D1225040

# Structure of Biological Membranes

## NOBEL SYMPOSIUM COMMITTEE (1976)

STIG RAMEL, *Chairman* • Executive Director, Nobel Foundation

ARNE FREDGA • Chairman, Nobel Committee for Chemistry

TIM GREVE • Director, Norwegian Nobel Institute (Peace)

BENGT GUSTAFSSON • Secretary, Nobel Committee for Medicine

LARS GYLLENSTEN • Member, Swedish Academy (Literature)

LAMEK HULTHÉN • Chairman, Nobel Committee for Physics

ERIK LUNDBERG • Chairman, Prize Committee for Economic Sciences

NILS-ERIC SVENSSON • Executive Director, Bank of Sweden Tercentenary Foundation

# Structure of Biological Membranes

Nobel Symposium, 34th

Edited by

Sixten Abrahamsson and Irmin Pascher
University of Göteborg
Göteborg, Sweden

Plenum Press · New York and London

Library of Congress Cataloging in Publication Data

Nobel Symposium, 34th, Skövde, Sweden, 1976.
Structure of biological membranes.

Held June 7-11, 1976.
Includes index.
1. Membranes (Biology)–Congresses. I. Abrahamsson, S. II. Pascher, Irmin, III. Title.
QH601.N6 1976 574.8'75 76-54955
ISBN 0-306-33704-5

Proceedings of the thirty-fourth Nobel Symposium on the Structure of the Biologica Membranes held in Skövde (near Göteborg), Sweden, June 7–11, 1976


A Division of Plenum Publishing Corporation
227 West 17th Street, New York, N.Y. 10011



Printed in the United States of America

# PUBLISHED NOBEL SYMPOSIA

1 • **Muscular Afferents and Motor Control**—*Edited by Ragnar Granit*
2 • **Prostaglandins**—*Edited by Sune Bergström and Bengt Samuelsson*
3 • **Gamma globulins**—*Edited by Johan Killander*
4 • **Current Problems of Lower Vertebrate Phylogeny**—*Edited by Tor Ørvig*
5 • **Fast Reactions and Primary Processes in Chemical Kinetics**—*Edited by Stig Claesson*
6 • **Problems of International Literary Understanding**—*Edited by Karl Ragnar Gierow*
7 • **International Protection of Human Rights**—*Edited by Asbjörn Eide and August Schou*
8 • **Elementary Particle Theory**—*Edited by Nils Svartholm*
9 • **Mass Motions in Solar Flares and Related Phenomena**—*Edited by Yngve Öhman*
10 • **Disorders of the Skull Base Region**—*Edited by Carl-Axel Hamberger and Jan Wersäll*
11 • **Symmetry and Function of Biological Systems at the Macromolecular Level**—*Edited by Arne Engström and Bror Strandberg*
12 • **Radiocarbon Variations and Absolute Chronology**—*Edited by Ingrid U. Olsson*
13 • **Pathogenesis of Diabetes Mellitus**—*Edited by Erol Cerasi and Rolf Luft*
14 • **The Place of Value in a World of Facts**—*Edited by Arne Tiselius and Sam Nilsson*
15 • **Control of Human Fertility**—*Edited by Egon Diczfalusy and Ulf Borell*
16 • **Frontiers in Gastrointestinal Hormone Research**—*Edited by Sven Andersson*
17 • **Small States in International Relations**—*Edited by August Schou and Arne Olav Brundtland*
20 • **The Changing Chemistry of the Oceans**—*Edited by David Dyrssen and Daniel Jagner*
21 • **From Plasma to Planet**—*Edited by Aina Elvius*
22 • **ESR Applications to Polymer Research**—*Edited by Per-Olof Kinell and Bengt Rånby*
23 • **Chromosome Identification Technique and Applications in Biology and Medicine**—*Edited by Torbjörn Caspersson and Lore Zech*
24 • **Collective Properties of Physical Systems**—*Edited by Bengt Lundqvist and Stig Lundqvist*
25 • **Chemistry in Botanical Classification**—*Edited by Gerd Bendz and Johan Santesson*
26 • **Coordination in the Field of Science and Technology**—*Edited by August Schou and Finn Sollie*
27 • **Super-Heavy Elements**—*Edited by Sven Gösta Nilsson and Nils Robert Nilsson*
28 • **Somatomedins and Some Other Growth Factors**— *Edited by Rolf Luft and Kerstin Hall*
29 • **Man, Environment, and Resources**—*Edited by Torgny Segerstedt and Sam Nilsson*
30 • **Physics of the Hot Plasma in the Magnetosphere**—*Edited by Bengt Hultqvist and Lennart Stenflo*
31 • **The Impact of Space Science on Mankind**—*Edited by Tim Greve, Finn Lied, and Erik Tandberg*
33 • **Molecular and Biological Aspects of the Acute Allergic Reaction**—*Edited by S. G. O. Johansson, Kjell Strandberg, and Börje Uvnäs*
34 • **Structure of Biological Membranes** — *Edited by Sixten Abrahamsson and Irmin Pascher*
36 • **Plasma Physics: Nonlinear Theory and Experiments** — *Edited by Hans Wilhelmsson*

Symposia 1-17 and 20-22 were published by Almqvist & Wiksell, Stockholm and John Wiley & Sons, New York; Symposia 23-25 by Nobel Foundation, Stockholm and Academic Press, New York; Symposium 26 by the Norwegian Nobel Institute, Universitetsforlaget, Oslo; Symposium 27 by Nobel Foundation, Stockholm and Almqvist & Wiksell International, Stockholm; Symposium 28 by Academic Press, New York; Symposium 29 by Nobel Foundation, Stockholm and Trycksaksservice AB, Stockholm; and Symposia 30, 31, 33, 34, and 36 by Plenum Press, New York.

# IN MEMORIAM

Dr. Hermann Träuble died on July 3, 1976 shortly after the Symposium. This is a great loss to his colleagues and to Science.

# Preface

Since 1965 the Nobel Foundation sponsors, through grants from the Bank of Sweden Tercentenary Fund, Symposia on subjects which are considered to be of central scientific importance and for which new results of a special interest have been reached. The aim of these Symposia is to bring together, by personal invitation, a limited number of leading scientists from various countries to discuss the current research situation within the field and to define the most urgent problems to be solved.

One of the most important fields in modern biomedical research concerns the structure and function of biological membranes. Research on this subject is very active and important scientific contributions appear at an increasing rate. It was therefore considered highly appropriate to devote Nobel Symposium 34 to the structure of membranes in order to get an expert summary of what is now known in the field.

The Symposium was held at Hotel Billingehus in Skövde (about 150 km from Göteborg), Sweden, from June 7 to 11, 1976. In addition to the grant from the Nobel Foundation financial support was received from the Nobel Institute of Chemistry of the Royal Academy of Sciences and from the Science Fund of Wilhelm and Martina Lundgren.

The Symposium was attended by some 50 scientists. The papers in this Volume had been distributed in advance to all participants. Therefore only summary presentations needed be given at the Symposium and the main emphasis was put on discussions.

The Organizing Committee for the Symposium was composed of S. Abrahamsson (chairman), K. Larsson, I. Pascher (secretary) and H. Virgin with an Advisory

Bord consisting of A. Engström, L. Ernster and G. Lundgren.

A special effort for rapid publication has been made by the contributors and publisher which is greatly appreciated.

S. Abrahamsson

I. Pascher

## Contents



* *Lecturer*

# MOLECULAR ARRANGEMENT AND CONFORMATION OF LIPIDS OF RELEVANCE TO MEMBRANE STRUCTURE

Sixten Abrahamsson, Birgitta Dahlén, Håkan Löfgren, Irmin Pascher & Staffan Sundell

Dept. of Structural Chemistry, Faculty of Medicine, University of Göteborg, P.O.B., S-400 33 Göteborg 33, Sweden

## INTRODUCTION

Intact membranes as well as membrane components have been studied extensively by various physical and chemical methods. Important data have accummulated but more specific structural information on the atomic level is still necessary in order to obtain a detailed understanding of lipid-lipid and lipid-protein interactions and of variations in structure and composition of lipids observed in different types of membranes. Only then will it be possible to explain the function of the different constituents and their significance for various membrane properties.

X-ray diffraction methods provide detailed structural information but require access to single crystals of pure and homogenious compounds. Studies of intact biological systems such as the membranes on the other hand will only give the overall molecular arrangement. The solid state is, of course, static in comparison with biological systems but it is known from many fields that structural features in crystals often reflect conditions in less ordered systems.

The lipid bilayer in the membranes varies in fluidity. In the gel state the molecules have such a close proximity that local order resembling that in the solid state most likely exists. As furthermore a conformational change in the hydrocarbon chain of one lipid molecule directly affects neighbouring chains in a cooperative way the arrangement of hydrocarbon chains in pure complex lipids should be of relevance to the bilayer structure.

Even if the conformation of a molecule is dependent on its environment, intramolecular forces give rise to preferred conformations. These manifest themselves by existing in different molecular surroundings and may also be apparent from energy calculations.

In the structure of glycerylphosphorylcholine (Abrahamsson & Pascher, 1966), for example, the two independent molecules of the asymmetric unit both show a characteristic *gauche* conformation about the $\overset{+}{N}$-C-C-O bond. This has been found to exist in a number of other related compounds as surveyed by Sundaralingam (1972) and also recently in an actual phospholipid, 1, 2-dilauroyl-(DL)-phosphatidylethanolamine (Hitchcock, Mason, Thomas & Shipley, 1974).

At this Department we are systematically studying by X-ray diffraction techniques lipids varying from fairly simple ones to complex membrane constituents. Special interest has been devoted to molecular packing behaviour such as the arrangement of hydrocarbon chains and in connection herewith the stacking of cholesterol skeleta and their accommodation into a lipid matrix. Furthermore major research efforts have concerned the correlation between structure and function of sphingolipids.

## ARRANGEMENT OF HYDROCARBON CHAINS

In lipids a number of modes of lateral packing of hydrocarbon chains with parallel axes are possible. As seen along the axes, the chains can either have their planes parallel or at approximately right angles to each other. The packing is usually described in terms of a subcell with the $c_s$ axis (~ 2.55 Å) along the chain direction and the other two axes representing repeat distances between chains. Symbols O, M and T denote orthorhombic, monoclinic and triclinic symmetry and $\perp$ and $\parallel$ mutual chain plane orientation. Sometimes a prime is used to differentiate subcells with otherwise similar symbols.

Some characteristics of T$\parallel$, O$\perp$, O$'\perp$ and O$\parallel$ were described by Abrahamsson, Ställberg-Stenhagen & Stenhagen (1963). Later other packings were discovered: M$\parallel$ (Abrahamsson & Westerdahl, 1963) and O$'\parallel$ (Abrahamsson & Ryderstedt-Nahringbauer, 1962) (Fig. 1). Segerman (1965) discussed the chain packings in terms of rows of identical chains and introduced a different nomenclature. This has not been used here as it does not account for new more complex chain arrangements.

In actual structures there are considerable random deviations from these idealized subcells. The O$\perp$ packing, however, shows principally important systematic variations. On heating, the short

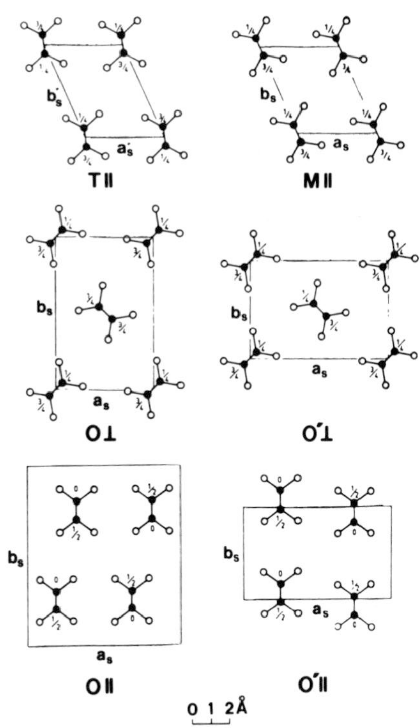


Fig. 1
Hydrocarbon chain packing subcells. Numbers at the carbon atoms represent fractional coordinates along the $c_s$ axis.

subcell axis, $a_s$, remains constant (~ 5 Å) or decreases slightly, whereas there is a marked expansion of the normally 7.40 Å $b_s$ axis. Similarly, substituents can be accommodated into the O⊥ matrix by the chain axes separating in the same direction. In 12-D-hydroxyoctadecanoic acid methyl ester (Lundén, 1976) hydroxyl groups and their hydrogen bond system are given enough space without local chain distorsion by expanding the $b_s$ dimensions to 7.87 Å (Fig. 2). This gives a cross section area per chain of 19.5 Å$^2$ as compared to 18.5 Å$^2$ in pure hydrocarbons. Though a quite different situation exists in sodium dodecyl sulphate (Sundell, 1976 a) in that the packing of the bulky polar groups leaves the chains with too much space, the chain matrix adapts in an analogous way by keeping one subcell axis nearly constant (5.1 Å) whereas the other expands to 8.2 Å giving a chain area of 20.9 Å$^2$. It should be noted that even though the chains in the latter case are considerably separated and consequently disordered with large thermal motion, a normal all anti-planar conformation is maintained in the chains (Fig. 3).

Near their melting point many lipids exist in the so called $\alpha$-form giving powder diffraction patterns implying a hexagonal symmetry of the carbon chains. The subcell dimensions corresponding to O⊥ are then 4.9 and 8.6 Å with a cross section of 21.1 Å$^2$. This is generally considered to be the chain arrange-

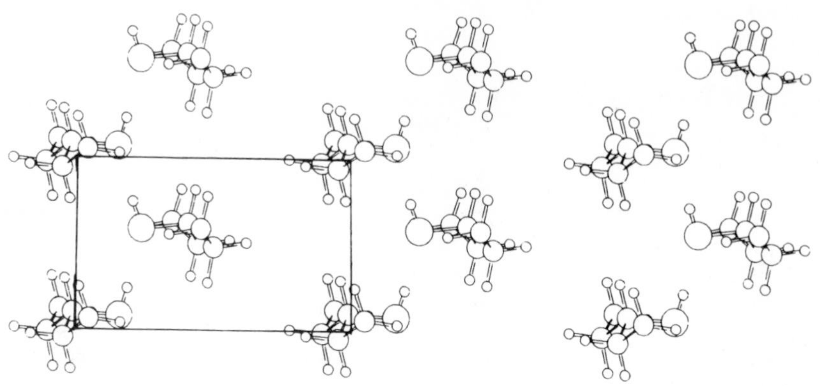

Fig. 2 The chain packing with accommodated hydroxyl groups in 12-D-hydroxyoctadecanoic acid methyl ester.

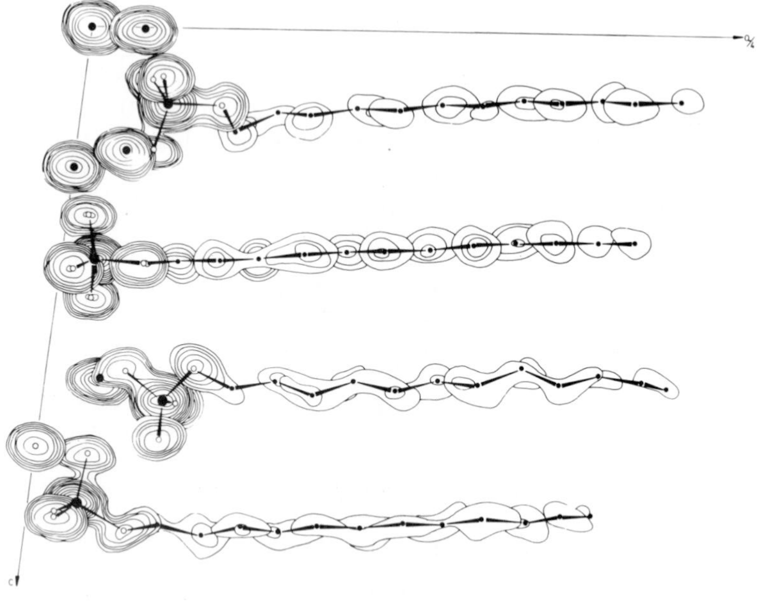

Fig. 3 Electron density map of sodium dodecyl sulphate showing the four molecules in the asymmetric unit.

ment in membrane lipid bilayers in the gel state. Even with this separation of the chain axes the hexagonal symmetry cannot be due to rotation of the chains about their axes (behaving as cylinders) but is rather caused by disorder. As already pointed out by Andrew (1950) in connection with an NMR-study, a completely synchronous motion of hexagonally arranged carbon chains with two-fold symmetry is impossible in a defined lattice. In fact, even moderate oscillations about the chain axes will lead to displacement of molecules within the lipid layer. In the super liquid state in monolayers (Harkins, 1944) in which a vigorous lateral movement in the tightly packed liquid layer is observed, the chains most likely do rotate completely.

Recently we have found two new types of chain packing in a cholesteryl ester (Abrahamsson & Dahlén, 1976 a, b) and a cerebroside (Pascher & Sundell, 1976) which appear to be of principal significance for the condensed state of the lipid matrix of membranes. In cholesteryl-17-bromoheptadecanoate the steroid skeleta determine the general packing and the chains cannot adopt positions suitable for any earlier known lateral arrangement. The chain axes are roughly hexagonally ordered as in the expanded O ⊥ case but the chains are closer packed with cross section area of only 19.3 $Å^2$. Both these conditions can be achieved simultanously only by a hybrid type of lateral stacking. The packing can formally be derived by a twinning operation on O ⊥ and can be described as being built up of pleated sheets in which all chains have parallel planes. Adjacent sheets have opposite tilt of the chain planes (Fig. 4). We

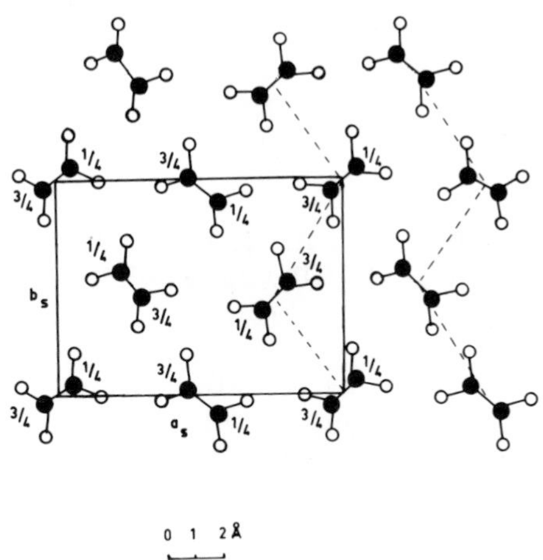


Fig. 4 Idealized subcell of cholesteryl-17-bromoheptadecanoate. Dotted lines indicate pleated sheets of chains with parallel planes.

have furthermore found that the very same chain arrangement also exists in 1, 2-dilauroyl-(DL)-phosphatidylethanolamine (Hitchcock, Mason, Thomas & Shipley, 1974).

The chain packing in a cerebroside discussed below is of a similar type in which the chain axes are roughly hexagonally arranged. (Fig. 5). It also contains pleated sheets in which the chains all have parallel planes. The tilt of the chain planes in relation to the sheet planes is however different from the packing described above.

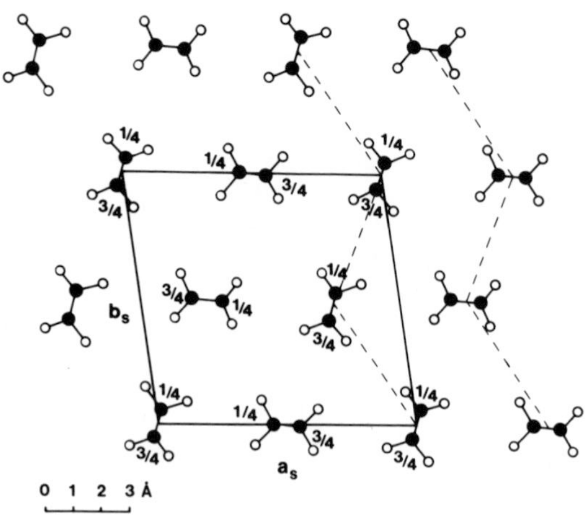


Fig. 5 Idealized subcell of β-D-galactosyl-N-(2-D-hydroxy-octadecanoyl)-D-dihydrosphingosine. Pleated sheets are indicated by dotted lines.

## CHOLESTEROL PACKING AND ACCOMMODATION INTO A HYDROCARBON MATRIX

Though the effect of cholesterol on fluidity and packing properties of phospholipids has been studied extensively, still no satisfactory model exists which explains more specifically the co-packing of hydrocarbon chains and steroid skeleta. Single crystal analyses of cholesteryl sulphate (Sundell, 1976 b, Fig. 6) and cholesteryl esters (Fig. 7) have therefore been performed. Cholesteryl sulphate is found as an abundant constituent in membranes of the starfish (Björkman et al., 1972) and in the brushborder membrane of intestine (Karlsson, unpublished). Cholesteryl esters are, apart from the attention they have attracted as con-

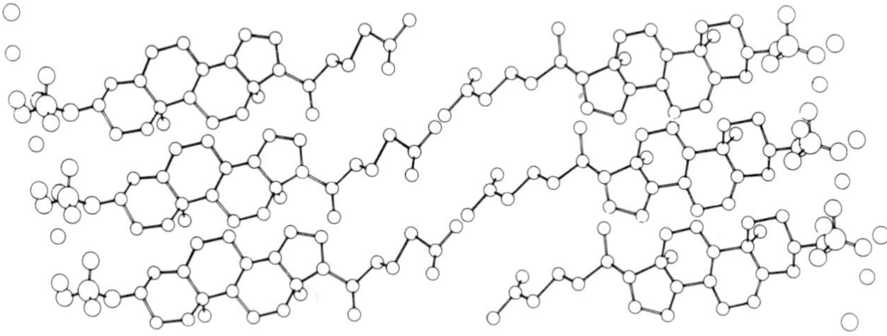

Fig. 6 Molecular packing of sodium cholesteryl sulphate.

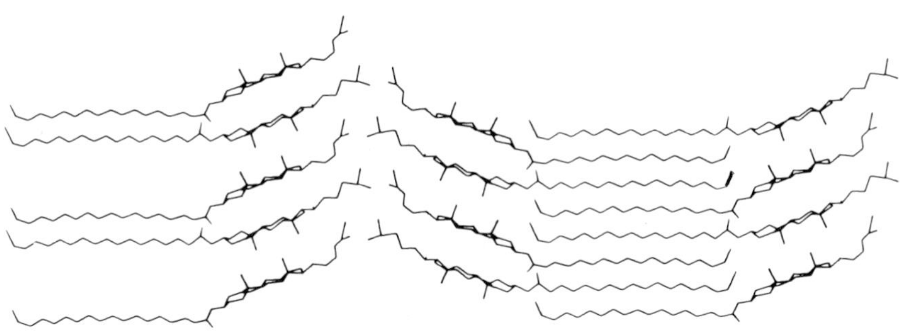

Fig. 7 Molecular packing of cholesteryl-17-bromoheptadecanoate.

stituents in lipoproteins and atherosclerotic lesions, of interest as they might provide direct information on cholesterol - hydrocarbon chain co-packing.

However, as is obvious from Fig. 7 cholesteryl-17-bromoheptadecanoate packs with alternating regions of steroid skeleta and hydrocarbon chains. This is also the case in cholesteryl myristate (Craven & DeTitta, 1976). From these analyses it is still possible to draw some conclusions about the co-packing of cholesterol and hydrocarbon chains.

The packing of cholesterol skeleta is principally different in the cholesteryl esters (Fig. 8) and cholesteryl sulphate (Fig. 9). In the esters the skeleta (area 36.7 $Å^2$) are arranged in double layers with the projecting methyl groups facing each other. A rather large translation in the direction of maximum extension of the skeleton allows the two methyl groups of one molecule to be accommodated into the space between the methyl groups of two molecules in the opposite layer half. The double layers thus have smooth contact surfaces. In cholesteryl sulphate only single layers exist and the methyl groups project into the space between two skeleta which then can have no direct contact sideways. This packing, however, is equally effective as that in the ester with an area per skeleton of 36.8 Å$^2$.

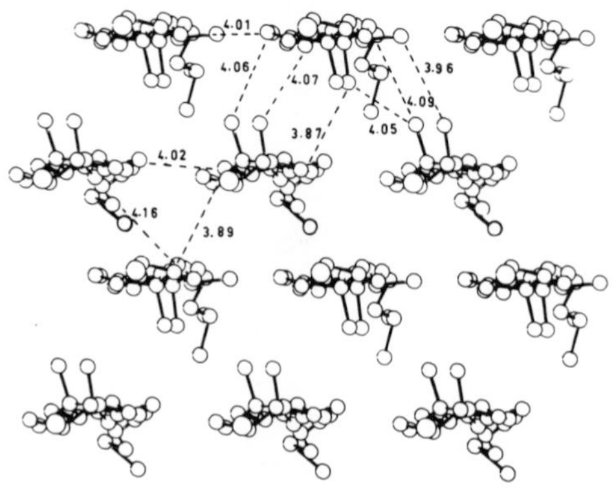


Fig. 8 Packing of cholesterol skeleta in cholesteryl esters as seen along the direction of maximum extension of the skeleton.

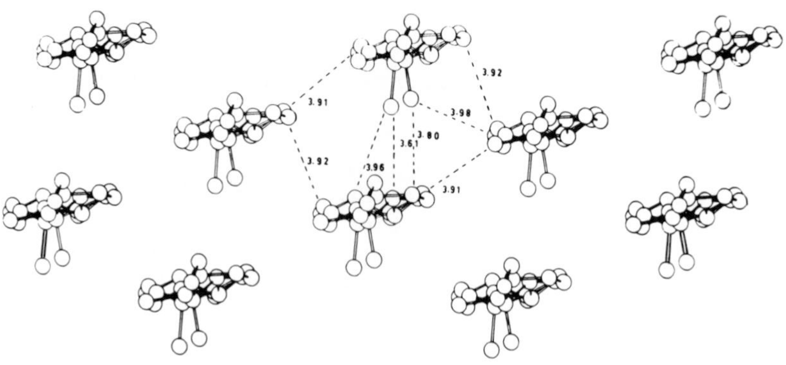


Fig. 9 Packing of cholesterol skeleta in sodium cholesteryl sulphate.

If the packing patterns of the hydrocarbon chains described above are compared with the two arrangements of cholesterol skeleta a good fit can be obtained where two chains superpose the area of one cholesterol molecule throughout the lattice. It is clear from Fig. 10 that cholesterol molecules can randomly replace two chains and vice versa without any major distorsions.

Such a random mutual replacement of various molecules in a lipid matrix requires that both the molecular areas and shapes are the same. This is, of course, rarely the case, as both polar groups, chain unsaturation etc. directly influence the effective cross section. Areas corresponding to that of cholesterol (37-38 Å$^2$) have been observed for saturated phosphatidylethanolamine in monolayers (Hayashi, Muramatsu & Hara, 1972) and in crystals (Hitchcock et al, 1974) but for phosphatidylcholine and sphingolipids (sphingomyelin and cerebrosides) the minimum packing area for saturated species are 42 Å$^2$ (Phillips & Chapman, 1968) and 40 Å$^2$ respectively (Löfgren, unpublished). The latter areas are not decreased to any extent in systems with cholesterol, which is known to have a marked condensing effect on expanded monolayers of phosphatidylcholines. However, a monolayer of an equimolar mixture of distearoyl-phosphatidylcholine, cerebroside (galactosyl-N-stearoyl-sphingosine) and cholesterol can be compressed to an molecular area of 37 Å$^2$ per molecule (Löfgren & Pascher, in preparation). This mixture appears representative for the lipid composition of the outer layer of the myelin membrane (O´Brien & Sampson, 1965) if an asymmetric lipid distribution is assumed. The formation of this highly condensed lipid layer can only be explained by a precise stereospecific fit between the three components.

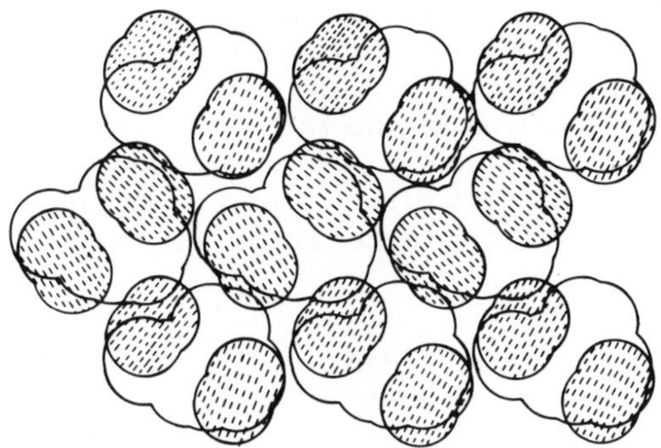

Fig. 10 Superposition of the hydrocarbon chain arrangement of cholesteryl-17-bromoheptadecanoate on the steroid skeleta packing of sodium cholesteryl sulphate. The contours are proportional to the van der Waals´ radii.

## SPHINGOLIPID STRUCTURE AND FUNCTION

Aspects on structure and function of sphingolipids are summarized by Karlsson (this volume). Sphingolipids appear to be mainly confined to the outer layer of surface membranes where they together with phosphatidylcholine and cholesterol occur in a molar ratio of approximately 1:1:1. They differ characteristically from glycerolipids in having donor groups for hydrogen bonds in their lipophilic ceramide part. In addition to an amide group they contain in their lipophilic ceramide part a varying number of hydroxyl groups which increases significantly in membranes which are exposed to pronounced physical and chemical stress.

To clarify the role of these polar groups in sphingolipids, series of different model compounds of high purity and well defined stereoconfiguration were synthesized and their properties and conformations studied by various physical methods (Abrahamsson et al., 1972).

### Ceramide

Fig. 11 shows the single crystal structures of a number of sphingolipids and sphingolipid components. From their common structural features the preferred conformation of ceramides was deduced (Fig. 12) and the functional significance of their hydroxyl groups for lipid-lipid interactions emphasized (Pascher, 1976). The rigid amide group, which serves as a link between the hydrocarbon chains stands perpendicular to the axis of the sphingosine chain. The hydrogen atom on carbon atom 2 of the sphingosine is thereby located in the amide plane. In order to bring the two carbon chains into packing contact one of them has to bend sharply. The conformations of the two fatty acid derivatives V and VI in Fig. 11 indicate that this requirement is met by a double *gauche* bend on both sides of the fatty acid α-carbon atom. The resulting stereospecific orientation of the fatty acid hydroxyl groups with its implications on physical properties could be confirmed by infrared spectroscopy as well as monolayer and thin-layer chromatographic behaviour of diastereomeric model ceramides (Fig. 13).

The orientation of the different hydrogen bond donors and acceptors allows lateral interaction with other lipid molecules which seems to be an important characteristic of sphingolipids.

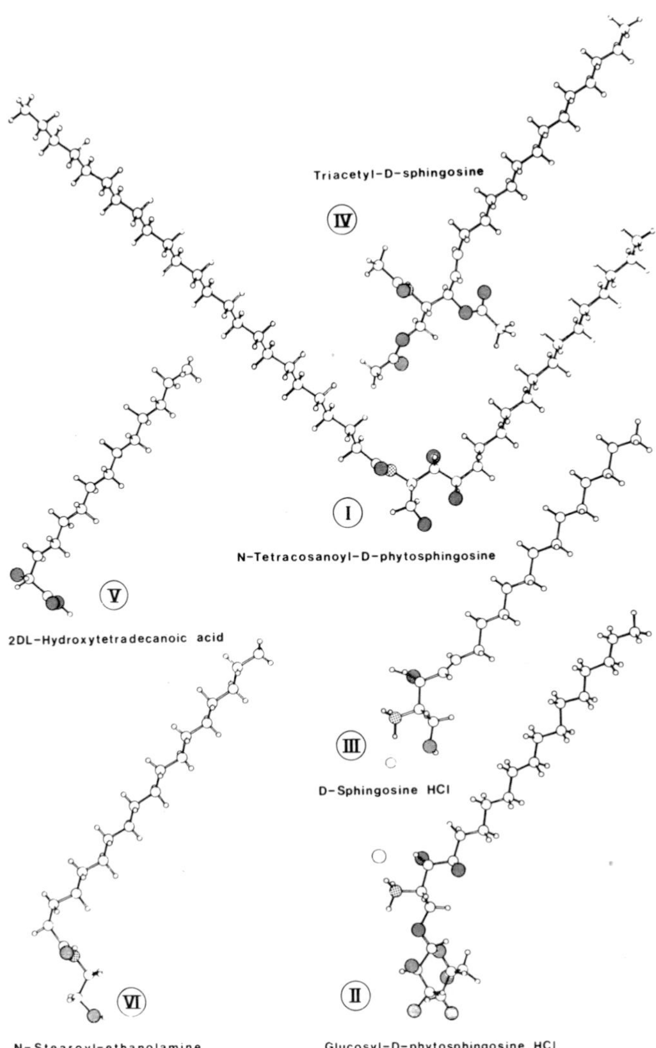


Fig. 11 Conformations of sphingolipid components

I N-Tetracosanoyl-D-phytosphingosine (Dahlén & Pascher, 1972)

II Glucosyl-D-phytosphingosine HCl (Abrahamsson, Dahlén & Pascher, 1976)

III D-Sphingosine HCl (Nilsson & Pascher, unpublished)

IV Triacetyl-D-sphingosine (O'Connell & Pascher, 1969)

V 2DL-Hydroxytetradecanoic acid (Dahlén, Lundén & Pascher, 1976)

VI N-Stearoyl-ethanolamine (Dahlén, Pascher & Sundell, 1976)

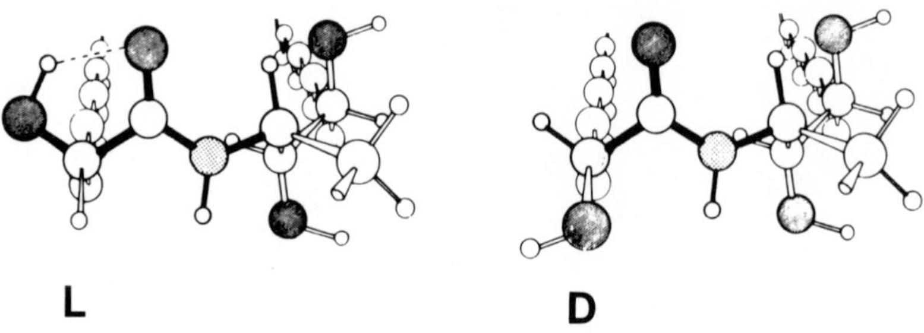


Fig. 12 Deduced conformation of ceramide. The models summarize the preferred conformational features observed in different ceramide components (cf. Fig. 11). The two molecules differ with respect to the configuration of the fatty acid hydroxyl group. They are seen with the amide group turned towards the viewer and with the hydrocarbon chains pointing away. All atoms connected with solid bonds are localized in or close to the plane of the amide group.

## Cerebroside

Cerebroside is the basic structure in all glycosphingolipids. Recently the crystal structure of β-D-galactosyl-N-(2-D-hydroxyoctadecanoyl)-D-dihydrosphingosine (Fig. 14) has been determined at this Department. The compound revealed two new unexpected and remarkable conformational features. Firstly, instead of an expected bend of the fatty acid, the parallel stacking of the hydrocarbon chains is accomplished by a sharp bend of the sphingosine chain as far up as at carbon atom 6 (Fig. 15). Secondly, due to another bend at carbon atom 1 of the sphingosine chain, the galactose ring adopts an almost perpendicular orientation towards the axes of the hydrocarbon chains and gives the molecule the shape of a shovel. This particular chain conformation of the ceramide part was first considered to be an adaption to the space requirements of the sugar groups but there are a series of experimental results which indicate that this chain bend, as well as the shape of the headgroup, represent preferred conformations of relevance also in a biological environment.

An isolated bend in the middle of a normal zig-zag chain is energetically unfavoured. Bends are usually formed in the vicinity of different functional groups which interrupt the regular chain packing. The most common long chain base - sphingosine - has a <u>trans</u> double bond in 4, 5 position. This excludes the formation of bends between carbon atoms 3 to 6, but would on the other hand naturally induce a bend at carbon atom 6. Such a <u>trans</u> double

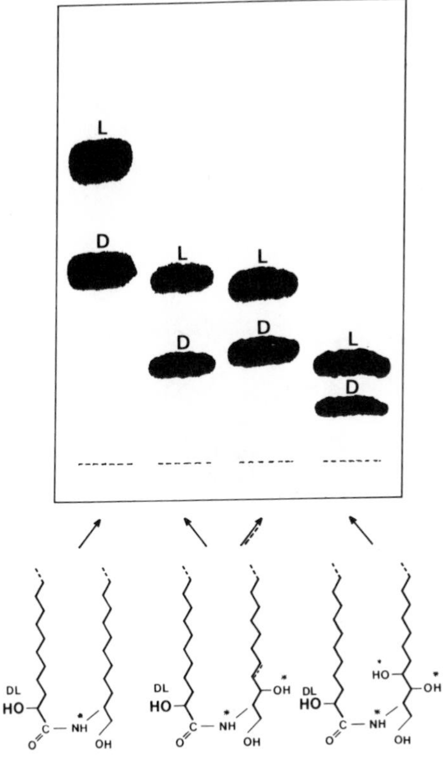


Fig. 13 Thin-layer chromatographic behaviour of diastereomeric pairs of ceramides which differ with respect to the configuration of the 2-hydroxyl group of the fatty acid. The increase in mobility of the unnatural L-stereoisomers reflects a decrease in polarity, consistant with the neutralisation of polar forces due to intramolecular hydrogen bonding (cf. Fig. 12).

bond would fit into the present structure without any change in conformation (Fig. 16). Thus it appears that the function of this trans double bond is to facilitate chain stacking and close packing of the lipid molecules. Monolayer studies on model ceramides with and without this double bond confirm this suggestion convincingly. While the saturated ceramide, N-octadecanoyldihydrosphingosine, can only be condensed at a relatively high lateral pressure the corresponding unsaturated ceramide, N-octadecanoylsphingosine, spontanously condenses at the water/air interface and forms a very well-packed lipid layer already at a lateral pressure of less than 10 dyn/cm (Fig. 16). The presence of a

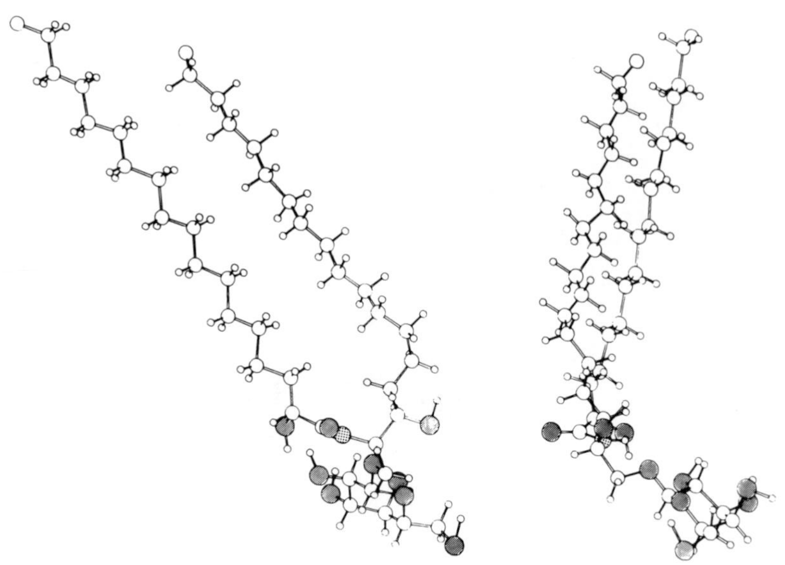

Fig. 14 The molecule of β-D-galactosyl-N-(2-D-hydroxyoctadecanoyl)-D-dihydrosphingosine as seen along the crystallographic b and a axes.

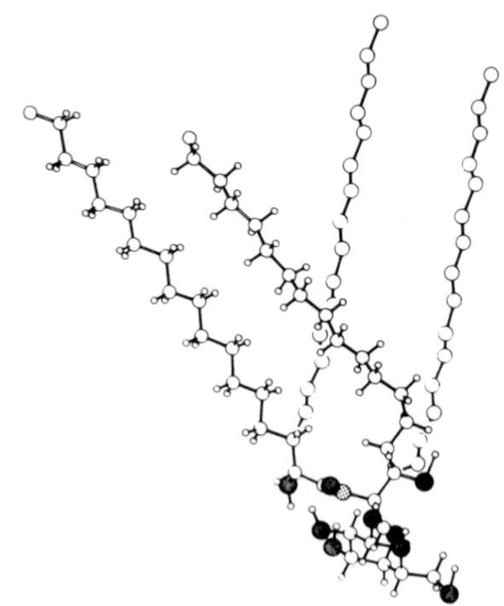

Fig. 15 Expected and observed chain stacking in cerebroside.

carboxyl group or a vinyl ether group of glycerolipids in a position structurally corresponding to that of the sphingosine trans double bond would similarly promote a chain bend.

There is another important implication of this special ceramide conformation. Due to the bend of the sphingosine chain, the fatty acid extends rather much over the end of the former. This effect would be even more pronounced in natural sphingolipids as fatty acids with a chain length of 24 carbon atoms are usually most abundant. The acids are predominantly saturated contrary to those of glycerolipids but there are also cis unsaturated species present. However among the latter only nervonic acid (C24:1) with a cis

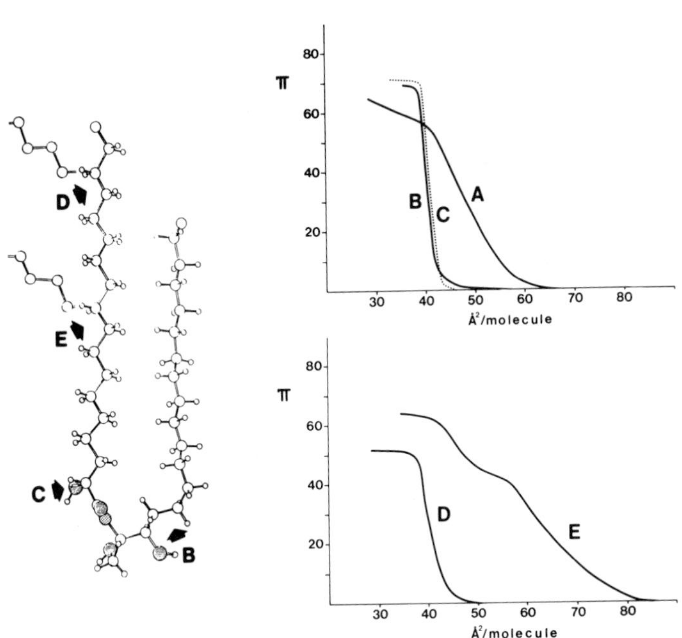


Fig. 16 Right: Surface pressure-area isotherms of ceramides
A: N-Octadecanoyl-D-dihydrosphingosine
B: N-Octadecanoyl-D-sphingosine
C: N-(2-D-Hydroxyoctadecanoyl)-D-sphingosine
D: N-cis-15-Tetracosenoyl-D-sphingosine and
E: N-cis-9-Octadecenoyl-D-sphingosine

Left: Conformation of ceramide as observed in cerebroside. The positions of the double bonds (B, D and E) and the hydroxyl group (C) of the different model compounds are indicated.

double bond in position 15 is found in any appreciable amount (Svennerholm & Ställberg-Stenhagen, 1968). The double bond of this acid should however not disturb the chain stacking in this particular ceramide conformation as it would be located above the end of the sphingosine chain (Fig. 16). Again monolayer studies of ceramides with a cis double bond in the fatty acid above and below the sphingosine chain end, according to this model show that this conformation of the crystalline compound is relevant also for ceramide in contact with an aqueous environment. Generally, groups such as double bonds and methyl branches (Karlsson & Samuelsson, 1973) in the fatty acid which might disturb the chain packing are found in positions close to the end of the sphingosine chain.

The remarkable shovel conformation of the galactose head group is determined by two important interactions. Primarily, there is an intramolecular hydrogen bond between the N-H group and the glycosidic oxygen (Fig. 17). The oxygen atom which serves as a link between ceramide and galactose thereby becomes fixed in a position in which the rigid sugar ring points away sideways from the ceramide part and is then only free to rotate about the bond between the glycosidic oxygen and the first carbon atom of the ring. This type of conformation-determining intramolecular interaction has also been observed in glucosylphytosphingosine (Fig. 10 II). Furthermore, the galactose ring turns the one of its sides towards the ceramide layer which allows the maximum number of intermolecular hydrogen bonds within the layer.

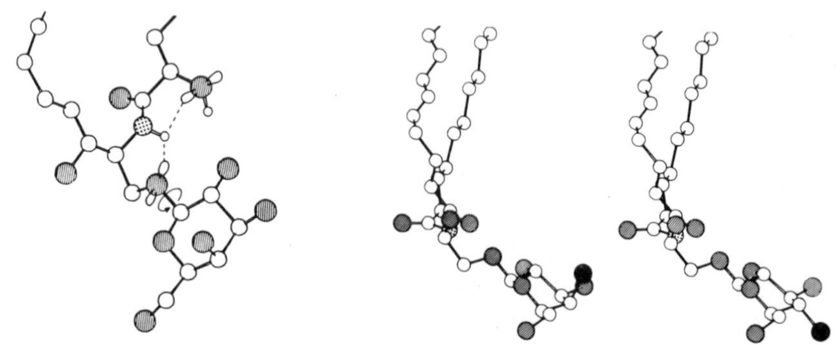

Fig. 17 Left: The intramolecular interactions in cerebroside locking the glycosidic oxygen and the hydroxyl group of the fatty acid. Right: The stereospecific differences between galactose and a similarly attached glucose.

All projecting groups on the galactose ring are equatorial with the exception of the axial hydroxyl group in position 4 which in this particular cerebroside conformation points almost perpendicularly back towards the ceramide layer (Fig. 17). In oligoglycosylceramides the oxygen in position 4 is usually involved in the glycosidic link to the next sugar. The galactose conformation in this cerebroside would not allow such an addition of another sugar residue. It is therefore of interest to note that galactose is the predominant carbohydrate in monoglycosylceramides whereas, with one single exception (digalactosylceramide with an unusual $\alpha$ (1→4) linkage between the sugar residues) all oligoglycosylceramides so far known have glucose instead of galactose in proximal linkage to ceramide (Wiegandt, 1970; Hakomori, 1973). These two carbohydrates resemble each other in all respects except for the configuration of the hydroxyl group in position 4 (Fig. 17). The substitution of galactose with glucose in the present structure would make the 4 hydroxyl group accessable for the addition of another sugar molecule.

As expected the molecular packing of cerebroside represents a typical lipid bilayer arrangement (Fig. 18). There is remarkably little polar interaction between sugar residues over the polar group contact planes. Only one hydrogen bond via a disordered molecule of ethanol extends between the double layers. The eight other possible hydrogen bonds of the molecule are formed within the lipid layer (Fig. 19). Due to the shape of the molecule the sugar residue is not in packing contact with its own ceramide part but with those of two neighbouring molecules.

The formation of lateral hydrogen bonds have been observed in ceramide monolayers and the implication of these bonds on membrane stability and permeability has been discussed (Pascher, 1976). A sugar residue, with its hydrogen bonds mainly within the lipid layer as described above, should further enhance lateral interaction. As shown in Fig. 20 the addition of galactose to a

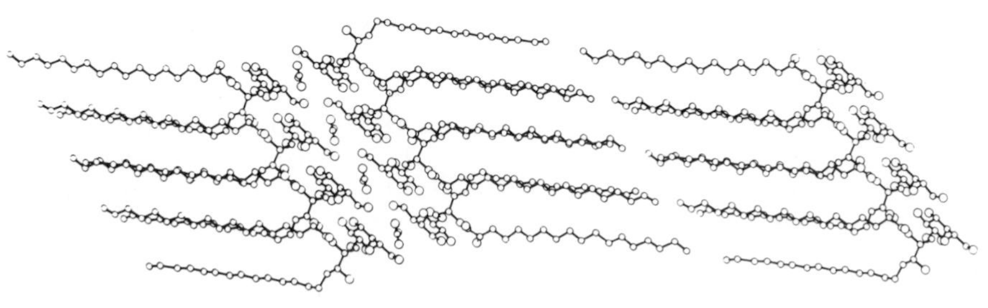

Fig. 18 Molecular packing of cerebroside.

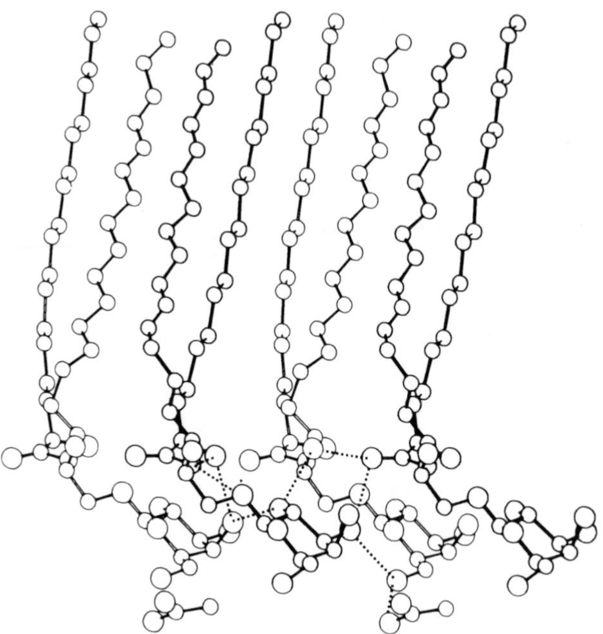

Fig. 19 Part of the hydrogen bond system (dotted lines) representative for one cerebroside molecule.

dihydroceremide, which alone gives an expanded monolayer, has a considerable condensing effect.

The condensation of sphingolipids is thus promoted by the following factors: hydroxyl groups in the ceramide part and the sugar residue and the 4, 5 trans double bond in the sphingosine chain. In cerebroside of myelin, which is considered to be one of the most densely packed membranes almost only unsaturated sphingosine and galactose are found combined with fatty acids of which almost 50 percent have a 2-hydroxyl group.

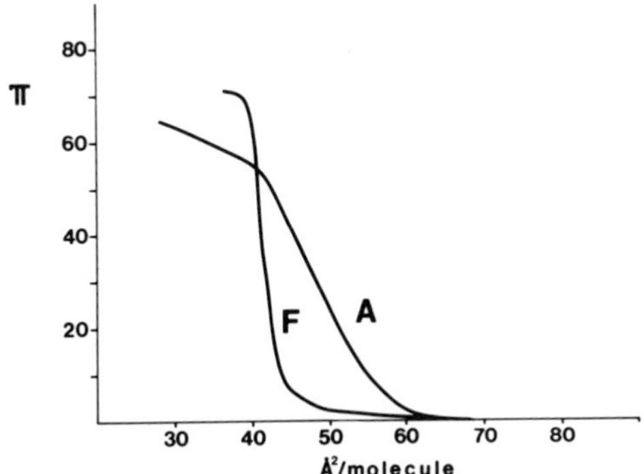


Fig. 20 Surface-pressure-area (π -A)-isotherms of
A: N-octadecanoyl-D-dihydrosphingosine and
F: β-D-galactosyl-N-octadecanoyl-D-dihydrosphingosine.

## IMPLICATIONS ON MEMBRANE STRUCTURE

From the present knowledge of the structural behaviour of lipids it is possible to indicate some general features of the surface membrane which are schematically illustrated by the model in Fig. 21. As mentioned earlier the main components of the outer layer appear to be phosphatidylcholine, cholesterol and sphingolipid which in the case of myelin is predominantly cerebroside. From a structural point of view it seems rational to define four functionally different regions in the layer: a liquid and a structural region of the hydrocarbon matrix and a structural and a surface functional region in the polar part.

The structural region of the polar part is characterized by lateral interaction mainly through hydrogen bonding. Sphingolipids are because of their donor and acceptor groups capable to form rigid networks of hydrogen bonds. In lipid mixtures these bonds can be directed to and terminated on the acceptor groups of glycerolipids. It is obvious that variations in structure and composition of the lipids will determine the extent of these polar lateral interactions and thus the stability and fluidity of the membrane.

The properties of the lipid layer is furthermore dependent on the structural hydrocarbon region. The width of this region is defined by the extension of the rigid steroid skeleton and the un-

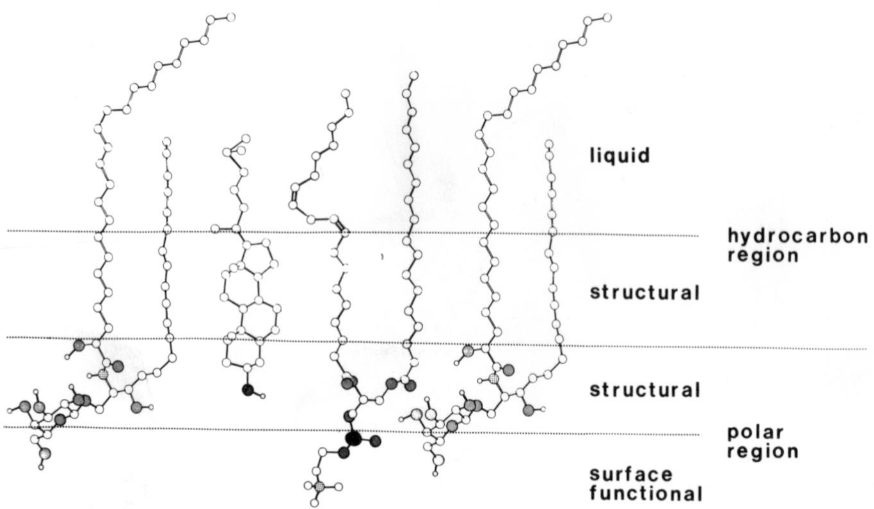


Fig. 21 Model schematically illustrating the functional regions in the lipid layer of surface membranes.

distorted part of the hydrocarbon chains of the lipids. The existence of both a polar and a nonpolar close packed region is of principal importance for the barrier properties of the membrane. Cholesterol, with its asymmetric structure, can pack both with saturated and unsaturated hydrocarbon chains. In addition it functions as a spacer to allow accommodation of bulky groups in the two adjacent regions.

Due to double bonds, branches and differences in chain length of the double chain lipids a close packing is impossible in the region above the hydrocarbon structural part. This leads to a liquid zone in the interior of the double layer. The two layer halves are thus structurally independent which is compatible with membrane asymmetry and dynamics.

There is finally a fourth region of headgroups which have specific functions and which do not directly influence membrane structure. The complex carbohydrate structures of blood group-active sphingolipids for instance would be part of the surface functional region, whereas galactose in cerebroside with its hydroxyl groups pointing back towards the lipid layer should mainly be considered as belonging to the polar structural region.

It is important to note that condensing effects in the polar and hydrocarbon structural regions will not always give rise to a close packed layer. This requires in addition a good steric fit between the various lipid molecules. In the monolayer described above of cerebroside, phosphatidylcholine and cholesterol in the same proportions as they appear in myelin, all these conditions are obviously fulfilled.

## ACKNOWLEDGMENTS

Grants in support of this Department were obtained from the Swedish Medical Science Research Council (B76-13X-6-12C), the Swedish Board for Technical Development, the Wallenberg Foundation and the U.S. Public Health Service (GM 11653).

## REFERENCES

Abrahamsson, S. and Ryderstedt-Nahringbauer, I., The crystal structure of the low-melting form of oleic acid, Acta Cryst., 15, 1261, 1962.

Abrahamsson, S., Ställberg-Stenhagen, S. and Stenhagen, E., The higher saturated branched chain fatty acids, in Progress in the Chemistry of Fats and other Lipids, Vol. 7, 1963.

Abrahamsson, S. and Westerdahl, A., The crystal structure of 3-thiadodecanoic acid, Acta Cryst., 16, 404, 1963.

Abrahamsson, S. and Pascher, I., Crystal and molecular structure of L-α-glycerylphosphorylcholine, Acta Cryst., 21, 79, 1966.

Abrahamsson, S., Pascher, I., Larsson, K. and Karlsson, K.-A., Molecular arrangements in glycosphingolipids, Chem. Phys. Lipids 8, 152, 1972.

Abrahamsson, S. and Dahlén, B., The crystal structure of cholesteryl-17-bromoheptadecanoate, J.C.S.Chem.Comm., 117, 1976 a.

Abrahamsson, S. and Dahlén, B., The crystal structure of cholesteryl-17-bromoheptadecanoate, submitted for publication 1976 b.

Abrahamsson, S., Dahlén, B. and Pascher, I., Molecular arrangements in glycosphingolipids. Crystal structure of glucosylphytosphingosine hydrochloride, Acta Cryst., in press, 1976.

Andrew, E.R., Molecular motion in certain solid hydrocarbons, J.Chem.Phys., 18, 607, 1950.

Björkman, L.R., Karlsson, K.-A., Pascher, I. and Samuelsson, B.E., The identification of large amounts of cerebroside and cholesterol sulphate in the sea star, asterias rubens, Biochim. Biophys. Acta 270, 260, 1972.

Craven, B.M. and DeTitta, G.T., Cholesteryl myristate: structures of the crystalline solid and mesophases, J. Chem. Soc. Perk II, in press, 1976.

Dahlén, B. and Pascher, I., Molecular arrangements in sphingolipids. Crystal structure of N-tetracosanoylphytosphingosine, Acta Cryst., B28, 2396, 1972.

Dahlén, B., Lundén, B.-M. and Pascher, I., The crystal structure of 2-DL-hydroxytetradecanoic acid, Acta Cryst., in press, 1976.

Dahlén, B., Pascher, I. and Sundell, S., The crystal structure of N-(2-hydroxyethyl)-octadecanamide, submitted for publication, 1976.

Hakomori, S.-I., Glycolipids of tumor cell membranes, Adv. Cancer Res., 18, 265, 1973.

Harkins, W.D., The surface of solids and liquids and the films that form upon them, in Colloid Chemistry, vol. 5, Reinhold Publishing Corp., New York, 1944.

Hayashi, M., Muramatsu, T. and Hara, I., Surface properties of synthetic phospholipids, Biochim. Biophys. Acta, 255, 98, 1972.

Hitchcock, P.B., Mason, R., Thomas, K.M. and Shipley, G.G., Structural chemistry of 1,2-dilauroyl-DL-phosphatidylethanolamine: Molecular conformation and intermolecular packing of phospholipids, Proc. Nat. Acad. Sci. U.S., 71, 3036, 1974.

Karlsson, K.-A. and Samuelsson, B.E., The structure of ceramide aminoethylphosphonate from the sea anemone, metridium senile, Biochim. Biophys. Acta, 337, 204, 1974.

Löfgren, H. and Pascher, I., Molecular arrangement in sphingolipids. Lateral interaction of ceramides in monolayers, to be published.

Lundén, B.-M., The crystal structure of 12-D-hydroxyoctadecanoic acid methyl ester, Acta Cryst., in press, 1976.

Nilsson, B. and Pascher, I., The crystal and molecular structure of the hydrochloride and hydrobromide of sphingosine. Submitted for publication, 1976.

O'Brien, J.S. and Sampson, E.L., Lipid composition of the normal human brain: grey matter, white matter, and myelin, J. Lipid Res., 6, 537, 1965.

O'Connell, A.M. and Pascher, I., The crystal structure of triacetylsphingosine, Acta Cryst., B25, 2553, 1969.

Pascher, I., Molecular arrangements in sphingolipids. Conformation and hydrogen bonding of ceramide and their implication on membrane stability and permeability, Biochim. Biophys. Acta, submitted for publication, 1976.

Pascher, I. and Sundell, S., Molecular arrangement of sphingolipids. The crystal structure of a cerebroside: β-D-galactosyl-N-octadecanoyl-D-dihydrosphingosine, in preparation.

Phillips, M.C. and Chapman, D., Monolayer characteristics of saturated 1,2-diacylphosphatidylcholine (lecithins) and phosphatidylethanol-amines of the air-water interface, Biochim. Biophys. Acta, 163, 301, 1968.

Segerman, E., The modes of hydrocarbon chain packing, Acta Cryst., 19, 789, 1965.

Sundaralingam, M., Molecular structures and conformations of the phospholipids and sphingomyelins, Ann. N.Y. Acad. Sci., 195, 324, 1972.

Sundell, S., The crystal structure of sodium dodecyl sulphate, to be published, 1976 a.

Sundell, S., The crystal structure of sodium cholesteryl sulphate, to be published, 1976 b.

Svennerholm, L. and Ställberg-Stenhagen, S., Changes in the fatty acid composition of cerebrosides and sulfatides of human nervous tissue with age, J. Lipid Res., 9, 215, 1968.

Wiegandt, H., Glycosphingolipids, Adv. Lipid Res. 9, 249, 1971.

# ENERGY CONSERVATION BY PROTON TRANSPORT THROUGH CHLOROPLAST MEMBRANES

M. Avron, U. Pick, Y. Shahak and Y. Siderer

Department of Biochemistry, Weizmann Institute of Science
Rehovot, Israel

The relation between electron transport, proton transport, proton gradient-formation and ATP synthesis is qualitatively and quantitatively investigated. Proton gradients have many properties which are in agreement with their being an intermediary energy pool between electron transport and ATP formation. Approximately three protons are required to transverse the chloroplast-vesicle-membrane to provide sufficient energy for the synthesis of an ATP molecule in the steady state. The system is shown to be fully reversible through the experimental demonstration of ATP driven proton gradients and reverse electron flow, and proton gradient driven reverse electron flow and reverse-electron-flow-luminescence.

## 1. Proton Uptake and Proton Gradient

Chloroplast membranes isolated from higher plants or algae show rapid and reversible electron transport dependent proton uptake from the medium into their innervesicular space (see Jagendorf, 1975). The number of protons taken up is a function of the natural internal buffering power, and can therefore be increased considerably by the addition of components which act as internal buffers (Avron, 1972).

In order to evaluate the possible contribution of the proton concentration gradients created during such proton uptake to the energy conserving system several methods were developed to measure the size of such gradients. Most of these methods depend on the equilibration of amines across the chloroplast membranes in response to the pH gradient (Fig. 1).

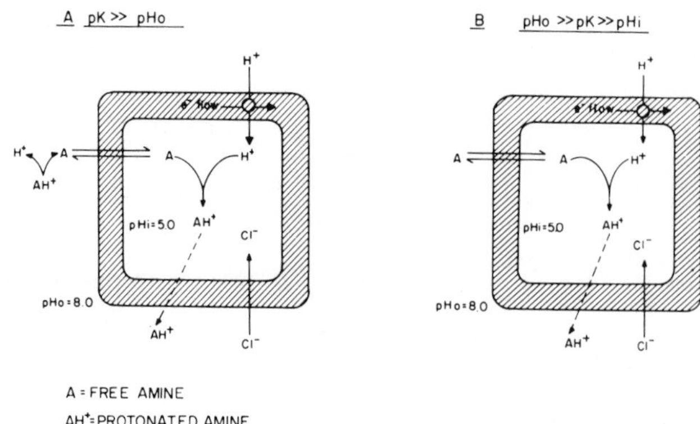


Fig. 1: Equilibration of amines across energised chloroplast membrane vesicles.

The non-protonated form of the amines is in all cases assumed to freely permeate the membrane and therefore its concentration inside equals its concentration outside. Amines with a pK more than 1 pH units above that of the reaction medium (fig. 1A) will be mostly in their protonated form in the medium. When a pH gradient exists across the vesicle due to proton uptake, the free amine inside will be protonated by the incoming protons pulling more free amine in, until at steady state the following relation will hold (Rottenberg et al., 1972):

$$\frac{(AH^+)out}{(AH^+)in} = \frac{(H^+)out}{(H^+)in} \qquad \text{equation 1}$$

The first method devised for measuring the transmembrane pH gradient utilized $^{14}C$-labeled methylamine (pK = 10.6), where under all experimental conditions employed the total amine concentration is essentially equal to the protonated amine concentration. Chloroplasts were illuminated in the presence of trace (1-10 $\mu$M), amounts of labelled methylamine, centrifuged in the light, and the $^{14}C$ content of pellet and supernatant determined. Independently the osmotic volume of the same chloroplast preparation was determined using $^{14}C$-labeled sorbitol which does not enter the chloroplast vesicle and therefore labels only the non-osmotic compartment. From these values the methylamine concentration inside and outside the vesicle can be determined and hence (equation 1) the ΔpH. Values of ΔpH approaching 3 were determined with this method (Rottenberg et al., 1972). Similar techniques using other amines have been employed (Portis and McCarty, 1974). The main disadvantage of such methods is the requirement for rather high chloroplast concentrations which results in undersaturation with light, and a consequent underestimation of the pH gradient.

This difficulty does not arise in the following methods, which follow the pH gradient continuously in the reaction mixture. Rottenberg and Grunwald (1972) added low concentrations of ammonium and followed the ammonium concentration in the medium directly with an ammonium-sensitive electrode. This is an excellent method but its usefulness is limited to media which contain no other ions to which the electrode responds.

Schuldiner et al. (1972) followed the fluorescence of a fluorescent amine, 9-aminoacridine, since it was previously demonstrated (Kraayenhof, 1970) that the fluorescence of such amines is totally quenched inside the chloroplast vesicles. From the light induced change in fluorescence and the osmotic volume of the vesicles the light induced ΔpH can be accurately and continuously followed. This is probably the most widely used method and with proper precautions (Fiolet et al., 1974, 1975) its reliability has been adequately established in a variety of biological and model systems (Deamer et al., 1972, Casadio et al., 1974, Avron, 1976; Graber and Witt, 1976; Chow and Hope, 1976).

Recently, we developed a new method based on a quantitative measurement of the extra proton uptake observed in illuminated chloroplasts upon addition of several amines which serve as internal buffers (fig. 1B, Crofts, 1958; Lynn, 1968; Nelson et al., 1971; Avron, 1972). It could be shown that such extra proton uptake, $\overline{\Delta H^+}$, relates to the internal proton concentration, $(H^+)_i$, by the following equation (Pick and Avron, unpublished):

$$(H^+)_i = \frac{\Delta H^+[\,(H^+)_o + K\,]^2}{(A_t)_o \cdot v \cdot K} \qquad \text{equation 2}$$

where $(H^+)_o$ refers to the external proton concentration; $(A_t)_o$ to the total amine concentration outside at steady state (which in most cases effectively equals the added amine concentration); v to the osmotic volume of the vesicles and K the amine dissociation constant. The type of data obtained is illustrated in fig. 2 for imidazole (pK = 6.9). The figure also illustrates the predicted finding (see equation 2) that the extra proton-uptake will be a (linear) function of the osmotic volume and will therefore decrease with an increase in osmolarity (in this case by the addition of sorbitol). Thus confirming the assumption underlying the derivation of equation 2, that the amines used are accumulated essentially freely in the internal volume of the vesicles, rather than being bound to the membrane.

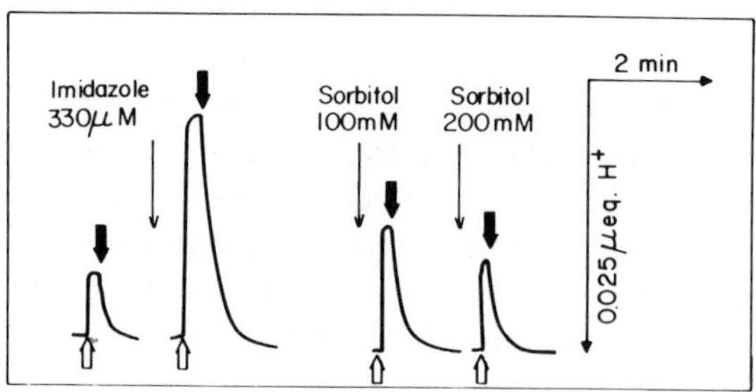


Fig. 2: Stimulation of proton uptake by imidazole.

The reaction mixture contained in 3 ml: KCl, 20 mM; Tricine, pH 7.9, 1 mM; pyocyanine, 15 μM, and chloroplasts containing 50 μg of chlorophyll. Light and dark arrows refer to onset and termination of illumination, respectively (Pick and Avron, unpublished).

Figure 3 illustrates the dependence of the stimulation of proton uptake upon the external amine concentration for seven amines differeing in pK from 4.5 (aniline) to 10 (9-aminoacridine). It can be seen that all exhibit the predicted (equation 2) linear dependence when used within a limited range of concentrations. Deviations from linearity have been traced to: (a) the well known uncoupling by amines at sufficiently high concentrations (e.g., $NH_4Cl$, 9-aminoacridine) (b) Swelling of the vesicles which occurs at higher concentrations when the internal osmolarity increases significantly (e.g. imidazole) and (c) Binding to the membrane which can account for a significant portion of the amine taken up at the lower concentration range (e.q. pyridine). This latter effect can be detected by the lack of response to increased external osmolarity (see fig. 2, and below).

Table 1 summarizes the internal pH values calculated from such data. It can be seen that similar values were obtained for the internal pH and the ΔpH with the different amines used and at all the tested pH values. Furthermore these values are similar to those obtained with the ammonium electrode method and the 9-aminoacridine fluorescence quenching method (table 2).

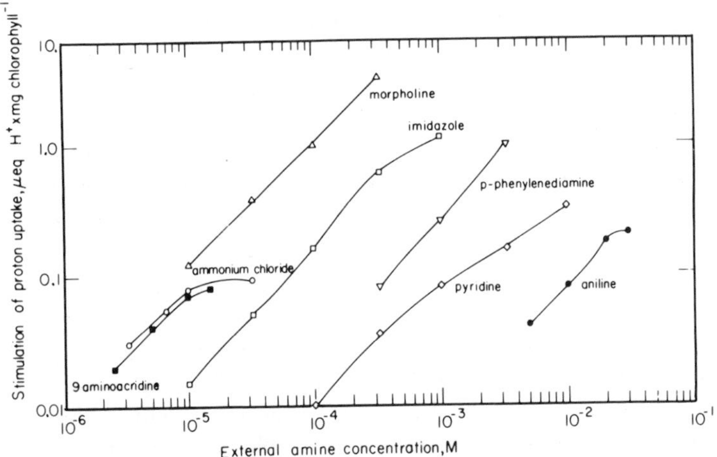


Fig. 3: Dependence of the stimulation of proton uptake by amines on the amine concentration.

The reaction mixture contained in 2.5 ml: KCl, 30 mM; tricine, pH 8.9, 0.5 mM; pyocyanine, 30 μM; and chloroplasts containing 50 μg of chlorophyll. The amine at the concentration indicated was added following three illumination cycles in its absence.

Table 1: The internal pH of illuminated chloroplasts at several external pH values, as calculated from the stimulation of proton uptake by several amines

| Amine | pK | External pH | | | |
|---|---|---|---|---|---|
| | | 8.8 | 7.9 | 7.0 | 6.1 |
| Ammonium chloride | 9.25 | - | 5.0 | - | 6.1 |
| morpholine | 8.4 | 5.2 | 4.8 | - | - |
| imidazole | 6.9 | 5.0 | 4.3 | 4.2 | - |
| p-phenylenediamine | 6.2 | 5.1 | 4.3 | 3.8 | 3.8 |
| pyridine | 5.2 | 5.0 | 4.0 | 3.7 | |
| aniline | 4.5 | 4.9 | 4.2 | 3.8 | 3.6 |
| Average $pH_i$ | - | 5.1 | 4.3 | 3.8 | 3.7 |
| Average ΔpH | - | 3.7 | 3.6 | 3.2 | 2.4 |

The use of membrane bound pH indicators for monitoring directly internal pH independent of internal volume measurements was attempted early by several investigators, and was abandoned because of difficulties in the quantitative evaluation of such data (Avron, 1966; Lynn, 1968; Mitchell, et al., 1968). Recently, Auslander and Junge (1975) suggested that under specified conditions neutral-red can serve as a reliable internal pH indicator. We checked whether this holds true also under the steady state illumination conditions employed with the other described techniques. Figure 4 (Pick and Avron, 1976) describes the type of observations made, and confirms the basic interpretation of Auslander and Junge. The light induced absorbance changes of neutral red were not affected by adding an external buffer, tricine, but were fully sensitive to the addition of the uncoupler SF-6847. We do not find bovine-serum-albumin (BSA) to be an ideal external buffer probably because it competes for neutral red binding with the membranes. Internal buffers like imidazole, at non-uncoupling concentrations (1 mM), slow down the kinetics of onset and, more clearly, decay of the neutral red absorbance change, while valinomycin accelerates both. However, this method was not found suitable for a quantitative evaluation of the pH gradient mainly because of the following phenomena: (a) direct binding studies indicated that about twice as much neutral red was bound to the chloroplast membranes in the light as in the dark; (b) essentially all of the bound neutral red responded in the light to the internal pH indicating that some of the bound neutral red molecules reorient themselves in the light to enable them to sense the internal pH; and (c) the low vs. high pH difference spectrum of the membrane bound neutral red differs markedly from that of the dye in solution.

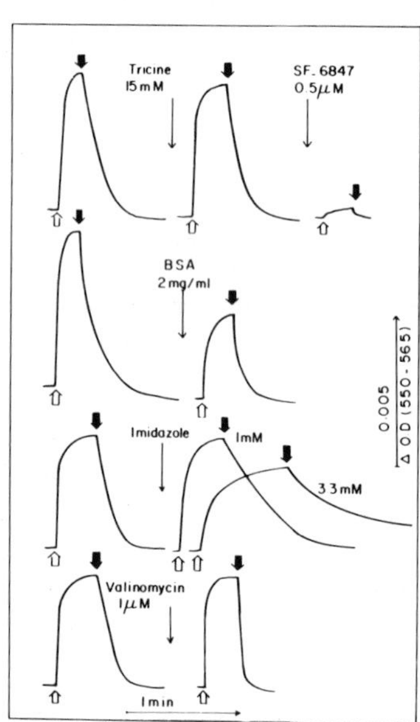


Fig. 4: The response of neutral red to the light induced pH gradient across chloroplast vesicles.

The reaction mixture contained in 3.0 ml: KCl, 40 mM; tricine, pH 7.9, 1 mM; pyocyanine, 15 μM; neutral red, 3.3 μM, and chloroplasts containing 25 μg chlorophyll. Measurements were made with an Aminco-Chance dual wavelength spectrophotometer.

Table 2 compares the ΔpH values as a function of medium pH as measured by different authors using several of the methods described. Except for the first two methods in which low values are obtained because of light limitation, reasonable agreement is observed. The values calculated from the distribution of N-(1-naphthyl)ethylenediamine are somewhat high possibly due to some internal binding.

Table 2: Comparison of ΔpH values determined by different methods in illuminated chloroplasts

| Method | External pH 6.0 | 7.0 | 8.0 | 9.0 | Reference |
|---|---|---|---|---|---|
| Methylamine distribution | 1.0 | 1.7 | 2.3 | 2.7 | Rottenberg *et al.*, 1972 |
| aniline distribution | - | 3.1 | 3.0 | - | Portis & McCarty, 1973 |
| Ammonium electrode | 1.7 | 2.5 | 3.5 | 3.5 | Rottenberg and Grunwald, 1972 |
| 9-aminoacridine quenching | 1.7 | 2.6 | 3.5 | 4.0 | Pick *et al.* 1974 |
| N-(1-naphthl)ethylenediamine quenching | 2.4 | 3.0 | 3.9 | 4.9 | Chow and Hope 1976 |
| Stimulation of proton uptake | 2.2 | 3.2 | 3.6 | 3.9 | Pick and Avron unpublished |

## 2. Proton Gradients and ATP Formation

The relation of the measured ΔpH values to ATP formation in chloroplasts was investigated using several approaches. It was observed that, as expected from the chemiosmotic hypothesis which postulates that ATP formation is driven by utilizing the pH gradient, the gradient observed during phosphorylation was smaller by about 0.5 pH units than that observed in the absence of phosphorylation (Pick *et al.*, 1973). When very low light intensities were used, the kinetics of the development of the pH gradient and of ATP synthesis could be simultaneously observed. As can be seen in Fig. 5 (Pick *et al.*, 1974b) the gradient developed immediately upon turning the light on, but a distinct lag in ATP formation was observed. On plotting the rate of ATP synthesis vs. the magnitude of ΔpH at any time point (fig. 5B), it is clear that no ATP formation could be observed before a threshold of ΔpH (about 2.3 units) was built up. Beyond that point the rate of ATP formation was sharply and linearly dependent upon further increase in ΔpH. Clearly, ATP formation and ΔpH are linked but their relation is not linear.

To quantitate the relation of proton movement across the membrane vesicle to ATP formation chloroplasts were permited to synthesize ATP at low light intensities (see fig. 5) until a steady state was achieved where no further ATP could be synthesized. At this point both the magnitude of the proton concentration gradient and of the phosphate potential were measured. In Table 3 (Avron, 1976) the results of such an experiment are presented. It is evident that the proton concentration gradient measured was thermodynamically capable

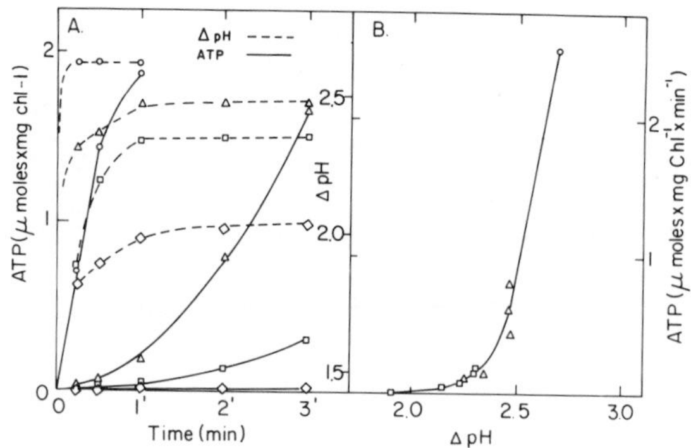


Figure 5: Time course of development of ΔpH and ATP formation at low light intensities

The reaction mixture contained in 3.0 ml: tricine-maleate, pH 7.2, 30 mM; sorbitol, 30 mM; $MgCl_2$, 1 mM; ADP, 1 mM; Pi, 1 mM (containing $3 \times 10^7$ cpm $^{32}P$). ATP, 100 μM, pyocyanine, 25 μM; 9-aminoacridine, 0.5 μM; and chloroplasts containing 75μg of chlorophyll. In A light intensities were in ergs x $cm^{-2}$ x $sec^{-1}$ 0—0; $4.5 \times 10^5$; Δ—Δ $6 \times 10^4$; □—□ $4 \times 10^4$; ◇—◇ $1.5 \times 10^4$.

of providing a sufficient driving force for ATP formation at the measured phosphate potential, if at least 3 protons transversed the membrane per ATP molecule synthesized. Membrane potential values were not considered in the calculation, since it has been amply demonstrated (Schroder et al., 1971; Rottenberg et al., 1972; Chow and Hope, 1976) that no significant membrane potentials exist in the steady state in chloroplasts.

Table 3: Analysis of ΔpH as a driving force for ATP formation in the steady state

| Light intensity | Phosphate potential | Electrochemical gradient | | |
|---|---|---|---|---|
| | | required assuming $2H^+$/ATP | $3H^+$/ATP | measured |
| ergsxcm$^{-2}$xsec$^{-1}$ | Kcalxmole$^{-1}$ | | milivolts | |
| $2 \times 10^5$ | 13.4 | 290 | 193 | 230 |
| $4 \times 10^4$ | 12.8 | 276 | 184 | 197 |
| $2 \times 10^4$ | 11.9 | 258 | 172 | 183 |

Reaction mixture in 3 ml: KCl, 20 mM; phosphate, pH 8.0, 2 mM; $MgCl_2$, 4 mM; ADP, 0.2 mM; phenazine methosulfate, 10 $\mu$M; 9-aminoacridine, 1 $\mu$M; and chloroplasts containing 54 $\mu$g of chlorophyll. It was illuminated for 10-20 minutes until a steady state phosphate potential was attained. Phosphate potential values were calculated from ADP content determination with pyruvate kinase and phosphoenolpyruvate. $\Delta G^{o'}$ of 7.8 Kcalxmole$^{-1}$ at pH 8.0, 10 mM $Mg^{+2}$ 25°C, 0.1 ionic strength was used (Rosing and Slater, 1972).

## 3. Proton Gradient and Reverse Electron Flow

We have previously shown that, under appropriate conditions, the coupled reaction between electron transport and ATP formation in chloroplasts is reversible (Rienits et al., 1974). Thus addition of ATP caused the reduction of "Q" and oxidation of cytochrome f (figures 6, 7). If proton gradients serve as intermediate energy pools between electron transport and ATP formation, it should be possible to demonstrate also proton-gradient dependent reverse electron flow. Indeed (Shahak et al., 1975, 1976) such a reaction was recently found in our laboratory (fig. 8).

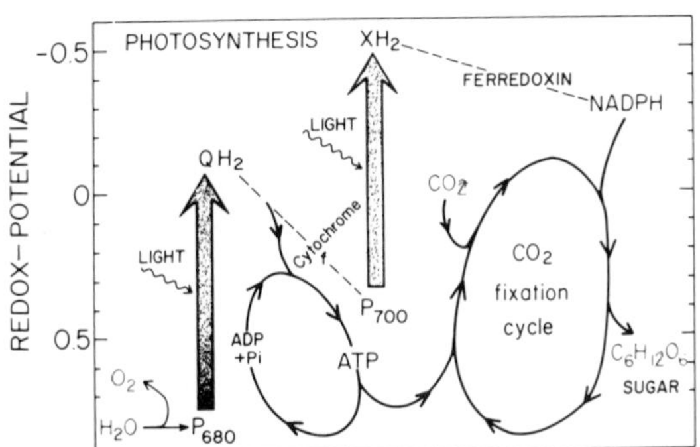


Fig. 6: An oversimplified scheme of photosynthesis indicating the site of coupling between electron transport and ATP formation.

Rapid addition of tris to a chloroplast preparation preequilibrated at pH 5.3 created a momentary pH gradient with the innervesicular space at pH 5.3 and the medium at pH 9.6. This momentary gradient served as a driving force for reverse electron flow, as indicated by the transient reduction of Q (Fig. 8). Reverse electron flow is clearly occurring since oxidizing the reduced electron carriers between the two photosystems by preillumination with far-red light which excites only the photoreaction sensitizing the $P_{700}$ to X electron transfer, (see fig. 6) severely inhibits the acid-base induced reduction of Q, and a following preillumination with green light (which excites also the second photoreaction, thus rereducing the electron carriers) fully restores the reaction (Fig. 9).

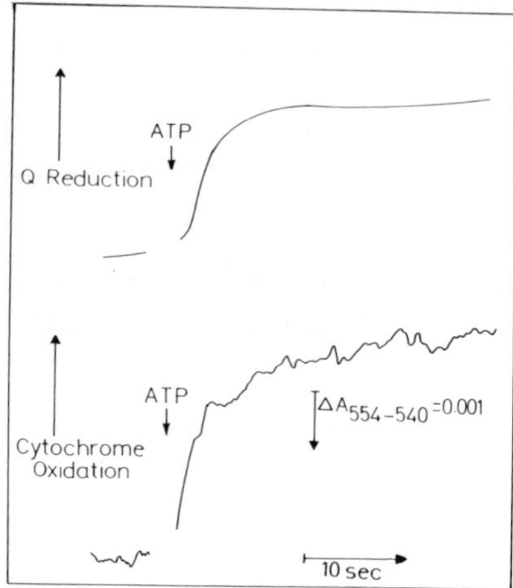


Fig. 7: ATP driven reduction of Q and oxidation of cytochrome f.

Q reduction was followed by the increase in chlorophyll fluorescence yield which accompanies it. Cytochrome f oxidation in a dual wavelength spectrophotometer at 554-540 nm. For further details see Rienits et al., 1974.

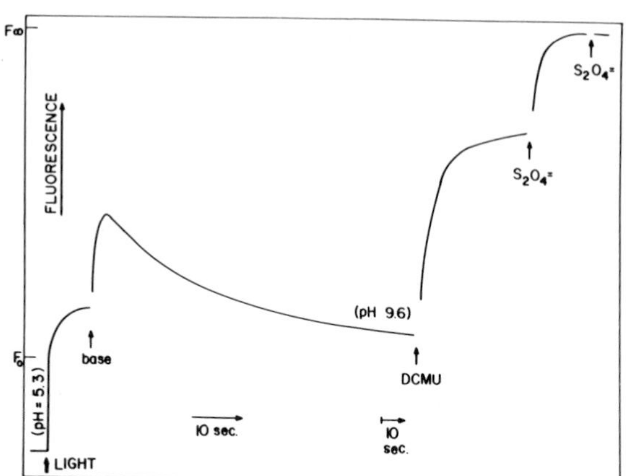


Fig. 8: Proton gradient drive reduction of Q

$F_o$ level indicates fully oxidized Q and $F_\infty$ level (observed on addition of DCMU and/or dithionite) fully reduced Q. Base refers to an injection of 0.2 ml of 1 M Tris at a predetermined pH to result in a final pH of 9.6 into a reaction mixture which contains in 2.0 ml: maleic acid, pH 5.3, 3 mM; $MgCl_2$, 10 mM; KCl, 30 mM; and chloroplasts containing 40 $\mu$g of chlorophyll. Where indicated 1.5 $\mu$M DCMU and a few grains of dithionite were added.

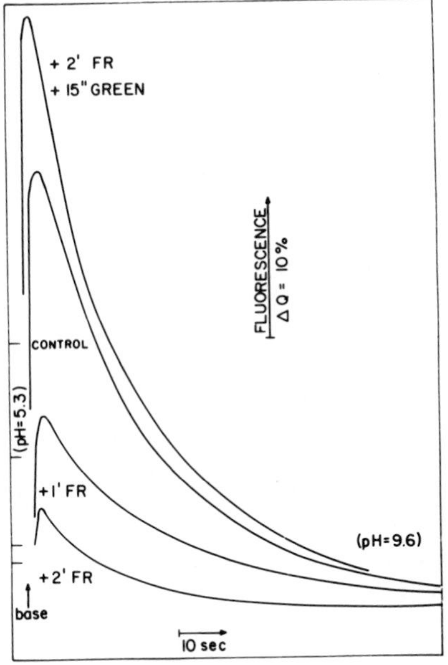


Fig. 9: Effect of far-red (FR) and green preillumination on the proton gradient driven reduction of Q.

Reaction conditions as described under Fig. 8.

During the last few months conditions were found under which proton gradient driven reverse electron flow was extended to include the photosystem. Thus, acid-base transition induced the emission of light photons from the photosystem. The experimental procedure and results are illustrated in Fig. 10. Following a preillumination period, the chloroplasts are placed in the dark during which their native luminescence decays essentially completely, the shutter in front of the photomultiplier is then opened to permit observation, Tris is injected and immediately a transient light emission is apparent. DCMU, which blocks electron flow between the site of coupling and Q, fully inhibits the reaction. Also, preillumination with far-red light (Fig. 11) to empty the electron donating pool, severely inhibits the reaction, clearly indicating that proton gradient driven reverse electron flow is leading to the observed luminescence under these conditions (see Fleischman and Mayne, 1973).

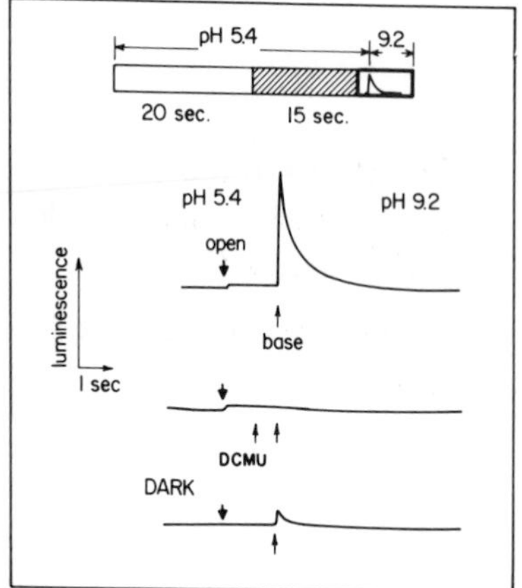


Fig. 10: Proton gradient driven reverse-electron flow luminescence.

The experiment was performed in the instrument designed by Malkin (Malkin and Hardt, 1971). Reaction mixture included in 2 ml: succinate, 3 mM; KCl, 30 mM; $MgCl_2$, 10 mM and chloroplasts containing 40 $\mu$g of chlorophyll. Final pH 5.4. 0.15 ml 0.5 M Tris and 0.1 ml 0.1 mM DCMU in 10% methanol were injected where indicated. Preillumination was with green light $8 \times 10^4$ ergs x $cm^{-2}$ x $sec^{-1}$

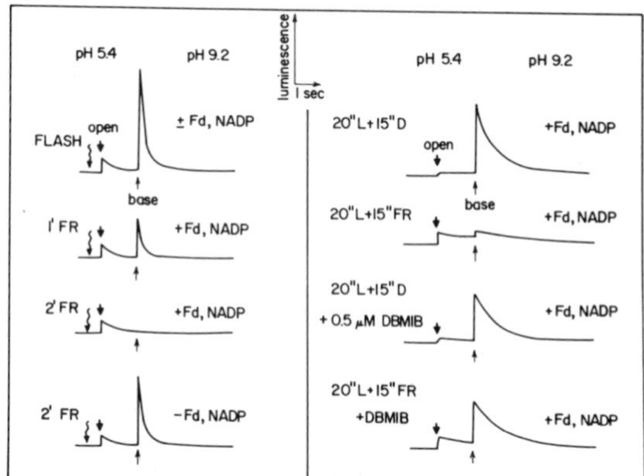


Fig. 11: Effect of far-red preillumination on proton gradient driven reverse-electron-flow-luminescence.

Conditions as described under Fig. 10, except that the wavy arrow indicates a saturating flash of green light with $t_{1/2} = 100\ \mu$sec , and that 95 $\mu$g ferredoxin and 0.25 mM NADP were added where indicated. Far-red illumination was through a 730 nm interference filter at an intensity of $10^4$ ergs x $cm^{-2}$ x $sec^{-1}$.

In conclusion, I have tried to show data from experiments with isolated chloroplasts which demonstrate the intimate qualitative and quantitative relation of proton gradients as an intermediary energy pool between electron transport and the ATP synthesizing apparatus. The system can be experimentally manipulated to show light induced electron transport, light induced proton gradient, proton gradient induced ATP formation, and in reverse, ATP induced proton gradient formation, ATP induced reverse electron transport, proton gradient induced reverse electron transport and proton gradient induced reverse-electron-transport-luminescence. About three protons seem to be necessary to drive the synthesis of an ATP molecule in the steady state. However, the mechanism through which this feat is achieved still eludes us, and remains as a challenge for the future.

## References


Avron, M., The relation of light induced reactions in isolated chloroplasts to proton concentrations. In: Proc. 2nd International Congress on Photosynthesis. Edited by G. Forti et al., pp. 861-871, N.V. Junk, The Hague, 1972.

Avron, M., Energy transduction in isolated chloroplast membranes, in: The Structural Basis of Membrane Function. Edited by Y. Hatefi and L. Djavani-Ohaniance, pp. 227-238, Academic Press, New York 1976.

Avron, M., Bromthymol blue as an internal pH indicator, Brookhaven Symposium 19, 243, 1966.

Auslander, W. and W. Junge, Neutral-red, a rapid indicator for pH changes in the inner phase of thylakoids. FEBS Letters 59, 310-315, 1975.

Casadio, R., A. Baccarini-Melandri and B.A. Melandri, Eur. J. Biochem. 47, 121-128, 1974.

Chow, W.S. and A.B. Hope, Light induced pH gradients in isolated spinach chloroplasts. Aust. J. Plant. Physiol. 3, 141-152, 1976.

Crofts, A.R., Ammonium uptake by chloroplasts and the high -energy state. In: Regulatory Function of Biological Membranes. Edited by J. Jarngelt, pp 247-263, Elsevier, Amsterdam, 1968.

Deamer, D.W., R.C. Prince, and A.R. Crofts, The response of fluorescent amines to pH gradients across liposome membranes. Biochim. Biophys. Acta 274, 323-335, 1972.

Fiolet, J.W.T., E.P. Bakker and K. Van Dam., The fluorescence properties of acridines in the presence of chloroplasts or liposomes. Biochim. Biophys. Acta 368, 432-445, 1974.

Fiolet, J.W.T., L.V.D., Haar, R. Kraayenhof, and K. Van Dam. On the stimulation of the light induced proton uptake by uncoupling aminoacridine derivatives in spinach chloroplasts. Biochim. Biophys. Acta 387, 320-334, 1975.

Fleischman, D.F. and B.C. Mayne, Chemically and Physically induced luminescence as a probe of photosynthetic mechanisms. Current topics in Bioenergetics 5, 77-105, 1973.

Graber, P. and H.T. Witt, Relation between the electrical potential, pH gradient, proton flux and photophosphorylation in the photosynthetic membrane. Biochim. Biophys. Acta 423, 141-162, 1976.

Jagendorf, A.T., Mechanism of Photophosphorylation, In: Bioenergics of Photo-Synthesis. Edited by Govindjee, pp. 423-492, Academic Press, New York, 1975.

Kraayenhof, R. Quenching of uncoupler fluorescence in relation to the "energized state" in chloroplasts. FEBS Letters 6, 161-165, 1970.

Lynn, W.S. Proton and electron poising and photophosphorylation in chloroplasts. Biochemistry 7, 3811-3820, 1968.

Lynn, W.S., Changes in internal proton concentration associated with photophosphorylation in intact and sonically treated chloroplasts. J. Biol. Chem. 243, 1060-1064, 1968.

Mitchell, P., J. Molye and L. Smith, Bromothymol blue as a pH indicator in mitochondrial suspensions, Eur. J. Biochem. 4, 9-19, 1968.

Malkin, S. and H. Hardt, Kinetics of various emission processes in chloroplasts, evidence for various reaction types. In: Proc. 2nd International Congress on Photosynthesis, Edited by G. Forti et al., pp. 253-269, 1972.

Nelson, N., H. Nelson, V. Naim and J. Neumann, Effect of pyridine on the light induced pH rise and post illumination ATP synthesis in chloroplasts. Arch. Biochem. Biophys. 145, 263-267, 1971.

Pick, U., H. Rottenberg and M. Avron, Effect of photophosphorylation on the size of the proton gradient across chloroplast-membranes. FEBS Letters, 32, 91-94, 1973.

Pick, U., H. Rottenberg and M. Avron, Proton gradients, proton concentrations and photophosphorylation. In: Proceedings 3rd International Congress on Photosynthesis. Edited by M. Avron, pp. 967-974, Elsevier, Amsterdam 1974b.

Pick, U., H. Rottenberg and M. Avron, The Dependence of photophosphorylation in chloroplasts on ΔpH and external pH, FEBS Letters 48, 32-36, 1974.

Pick, U. and M. Avron. Neutral red response as a measure of the pH gradient across chloroplast membranes in the light. FEBS Letters, In press, 1976.

Portis, A.R. and R.E. McCarty. On the pH dependence of the light induced hydrogen ion gradient in spinach chloroplsts. Archives Biochem. Biophys. 156, 621-625, 1973.

Portis, A.R. and R.E. McCarty, Effect of adenosine nucleotides and of photophosphorylation on $H^+$ uptake and the magnitude of the $H^+$ gradient in illuminated chloroplasts. J. Biol. Chem. 249, 6250-6254, 1974.

Rienits, K.G., H. Hardt, and M. Avron. Energy dependent reverse electron flow in chloroplasts. Eur. J. Biochem. 43, 291-298, 1974.

Rosing, J. and E.C. Slater, The value of $\Delta G^o$ for the hydrolysis of ATP. Biochim. Biophys. Acta 267, 275-290, 1972.

Rottenberg, H. and T. Grunwald, Determination of ΔpH in chloroplasts. Eur. J. Biochem. 25, 71-74, 1972.

Rottenberg, H., T. Grunwald and M. Avron. Determination of ΔpH in chloroplasts. Eurp. J. Biochem. 25, 54-63, 1972.

Schroder, H., H. Muhle and B. Rumberg, Relationship between ion transport phenomena and phosphorylation on chloroplasts. In Proceedings 2nd International Congress on Photosynthesis. Edited by G. Forti et al., pp. 919-930, W.Junk, The Hague, 1971.

Schuldiner, S., Rottenberg, H. and M. Avron, Determination of ΔpH in chloroplasts - Fluorescent amines as a probe for the determination of ΔpH in chloroplasts. Eurp. J. Biochem. 25, 64-70, 1972.

Shahak, Y., H. Hardt and M. Avron, Acid-base driven reverse electron flow in isolated chloroplasts. FEBS Letters 54, 151-154, 1975.

Shahak, Y., U. Pick and M. Avron, Energy dependent reverse electron flow in chloroplasts. In: Proc. 10th FEBS Meeting, Edited by P. Desnuella and A.M. Michelson, Vol. 40, pp. 305-314, Elsevier, Amsterdam, 1976.

# ENERGY TRANSDUCTION IN THE CHROMATOPHORE MEMBRANE

Margareta Baltscheffsky

Department of Biochemistry, Arrhenius Laboratory

University of Stockholm, Stockholm, Sweden

## ABSTRACT

In chromatophore membranes from the photosynthetic bacterium _Rhodospirillum rubrum_ inorganic pyrophosphate (PPi), which can be formed as an alternative end product to ATP in photophosphorylation, may also be utilized as energy donor in the dark in reversed energy conversion reactions. The light-induced energy conversion is reflected in a biphasic absorbance change of endogenous membrane-bound carotenoids. Evidence is presented that the slow phase of this absorbance change is in response to the presence and activity of the ATP-synthesizing enzyme. The present widespread use of the total carotenoid absorbance change as an indicator of a transmembrane potential is discussed and questioned. Alternative interpretations are given on the basis of our findings concerning different characteristics of two phases of the total carotenoid absorbance change.

## INTRODUCTION

Chromatophores, the small vesicular structures obtained from the inner membrane of photosynthetic bacteria, are capable of carrying out the whole process of harvesting light energy and converting this energy into a biologically useful form in the synthesis of the energy-rich compound ATP. How is this done? A great help on the way to get a better understanding of this process has been offered the investigator by certain special features present in the chromatophores from some photosynthetic bacteria. One of these is the ability to form and utilize inorganic pyrophosphate

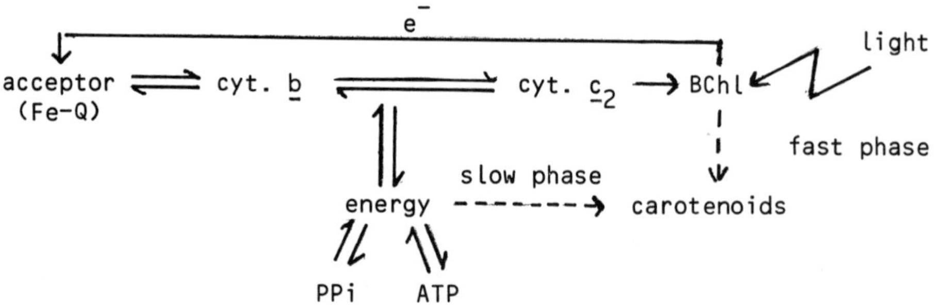


Fig. 1. Tentative minimum scheme for the light induced electron transport and phosphorylation in *R. rubrum* chromatophores

(PPi) as an alternative energy reservoir and source. Another one is the presence of a built-in membrane probe responding to the energetic state, the carotenoids. The combination of these two features in *Rhodospirillum rubrum* has provided new opportunities to study the energy conversion reactions of this photosynthetic system thus extending the possibilities as compared to other energy converting systems. The information obtained has given us new insight into the mechanism of bacterial photophosphorylation. More generally, it may also contribute to the current concepts on the still elusive details of mechanism of cellular electron transport-coupled phosphorylation.

An overall tentative scheme for the energy conversion reactions in chromatophores from *Rhodospirillum rubrum* is given in Fig. 1.

## PPi AS AN ENERGY DONOR

### PPi as an Alternative End Product

The finding that inorganic pyrophosphate was formed at the expense of light energy as an alternative pathway to ATP in the absence of phosphate acceptor (H. Baltscheffsky et al., 1966) formed a basis for the exploration of the properties of PPi as energy donor. The synthesis of PPi was shown to be dependent on electron transport, as is the ATP-synthesis, and was inhibited by uncouplers in the same concentration range as that inhibiting ATP-synthesis. One significant and experimentally useful difference between phosphorylation to PPi and ATP was found, in that the energy transfer inhibitor oligomycin had no inhibitory effect on the PPi-synthesizing system, even at concentrations where the ATP-synthesis was totally inhibited. In fact, oligomycin induced a marked stimulation of PPi-synthesis at those concentrations, as if an energy leak through the potential ATP-pathway was cut off (H. Baltscheffsky and von Stedingk, 1966). It has also been shown that the PPi-synthesis is saturated at lower light intensities than the ATP-synthesis. The rate of PPi-formation at these low intensities is equal to that of ATP-formation, whereas it is only about one tenth at the higher intensities that saturate the ATP-synthesis (Guillory and Fisher, 1972).

## Localization of the Two Terminal Photophosphorylation Enzymes

The topology of the membrane with respect to the energy conservation system in chromatophores is a question of interest, perhaps specially since in R. rubrum chromatophores we have the two terminal energy-rich products, ATP and PPi, and since all the available evidence indicates that the synthesis of these two products is catalyzed by two different membrane-bound enzymes.

An early investigation by Löw and Afzelius (1965) showed that chromatophores from R. rubrum in the electron microscope display typical knoblike structures on the outside, similar to those which had earlier been described in submitochondrial particles and in chloroplast membranes, and identified as the ATPase enzymes $F_1$ and $CF_1$ by the elegant work of Racker and his associates (Kagawa and Racker, 1966; Lien and Racker, 1971). In chromatophores from Rhodopseudomonas spheroides the identification of the knoblike structures with the ATPase has been achieved by Reed and Raveed (1972). The ATPase activity can relatively easily be detached from the chromatophores and reconstituted with the depleted membranes (Johansson, 1972; Johansson et al., 1972). Where is then the PPi-ase localized? For this we have not yet been able to obtain any direct evidence, but a number of observations are consistent with the concept of a membrane-bound, integral protein, dependent on a hydrophobic environment for its activity. All attempts to isolate this enzyme have so far failed. Treatments, that completely remove or inactivate the ATPase, i.e. sonication in the presence of EDTA (Table 1) or washing with 1 M LiCl (Fisher and Guillory, 1972) only marginally affect the PPiase. Also, treatments that can be considered as interfering with the membrane lipid structure, such as addition of ethanol and butanol (Baltscheffsky et al., 1966) or incubation with phospholipase A (Klemme et al., 1971) interfere at lower concentrations or milder conditions with the PPiase than with the ATPase. The inactivation after treatment with phospholipase is reversed by the addition back of phospholipids. The general picture that seems to emerge is that of an enzyme residing in the membrane, possibly at the protein lipid interphase, dependent on a lipid environment for its activity.

TABLE 1

Effect of Sonication in 1 mM EDTA on Chromatophore-Bound ATPase and PPiase

| Sonication time | % remaining activity ATPase | PPiase |
|---|---|---|
| 0 | 100 | 100 |
| 3 x 30 s | 3.2 | 85 |
| 10 x 30 s | 0 | 89 |

Reaction mixture for ATPase and PPiase contained in a final volume of 2 ml: 50 mM Tris-HCl, pH 8.0, 5 mM $MgCl_2$, 5 mM ATP or PPi and chromatophores equivalent to 25 μM Bchl. Reaction time 10 min.

## PPi as energy donor

PPi has been shown to be an efficient energy donor in this photophosphorylating system. When added in the dark to chromatophores it is capable of driving various energy requiring reactions, amont those reduction of cytochrome b by electrons from cytochrome $c_2$ (Fig. 2) and an absorbance change of endogenous membrane-bound carotenoids (Baltscheffsky, 1967a; 1967b). For these two reactions the "apparent efficiency" of PPi as an energy donor is at least twice that of ATP. PPi is usually hydrolyzed twice as fast as ATP. In Fig. 3 it is seen that the carotenoid absorbance change is both larger in extent and rises faster with PPi than with ATP. The dependence of the energy-requiring cytochrome b reduction upon the hydrolysis of PPi has been established in experiments where the spectrophotometric monitoring of a PPi-induced reduction-oxidation cycle has been combined with simultaneous analysis of the PPi-concentration at different points. The "turning point", where the net reoxidation starts is at 25 μM PPi (Fig. 4). This is the same concentration that earlier (Baltscheffsky, 1967) has been shown to be required for the maximum extent of cytochrome reduction or carotenoid absorbance change.

In chromatophores both from R. rubrum and from some other species of photosynthetic bacteria it has been shown that PPi can serve as an energy donor for the energy requiring transhydrogenase as well as for succinate-linked pyridine nucleotide reduction (Keister and Yike, 1967a; 1967b; Jones and Saunders, 1972). In these cases the rate of the ATP-driven reaction is usually higher than the rate of the PPi-driven reaction. The stoichiometry

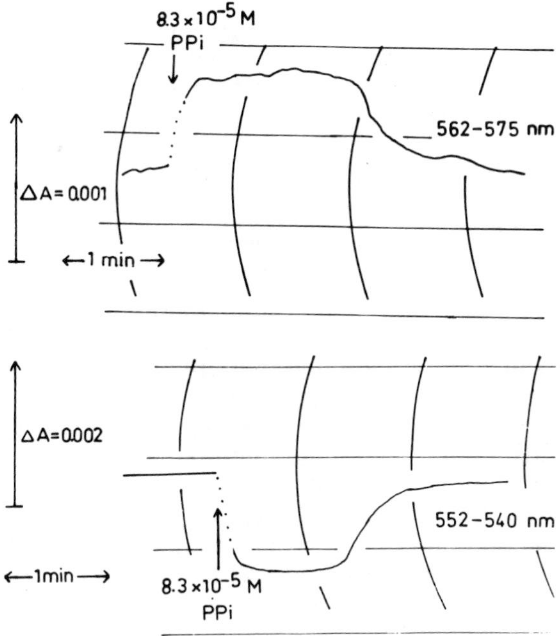


Fig. 2. PPi-induced reversed cross-over between cytochromes b and $c_2$.
The reaction mixture contained in a final volume of 2.5 ml: 0.2 M glycyl-glycine buffer, pH 7.4, 5 mM $MgCl_2$ and chromatophores from *R. rubrum* strain G-9 equivalent to 125 μM Bchl.

for the transhydrogenase is close to 1 ATP per NADPH formed whereas for PPi this value is 6-8 PPi per NADPH.

## PPi as Energy Donor for ATP Synthesis

The presence of two separate terminal systems for energy conservation in the chromatophore membrane offers a unique possibility to study the phosphorylation reactions at the level of the terminal enzymes.

We found spectrophotometrically that chromatophores energized in the dark with PPi respond to the addition of ADP in the same

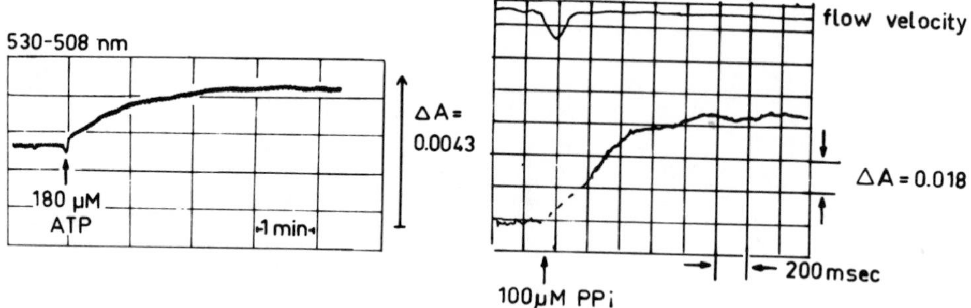


Fig. 3. ATP- and PPi-induced carotenoid absorbance change. Reaction mixture contained for the experiment with ATP: 0.2 M glycyl-glycine buffer, pH 7.4, 2 mM $MgCl_2$ and chromatophores equivalent to 37 μM Bchl in a final volume of 1.5 ml.
The experiment with PPi was performed in a stopped flow apparatus from the Johnson Research foundation. The reaction mixture contained: 0.2 M glycyl-glycine buffer, pH 7.4, 5 mM $MgCl_2$ and chromatophores equivalent to 73 μM Bchl in a total volume of 15 ml.

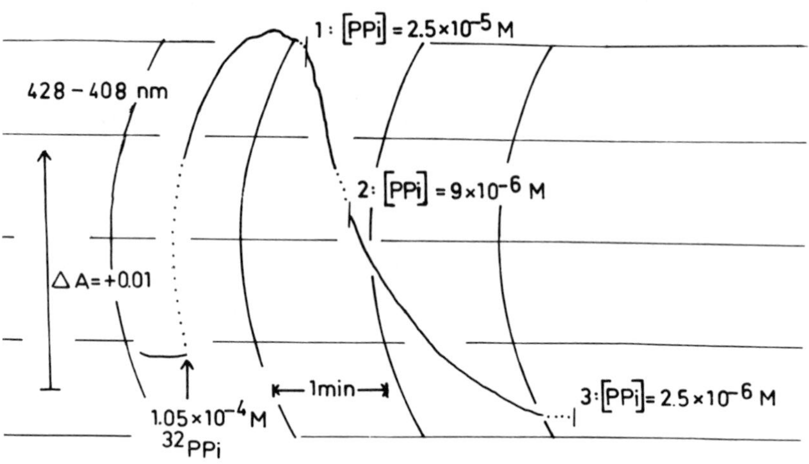


Fig. 4. Simultaneous measurement of PPi concentration and cytochrome b cycle. The reaction mixture contained in a starting volume of 3.5 ml: 0.2 M glycyl-glycine buffer, pH 7.4, 2.5 mM $MgCl_2$, and chromatophores corresponding to 125 μg Bchl. At the points marked 1, 2 and 3 aliquots of 1 ml were withdrawn and analyzed for their PPi content, by determining $^{32}PPi$ and $^{32}Pi$. (From Baltscheffsky, 1969b).

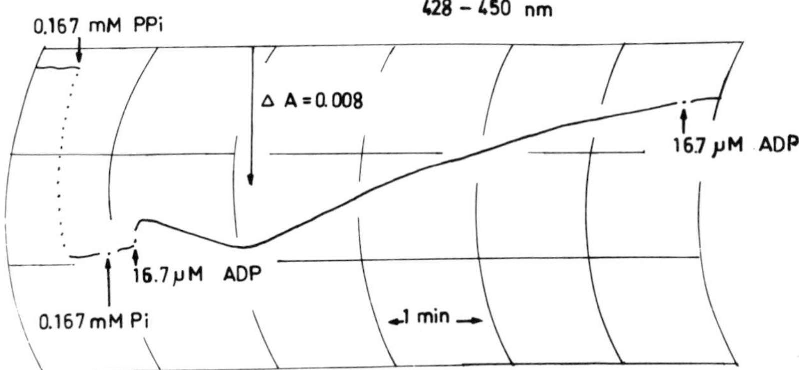


Fig. 5. ADP induced change on dark, PPi energized redox state of b-type cytochrome. Reaction mixture contained in a final volume of 1.2 ml: 0.2 M glycyl-glycine buffer, pH 7.5, 1.25 mM $MgCl_2$, 83 μM Na-succinate and chromatophores equivalent to 43 μM bacteriochlorophyll. (From Baltscheffsky and Baltscheffsky, 1972).

way as if they were energized by illumination (M. Baltscheffsky and H. Baltscheffsky, 1970, 1972). Fig. 5 shows that cytochrome b reduced in the dark by energization with PPi becomes partly reoxidized by addition of ADP which causes an oxidation-reduction cycle, apparently concomitant with its phosphorylation to ATP. This reaction is inhibited by oligomycin but not by antimycin, showing that the electron transport chain is not involved in this energy transfer from PPi to ATP. In fact, it is the case of only energy transfer, as phosphate emanating from PPi does not appear to participate in the reaction, since addition of exogenous Pi is necessary for obtaining the effect of ADP.

Keister (1971) has investigated this PPi-driven ATP synthesis in some detail and found that the stoichiometry is about 10 PPi hydrolyzed per ATP formed. However, this stoichiometry is not altered by the simultaneous occurrence of another energy driven reaction, for example $NAD^+$ reduction with succinate as electron donor. This indicates that the energy from PPi only partially is utilizable for ATP-synthesis and that there probably is no direct interaction between the two terminal enzymes.

## THE CAROTENOID ABSORBANCE CHANGE AS A MEMBRANE PROBE

In early studies of the light induced redox changes of cytochromes it was found that, in *R. rubrum* chromatophores or whole cells, there were light induced absorbance changes between 500-600 nm which were too large to be attributable to the α-bands of cytochromes. These absorbance changes were found to be due to carotenoids. The extent of these light-induced changes was decreased under phosphorylating conditions (Smith *et al.*, 1960).

In the course of studying the effect of reversed phosphorylation on the redox state of endogenous cytochromes, by addition of PPi or ATP in the dark, I observed the same type of carotenoid absorbance changes as were induced in the light (Baltscheffsky, 1967, 1969). These changes were inhibited by uncouplers and, when induced by ATP, also inhibited by oligomycin. In contrast to the light-induced change, the dark energization was not inhibited by the electron transport inhibitor antimycin A, indicating a direct interaction between the energy-rich state and the carotenoids rather than an interaction *via* the electron transport system.

The light induced carotenoid absorbance change when induced by a short xenon flash shows in *R. rubrum* chromatophores biphasic kinetics (Fig. 6a). The fast phase, which is not time-resolved in

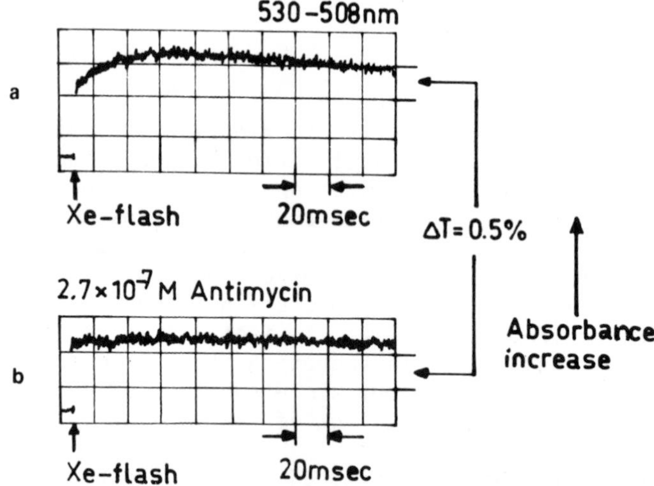


Fig. 6. Flash-induced carotenoid absorbance change at 530-508 nm. Chromatophores equivalent to 41 μM bacteriochlorophyll suspended in 0.2 M glycylglycine buffer (pH 7.4). Illumination source was a single saturating Xe flash of 5 μs (full width at half height) passed through double layers of Wratten 88A gelatine filters. (From Baltscheffsky, 1974)

this figure, has a rise time shorter than 0.2 µsec whereas the slower phase has a half rise time of about 10 msec. The slower phase is completely abolished by antimycin A (Fig. 6b), and thus seems to respond to some event related to coupled electron flow in the cytochromes b-c region. This inhibition by antimycin of the slow phase makes it possible to study the two phases separately even under continuous illumination and reveals that there are a number of differences in response to various influences between the two phases. Since in the literature it has been repeatedly assumed that all of the carotenoid absorbance change occurs in response to a transmembrane potential (Jackson and Crofts, 1969, 1971; Jackson et al., 1975) it would appear relevant to consider these differences in some detail.

## Light Intensity Dependence

A lack of antimycin sensitivity of the light induced absorbance change at low light intensity led us to study this change as

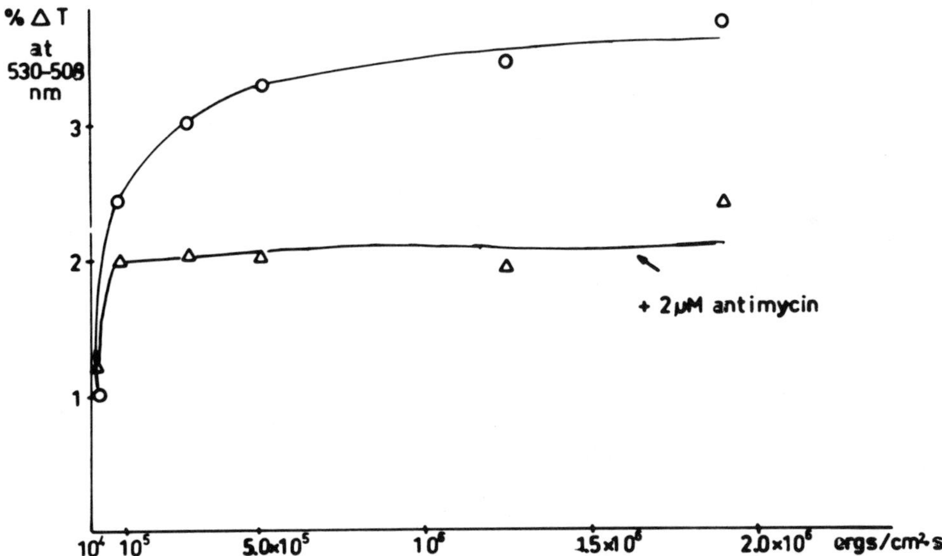


Fig. 7. Carotenoid absorbance change as a function of light intensity. The reaction mixture contained in a final volume of 1.5 ml: 0.2 M glycyl-glycine buffer, pH 7.4, 5 mM $MgCl_2$ and chromatophores equivalent to 40 µM Bchl. The intensity of the actinic light was measured with an YST-instruments light meter.

a function of light intensity in the absence and presence of this inhibitor. These studies were performed with steady state illumination to facilitate measurements of the light intensity. The actinic light was turned on for approximately 30 sec for each measuring point. As is seen in Fig. 7, there is an about tenfold difference in the light saturation in the presence and absence of the inhibitor, showing that the part of the carotenoid absorbance change that corresponds to the fast phase after single flashes saturates at much lower intensity than does the total change (fast + slow), which must mean that the high light requirement is in the slow phase. If we compare these light saturation curves with those of other light induced reactions such as photophosphorylation (Fig. 8) or the light induced proton uptake (Fig. 9) we find that both these reactions are saturated at intensities comparable with that of the slow phase. So it appears that the slow phase may be more intimately connected than the fast one with the events that are normally identified with energy coupling, in agreement with our earlier claims (Baltscheffsky, 1971).

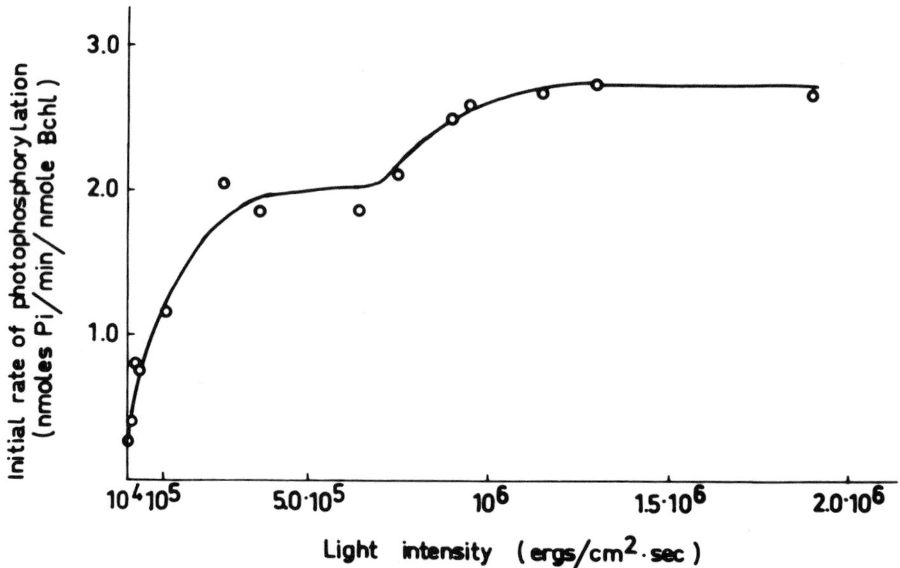


Fig. 8. Photophosphorylation as a function of light intensity. The rate of photophosphorylation was determined according to the pH-method of Nishimura et al., 1962). The reaction mixture contained in a final volume of 3 ml: 0.1 M KCl, 4.7 mM $KH_2PO_4$, 7 mM $MgCl_2$, 0.23 mM Na-succinate and chromatophores equivalent to 19 µM. The starting pH was 7.40 and ADP was added to a final concentration of 3.3 mM.

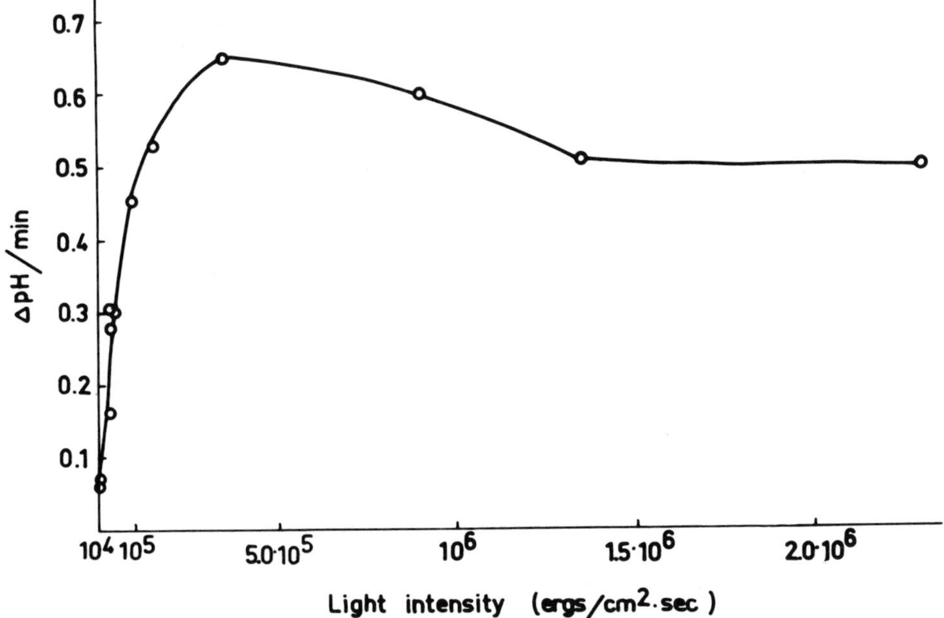


Fig. 9. Light induced $H^+$ uptake as a function of light intensity. The reaction mixture contained in a final volume of 3 ml: 0.1 M NaCl, 0.016 mM Na-succinate and chromatophores equivalent to 27.5 µM.

With this in mind, we have also studied the effects of various agents and treatments on both the fast and slow phases of the carotenoid absorbance change.

## Effects of ADP, ATP and PPi

The extent of inhibition by antimycin on the light induced change was found to be the same as that obtained by the presence of ADP under continuous illumination. If only a low concentration of ADP was added, the inhibition was reversed when the added ADP had been phosphorylated to ATP (Fig. 10). This could be rationally explained, although in my opinion erroneously, by the assumption that phosphorylating conditions cause a fast break-down of the membrane potential and thus an increased rate of decay of the light induced carotenoid change, which under continuous illumination would appear as a decrease in extent. Such an increase in the rate of decay has also recently been reported and studied in detail

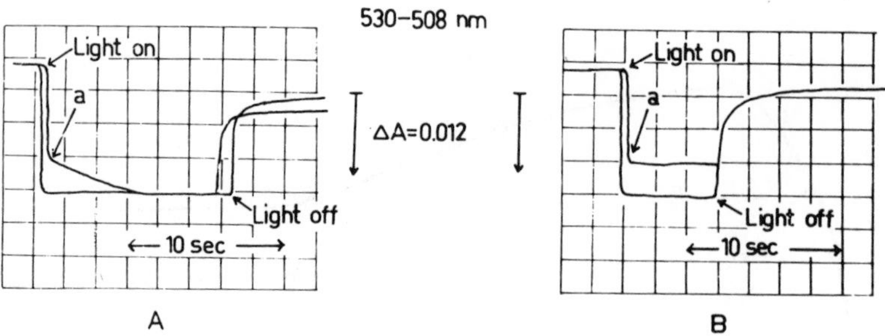


Fig. 10. The reaction mixture contained: 0.2 M glycyl-glycine, pH 7.4; 5 mM $MgCl_2$ and 1 mM Na-succinate and 31 μM Bchl. A: Trace a contained initially 83 μM ADP and 5 mM Pi; B: Trace a contained 0.5 μM antimycin.

(Saphon et al., 1975a, 1975b; Jackson et al., 1975) using a single beam spectrophotometer with a signal averaging system and multiple flashes, and given the above mentioned interpretation. We have instead used a dual beam spectrophotometer and single flashes, which does not give us time resolution enough to study the rise kinetics of the fast phase but is perfectly adequate for following the slow phase. This gives us the advantage to be able to distinguish events after one single flash from those after a number of flashes.

If we add ADP and Pi to a chromatophore suspension and activate the system by one single flash we find that the primary effect of ADP + Pi is to inhibit the slow phase with no apparent acceleration of the decay, as is seen in Fig. 11. If we continue the flash activation we see that already after the second flash there is a slight acceleration of the decay which is even more pronounced after the third flash (Fig. 12). The relative initial rates of the decay are 1:2:3.3 after the 1st, 2nd and 3rd flashes, respectively. This gradual increase of the rate of decay as ATP is being formed would rather point to that ATP, and not ADP + Pi, is responsible for the effect. That this is so can be seen in Fig. 13a. An accelerating effect by ATP on the rate of decay was observed but not commented upon by Saphon et al. (1975b). The acceleration is even more pronounced if PPi is added in place of ATP, Fig. 13b. The effect of PPi was first reported some years ago (M. Baltscheffsky, 1970), and, as is seen in Fig. 13b, it apparently is related to

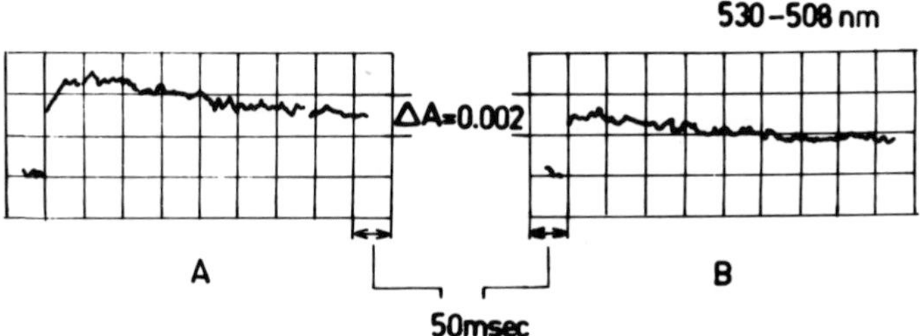


Fig. 11. Effect of ADP + Pi on light induced carotenoid absorbance change.
Light activation was by a Xe-flash of 0.5 msec duration.
Reaction mixture contained in a final volume of 2.0 ml: 0.2 M glycyl-glycine buffer, pH 7.4, 5 mM $MgCl_2$ and 0.3 mM Na-succinate.
In B was added 0.2 mM ADP and 0.5 mM Pi, final concentrations.

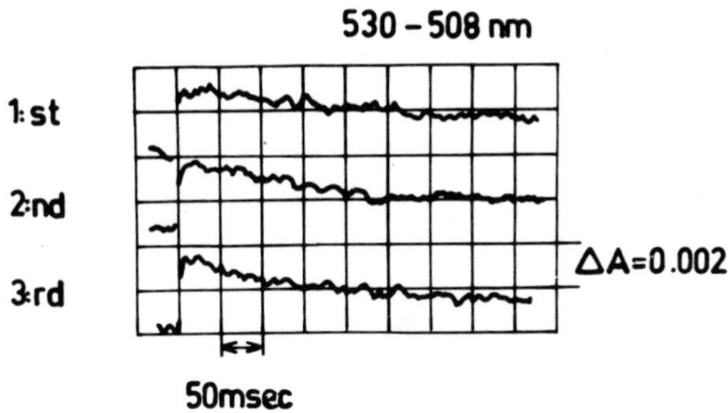


Fig. 12. Effect of multiple flashes on carotenoid absorbance change under phosphorylating conditions. Experimental conditions as in Fig. 11B.

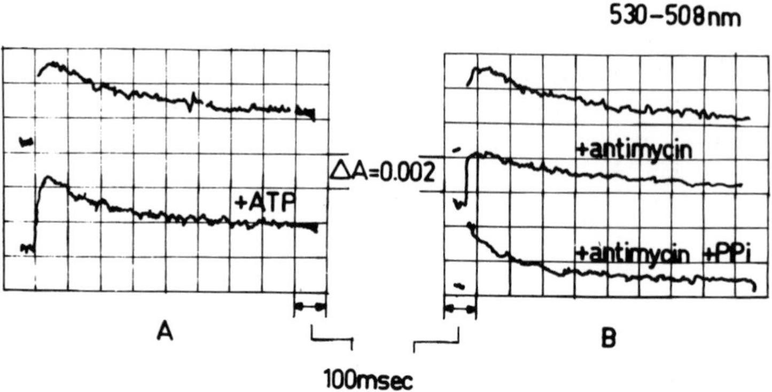


Fig. 13. ATP- and PPi-induced acceleration of the decay rate of light induced carotenoid absorbance change. Experimental conditions as in Fig. 11A. Additions were where indicated: 0.2 mM ATP, 0.2 mM PPi and 1.5 μM antimycin.

the fast phase, since this experiment was performed in the presence of antimycin A.

Recently, Girault and Galmiche (1976) have found that in spinach chloroplasts the decay rate of the 520 nm absorbance change is governed rather by the interaction of ATP with the coupling factor than by photophosphorylation, which is in accordance with the above data. Addition of ADP + Pi also in this system causes a strong inhibition of the extent of the 520 nm signal, both under continuous illumination (Baltscheffsky and Hall, 1974), and after a group of short flashes (Girault and Galmiche, 1976).

The specific inhibitory effect of ADP + Pi on the slow phase can be envisaged to be connected with the binding of the substrate to its enzyme, in this case the ATPase. If this is the case, one might expect to find an effect on the slow phase by removing the ATPase from the chromatophore membrane.

Chromatophores can easily be depleted from the ATPase by sonication, leaving a relatively intact membrane behind. These depleted membranes show a light induced $H^+$ uptake which is 80-90 % of that

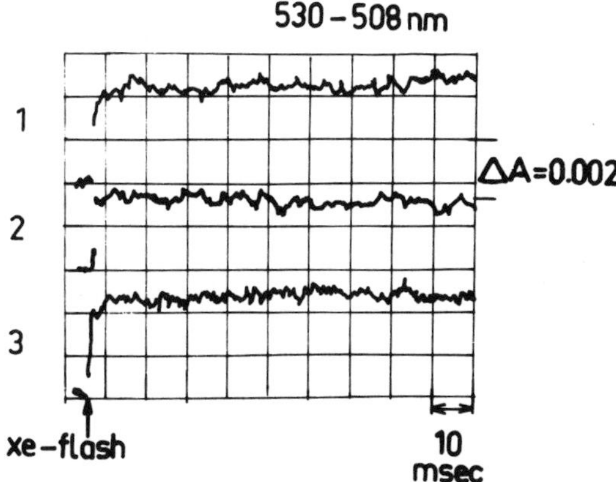


Fig. 14. The effect on the light induced carotenoid absorbance change of depletion from and reconstitution with the coupling factor ATPase. 1, untreated chromatophores; 2, depleted chromatophores; 3, reconstituted chromatophores.
Reaction mixture contained in a final volume of 2.3 ml: 0.2 M glycyl-glycine buffer, pH 7.4, 5 mM $MgCl_2$ and chromatophores equivalent to 53 μM Bchl.
In 3, reconstitution was performed by 30 min preincubation at room temperature with 0.5 ml supernatant fluid. Light activation as in Fig. 6.

in the untreated chromatophores (Johansson, personal communication). The photophosphorylation declines to 10 % or less, as does the ATPase activity, indicating that most of the coupling factor indeed has been physically removed from the membranes. A $Ca^{++}$-stimulated ATPase activity is found in the soluble fraction after sonication and centrifugation and by adding this fraction back to the depleted membranes it is possible to achieve complete restoration of photophosphorylation (Johansson, 1972).

Fig. 14 shows that the depletion of chromatophores results in a loss of specifically the slow phase of the light induced carotenoid change as has been briefly reported earlier (Baltscheffsky, 1974). This is not combined with any significant increase in the rate of decay of the fast phase which would have indicated that the depleted membranes had become "leaky" and unable to maintain a membrane potential. The fact that the $H^+$-uptake is nearly unimpaired also indicates intactness. Under continuous illumination the carotenoid change in the depleted membranes is not sensitive to antimycin. Reconstitution with a sufficient amount of the soluble fraction to restore photophosphorylation also results in restoration of the slow phase after a single flash (Fig. 13) and of the antimycin sensitivity under continuous illumination. Similar to these effects on the endogenous membrane probe, the carotenoid, are the results with added membrane probes. Added ANS (8-anilino-naphthalene-1-sulfonic acid) which has been shown to give energy-linked responses in a number of phosphorylating systems, shows a decreased light induced response which is restored upon reconstitution with the ATPase (Johansson *et al.*, 1971) and the likewise energy-responding probe merocyanin V (Chance and Baltscheffsky, 1975) has recently been shown to perform in the same way (Chance and Baltscheffsky, unpublished experiments).

## Carotenoid Changes at Lower Temperatures

Very recently, in collaboration with B. Chance a study of the light induced carotenoid absorbance change at different temperatures was begun. This investigation of the light-induced carotenoid absorbance change as a function of temperature over a wide range, in chromatophores from three different species of photosynthetic bacteria, has already provided some quite surprising results.

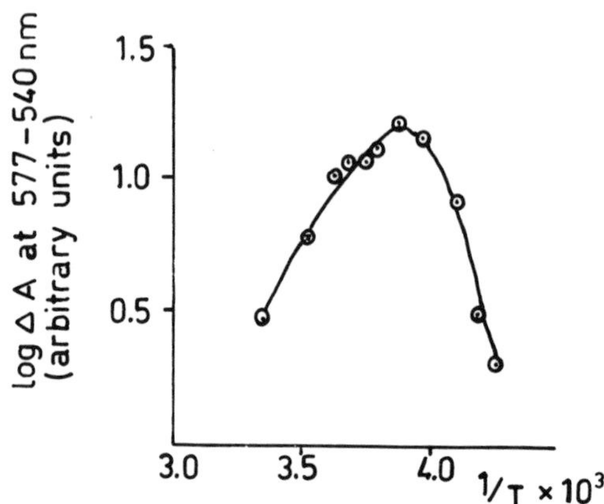


Fig. 15. The light induced carotenoid absorbance change in chromatium chromatophores as a function of temperature. Reaction mixture contained in a volume of 1 ml: 3.5 mM Tris-$SO_4$, pH 7.5, 140 mM mannitol, 4 μM merocyanine V, 40 % ethylene glycol and chromatium chromatophores equivalent to 50 μM.

Chromatophores from Rhodopseudomonas spheroides showed a maximum extent from 25°C - 0°C with a fairly sharp decline below 0°C. In R. rubrum chromatophores there is a small increase as the temperature is lowered from 25°C, the maximal extent being around + 5°C, then followed by a gradual decline. Chromatophores from Chromatium vinosum have been considered as devoid of any significant carotenoid absorbance change at room temperature. As seen in Fig. 15 this is indeed so, but as the temperature is lowered, the extent becomes markedly increased. The maximum is not reached until - 19° C, a temperature at which chromatophores from the two other species of bacteria almost had lost their carotenoid responses. At this low temperature the light induced change was partly sensitive to antimycin, and completely inhibited by the uncoupler FCCP, thus still responding in an energy-linked fashion. So far these studies have been performed only under continuous illumination but the slow kinetics of the changes at low temperatures permit a time resolution even under these conditions.

The limited number of reactions that are active at - 19°C considerably narrows the possible source of the light-induced absorbance change. The primary charge separation in the reaction center as well as the cyclic electron transport are known to be active, and work is now in progress to test the possible relation of both the onset and the decay of the light-induced carotenoid absorbance change to such events at low temperatures.

## Membrane Potential or Local Charges?

It seems that the widespread use of the carotenoid absorbance change as a calibrated measure of a transmembrane potential in photosynthetic systems may not always be warranted in the coupled membrane capable of active phosphorylation. The usual way of calibrating the extent of the light induced carotenoid change to the size of the membrane potential is by inducing a diffusion potential with $K^+$ ions in the presence of valinomycin (Jackson and Crofts, 1969).

Our data show that it is necessary to distinguish between the slow and fast phases of the light induced carotenoid absorbance change. I will now attempt to give more detailed interpretations of the results.

It is not difficult to visualize that the energization of the membrane, which may well be a formation of an electrochemical gradient or a protein conformational change, or both, may result in local charge redistributions in the membrane. Conformational changes in the ATPase molecule from another photosynthetic bacterium, Rhodopseudomonas capsulata, have been shown to occur upon energization (Melandri *et al.*, 1974). Such changes may well affect the distribution of charges adjacent to carotenoid molecules close to the ATPase in a specific manner, which could give rise to an electrochromic absorbance change of those carotenoids. In agreement with this reasoning, removal of the coupling factor ATPase, as well as binding of ADP + Pi to it, would exert the inhibitory action on the light induced slow phase by preventing such local charge redistributions. The same would occur when antimycin A prevents energization. If the slow phase, on the other hand, would have been caused by the formation of a membrane potential, utilized to drive ATP-synthesis, one would have expected that phosphorylating conditions should be reflected as a fast break-down after an initial spike, rather than the occurring total inhibition.

The acceleration of the decay of a flash-induced fast absorbance change in the presence of ATP or PPi (Fig. 13) can be explained as follows, in accordance with existing models for excitable membranes. The presence of either phosphate compound causes the formation of a membrane potential, reflected in an initial carotenoid absorbance change in the dark. An increase of this potential (by a subsequent light flash) will result in a faster breakdown of the potential if the capacity of the membrane to maintain a potential is limited in this potential range.

In conclusion, the slow phase of the light-induced carotenoid absorbance change, inhibited by ADP + Pi or antimycin, is closely linked to the phosphorylation reactions but does not seem to be in response to a transmembrane potential, as discussed above. In contrast, the fast phase, as well as the changes induced in the dark by ATP or PPi appear to be in response to a transmembrane potential, but this potential is apparently not utilized under physiological conditions as driving force for ATP-synthesis. These data are not in agreement with the proposed role of a light induced membrane potential as a significant contribution to the electrochemical gradient which has been postulated as the energy source for ATP-synthesis (Jackson and Crofts, 1969, 1971; Jackson *et al.*, 1975) in photosynthetic bacteria.

The above results would seem to necessitate a focussing of the discussion on the mechanism of bacterial photophosphorylation to the events reflected by the slow phase of the light-induced carotenoid absorbance change rather than the sum of both fast and slow phases. I would like to suggest that these results should be considered also when evaluating the role of membrane potentials in other energy transducing membranes, such as, for examples, chloroplasts and mitochondria.

## ACKNOWLEDGMENTS

This work has been supported by the Swedish Natural Science Research Council. The excellent technical assistance of Ms. Rosa Johansson is gratefully acknowledged.

## REFERENCES

Baltscheffsky, H. and L.-V. von Stedingk, Bacterial photophosphorylation in the absence of added nucleotide. A second intermediate stage of energy transfer in light-induced formation of ATP, Biochem. Biophys. Res. Commun. 22, 722, 1966.

Baltscheffsky, H., L.-V. von Stedingk, M.-W. Heldt and M. Klingenberg, Inorganic pyrophosphate: Formation in bacterial photophosphorylation, Science 153, 1120, 1966.

Baltscheffsky, M., Inorganic pyrophosphate and ATP as energy donors in chromatophores from Rhodospirillum rubrum, Nature 216, 241, 1967.

Baltscheffsky, M., Inorganic pyrophosphate as an energy donor in photosynthetic and respiratory electron transport phosphorylation systems, Biochem. Biophys. Res. Comm. 28, 270, 1967b.

Baltscheffsky, M., Energy conversion-linked changes of carotenoid absorbance in Rhodospirillum rubrum chromatophores, Arch. Biochem. Biophys. 130, 646, 1969a.

Baltscheffsky, M., Reversed energy conversion reaction of bacterial photophosphorylation, Arch. Biochem. Biophys. 133, 46, 1969b.

Baltscheffsky, M., Discussion comment in Electron Transport and Energy Conservation, edited by J.M. Tager, S. Papa, E. Quagliariello and E.C. Slater, pp. 419-421, Adriatica Editrice, Bari, 1970.

Baltscheffsky, M., Reversible energization in photosynthesis as measured with endogenous carotenoid, in Dynamics of Energy Transducing Membranes, edited by L. Ernster, R.W. Estabrook and E.C. Slater, pp. 365-376, Elsevier, Amsterdam, 1974.

Baltscheffsky, M., H. Baltscheffsky and L.-V. von Stedingk, Light-induced energy conversion and the inorganic pyrophosphatase reaction in chromatophores from Rhodospirillum rubrum, Brookhaven Symp. Biol. 19, 246, 1966.

Baltscheffsky, M. and H. Baltscheffsky, Coupling and control at the cytochrome level of bacterial photosynthetic electron transport, Abstract, Wenner-Gren Symposium on Oxidation Reduction Enzymes, p. 39, Stockholm, 1970.

Baltscheffsky, M. and H. Baltscheffsky, Coupling and control at the cytochrome level of bacterial photosynthetic electron transport, in Oxidation Reduction Enzymes, edited by Å. Åkesson and A. Ehrenberg, pp. 257-262, Pergamon Press, Oxford, 1972.

Baltscheffsky, M. and D.O. Hall, Photophosphorylation and the 518 nm absorbance change in tightly coupled chloroplasts, FEBS Letters 39, 345, 1974.

Casadio, R., A. Baccarini-Melandri, D. Zannoni and B.A. Melandri, Electrochemical proton gradient and phosphate potential in bacterial chromatophores, FEBS Letters 49, 203, 1974.

Chance, B. and M. Baltscheffsky, Carotenoid and merocyanine probes in chromatophore membranes, in Biomembranes, vol. 7, edited by H. Eisenberg, E. Katchalsky-Katzir and L.A. Manson, pp.

33-55, Plenum Press, New York, 1975.

Girault, G. and J.M. Galmiche, Nucleotides effect on the decay kinetics of the 520 nm absorbance change in tightly coupled chloroplasts, Biochem. Biophys. Res. Commun. 68, 724, 1976.

Guillory, R.J. and R.R. Fisher, Studies on the light-dependent synthesis of inorganic pyrophosphate by Rhodospirillum rubrum chromatophores, Biochem. J. 129, 471, 1972.

Jackson, J.B. and A.R. Crofts, The high energy state in chromatophores from Rhodopseudomonas spheroides, FEBS Letters 4, 185, 1969.

Jackson, J.B. and A.R. Crofts, The kinetics of light induced carotenoid changes in Rhodopseudomonas spheroides and their relation to electrical field generation across the chromatophore membrane, Eur. J. Biochem. 18, 120, 1971.

Jackson, J.B., S. Saphon and H.T. Witt, The extent of the stimulated electric potential decay under phosphorylating conditions and the $H^+$/ATP ratio in Rhodopseudomonas spheroides chromatophores following short flash excitation, Biochim. Biophys. Acta, 408, 83, 1975.

Johansson, B.C., A coupling factor from Rhodospirillum rubrum chromatophores, FEBS Letters 20, 339, 1972.

Johansson, B.C., M. Baltscheffsky and H. Baltscheffsky, Coupling factor capabilities with chromatophore fragments from Rhodospirillum rubrum in Proceedings of the II-nd International Congress on Photosynthesis Research, Stresa, 1971, edited by G. Forti, M. Avron and A. Melandri, pp. 1203-1209, Junk Publishers, The Hague, 1972.

Jones, O.T.G. and V.A. Saunders, Energy-linked electron transfer reactions in Rhodopseudomonas viridis, Biochim. Biophys. Acta 275, 427, 1972.

Kagawa, Y. and E. Racker, Partial resolution of enzymes catalyzing oxidative phosphorylation X. Correlation of morphology and function in submitochondrial particles, J. Biol. Chem. 241, 2475, 1966.

Keister, D.L. and N.J. Minton, Energy-linked reactions in photosynthetic bacteria. VI. Inorganic pyrophosphate-driven ATP synthesis in Rhodospirillum rubrum, Arch. Biochem. Biophys. 147, 330, 1971.

Keister, D.L. and N.J. Yike, Energy-linked reactions in photosynthesic bacteria. I. Succinate-linked ATP-driven $NAD^+$ reduction by Rhodospirillum rubrum chromatophores, Arch. Biochem. Biophys. 121, 415, 1967a.

Keister, D.L. and N.J. Yike, Energy-linked reactions in photosynthetic bacteria. II. The energy-dependent reduction of oxidized nicotinamide-adenine dinucleotide phosphate by chromatophores of Rhodospirillum rubrum, Biochemistry 6, 3847, 1967b.

Klemme, B., J.-H. Klemme and A. San Pietro, PPase, ATPase, and photophosphorylation in chromatophores of Rhodospirillum rubrum: Inactivation by phospholipase A. Reconstitution by phospholipids, Arch. Biochem. Biophys. 144, 339, 1971.

Lien, S. and E. Racker, Partial resolution of the enzymes catalyzing photophosphorylation, VII. Properties of silicotungstate-treated subchloroplast particles, J. Biol. Chem. 246, 4298, 1971.

Löw, H. and B. Afzelius, Subunits of the chromatophore membranes in Rhodospirillum rubrum, Exp. Cell Res. 35, 431, 1965.

Melandri, B.A., E. Fabbri, E. Firstater and A. Baccarini-Melandri, Allotopic properties and energy dependent conformational changes of bacterial ATPase, in Membrane Proteins in Transport and Phosphorylation, edited by G.F. Azzone, M.E. Klingenberg, E. Quagliariello and N. Siliprandi, pp. 55-60, North-Holland, Amsterdam, 1974.

Nishimura, M., T. Ito and B. Chance, Studies on bacterial photophosphorylation. III. A sensitive and rapid method of determination of photophosphorylation. Biochim. Biophys. Acta 59, 177, 1962.

Reed, D.W. and D. Raveed, Some properties of the ATPase from chromatophores of Rhodopseudomonas spheroides and its structural relationship to the bacteriochlorophyll proteins, Biochim. Biophys. Acta 283, 79, 1972.

Saphon, S., J.B. Jackson, V. Lerbs and H.T. Witt, The functional unit of electrical events and phosphorylation in chromatophores from Rhodopseudomonas spheroides, Biochim. Biophys. Acta 408, 58, 1975a.

Saphon, S., J.B. Jackson and H.T. Witt, Electrical potential changes, $H^+$ translocation and phosphorylation induced by short flash excitation in Rhodopseudomonas sphaeroides chromatophores, Biochim. Biophys. Acta 408, 67, 1975b.

Smith, L., M. Baltscheffsky and J.M. Olson, Absorption spectrum changes observed on illumination of aerobic suspensions of photosynthetic bacteria, J. Biol. Chem. 235, 213, 1960.

# MONOMOLECULAR FILMS AND MEMBRANE STRUCTURE

D.A. Cadenhead

Department of Chemistry

State University of New York at Buffalo
Buffalo N.Y. 14214

## ABSTRACT

The liquid condensed and liquid expanded physical states in an insoluble monomolecular film at the air-water interface and the gel and liquid crystelline states in fully hydrated bilayers are discussed and compared. It is postulated that, under reasonable restraints, the monolayer may be regarded as an adequate model membrane system. The interfacial behavior of a number of fatty acid spin-label and fluorescent probes are then outlined both for pure and mixed monomolecular films. As in bilayers and membranes these probes exhibit perturbation and immiscibility phenomena which should be taken into account when investigating membrane structure. While the perturbations of any individual probe may be small, they appear to be significant where bilayer profiles are concerned.

## INTRODUCTION

The monomolecular film has been regarded for some time as a model membrane system. It clearly constitutes half of bilayer and the bilayer appears to be an integral part of most, if not all, membranes. While it has seen less application, than the liposome, nevertheless, it provides a simple, direct method of obtaining information at the molecular level on the packing and conformation of membrane lipids under membrane like conditions. In the first part of this presentation I will consider some of the important assumptions made in utilizing the monolayer as a model membrane system. In the second I will discuss one of the aspects with which my group has been concerned: the interfacial behavior of molecular probes and how such behavior concerns the interpretation of molecular structure in membranes

## SECTION I: THE MONOLAYER AS A MODEL MEMBRANE

Ever since the original monomolecular film studies of erythrocyte membrane lipids of Gorter and Grendal,[1] insoluble monolayers have been used as model membranes. Both the air-water and oil-water interfaces were considered for such studies, however, in spite of an "intuitive" preference for the latter,[2] the much greater experimental ease of handling such films at the air-water interface, has led to the overwhelming selection of this interface. Recent work by Mingins and co-workers of lecithins and ethanolamines at the oil-water interface,[3] have adequately demonstrated that typical membrane lipid films would exist only in a fully expanded state at an oil-water interface. Thus at 37°C, films of distearoyl lecithin (DSPC) are fully gaseous expanded (see Figure 1 for an explanation of such terms) and below 40 dynes/cm still occupy an area/molecule greater than 60 Å$^2$, a situation inconsistent with both x-ray[4] and electron diffraction[5] data for lipid bilayers. The highly expanded state of membrane lipids at this interface clearly reflects a sharp reduction of chain-chain interactions through oil penetration between individual film molecules, particularly at low pressures. At high pressures however oil may be squeezed out and the presence of chain-dependent phase changes may be observed,[3,6] contrary to the predictions of Davies.

In contrast, the correspondence between lipid monolayers at the air-water interface and lipid bilayers appears to be much better documented. From the beginning it has been clear that those phospholipids of scientific interest because of their ability to undergo phase changes were also of considerable biological significance.[8] It was left to Phillips and Chapman,[9] however to first point out the correspondence between the gel-liquid crystalline transition in hydrated lecithin bilayers and the liquid expanded-liquid condensed phase transition in monolayers. They predicted that the critical temperature for this latter transition in monomolecular films would occur at approximately 41°C for dipalmitoyl lecithin (DPPC) in agreement with the value of 41.5°C for the gel-liquid crystalline transition. Subsequently Hui et al. confirmed this and, by comparing molecular packing in wet bilayers and monolayers, established an equivalent packing pressure for monolayers of DPPC at 41.5°C of 47 dynes/cm.[5] This value agrees well with one of 50 dynes/cm estimated from interfacial tensions by Nagle.[10]

Below 41.5°C an excellent correspondence exists between the liquid or solid condensed monolayer and the bilayer.[5] The anticipated correspondence between the liquid expanded state and the

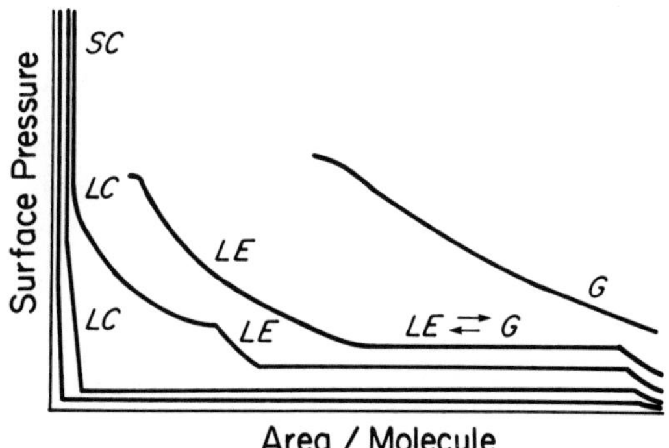


Figure 1. A schematic representation (not to scale) of the various physical states of monomolecular films SC, solid condensed; LC, liquid condensed; LE, liquid expanded; G, gaseous.

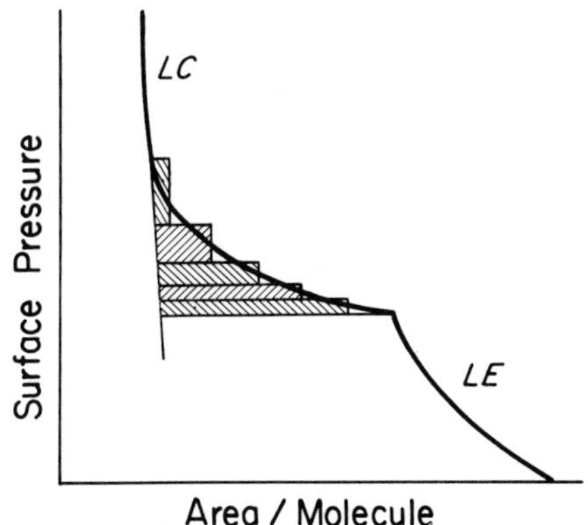


Figure 2. The liquid expanded-liquid condensed phase change below the critical temperature. The hatched regions indicate how hypothetical domain structure, undergoing first order phase changes, can result in the overall phase change being second order.

bilayer above 41.5°C has not been fully verified because of diffuse electron diffraction in bilayers and film collapse in monolayers. Nevertheless the correspondence exists at 41.5° and 47 dyne/cm and it would seem reasonable to anticipate that at lower temperatures, and the equivalent lower pressures, the liquid expanded state, close to the transition, would still approximate to a liquid crystalline lipid bilayer, particularly at pressures of 20 dynes/cm or more, where areas/molecule approach 60 Å$^2$. This correspondence found for lecithins,[5,9] can be extended to ethanolamines and mono-glycerides,[11] and it would seem reasonable to propose that a similar correspondence of physical states exists for all water-dispersible amphipathic lipids.

Unfortunately while the liquid condensed, liquid expanded states and their phase transition have helped establish a correspondence between monolayers and bilayers the nature of the transition, at a molecular level, is not well understood. In appearance the monolayer transition ranges from near first order to second order and is usually regarded as a degenerative first order[9] or, more recently, as a 3/2 order phase transition.[12] The non-zero slope of the transition (see Figure 1) is quite reproducible over a wide range of compressional velocities though there does appear to be a small compressional-decompressional hysteresis associated with it. Thus the use of very slow film compressional velocites does not eliminate the positive slope of the transition suggesting that the transition is not first order.

Historically the first attempt to treat this transition theoretically was made by Langmuir in his now classic paper on duplex films.[13] Langmuir postulated that at the onset of the transition only "free" molecules were present, while at its completion there would be only "micelles". The number of molecules in a micelle would remain fixed for a given compound and only the number of micelles could vary. For differing molecules the number of molecules in a micelle could vary from 5 to 60 with the micelle surface pressure contribution increasing with decreasing molecular content. The theory has the problem that above the initial transition pressure all "free" molecules would be unstable with respect to micelle formation.

An alternate theory proposed by Kirkwood postulated free rotations for rigid-rotor like molecules giving rise to hindered rotation on close packing and a subsequent second order phase transition.[14] Apart from the obvious problem of treating flexible amphipathic molecules as rigid rotors, the theory suffers from the fact that, for all surfactant molecules so far studied, significant molecular interactions and hindered rotation occur long before a close-packed state is achieved. What should be emphasized is that this phase transition has been observed <u>only</u> for flexible amphipathic molecules.[15]

Among the more recent attempts of interest to explain this transition in both bilayers and monolayers have been those of Nagle.[10,12,16] Nagle has been able to demonstrate that although monolayers and bilayers behave differently, use of the same model (co-operatively interacting hydrocarbon chains) can explain the characteristics of both.[10] Introduction of a minor role for polar group interactions[17] led to an improved though not perfect correspondence for the experimental and theoretical transition temperature.[10] In spite of this latter modification, in the interests of an exact solution, the model did not taken into account per se the existence of micelles or domains for which considerable evidence exists in bilayers,[18,19] at least in the condensed state.

One way to explain the liquid expanded-liquid condensed phase change would be to assume that, although difficult to observe experimentally, domains already exist in the liquid expanded state. An extremely sharp domain size distribution would lead to a near first order transition. As the domain size distribution broadened an apparent second order phase transition would be obtained as envisaged in Figure 2, with differing domains undergoing the transition at slightly differing pressures. The sharp increase in slope usually observed as the transition nears completion could be due to a cooperative domain transition mechanism. This theory could explain the enhanced water evaporation at the transition point observed by LaMer and coworkers[20] through the formation of a pronounced two-dimentional defect structure, an effect resembling the sodium permeability studies of Papahadjopoulos,[21] and van Deenen.[22] It would also explain the monolayer compressional-decompressional hysteresis and the reproducebility of the isotherm.

In all of the above discussion we have paid no attention to mixed monolayers, an aspect of prime importance when considering the monomolecular film as a model membrane. In such films, as in bilayers, distinction of expanded and condensed states may become blurred and the question of misability is of considerable importance. Typically Gershfeld and coworkers,[23,24,25] have pointed out that many cholesterol-phospholipid systems are immisable at low pressures and may be immisable under high (membrane-like) pressure conditions. They suggested that this throws into considerable doubt the now widely assumed ability of cholesterol to condense expanded phospholipids.[26]

Recently we investigated this situation for a typical mixed monolayer system of this type: DPPC-cholesterol. After establishing the stability of the individual components under non-oxidizing conditions for long periods of time, we added just sufficient cholesterol to DPPC to suppress the liquid expanded-liquid condensed phase charge, then awaited the reemergence of this phase change while holding the film at a well defined surface pressure. We found that at low (near zero) pressure conditions, perceptible segregation had taken place in thirty minutes, at 5-10 dynes/cm the equivalent time was 5-7 hours, while above 20 dynes/cm no segregation was detected in 16 hours. We interpreted these results as meaning that under membrane-like conditions DPPC and cholesterol constitute a long-term metastable mixture, a situation consistent with that in bilayers and membranes.[27,28] This is not a serious matter when the shorter turnover time of cholesterol in membranes is taken into account.[29] It would appear that such films are sufficiently metastable to constitute reasonable model membranes. Details of this work will be published elsewhere.[30]

In summary, monomolecular films at the air-water interface do appear to represent a satisfactory model membrane system in which the liquid expanded and liquid condensed states may be compared with the gel and liquid crystalline respectively in bilayers. The correspondence between the gel-liquid crystalline transition in bilayers and the liquid condensed-liquid expanded critical temperatures appears to hold not only for lecithins but for all water dispersible lipids. Energetically the monolayer and bilayer are similar though not identical, and the same theoretical models can be used to treat the behaviour of both systems.

## SECTION 2: MOLECULAR PROBES AND MEMBRANE STRUCTURE

The past few years we have seen large advances in our knowledge of membrane structure through the use of molecular probes, particularly spin-label[31] and fluorescent[32] probes. With the growth of the usage of such probes has come the realization that they should be as close a resemblance as possible to one or other membrane components in order to minimize any possible perturbation and to assist in establishing the precise location of the probe within the membrane. We will demonstrate that monolayer studies can provide information on both these points.

### a) Fatty-Acid Spin-Label Probes (Pure Films)

In our initial studies with spin-labelled compounds we learned that the substitution of an oxizolidine ring for a membrane lipid polar group could provide a probe molecule with very similar behaviour to that of the original "parent" lipid. Thus 3-nitroxide cholestane in both pure and mixed films behaved in a very similar way to cholesterol[33] even to the ability to condense expanded lipids.[34] In contrast, when the oxizolidine ring is added to a hydrocarbon chain the behaviour of the molecule in a monolayer can be dramatically affected, as is shown in Figure 3 for a series of fatty acid nitroxide spin-label probes[35] where the oxazolidine ring is shifted from the 5- to the 16- position.

At all times the probe isotherms are significantly more expanded than their "parent" fatty acid. With the oxazolidine ring at or beyond the 8-position the isotherm becomes gaseous expanded, behavior typical of a bipolar rather than a monopolar molecule. This behavior is best understood by regarding the oxazoline ring as a weakly polar group. At low pressures both the ring and the carboxyl (ester) group occupy the interface. As the film is compressed the nitroxide group is gradually forced from the substrate until the film molecules collapse or adapt an erect conformation.[36] What is more important, however, is that, as the position of the probe group is changed, the interfacial behavior of the molecule is greatly affected. The assumption generally made that changes seen by a series of probe molecules of this type[37,38] result solely from a change in the probe environment are thrown into doubt, and is not surprising that spin-labels and deuterium resonance techniques predict somewhat different segmental order parameters in a lipid bilayer.[39]

## b) Myristic Acid - 12 Nitroxide Stearic Acid

While pure films may indicate that a given probe may give rise to problems, it is only by examining mixed monolayers with a host lipid that this can be established. Figure 4 shows the resultant isotherms for a series of mixed monolayers of 12-nitroxide stearic acid with a simple host lipid, myristic acid, at room temperature. Under these circumstances the myristic acid exists in both an expanded and a condensed state and we may examine the effects of a change in the physical state of the host on the behavior of the probe. Only the primary finding are presented here, a more detailed presentation will be published later.

The isotherms in Figure 4 show a gradual transition from the isotherm of myristic acid to that of 12-nitroxide stearic acid. At low probe concentrations ($<$5 mole %) the myristic acid isotherm is only slightly affected yet the liquid expanded - liquid condensed phase change is perceptible shifted, a shift that resembles that brought about by an increase in temperature. By extrapolating the effect at the lowest probe concentrations measured ($\sim$2 mole %) to a concentration that the probe "sees", we can quantify the perturbation. Because such probes can detect cooperative lipid motion they must "see" beyond their nearest neighbours.[17] It seems likely however, as with other interactions,[40] that such effects fall off rapidly resulting in small "halos". An equivalent host-lipid:probe molar ratio of between 12:1 an 15:1 seems reasonable and for this system would produce an equivalent temperature

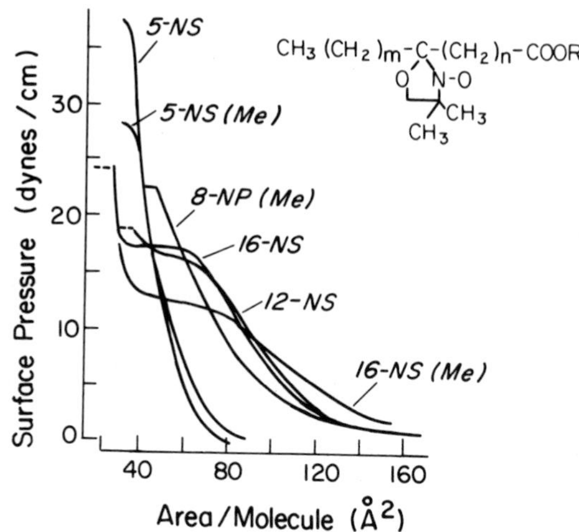


Figure 3. The isotherms at 23°C for a series of fatty acid spin-label probes. 5-NS, 5-nitroxide stearic acid; 5-NS (Me), 5-nitroxide stearic acid, methyl ester. The other probes are similarly abbreviated. All data were continuously recorded.

increase of between 2-4°C.

A second effect, particularly observable here with the 48.4 mole % 12-NSA film, but also detectable for other probe concentrations, is the shift to low areas/molecule ($< 20Å^2$) above the transition usually ending in a short condensed region. This clearly indicates that as the film condenses, the probe laterally segregates and is partially lost from the film leaving a myristic acid rich residue to undergo further condensation. Both perturbation and partial immiscibility effects have been observed for such probes in bilayers[41] and membranes.[42]

Further insight into the data may be obtained by their representation in mean molecular area versus compositional plots as a function of surface pressure. Before doing so however it is important to understand precisely what such a plot shows. As is indicated in Figure 5, locating a probe molecule within a lipid assembly may be considered in two parts: the insertion of the probe and the subsequent accommodation, positive or negative. The mean molecular area plot, comparing as it does, the ideal and actual areas occupied, describes only the subsequent accommodation.

Perhaps the most obvious feature of the mean molecular area plots for the myristic acid - 12-NSA system at 21°C (Figure 6) is the negative deviation from ideality at high probe concentrations.

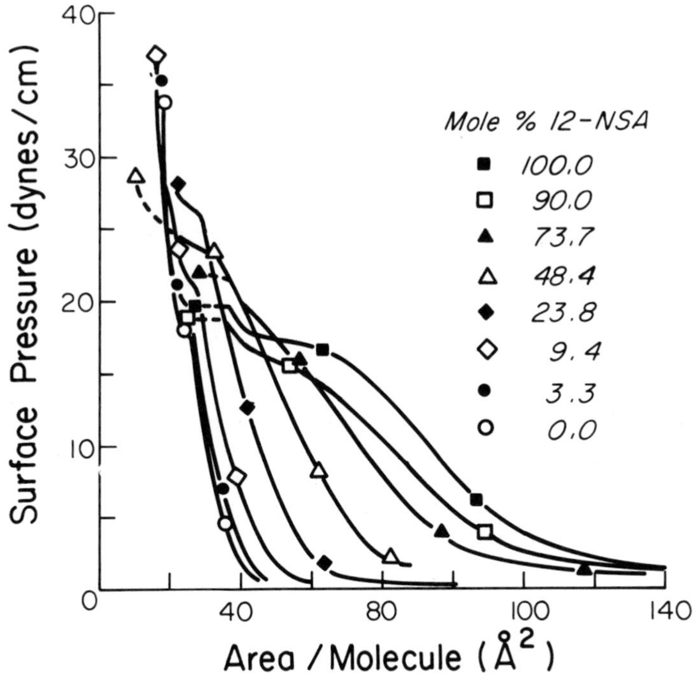


Figure 4. Isotherms at 20.8°C for the mixed film system myristic acid - 12-nitroxide stearic acid for the compositions indicated. All data were continuously recorded. The broken lines indicate a collapse phase.

At low pressures we appear to have ideal mixing. The negative deviation at about 90 mole 12-NSA increases with increasing pressure but then dissapears at about 18 dynes/cm. Reference to Figure 4 shows that what we are observing is an erection of 12-NSA from a bent to an erect conformation with the addition of only small amounts of host lipid (myristic acid). This confirms that, for this system, except for almost pure probe films, the oxazolidine ring is located in the hydrophobic portion of the lipid agregate.

Comparison with ideal behavior (solid lines, Figure 6) at say 16 dynes/cm reveals large negative deviations at high probe concentrations with slight positive deviations over the remainder of the concentration range. If however we approximate that the composition at the minimum of the negative deviation represents nearly pure erect probe it is clear that deviations for the

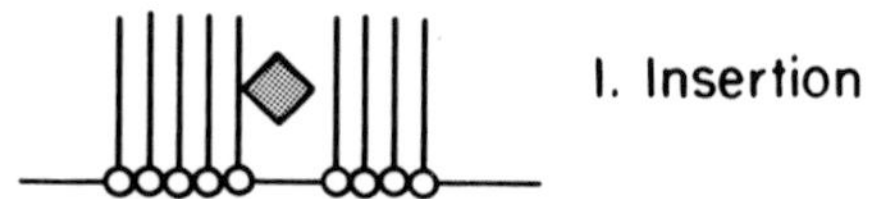


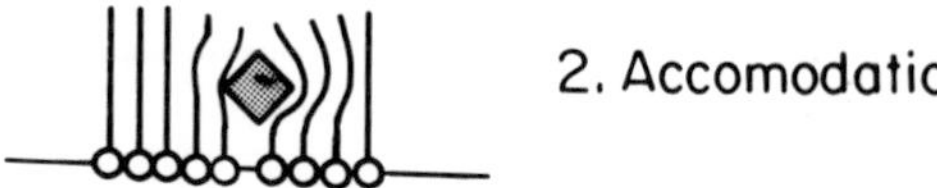


Figure 5. Schematic depiction of the perturbation of a lipid monolayer by a probe molecule. The perturbation is envisaged in two stages: the initial insertion producing a disruption of chain packing related to the probe size and the host lipid accommodation (depicted here as positive). The accommodation (positive or negative) is dependent on the nature of the probe and the surrounding host lipid environment. The net perturbation is taken as the summation of both effects.

system myristic acid - **erect** 12-NSA are strongly positive over the entire concentration range.

At high myristic acid concentrations, where the mixed film behavior is dominated by that component, negative, then positive deviations appear between 17 and 20 dynes/cm. Here we are observing the transient effect of the small myristic acid LE-LC phase change, followed by what is best regarded as immiscibility. Further evidence of the immiscibility of 12-NSA in the condensed state of myristic acid can be seen in the fact that while the 21 dyne/cm data extrapolate to approximately $33A^2$/12-NSA molecule, data above this pressure, where the myristic acid is fully condensed, extrapolate to zero area.

## c) Mixed Films of Dipalmitoyl Lecithin and 12-Nitroxide Stearic Acid

Studies for this system reported here were carried out at 39°C, at which temperature DPPC is liquid expanded until approximately 40 dynes/cm. Because of this the mean molecular area plots shown in Figure 7 do not reflect a liquid expanded - liquid condensed phase change for DPPC. The negative deviations found for myristic acid-12-NSA at lower pressures ($<$ 14 dynes/cm) are also present here, however they occur over a wider concentrational

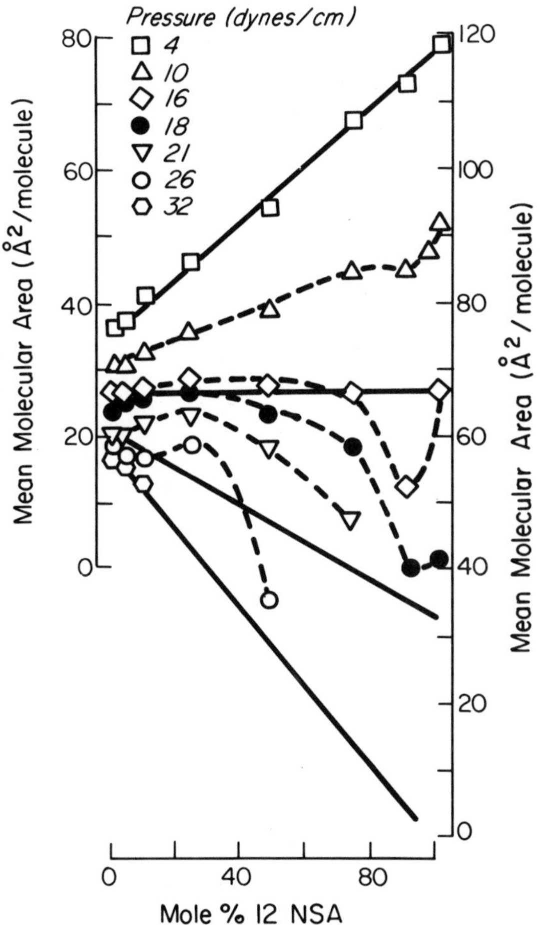


Figure 6. Mean molecular area Vs composition plots for the system myristic acid - 12-nitroxide stearic acid at 20.8°C at the surface pressures indicated. Solid lines represent ideal behavior, broken lines actual behavior. The left and right hand ordinates are displaced for clarity.

range and suggest that DPPC is less capable of erecting 12-NSA than myristic acid. At higher pressures, under membrane-like conditions, deviations from ideality become positive. This means, once again, that the bulk of the probe molecules must be in an erect position with the oxazolidine ring in the hydrophobic

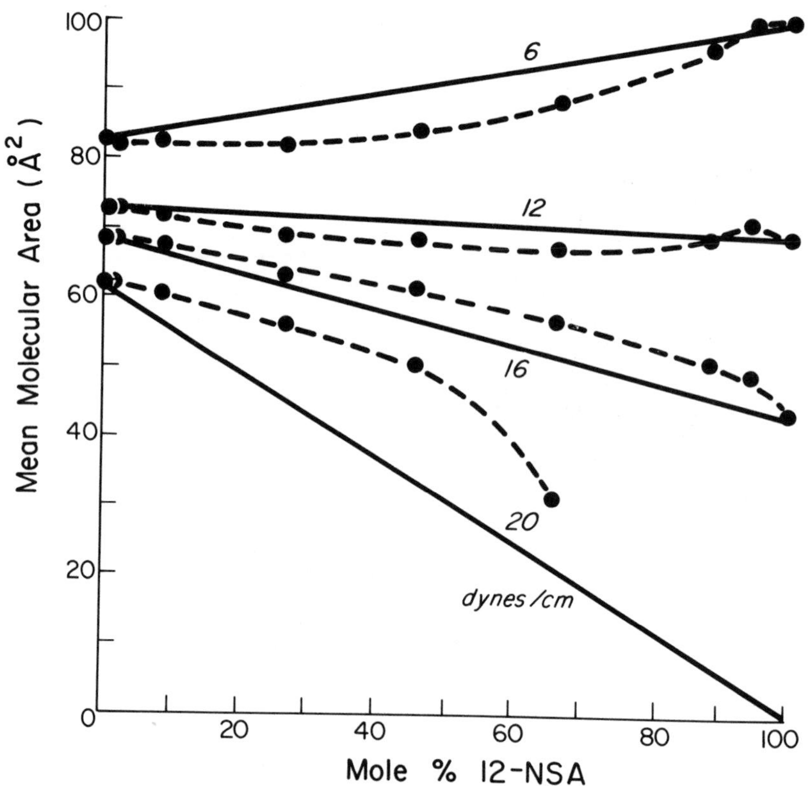


Figure 7. Mean molecular area Vs composition plots for the system dipalmitoyl lecithin - 12-nitroxide stearic acid at 39°C at the surface pressures indicated. Solid lines represent ideal behaviour, broken lines actual behaviour.

region and that only a small percentage could retain a bent conformation. The existance of such a bent conformation in bilayers would readily give rise to an ESR liquid line. Unfortunately the same effect is also produced by dissolved probe molecules.[43] These positive deviations from ideality, taken in conjunction with a DPPC, liquid expanded-liquid condensed phase transition shift, indicate that, not only does 12-NSA perturb the packing of DPPC, but that the polar nature of the oxazolidine ring in the hydrophobic region enhances the effect (a negative accommodation as envisaged in Figure 5). This dependence of the overall perturbation on

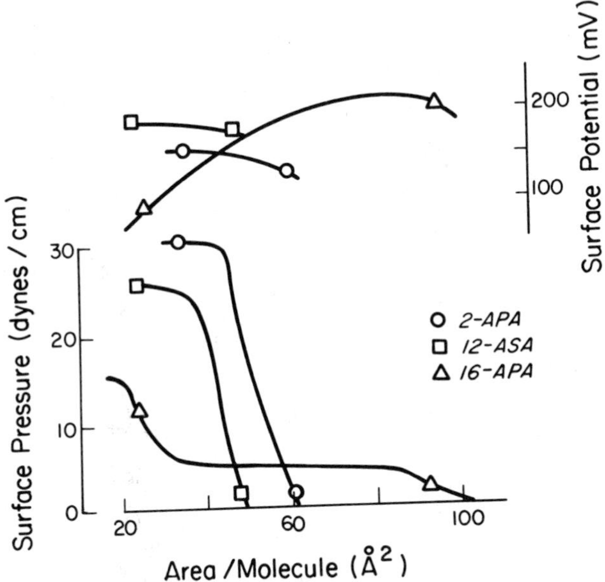


Figure 8. Isotherms (left hand ordinate) and surface potentials (right hand ordinate) for 2-anthroyl palmitic (2-APA), 12-anthroyl stearic acid (12-ASA) and 16-anthroyl palmitic acid (16-APA). Data were recorded continuously.

both the size and nature of the probe has been previously emphasized.[44]

We have also completed studies of the systems DPPC-5-NSA and DPPC-16-NSA both at 23° and 39°C. The results are qualitatively similar to those reported above for DPPC-12-NSA at 39°C, however they show quantitative differences that reflect the changing location of the probe group.

## d) Pure Films of Anthroyl Fatty Acid Probes

The surface pressures and potentials for three fluorescent probes [2-anthroyl palmitic acid (2-APA), 12-anthroyl stearic acid (12-ASA) and 16-anthroyl palmitic acid (16-APA)] are given in Figure 8 as a function of area/molecule at 23°C, the isotherm for 12-ASA was previously published when it was contrasted with those of 12-NSA and stearic acid.[44] 2-APA was supplied by Dr. R.A. Badley of Unilever Research, England, while 16-APA was synthesized by Mr. B.M.J. Kellner of our research group. Once again only a summary

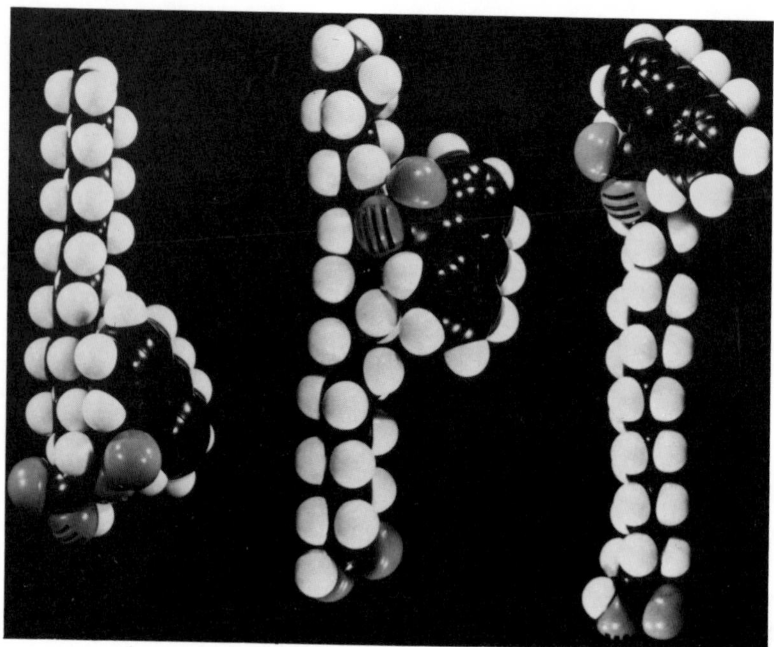

Figure 9. CPK molecular models of (left) 2-anthroyl palmitic acid, (center) 12-anthroyl stearic acid and (right) 16-anthroyl palmitic acid.

of our findings will be reported here.

The isotherms of Figure 8 are best understood in terms of the CPK molecular models shown in Figure 9. The stable film formed by 12-ASA, shows a fully condensed isotherm over the entire pressure range with a collapse at about 25 dynes/cm. This reflects the ability of the anthracene group to pack parallel to the alkane chain, screening the ester linkage, and actually permitting enhanced molecular interactions when compared to stearic acid.

Initially we felt that locating the anthroyl group at the 2-position would substantially weaken the primary polar group (carbonyl) interaction with the substrate and the film would readily collapse. In fact, as Figure 8 and 9 indicate, while locating the anthroyl group at the 2-position does lead to a looser packing (as does the shorter alkane chain), the film is even more stable than 12-ASA, collapsing at about 30 dynes/cm. It would seem that the carboxyl and ester groups combine to form a single polar group for this molecule. 16-APA contrasts with the other two probes both in the isotherm and surface potential. Located at the 16-position

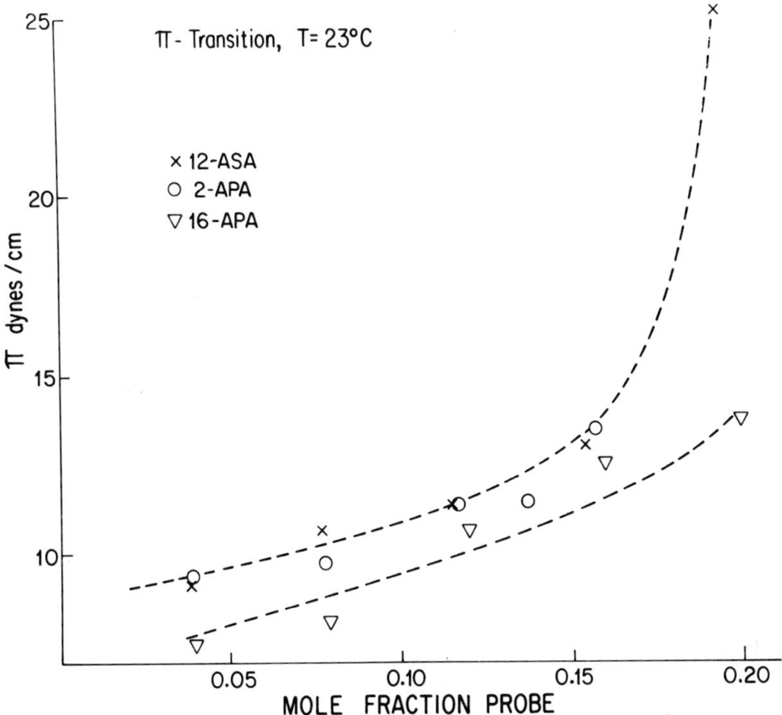


Figure 10. The surface pressure at the onset of the liquid expanded-liquid condensed phase transition for dipalmitoyl lecithin at 23°C as a function of concentration of 2-anthroyl palmitic acid (O), 12-anthroyl stearic acid (X) and 16-anthroyl palmitic acid (▽).

the anthroyl group can orient such that the ester group can enter the substrate with the anthracene just above the interface and the molecule behaves as a bipolar molecule at low pressures. With increasing pressure the weaker ester group is forced out of the substrate (at about 85 Å$^2$/molecule) producing a sharp knee in the isotherm to be followed by a short condensed region collapsing at approximately 15 dynes/cm. Such an orientation is not possible for 12-ASA (Figure 9), where the ester group cannot be inserted in the substrate without also immersing the anthracene.

### e) Anthroyl Fatty Acid Probes (Mixed Films with DPPC)

As has been previously pointed out, a shift in the liquid expanded-liquid condensed phase change can be used to evaluate a net perturbation. Figure 10 reveals the effects of these three

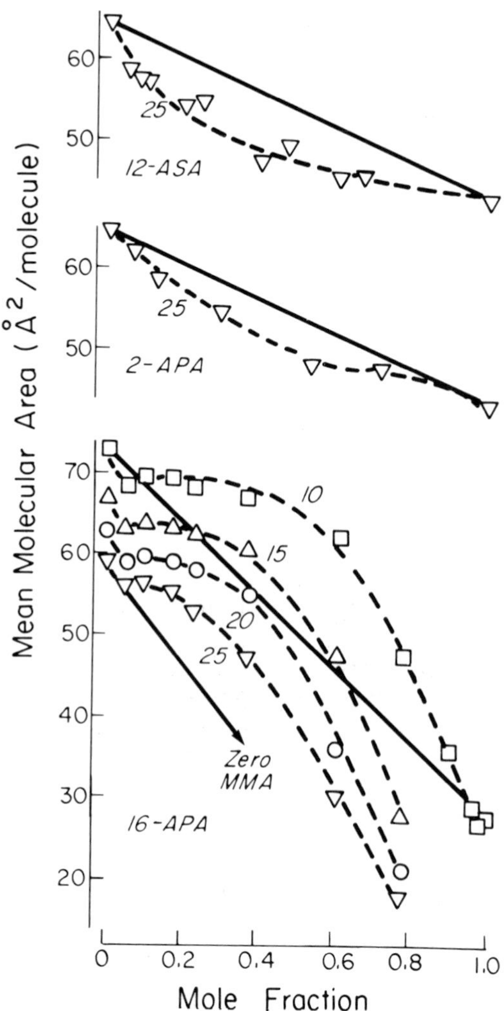


Figure 11. Mean molecular area Vs composition plots for 2-anthroyl palmitic acid, 12-anthroyl stearic acid and 16-anthroyl palmitic acid at the surface pressures indicated. Solid line represent ideal behavior, broken lines represent actual behavior.

probes on the phase change at 23°C. At low probe concentrations 16-APA has little or no effect, while 2-APA and 12-ASA produce a small increase. At higher probe concentrations 2-APA has caused the transition to vanish, 12-ASA has produced a 15 dyne/cm shift while that for 16-APA is about 5 dynes/cm. Thus at approximately 0.2 mole fraction probe the net perturbation decreases 2-APA > 12-ASA > 16-APA.

Both 2-APA and 12-ASA show negative deviations from ideality in their mean molecular area plots (Figure 11), while those for 16-APA are near zero at low probe concentrations but become positive at higher concentrations. What these data signify is that for both 2-APA and 12-ASA the mixed DPPC systems are capable of significant accommodation of these probes. For 16-APA the accommodation is either zero or negative. It would seem that the relatively unscreened polar ester group and/or poor packing of the anthracene in the hydrophobic region enhances the initial perturbation.

### f) A Monolayer-Bilayer Comparison

Having studied the effect of 2-APA, 12-ASA and 16-APA on the phase transition for DPPC in mixed monomolecular films it seemed rational to compare the results with fluorescence polarization studies on DPPC vesicles along the lines of the procedures used by Vanderkooi *et al.*[45] or Jacobson and Papahadjopoulos.[46] The actual studies were carried out by Dr. K. Jacobson of Rosewell Memorial Center Research Institute in Buffalo, New York. Once again I will only report the primary findings because of a more detailed publication to follow.

Using a 1:400 molar ratio of probe: DPPC the DPPC gel-liquid crystalline phase change occurred at 41.5°C (16-APA), 40° (12-ASA) and 39° (2-APA). Since 41.5° is the accepted transition temperature using differential scanning calorimetry, the order of decreasing perturbation would appear to be 2-APA > 12-ASA > 16-APA in agreement with the monolayer finding. In addition a pretransition point was detected readily by the 16-APA at about 32°, barely detected by the 12-ASA and was not identifiable with the 2-APA. The pretransition has been associated with a change in tilt of lipid hydrocarbon chains,[47] and these data would seem to provide confirmation of this since they emphasize the transition is best seen in the hydrophobic region. It seems more likely however that if the probes perturb the gel-liquid crystalline transition they will also affect the pretransition[46] and that the degree of difficulty in observing the pretransition would decrease in the order 2-APA > 12-ASA > 16-APA. It would also seem reasonable that in a gel-like state any transition would have to affect *both* polar and non-polar regions.

### g) Summary

A parallel between monolayer and bilayer studies can be established. In this case we have cosidered the behavior of both spin-label and fluorescent probes and found similar phenomena. Monolayer studies of the liquid expanded-liquid condensed phase change shift provide a measure of the net perturbation (insertion plus accommodation). Mean molecular area plots provide information on the accommodation only. For fatty acid spin-label probes the initial perturbation is enhanced by the polar nature of the oxazolidine ring in the hydrophobic region. For the anthroyl derivatives accommodation is significant for both 2-APP and 12-ASA but not for 16-APA. The latter molecule, however, shows the least net perturbation effect.

In evaluating such perturbations each system must be studied individually. It is clear however that the magnitude of most perturbations is small and does not prevent semi-quantitative data being obtained. More serious is the effect of changing the probe location on the hydrocarbon chain. For both spin label and fluorescent probes the interfacial behavior can change rapidly. The alternative to such studies would seem to be the selection of a "non-perturbing" technique such as $^{13}C$ or deuterium-NMR studies. It is interesting to note that in this regard the total deuteration of DPPC brings about an apparent 5°C enhancement of fluidity.[48] This suggests that the substitution of a single deuterium atom must have a negligible effect.

## ACKNOWLEDGMENTS

I would like to acknowledge the financial assistance of the U.S. National Heart and Lung Institute through grant No. HL 12760-06, the work of my students Friedel Müller-Landau and Benjamin M.J. Kellner, the assistance of Dr. Kenneth Jacobson and the helpful comments of Dr. M.C. Phillips.

## REFERENCES

1. Gorter, E. and Grendel, F., J. Exptl. Med. 41 (1925) 439.

2. Hutchinson, E., in Monomolecular Layers (Ed. Sobotka, H.) American Association for the Advancement for Science (1954). p. 161-174.

3. Taylor, J.A.G., Mingins, J., Pethica, B.A., Tan, B.Y.J. and Jackson, C.M., Biochim. Biophys. Acta 323 (1973) 157.

4. Williams, R.M. unpublished results.

5. Hui, S.W., Cowden, M., Papahadjopoulos, D. and Parsons, D.F., Biochim. Biophys. Acta 382 (1975) 265.

6. Demel, R.A. unpublished results.

7. Davies, J.T., Proc. Roy. Soc. 208A (1951) 224.

8. Cadenhead, D.A., Demchak, R.J. and Phillips, M.C., Kolloid Z. u. Z. Polymere 220 (1967) 59.

9. Phillips, M.C. and Chapman, D., Biochim. Biophys. Acta. 163 (1968) 301.

10. Nagle, J.F., In press. J. Membrane Biol. (1976).

11. Phillips, M.C. and Cadenhead, D.A. unpublished work.

12. Nagle, J.F., J. Chem. Phys. 63 (1975) 1255.

13. Langmuir, I., J. Chem. Phys. 1 (1933) 756.

14. Kirkwood, J.G. in Surface Science (American Ass. for the Advancement of Science, Washington, D.C. (1943) Publication No. 21, p. 157.

15. Cadenhead, D.A. and Demchak, R.J., J. Chem. Phys. 49 (1968) 1372.

16. Nagle, J.F., J. Chem. Phys. 58 (1973) 252.

17. Phillips, M.C., Cadenhead, D.A., Good, R.J. and King, H.F., J. Colloid Interface Sci. 37 (1971) 437.

18. Hui, S.W. and Parsons, D.F., Science 190 (1975) 383.

19. Hui, S.W., Parsons, D.F. and Cowden, M., Proc. Nat. Acad. Sci. (US) 71 (1974) 5068.

20. LaMer, V.K. in Retardation of Evaporation by Monolayers: Transport Processes, Academic Press, N.Y. (1962).

21. Papahadjopoulos, D., Jacobson, K., Nir, S. and Isac, T., Biochim. Biophys. Acta 311 (1973) 330.

22. Blok, M.C., Van der Neut-Kol, E.C.M., van Deenen, L.L.M. and de Gier, J., Biochim. Biophys. Acta 406 (1975) 187.

23. Pagano, R.E. and Gershfeld, N.L., J. Colloid Interface Sci. 44 (1973) 382.

24. Gershfeld, N.L. and Pagano, R.E., J. Phys. Chem. 76 (1972) 1244.

25. Tajima, K. and Gershfeld, N.L., Adv. Chem. Ser. 144 (1975) 165.

26. Phillips, M.C., Prog. Surface Membrane Sci. 5 (1972) 139.

27. Enholm, C. and Zilversmit, D.B., J. Biol. Chem. 248 (1973) 1719.

28. Bruckdorfer, K.D., Edwards, P.A. and Green, C., Eur. J. Biochem. 4 (1968) 506.

29. Hagerman, J.A. and Gould, R.G., Proc. Soc. Exp. Biol. Med. 78 (1951) 329.

30. Cadenhead, D.A., Kellner, B.M.J. and Phillips, M.C., J. Colloid Interface Sci. In press (1976).

31. Smith, I.C.P. in Biological Applications of Electron Spin Resonance (L. Rothfield Editor) Chap. 3. Academic Press, N.Y. 1971.

32. Waggoner, A.S. and Stryer, L., Proc. Nat. Acad. Sci (US) 67 (1970) 579.

33. Cadenhead, D.A., Demchak, R.J., and Müller-Landau, F., Ann. N.Y. Acad. Sci. 195 (1972) 218.

34. Cadenhead, D.A. and Phillips, M.C., Adv. Chem. Ser. 84 (1968) 131.

35. For details see Cadenhead, D.A. and Müller-Landau, F., Adv. Chem. Ser. 144 (1975) 294.

36. Cadenhead, D.A. and Müller-Landau, F., J. Colloid Interface Sci 49 (1974) 131.

37. McConnell, H.M., McFarland, B.G., Ann. N.Y. Acad. Sci 195 (1972) 207.

38. Griffith, O.H., Dehlinger, P.J., Van, D.P., J. Membrane Biol. 15 (1974) 159.

39. Seelig, J. and Niederberger, W., Biochemistry 13 (1974) 1585.

40. Kleemann, W. and McConnell, H.M., Biochim. Biophys. Acta 419 (1976) 206.

41. Hubbell, W.L. and McConnell, H.M., J. Am. Chem. Soc. 93 (1971) 314.

42. Bieri, V.G., Wallach, D.F.H. and Lin, P.S., Proc. Nat. Acad. Sci (U.S.A.) 71 (1974) 4797.

43. Butler, K.W., Tattrie, N.H. and Smith, I.C.P., Biochim. Biophys. Acta 363 (1974) 351.

44. Cadenhead, D.A., Kellner, B.M.J. and Müller-Landau, F., Biochim. Biophys. Acta 382 (1975) 253.

45. Vanderkooi, J. Fischkoff, S., Chance, B. and Cooper, R.A., Biochemistry 13 (1974) 1589.

46. Jacobson, K. and Papahadjopoulas, D., Biochemistry 14 (1975) 152.

47. Rand, R.P., Chapman, D. and Larssun, K., Biophysical J. 15 (1975) 1117.

48. Phillips, M.C. Personal communication.

# PHASE TRANSITIONS, PROTEIN AGGREGATION AND MEMBRANE FLUIDITY

D. Chapman and B.A. Cornell

Department of Chemistry, Chelsea College

University of London, London, SW3 6LX

## INTRODUCTION

Considerable research over the past ten years has increased our understanding of cell membrane structure. Whilst there appear to be variations from one membrane to another in the detailed organisation of the lipid and protein components it is now generally accepted that a common theme of biological membrane structure is a lipid bilayer. The lipid is often in a "fluid" condition (1) although there are membrane systems where the lipid is relatively immobile due to the presence of cholesterol (2) or protein (3) and in some instances, both. With unsupplemented and supplemented membranes, at least some of the lipid at the growth temperature of the cell is in a "crystalline" packing arrangement (4,5).

Because of the occurrence of the lipid bilayer structure in cell membranes the phase transition which occurs with lipid systems, i.e. from fluid to crystalline chain packing (6,7) has attracted much attention.

The emerging structure of cell membranes includes protein both within the lipid bilayer and on the bilayer surface. Combinations of these arrangements, including proteins which span the membrane have been postulated for certain membrane systems (8,9). Apparently the proteins of some cell membranes can rapidly rotate as well as move laterally, e.g. rhodopsin (10), whereas with other membranes, e.g. bacteriorhodopsin (11), the protein appears to be fixed and relatively immobile at its position in the membrane. Protein aggregation within the plane of the lipid bilayer has also been postulated and related to pinocytotic activity (12), the cell growth cycle (13) and to temperature effects (14).

In the present paper we discuss these lipid phase transitions and by examining a model system containing a hydrophobic polypeptide attempt to provide insight into some of the factors which could be associated with protein aggregation within cell membranes.

## LIPID PHASE TRANSITIONS AND AN INTRINSIC POLYPEPTIDE (GRAMICIDIN A)

The endothermic phase transitions of lipid systems have been studied extensively and the marked transition where the lipid chains "melt" examined in some detail (15, 16). It is known that the transition temperature depends upon the head group, the hydrocarbon chain length and the degree and type of unsaturation present (17). The presence of cations and anions has been shown to shift the lipid transition temperature (18,19).

Binary mixtures of lipids have been studied and in some cases a migration of the lipid molecules occurs so as to give crystalline regions of the two separated compounds. In other cases co-crystallisation takes place resulting in a broad range of transition temperatures on reheating (20). Various techniques have been used to investigate this behaviour (21).

In some of our recent studies (22) we have examined the behaviour of a hydrophobic polypeptide Gramicidin A as a model for intrinsic proteins which span cell membranes. Gramicidin A is thought to form a cation selective transmembrane channel (≃ 37 Å in length) through the association of two such molecules via a hydrogen bond between their terminal formyl groups (23). The solubility of Gramicidin A in water is negligible and with the exception of the glycine group in position 2 all of the amino acid residues are hydrophobic. The structure of Gramicidin A proposed by Urry (23) is based upon a $\Pi_{(L,D)}$ helix with 4.4 residues per turn. A number of interesting features are observed with this model system (22).

### Below the Lipid Phase Transition Temperature

At low polypeptide content the "pre-transition" endotherm of the lipid is removed. This "pre-transition" endotherm has recently been associated in the case of dipalmitoyl lecithin with a change from a tilted chain condition to a vertical chain condition (24). This indicates that small amounts of the polypeptide affect the organisation of the lipid chains.

Furthermore, the rippled appearance observed by electron microscopy when pure dimyristoyl lecithin is quenched from below its transition temperature is eliminated. The latter observation shows that the interpretation sometimes made of the rippled appearance of freeze fracture electron micrographs of lipid surfaces must be treated with caution. A number of authors have used the presence and absence of this surface feature to indicate areas within cell membranes that were rigid or fluid at the temperature from which they were quenched (25). Clearly the lipid in the bilayer can be rigid but owing to the presence of small amounts of polypeptide, cholesterol or ions, the chains are in a vertical configuration which does not give the rippled appearance.

Small amounts of the polypeptide cause the main lipid endotherm to broaden and produce a loss in the transition enthalpy. This demonstrates that the presence of a transchannel polypeptide (or intrinsic protein) will smear out the lipid transition. This is relevant to recent discussions of boundary layer effects associated with various protein lipid systems (26). If it is assumed that the boundary layer lipid surrounding the polypeptide has its transition spread out such that it does not contribute to the observed enthalpy of the transition, the associated enthalpy loss is in good agreement with the experimental values (see Fig.1).

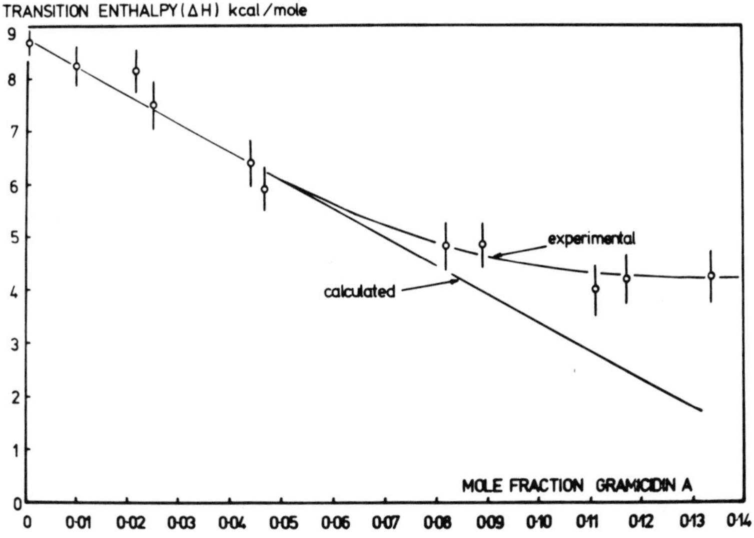


Figure 1 The enthalpy of the main endothermic calorimetric peak as a function of Gramicidin A concentration for DPL. The straight line is calculated as described in the text.

When the concentration of Gramicidin A is greater than 20 lipids per polypeptide an aggregation process occurs. This is seen from the enthalpy data, which deviates markedly from those expected at this value (Fig. 1). Further evidence is the associated deviation from the linear increase in transition width as the polypeptide content is raised above this level. This raises the question as to why such aggregation occurs both in the present study and with proteins in certain membrane systems when the temperature is lowered so that lipid chain crystallisation occurs (25).

Some progress towards such an explanation may be made by examining the packing of circles with diameters related to the mean diameters of the Gramicidin A and a single lipid hydrocarbon chain respectively.

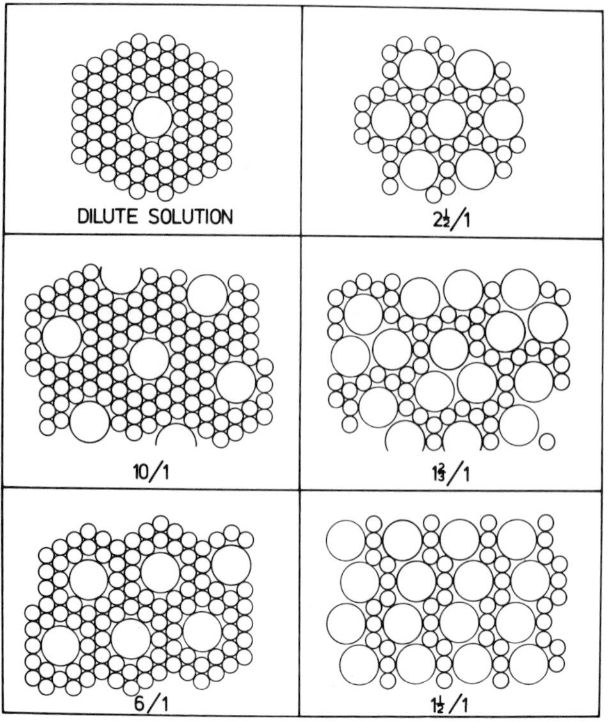


Figure 2 The packing of large (polypeptide) and small (lipid chains) circles in a hexagonal array. The larger diameter corresponds to the critical size of 11.8 Å for a small circle diameter of 4.8 Å. The figures shown below each arrangement correspond to the ratio of lipid to polypeptide in the continuous two dimensional array.

Unless the diameter of the larger circles shown in Fig.2 correspond to certain critical values it is not possible to completely fill a two dimensional area by surrounding them with the smaller diameter circles without producing a considerable disruption to the hexagonal packing. This disorder will take the form of packing faults adjacent to, and radiating outward from the larger circles (Fig.3). These packing faults will close after a number of molecular distances (dependent upon the elasticity assumed for the lipid matrix) to form dislocations which will ultimately vanish allowing the lipids to return to the original hexagonal packing. Many investigations of other solid state systems has established that the direction along which such dislocations occur is almost always that along which the atoms are most closely packed. In a two dimensional hexagonal matrix these are the directions of the three hexagonal axes, mutually separated by 120°.

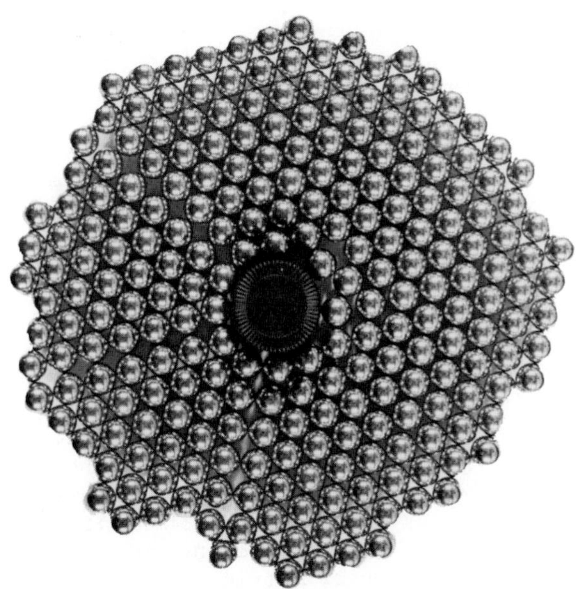

Figure 3 A typical pattern of packing faults which occurs when a cylinder of non-critical size is introduced into a ball bearing raft. By viewing this picture at low angle it can be seen that packing faults lie along the hexagonal axes of the close packed array of ball bearings. In a real molecular packing situation as occurs in the crystalline phase of a phospholipid these packing faults will close, due to the elasticity of the real system, to form dislocations along the same set of axes.

At low concentrations of polypeptide, the disruptive effect of the imperfect packing arrangements will be absorbed as dislocations surrounding the polypeptide. However, as the lipid content is lowered the width of the dislocations begins to approach the average spacing of the polypeptides. Assuming that the sample is cooled evenly throughout the transition, when the hydrocarbon chains crystallize the Gramicidin A will become trapped within the dislocations which spontaneously form due to the thermal fluctuations in the crystallisation process. Until the concentration of the polypeptide is high enough to produce a dislocation density in excess of that which occurs spontaneously in the pure lipid, the "zone refining" of the polypeptide into aggregates is prevented by the entrapment of the polypeptide into naturally occurring packing faults. At Gramicidin A concentrations significantly above this level, the effect of the lipid crystallization is to produce localised regions distributed throughout the plane of the bilayer in which the polypeptide to lipid ratio is relatively high.

It is suggested that geometrical considerations similar to these are necessary for an understanding of the lipid protein architecture of biomembranes and reconstituted systems. It is anticipated that regardless of the actual cross sectional shape of the polypeptide or transmembrane protein, if it does not permit a hexagonal packing of the surrounding lipid chains the system will be unstable to phase transitions. Should the hydrocarbon chains in such a system crystallize due to the temperature being lowered below the pure lipid phase transition, the system will be driven towards an aggregation of the protein and the lipid.

When the polypeptide Gramicidin A is present in a mixed lipid system (containing equimolar amounts of dilauroyl and dipalmitoyl lecithin) which is cooled at a rate of 5 K per minute the higher melting lipid crystallises out and appears to exclude the polypeptide from this lipid region. The polypeptide therefore occurs predominantly in the lower melting phase (as indicated by the absence of an observable enthalpy for the low melting lipid) rather than in the higher melting phase. This is interesting with regard to the interpretation of the calorimetric heating curves of cell membranes (27). If a membrane is cooled slowly prior to a calorimetric heating run it is clearly quite possible for some proteins to be squeezed out of the high melting lipid regions. Upon a subsequent heating run the scanning calorimetric curves will then correspond to melting of the lipids of a "zone refined" membrane. Experiments showing the existence of breaks or discontinuities in Ahrrenius plots of enzyme activity (28) with membranes of Acholeplasma laidlawii also need to be considered in the light of movement of some proteins into aggregated regions upon cooling the membranes.

Another feature of the calorimetric curves is the absence of a peak or endotherm associated with a "denaturation" of the polypeptide. This raises the question as to whether "intrinsic" proteins will show denaturation characteristics upon heating. Heat denaturation of proteins arises when hydrophobic groups from the interior of the protein move into a water environment.

## Above the Lipid Phase Transition Temperature

Above the lipid transition temperature when the concentration of the Gramicidin A is increased to greater than five lipids per polypeptide, two lipid regions occur. One corresponds to the relatively fluid lipid region normally observed at these temperatures and the other to a relatively rigid lipid region. The latter is thought to be associated with clusters of the polypeptide in which some of the lipid is entrapped between two or more molecules of the Gramicidin A.

The packing arrangements which occur between the lipids and the polypeptides within a randomly distributed close packed array of phospholipid molecules have been studied using a simple Monte Carlo model. These studies provide a conceptually useful picture of the packing arrangements of proteins and lipids in a fluid biomembrane. A simulated random array is shown in Fig.4.

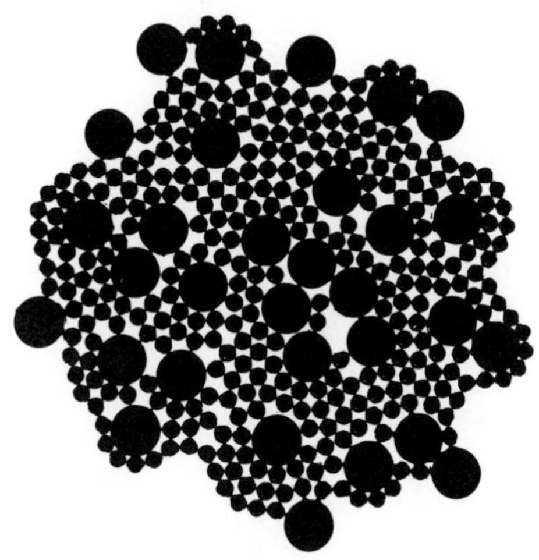

Figure 4 A simulation of a randomly distributed array of phospholipid chains and cylindrical polypeptides seen in two dimensions. (lipid mole to polypeptide ratio 5:1).

It can be shown for example that a relatively long term order may be produced between lipids, polypeptides and by implication, proteins which is related simply to their local concentration in membranes. It is also important to consider these effects when carrying out experiments with systems of low lipid content and high protein content when attempting to study boundary layer effects (26).

Further studies of the organisation of polypeptides and proteins in the fluid lipid bilayer structure seem worthy of attention.

## New Developments

The term lipid fluidity or membrane fluidity is now widely adopted. It was introduced to encompass the bewildering complexity of fatty acid chain lengths and unsaturation found to occur in cell membranes (15).

The modulation of lipid fluidity of cell membranes by cholesterol and other molecules has been studied and discussed by many workers. There have also been many methods adopted to alter lipid fluidity including genetic, nutritional and thermal manipulation as well as drugs and inhibitors. Our recent studies with Dr. P.J. Quinn have shown that the hydrogenation of the unsaturated double bonds of the lipid molecules within both model and natural biomembranes can be accomplished using a homogeneous catalyst. This provides the basis for a new technique for the modulation of membrane fluidity which may have many future applications.

## ACKNOWLEDGMENTS

We wish to acknowledge interesting discussions with Mr. A.W. Eliasz and Dr. W.E. Peel.

We also acknowledge a Senior Wellcome Trust Research Fellowship (to D.C.) and support from the Nuffield Foundation (to B.A.C.).

## REFERENCES

1. Chapman, D. Ann.N.Y.Acad.Sci. 137, 745 (1966).

2. Ladbrooke, B.D., Williams, R.M. & Chapman, D. Biochim. Biophys.Acta 150, 333 (1968).

3. Esser, A.F. & Lanyi, J.K. Biochemistry 12, 1933 (1973).

4. McElhaney, R.N. J.Mol.Biol. 84, 145 (1974).

5. Fox, F.C. & Tsukagoshi, T. Membrane Research, p.145 Academic Press (1972).

6. Chapman, D., Williams, R.M. & Ladbrooke, B.D. Chem.Phys.Lipids 1, 445 (1967).

7. Chapman, D. Q.Rev.Biophysics 8, 185 (1975).

8. Bretscher, M.S. & Raff, M.C. Nature 258, 43 (1975).

9. Steck, T.L. J.Cell.Biol. 62, 1 (1974).

10. Cone, R.A. Nature 235, 39 (1972).

11. Naqvi, K.R., Gonzalez-Rodriguez, J., Cherry, R.J. & Chapman, D. Nature 245, 249 (1973).

12. Orci, L. & Perrelet, A. Science 181, 868 (1973).

13. Scott, R.E., Carter, R.L. & Kidwell, W.R. Nature 233, 219 (1971).

14. James, R. & Branton, D. Biochim.Biophys.Acta 323, 378 (1973).

15. Chapman, D., Byrne, P. & Shipley, G.G. Proc.Roy.Soc. 290A, 115 (1966).

16. Chapman, D. The Structure of Lipids. Methuen, Lond. (1965).

17. Chapman, D. Biological Membranes (ed. D. Chapman) Vol.1, Academic Press (1968).

18. Cater, B.A., Chapman, D., Hawes, S. & Saville, J. Biochim. Biophys.Acta 363, 54 (1974).

19. Chapman, D., Urbina, J. & Keough,K. J.Biol.Chem. 249, 2512 (1974).

20. Ladbrooke, B.D. & Chapman, D. Chem.Phys.Lipids 3, 304 (1969).

21. Shimshick, E.J. & McConnell, H.M. Biochemistry 12, 2351 (1973).

22. Cornell, B.A., Chapman, D. & Eliasz, A.W. (awaiting publication)

23. Urry, D.W. Proc.Nat.Acad.Sci. 68, 672 (1971).

24. Rand, R.P., Chapman, D. & Larsson, K. Biophys.J. 15, 1117 (1975).

25. Kleeman, W., Grant, C.W.M. & McConnell, H.N. J.Supramolec.Structure 2, 609 (1974).

26. Jost, P.C., Griffith, O.H., Capaldi, R.A. & Vanderkooi, G. Proc.Nat.Acad.Sci. 70, 480 (1973).

27. Steim, J.M., Tourtellotte, M.E., Reinert, J.C., McElhaney, R.N. & Rader, R.L. Proc.Nat.Acad.Sci. 63, 104 (1969).

28. de Kruyff, B., van Kijk, P.W.M., Goldbach, R.W., Demel, R.A. & van Deenen, L.L.M. Biochim.Biophys.Acta 330, 269 (1973).

# BIOSYNTHESIS AND TRANSPORT OF MICROSOMAL MEMBRANE GLYCOPROTEINS

Gustav Dallner

Dept. of Pathology at Sabbatsberg Hospital,
Karolinska Institutet, and Dept. of Biochemistry,
Arrhenius Laboratory, University of Stockholm
Stockholm, Sweden

There are a large number of proteins, both in the extracellular space and in intracellular membranes, which contain covalently bound carbohydrate. Of the various membranes in hepatocytes, the plasma membrane has been most extensively studied (1); but it has also been established that the outer and inner mitochondrial membranes (2), lysosomes (3), nuclear membranes (4), and microsomes (5) also contain glycoproteins.

Experimental evidence demonstrates that microsomal glycoproteins are constitutive membrane components and that their presence cannot be explained by adsorption of nonmembranous proteins to the vesicle surface or by entrapment of secretory proteins in the vesicle lumen (6). In contrast to our knowledge about the glycoproteins of the plasma membrane, the functions of microsomal glycoproteins are not known. It is possible that some enzymes contain short oligosaccharide chains, but this does not seem sufficient to explain the large quantities of protein-bound sugar present in microsomes.

With respect to the biogenesis of microsomal membrane glycoproteins, it has turned out to be difficult to find a simple explanation for the microsomal localization of these proteins. Chemical and gas chromatographic analyses have conclusively demonstrated the presence of terminal sialic acid residues in both rough and smooth microsomes, fractions which completely lack the transferase for sialic acid (5,7). Consequently, the biogenetic sequence would seem to involve transport of a part of or the completed glycoprotein molecule from the site at which sialic acid is added to the membrane of the endoplasmic reticulum.

## Protein-Bound Sugar Residues in Microsomes

Table I. Sugar residues of liver microsomal membranes

| Fraction | Mannose | Galactose | Glucosamine | NANA |
|---|---|---|---|---|
| | | µg/mg phospholipid | | |
| Rough microsomes | 10.3 | 3.3 | 8.9 | 4.9 |
| Smooth I microsomes | 11.2 | 5.9 | 12.1 | 11.3 |
| Smooth II microsomes | 16.4 | 13.8 | 30.1 | 10.0 |
| Golgi membranes | 6.7 | 10.1 | 23.3 | 8.3 |

Data taken from ref. 7.

Washed microsomal and Golgi membranes contain the neutral sugars mannose and galactose, the amino sugar glucosamine, as well as N-acetyl neuraminic acid (NANA) (Table I). On a phospholipid basis the content of both neutral sugars varies greatly between the subfractions, with the ratio of mannose to galactose being highest in rough microsomes, lower in smooth I, lower still in smooth II, and lowest in Golgi membranes. Relatively large amounts of glucosamine are present in all the subfractions and the distribution pattern of this sugar is similar to that found for galactose. About 5 µg/mg phospholipid of protein-bound NANA is found in rough microsomes; 1.7 times this specific content is recovered in Golgi membranes; and the amount of NANA per mg phospholipid in smooth I and II microsomes is about twice as much as that in rough microsomes.

## Asymmetric Distribution of Glycoproteins

Trypsin treatment removes about 40 % of microsomal protein from the surface of intact microsomal vesicles, including NADPH-cytochrome c reductase and cytochrome $b_5$ (8) (Fig. 1). Such proteolysis also liberates about half of the protein-bound mannose and galactose.

Deoxycholate (DOC) under appropriate conditions (0.05 % DOC, 0.5 M KCl, 4 mg protein/ml) does not solubilize membrane components but makes microsomal vesicles permeable to macromolecules such as enzymes. When trypsin is introduced into the intramicrosomal space using DOC, 10 % of the total microsomal protein is solubilized from the inner surface, together with the solubilization of inactivation of nucleoside disphosphatase, esterase, β-glucuronidase and glucose-6-phosphatase. This treatment removes an additional 20 % of protein-bound mannose; in contrast, no galactose is bound to the protein on the inner surface which is accessible to trypsin.

A similar asymmetric distribution is also seen for microsomal NANA (Table II).

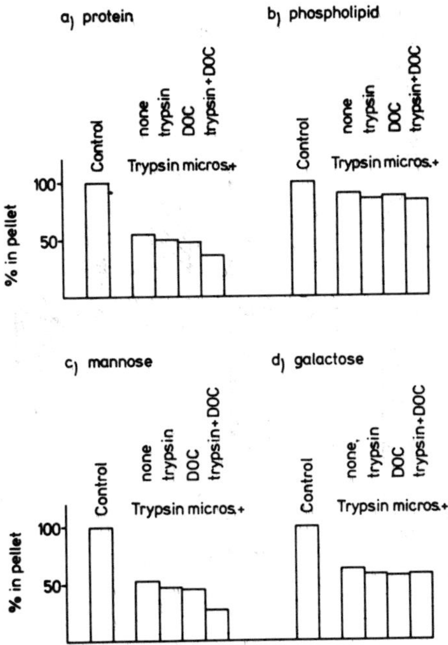


Fig. 1. Intramembranous localization of neutral sugars in microsomal membranes. Data taken from ref. 7.

Table II. Effect of neuraminidase treatment on microsomal NANA

| Treatment | NANA content in | |
|---|---|---|
| | Rough microsomes | Smooth microsomes |
| | µg/mg PL | |
| None | 5.9 | 13.9 |
| Neuraminidase | 2.8 | 12.7 |
| Pronase | 4.5 | 12.5 |
| Pronase + neuraminidase | 1.9 | 8.7 |
| 0.05 % DOC | 5.8 | 13.5 |
| 0.05 % DOC + neuraminidase | 2.1 | 10.1 |

The various treatments were performed as described earlier (9). Data taken from ref. 9.

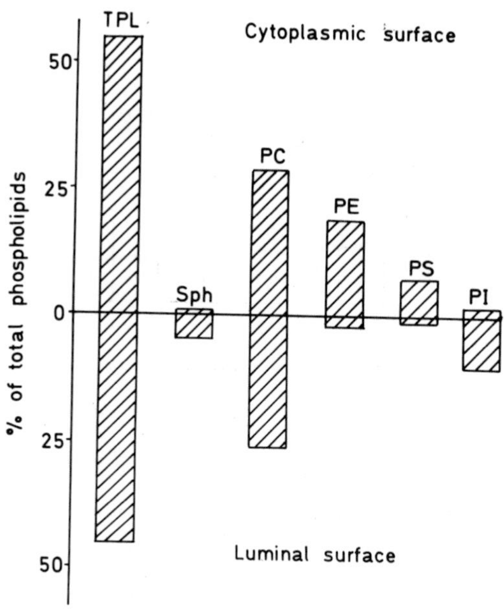


Fig. 2. Tentative distribution of various phospholipids in microsomal membranes. Data taken from ref. 8.

Neuraminidase treatment of intact rough vesicles removes more than half of the NANA but the same treatment liberates only a small amount of NANA from smooth microsomes. Treatment with both protease and neuraminidase solubilizes an additional amount of NANA from intact microsomal vesicles, indicating that a part of the protein-bound NANA on the outer surface is "shielded". When neuraminidase was introduced into the vesicles, only an additional 10-20 % of the original NANA could be released from both rough and smooth microsomes.

## Phospholipid Asymmetry

The distribution of phospholipids in microsomal membranes was studied using phospholipase $A_2$, an approach previously used on erythrocytes by Verkleij and his coworkers (10). Phospholipase $A_2$ treatment of intact microsomes hydrolyzed all of the phosphatidylethanolamine and -serine and 55 % of the phosphatidylcholine. It appears from this observation that these lipids are localized in the outer half of the bilayer (Fig. 2). Phosphatidylinositol, 45 % of the phosphatidylcholine and probably all of the sphingomyelin are thus assigned to the inner half of this bilayer.

## Localization of Membrane Glycoprotein Synthesis

Treatment of intact microsomal vesicles with trypsin, followed

Table III. Effect of trypsin pretreatment on the in vitro incorporation of mannose and glucosamine into cytoplasmic membranes

| | GDP-mannose-$^{14}$C | | | UDP-GlcNAc-$^{14}$C | | |
|---|---|---|---|---|---|---|
| | $LI_1$ | $LI_2$ | Protein | $LI_1$ | $LI_2$ | Protein |
| | % of non-treated | | | | | |
| Rough microsomes | 58 | 60 | 91 | 78 | 104 | 55 |
| Smooth microsomes | 42 | 50 | 22 | 101 | 103 | 95 |
| Golgi membranes | 35 | 29 | 44 | | | 60 |

Trypsin treatment before incubation was performed using 50 µg trypsin/mg protein at 30°C for 10 min. $LI_1$ = lipid intermediate 1 (dolichol phosphate); $LI_2$ = lipid intermediate 2 (dolichol pyrophosphate oligosaccharide).

by incubation with GDP-mannose-$^{14}$C and UDP-N-acetylglucosamine-$^{14}$C was carried out using rough and smooth microsomes and Golgi membranes (Table III). In most cases the activities of the enzymes transferring sugar residues from sugar nucleotides to dolichol monophosphate, from dolichol monophosphate to dolichol pyrophosphate-oligosaccharide, and finally from the second lipid intermediate to the protein acceptors were affected by trypsin treatment and sugar transfer was decreased to various degrees.

The isolated intact vesicles were also treated with trypsin after incubation *in vitro* with sugar nucleotides or after labeling *in vivo* with radioactive mannose and glucosamine (Table IV). In both cases part of the protein acceptor and also part of the lipid intermediates were removed. These experiments indicate that part of the transferase system, as well as the proteins which contain oligosaccharide chains under completion are situated at the cytoplasmic surface of the endoplasmic reticulum and Golgi membranes.

## Time Course of NANA Labeling

Pulse labeling with glucosamine-$^{3}$H *in vivo* resulted in a high rate of incorporation of NANA into proteins of the Golgi membranes and of the cytosol within the first 30 min, followed by a rapid decrease (Fig. 3). The time course for incorporation of NANA into glycoproteins of rough and smooth microsomes was very different; the radioactivity increased slowly and reached a maximum after 3 h. This pattern of incorporation suggests the possibility that glycoproteins of the cytosol and Golgi may later be incorporated into the membrane of the endoplasmic reticulum.

Table IV. Removal of Mannose and Glucosamine Incorporated *In Vitro* and *In Vivo* by Trypsin

| | Rough | | | Smooth | | | Golgi | | |
|---|---|---|---|---|---|---|---|---|---|
| | $LI_1$ | $LI_2$ | Protein | $LI_1$ | $LI_2$ | Protein | $LI_1$ | $LI_2$ | Protein |
| | % of non-treated | | | | | | | | |
| *In vitro* incubation | | | | | | | | | |
| GDP-mannose-$^{14}C$ | 43 | 53 | 57 | 70 | 68 | 98 | 54 | 68 | 76 |
| UDP-GlcNAc-$^{14}C$ | 82 | 83 | 83 | 51 | 43 | 75 | 62 | 90 | 74 |
| *In vivo* injection | | | | | | | | | |
| mannose-$^3H$ | 73 | 12 | 83 | | | 66 | | | |
| glucosamine-$^3H$ | 98 | 64 | 99 | | | 76 | | | |

Mannose-$^3H$ (100 µCi/100 g body weight) and glucosamine-$^3H$ (150 µCi /100 g) were injected into the portal vein 5 min before decapitation. Otherwise as in Table III.

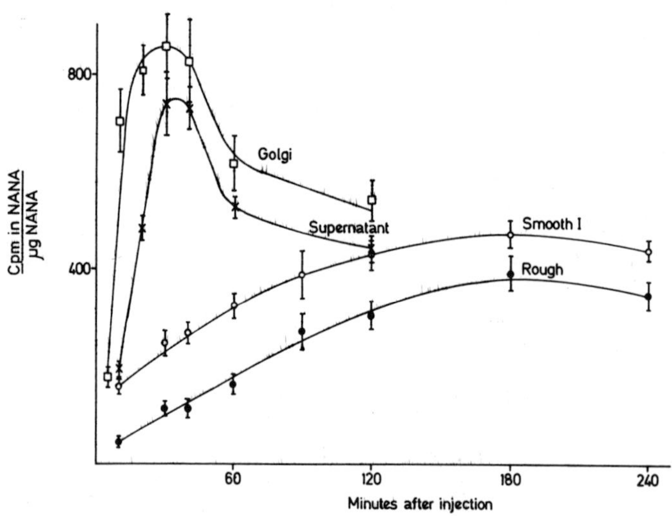


Fig. 3. Incorporation of glucosamine-$^3H$ *in vivo* into protein-bound NANA of different subcellular fractions. Data taken from ref. 6.

## Incorporation of Cytosolic Lipoprotein into Microsomes

In order to test the possibility that glycoprotein components of the cytosol may be transferred into microsomal membranes, three fractions were prepared from the liver supernatant of rats injected with glucosamine-$^3$H and leucine-$^{14}$C: a) the total protein in the void volume from a Sephadex G-25 column (G-25 pool); b) the G-25 pool partially purified on a Sephadex G-100 column (G-100 pool); and c) a supernatant lipoprotein (LP fraction) obtained by flotation of the G-25 pool in a KBr solution with a density of 1.21. The increasing purification of the sialoprotein in question is demonstrated by the dramatic increase of the specific amount of protein-bound NANA that can be incorporated into microsomes from the G-25 pool to the G-100 pool and finally to the LP fraction (Table V). Upon incubation of rough microsomes *in vitro* with each of the supernatant fractions both labels were incorporated, and this incorporation was highest with the LP fraction. The radioactivity incorporated into the microsomes could not be removed by various washing procedures nor using a low concentration of detergent.

The lipoprotein complex isolated by the flotation procedure (LP fraction) contains two peptides as revealed by electrophoresis in the presence of SDS, one with a molecular weight around 70,000 daltons and one around 10,000, and both contain sugar moieties (Fig. 4). The gel electrophoretic pattern of the three blood lipoproteins, VLDL, LDL and HDL, are different from that of the supernatant lipo-

Table V. Incorporation of glycoproteins from different fractions into rough microsomes

| | G-25 pool | G-100 pool | LP fraction |
|---|---|---|---|
| Amount of NANA (μg NANA/mg protein) | 0.30 | 0.96 | 38.1 |
| Total radioactivity in the incubation mixture | | | |
| GlN-$^3$H, cpm | 3,610 | 7,790 | 11,425 |
| Leucine-$^{14}$C, cpm | 4,410 | | 4,277 |
| Radioactivity transferred to microsomes | | | |
| GlN-$^3$H, cpm | 843 | 2,160 | 4,650 |
| GlN-$^3$H, % | 23.3 | 27.8 | 40.7 |
| Leucine-$^{14}$C, cpm | 554 | | 840 |
| Leucine-$^{14}$C, % | 12.5 | | 19.6 |

Rough microsomes were incubated with G-25 pool, G-100 pool and LP fractions as described earlier (11). Data taken from ref. 11.

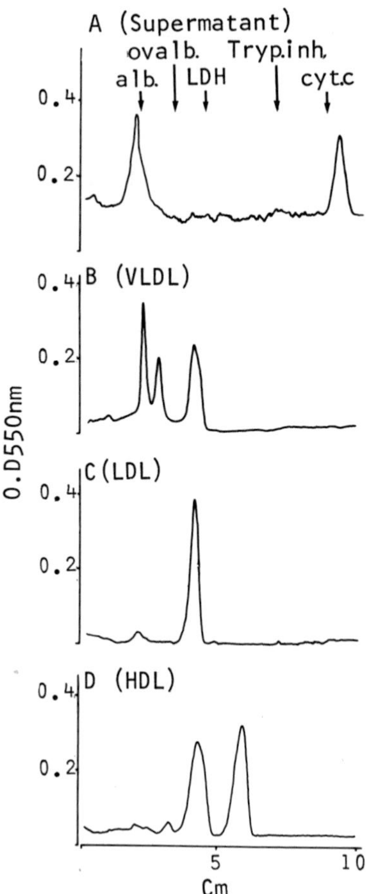


Fig. 4. SDS-gel electrophoretic patterns of supernatant and serum lipoproteins. A) Supernatant LP; B) VLDL; C) LDL; D) HDL. Data taken from ref. 11.

protein; and it is therefore improbable that the presence of the lipoprotein in the supernatant is due to contamination from serum proteins.

The gel electrophoretic pattern of rough microsomes after incubation with LP is shown in Fig. 5. The protein peak in fraction 9 is characterized by the highest levels of glucosamine-$^{3}H$ and leucine-$^{14}C$, suggesting that the whole glycoprotein unit is incorporated.

Some of the features of the cytoplasmic lipoprotein which is incorporated most readily into microsomal membranes under in vitro conditions are summarized in Table VI.

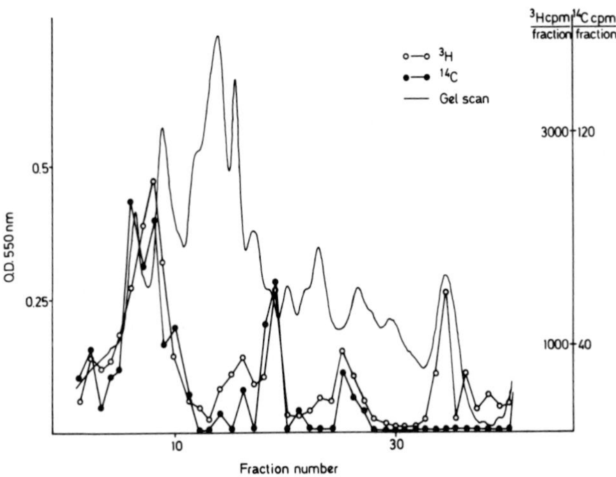


Fig. 5. SDS-gel electrophoretic pattern of rough microsomes after incubation *in vitro* with double-labeled supernatant LP. Data taken from ref. 11.

Table VI. Properties of the cytoplasmic lipoprotein complex

| | |
|---|---|
| Equilibrium density | 1.07-1.14 g/ml |
| Sedimentation coefficient | 5.3 S |
| Calculated molecular weight | 270,000 daltons |
| Peptides after SDS-gel electrophoresis | 11-13,000 and 67-69,000 daltons |
| Oligosaccharide composition | mannose, galactose, glucosamine, NANA |
| Lipid composition (44 % by weight) | |
| neutral lipids (50 %) | cholesterol, triglycerides |
| phospholipids (50 %) | phosphatidylcholine<br>lysophosphatidylcholine<br>sphingomyelin<br>phosphatidylethanolamine<br>phosphatidylinositol |

Data taken from ref. 12.

## Proteins Released from Golgi Vesicles

The possibility that the supernatant LP is released from Golgi membranes was suggested by the *in vivo* labeling experiments

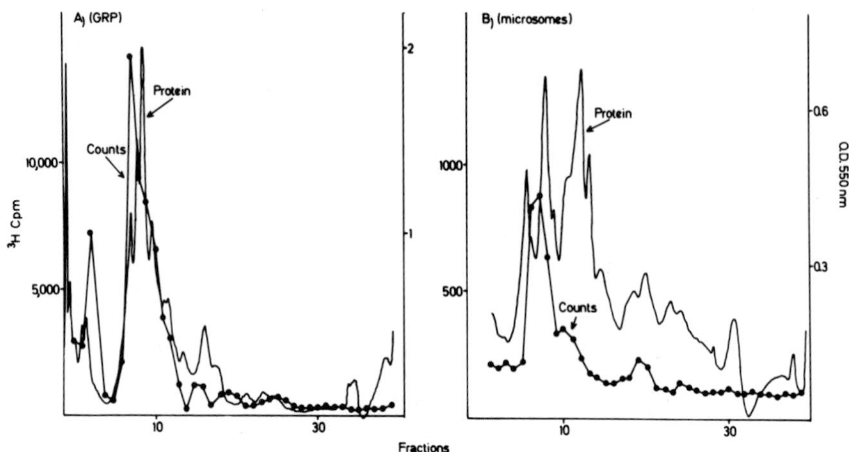


Fig. 6. Incubation of microsomes with proteins released from Golgi vesicles. A) GRP before, and B) microsomes after incubation. Data taken from ref. 13.

(Fig. 3). For this reason the Golgi fraction was prepared from rats which had been previously injected with glucosamine-$^3$H. When incubated in sucrose, the Golgi vesicles released four major weakly acidic glycoproteins (Fig. 6A). This protein released from the Golgi (GRP) was readily incorporated into microsomes upon incubation *in vitro*. When microsomes were analyzed by SDS-gel electrophoresis after such incubation, protein with high specific radioactivity appeared in only one region of the gel, around fraction 9 (Fig. 6B). The properties of GRP proved to be very similar to that of the supernatant LP in several respects. 30 % of the total

Table VII. Incorporation of total GRP into microsomes, mitochondria and Golgi membranes

| Fraction | Glucosamine-$^3$H incorporated | |
|---|---|---|
| | total cpm | % |
| GRP | 183,000 | |
| Total microsomes | 56,880 | 31.1 |
| Rough microsomes | 54,230 | 29.6 |
| Mitochondria | 10.490 | 5.7 |
| Golgi | 15.400 | 8.4 |

The various fractions (5 mg protein) were incubated with 0.3 mg GRP protein, centrifuged, washed, and the radioactivity in the pellet determined. Data taken from ref. 13.

radioactivity in the GRP fraction could be incorporated into total and rough microsomes by incubation (Table VII). Incorporation into liver mitochondria was poor and reincubation of GRP with the total Golgi fraction resulted in incorporation which was less than one-third of that obtained with microsomes. These findings indicate that the glycoproteins are designed to be incorporated into specific membranes.

## CONCLUSION

The results described above indicate that microsomal membrane glycoproteins may be located on the cytoplasmic surface of the rough endoplasmic reticulum after synthesis of the protein moiety and that they are transported to smooth membranes and subsequently to the Golgi system and are probably found in a similar lipid environment throughout different stages of this transport process. During this transport the oligosaccharide chain is completed. The protein is released from Golgi membranes into the cytoplasm as a lipoprotein complex. This complex is readily incorporated into microsomal membranes and may represent a pathway for the renewal of microsomal membrane glycoproteins.

## ACKNOWLEDGMENTS

The work described in this paper was carried out in collaboration with Drs. F. Autuori, A. Bergman, Å. Elhammer, L. Eriksson, O. Nilsson, H. Svensson, L. Winqvist and was supported by grants from the Swedish Medical Research Council.

## REFERENCES

1. Kawasaki, T., and I. Yamashina (1971). Biochim. Biophys. Acta 225, 234.
2. Martin, S.S., and H.B. Bossmann (1971). Exp. Cell Res. 66, 59.
3. Goldstone, A., and H. Koenig (1974). FEBS Lett. 39, 176.
4. Kawasaki, T., and I. Yamashina (1972). J. Biochem. 72, 1517.
5. Miyajima, N., M. Tomikawa, T. Kawasaki, and I. Yamashina (1969). J. Biochem. 66, 711.
6. Autuori, F., H. Svensson, and G. Dallner (1975). J. Cell Biol. 67, 687.
7. Bergman, A., and G. Dallner (1976). Biochim. Biophys. Acta, in press.
8. Nilsson, O., and G. Dallner (1975). FEBS Lett. 58, 190.
9. Winqvist, L., L. Eriksson, G. Dallner, and B. Ersson (1976). Biochem. Biophys. Res. Comm. 68, 1020.

10. Verkleij, A.J., R.F.A. Zwaal, B. Roelofsen, P. Comfurius, D. Kastelijn, and L.L.M. van Deenen (1973). Biochim. Biophys. Acta 323, 178.
11. Autuori, F., H. Svensson, and G. Dallner (1975) J. Cell Biol. 67, 700.
12. Svensson, H., Å. Elhammer, F. Autuori, and G. Dallner. In manuscript.
13. Elhammer, Å., H. Svensson, F. Autuori, and G. Dallner (1975). J. Cell Biol. 67, 715.

# SOME TOPOLOGICAL AND DYNAMIC ASPECTS OF LIPIDS IN THE ERYTHROCYTE MEMBRANE

L.L.M. van Deenen, J. de Gier, L.M.G. van Golde, I.L.D. Nauta, W. Renooy , A.J. Verkleij and R.F.A. Zwaal

Laboratory of Biochemistry and Laboratory of Veterinary Biochemistry, University of Utrecht, The Netherlands

## ABSTRACT

This paper reviews the use of phospholipases to determine the topology of phospholipids in the erythrocyte membrane. In mammalian red cell membranes the choline-containing phospholipids, sphingomyelin and lecithin are predominantly located at the exterior region whereas phosphatidylethanolamine and phosphatidylserine are found preferentially in the inner monolayer. A topological asymmetry of phospholipid metabolism was detected in rat erythrocyte membranes. Exchange of phospholipids between plasma lipoproteins and intact red cells occured at the exterior region of the lipid bilayer. The incorporation of fatty acids into phospholipids appeared to be located at the interior side of the membrane. The data indicate that a translocation of lecithin species between the two pools or sides of the membrane may be possible.

Electron microscopic studies demonstrated that the enlargement of erythrocytes in patients with cholestasis is due to a fusion of abnormal plasma lipoprotein with the red cell membrane, resulting in an increased content of lecithin and cholesterol.

## 1. PHOSPHOLIPID ASYMMETRY

The localization of membrane phospholipids can be investigated by the use of phospholipases or non-permeant group-specific labels. The latter group of reagents, used to date, are specific for amino-groups, and can therefore only provide direct information on the disposition of amino-containing phospholipids like phosphatidylethanolamine and phosphatidylserine. On the other hand, selecting a proper combination of phospholipases provides a tool with which in principle all available phospholipids can be attacked, so that information on the localization of each phospholipid class can be obtained (Zwaal *et al.*, 1973). A serious drawback might be that application of both methods results in a chemical alteration of the phospholipid molecules with a concomitant change in membrane structure.

The general strategy to reveal asymmetric phospholipid distributions in erythrocyte membranes, using pure phospholipases as tools, can be outlined as follows: (i) enzymatic breakdown of phospholipids, obtained by the action of phospholipases on intact red cells, is restricted to those phospholipid molecules which are located at the outside of the membrane, provided that no hemolysis of the cells occurs; (ii) enzymatic action on non-sealed ghosts will lead to hydrolysis of phospholipids at either side of the membrane and will in addition provide information about the availability of the phospholipids to the phospholipases; (iii) trapping of phospholipases (without cofactor) within resealed ghosts followed by addition of a cofactor, will result in enzymatic hydrolysis of those phospholipids which are present at the membrane interior, as long as no lysis of the resealed cells occurs.

When intact human erythrocytes are treated successively with phospholipase $A_2$ (*Naja naja*) and sphingomyelinase C (*Staphylococcus aureus*) no hemolysis occurs, although the osmotic stability of the cells is strongly impaired. This subsequent action, though not producing breakdown of phosphatidylserine, degrades 75% of phosphatidylcholine, 20% of phosphatidylethanolamine and 82% of sphingomyelin. It has been concluded that this phospholipid fraction, which comprises nearly 50% of the total phospholipids, forms the outer half of the lipid bilayer of the human red cell membrane. The conclusion that sphingomyelin is located in the outer monolayer of the membrane was also supported by freeze-etch electronmicroscopy. Sphingomyelinase action (either alone or in combination with phospholipase $A_2$) on intact erythrocytes produces formation of small spheres (probably ceramide droplets) with diameters of 75 Å or 200 Å on the outer fracture face with corresponding pits on the inner fracture face (Verkleij *et al.*, 1973). The combined action of phospholipase $A_2$ and sphingomyelinase on non-sealed ghosts produces complete breakdown of all the major phospholipid classes, including phosphatidylserine.

Due to its absolute requirement for $Ca^{2+}$, pancreatic phospholipase $A_2$ can be satisfactorily trapped within resealed ghosts in the presence of EDTA. Subsequent activation of the enzyme by addition of $Ca^{2+}$ produces substantial breakdown of glycero-phospholipids before the cells start to lyse (Fig. 1). Just before the onset of lysis, approximately 25% of phosphatidylcholine, 50% of phosphatidylethanolamine, and some 65% of phosphatidylserine were found to be converted into their respective lyso-derivatives (Zwaal

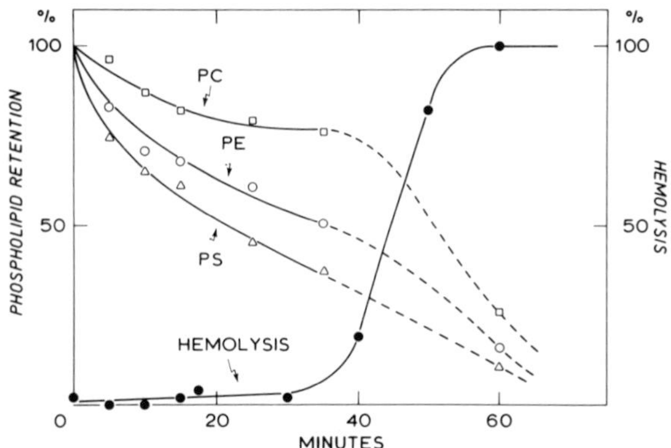


Figure 1 Action of phospholipase $A_2$ at the inside of the erythrocyte membrane.
Percentage retention of glycerophospholipids after trapping of phospholipase without $Ca^{++}$ inside resealed human erythrocyte ghosts followed by the addition of $Ca^{++}$ to start enzymatic hydrolysis at t = 0.
PC, lecithin; PE, phosphatidylethanolamine; PS, phosphatidylserine.

*et al.*, 1975). Apparently, the accumulation of lyso-compounds in the membrane reaches a critical level resulting in lysis of the resealed cell. Unlike the phosphatidylcholine degradation at the inside, the breakdown of phosphatidylethanolamine and phosphatidylserine is not completely complementary to the results obtained on intact erythrocytes. Nevertheless, the trapping experiments give strong support to the non-uniform distribution of phospholipids between outside and inside of the red cell membrane.

Recently, quantitative data from other laboratories concerning phospholipase action on intact red cells (Gul and Smith, 1974) and sealed inside-out ghosts (Kahlenberg *et al.*, 1974) also seem to support the asymmetric phospholipid distribution. Moreover, reaction of intact erythrocytes with relatively non-permeant reagents like formylmethionyl-sulphone-methylphosphate fails to tag phosphatidylethanolamine and phosphatidylserine, whereas both phospholipids are readily labelled in non-sealed ghosts (Bretscher, 1972). Using trinitrobenzenesulphonate, Gordeski and Marinetti (1973) also failed to label phosphatidylserine in intact cells, but some phosphatidylethanolamine was found to react with this relatively non-permeant probe.

In addition to revealing a non-random distribution of phospholipids in the membrane, phospholipases can also provide information about the compression state of the lipids in the native membrane.

This is based on the observation that one group of phospholipases produces phospholipid breakdown in intact erythrocytes, whereas the other group of phospholipases fails to exert its action on intact cells (Zwaal *et al.*, 1975). These two groups of enzymes can also be distinguished in studies dealing with the activity of phospholipases towards monomolecular films of phospholipids, spread at an air-water interface at various initial surface pressures (Demel *et al.*, 1975). Those phospholipases which fail to exert their action on intact cells are also unable to hydrolyse lipids when injected under a monolayer of choline-containing phospholipids spread at an initial surface pressure above 31 dynes/cm. On the other hand, those phospholipases which are able to attack the intact red cell membrane can produce phospholipid degradation in monolayers with an initial surface pressure of at least 34 dynes/cm. From these observations it is concluded that the packing of the phospholipids at the exterior layer in the intact erythrocyte membrane is comparable with a lateral surface pressure of 31-34 dynes/cm. This surface pressure is considerably higher than the pressure used in the classic experimental design of Gorter and Grendel (1925), which initiated many of the current concepts about the molecular structure of biomembranes. They measured the surface area of the lipids, extracted from erythrocyte membranes, by spreading them from a benzene solution onto a Langmuir Adam trough at low surface pressures of about 2 dynes/cm. The results were found to fit in well with the idea that the number of lipid molecules per cell is sufficient to accomodate the erythrocyte with a lipid-bilayer membrane.

Bar *et al.* (1966) reinvestigated this subject, carefully avoiding the errors made in the past, and found as could be expected that the ratio of lipid monolayer area to red cell surface area is strongly dependent on the film pressure employed. A ratio of film area to cell area of 2 is indeed possible at low monolayer surface pressures. At a lateral film pressure of 31-34 dynes/cm - as deduced above - this ratio appeared to be in the order of 1.5. This might imply that the number of lipid molecules per cell is sufficient to form a bilayer with a surface area occupying approximately 75% of the total membrane area (Fig. 2). This value, inaccurate as it may be, is in good agreement with the surface area occupied by the intrinsic membrane proteins, if one assumes that these proteins interrupt the lipid bilayer and are represented by the membrane intercalated particles observed with freeze-fracture electron microscopy. The number of particles per cell multiplied with their average surface area accounts for a total particle area per cell of approximately 30 square micrometer, which corresponds to the remaining 25% of the total cell surface area. Although it is realized that this estimation is subject to several oversimplifications, the results can be taken to support the concept that red cell membranes are essentially lipid-bilayers interrupted by membrane penetrating proteins.

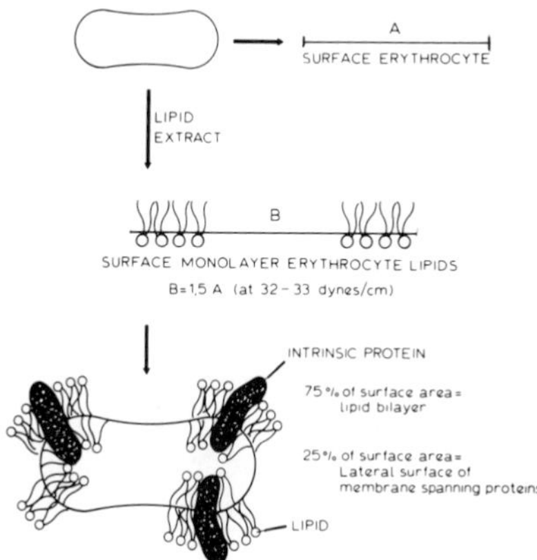


Figure 2 Surface area of the human erythrocyte membrane occupied by lipids and proteins.

## 2. TOPOLOGY OF PHOSPHOLIPID METABOLISM

Circulating mature erythrocytes from mammals are limited in their lipid metabolism, although some biosynthetic and catabolic reactions occur which, in combination with exchange processes, contribute to a given dynamic state of the erythrocyte membrane. Some of the reactions are summarized in figure 3; for a recent review compare van Deenen and de Gier, 1974. Two mechanisms may play a ma-

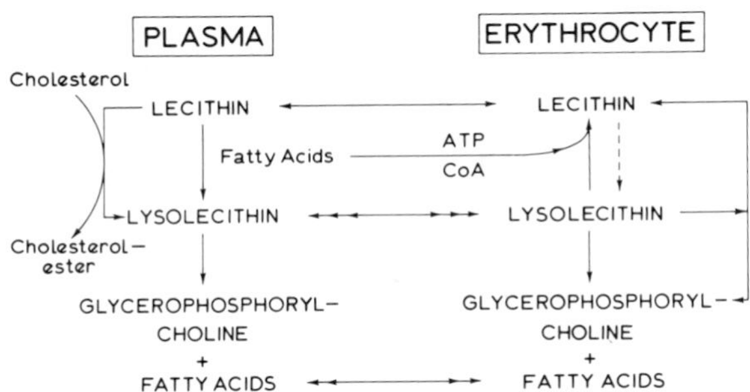


Figure 3 Pathways for renewal of phospholipids in the erythrocyte membrane.

jor role in the renewal of phospholipids of the erythrocyte membrane: (i) incorporation of fatty acids into lysophospholipids, (ii) and exchange of phospholipids between serum lipoproteins and the erythrocyte membrane. Recent evidence supports the concept that two metabolically distinct pools contribute to phospholipid turnover in the red cell membrane and that the two mechanisms occur in different compartments of the membrane. Renooy *et al.* (1974) studied the incorporation of fatty acids in human erythrocytes and used phospholipase $A_2$ from *Naja naja* as a tool to establish in which lecithin pool of the intact erythrocyte membrane the labelled fatty acids were recovered. When the lecithin present in the membrane exterior is converted into lysolecithin the majority of labelled lecithin appeared not to be degraded. A possible explanation is that the enzymes involved in the formation of acyl-CoA esters and the acyltransferases are located at the inside of the membrane and that the labelled lecithin molecules are located in the interior monolayer of the membrane. Recently these studies were extended to rat erythrocytes so as to include a study of the exchange process. The phospholipid distribution in rat erythrocytes as determined by means of phospholipase action is rather similar to that proposed for human erythrocytes (Fig. 4). Phosphatidylserine and phosphatidylethanol-

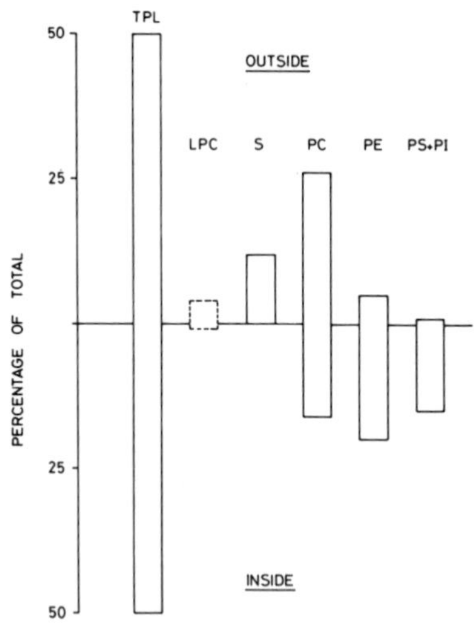


Figure 4 Distribution of phospholipids between inner and outer layer of the rat erythrocyte membrane.
TPL, total phospholipid; LPC, lysolecithin; S, sphingomyelin; PC, phosphatidylcholine; PE, phosphatidylethanolamine; PS, phosphatidylserine.

amine are mainly located at the inside of the membrane. Rat erythrocytes contain more lecithin and less sphingomyelin than human erythrocytes; the amount of choline-containing phospholipids in the outside monolayer of the membrane is about the same in both types of erythrocytes. When rat erythrocytes are incubated with radioactive fatty acids the highest total and specific activity was observed in lecithin. After treatment of the labelled erythrocytes with phospholipases, the radioactivity remained for the greater part unchanged, indicating that the incorporation of fatty acids occurs in a pool not accessible to phospholipases added at the outside of the intact erythrocyte.

After incubation of erythrocytes with $^{32}$P-labelled plasma an uptake of $^{32}$P-phospholipids by exchange is observed in the erythrocyte membrane. Labelled lysolecithin, lecithin and sphingomyelin were detected in the membrane. It was demonstrated by Renooy *et al.* (1976) that the labelled lecithin, now present in the membrane erythrocyte can be degraded by the action of phospholipase $A_2$ from *Naja naja*. This indicates that exchange of lecithin between plasma lipoproteins and rat erythrocyte membrane occurs mainly at the outer monolayer of the membrane, this by contrast to the labelling obtained after incubation with radioactive fatty acids. Furthermore, it was found that the specific radioactivity of the lecithin fraction, which can be hydrolyzed by phospholipases at the outside of the erythrocyte, decreases with incubation times. This result suggested a transfer of lecithin from the degradable to the non-degradable lecithin compartment of the membrane. Therefore, rat erythrocytes were labelled by incubating non-radioactive red cells for a given period in $^{32}$P-labelled plasma; the plasma was removed by washing and the incubation continued in a buffer, thus excluding a further contribution by the exchange process. After different incubation times the distribution of $^{32}$P lecithin between the both pools was measured by subjecting the cells to action of phospholipase $A_2$ and sphingomyelinase. As demonstrated in figure 5, the to-

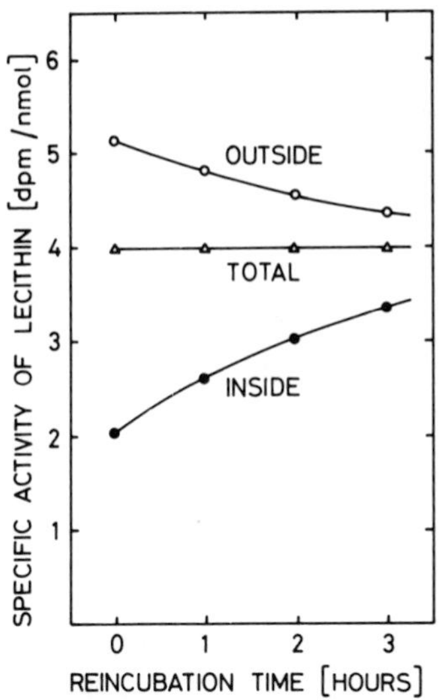


Figure 5 Transfer ($^{32}$P) lecithin between the pools susceptible (outside) and non-susceptible (inside) to phospholipase action on intact rat erythrocyte membrane. Labelling was obtained by incubated red cells in $^{32}$P-labelled plasma for 2 h at 37$^{o}$C. The plasma was removed and the labelled erythrocytes were reincubated in a buffer at 37$^{o}$C, for time periods indicated.

tal radioactivity of lecithin in the membrane remains constant throughout this reincubation period. The specific radioactivity of the lecithin fraction which can be degraded in the intact erythrocyte by phospholipase action decreases with time, with a corresponding increase of radioactivity of the lecithin fraction which is not being degraded by adding phospholipases to intact cells. These results suggest a transfer of lecithin molecules between two pools which may have a different location within the membrane. Both fractions are denoted in figure 5 as outside and inside respectively, because the exchange process is likely to occur initially at the exterior region of the membrane. The most simple explanation may be that the exchange of lecithin occurring at the outside of the membrane is followed by a trans-membrane movement to the inner part of the membrane. This redistribution occurring with a half-time rate of 4.5 h* might provide an explanation for the fact that no significant differences have been found between the molecular species of lecithin at either side of the membrane.

As discussed above, incubation of intact erythrocytes with radioactive fatty acids resulted in a labelling of the lecithin pool which is not hydrolysed by phospholipases added at the outside. Current experiments of Renooy demonstrate that with increasing incubation times more of the labelled lecithin becomes available for enzymatic attack. These results also suggest a transfer of lecithin from an inside to an outside pool of the membrane, occurring at a rate comparable to that observed with respect to the transfer of $^{32}$P-labelled lecithin incorporated by the exchange process. Thus it is possible that the two metabolically different pools of lecithin are in equilibrium by a translocation mechanism. Further experiments are needed to prove that this translocation process occurs indeed across the membran

## 3. FUSION OF LP-X AND ERYTHROCYTE MEMBRANE IN CHOLESTASIS

A complex exchange equilibrium of the phospholipids and cholesteral exists between the red cell membrane and serum lipoproteins (Fig. 3), and a disturbance of this equilibrium may alter the lipid composition of the red cell membrane. In hepatobiliary diseases an increase of the mean erythrocyte diameter has been observed, and chemical analysis have shown that this increase in cell diameter and surface area is associated with an increased content of cholesterol and phospholipid (for a review see Cooper, 1970). Recent investigations have demonstrated that in patients with intra or extra hepatic cholestases an abnormal plasma lipoprotein occurs (LP-X) which contains predominantly cholesterol and lecithin (ratio 1:1) and about 6% protein (Seidel *et al.*, 1970). It was demonstrated by Hamilton *et al.* (1971) that LP-X occurs as small vesicles bounded by a lipid bilayer and it can be visualized that this lipoprotein is responsible for the increased lipid content of erythrocytes in patients with cholestasis. Using freeze-etch electron micros-

*Recently comparable rates were observed by Bloj and Zilversmit, Biochemistry, 15(1976)1277, and Rousselet *et al.*, Biochim. Biophys. Acta, 426(1976)357.

Figure 6 The fusion of abnormal plasma lipoprotein (LP-X) and the erythrocyte membrane in patients with cholestasis.
a) Normal erythrocyte, inner fracture face (IFF)
b) Initial stage of the enlargement of an erythrocyte in patients with intra or extra hepatic cholestasis. Note the smooth structures of about 400 Å on the inner fracture face and at the etch face.
c) Thin sections of b
d) End stage of the enlargement of an erythrocyte in patients with intra and extra hepatic cholestasis

copy Verkleij *et al.* (1976) could detect changes in the lipid-intrinsic protein ratio in the red cell membranes from patients with cholestasis. The intra membraneous protein particles are not homogeneously distributed on the fracture faces, and the particle density on the inner fracture faces of the abnormal erythrocyte is significantly lower. (Compare figure 6a and d). In the pathologically enlarged cells a particle density dilution factor of 1.25 has been observed. At the early stage of cell enlargement the cell diameter is still normal and no abnormality was detected in the particle distribution on the fracture faces. However, at this stage smooth areas of 400-800 Å are frequently present at the fracture faces and the etch face. (Fig. 6b). Similar alterations could be induced *in vitro* after incubation of normal erythrocytes with serum containing the abnormal protein LP-X. After a two minutes incubation, freeze-etching showed similar smooth structures as observed in the early stage of the erythrocyte enlargement *in vivo*. At this stage the particle distribution is still normal, but after two hours incubation, particle aggregation and a significant decrease in partacle density on the inner fracture face was visible (Verkleij *et al.* 1976). The smooth structures could also be observed after incubation of erythrocytes with purified LP-X. Both *in vivo* and *in vitro* two characteristic freeze-etch stages were distinguished: (i) the initial stage having a normal erythrocyte size and normal particle distribution on the fracture face, but with the presence of characteristic smooth structures; (ii) the end stage which is characterized by an erythrocyte enlargement and a dilution of protein particles on the fracture faces. Thin sections at the initial stage showed both *in vitro* and *in vivo* adhesion of LP-X of vesicles at the surface of the erythrocyte membrane (Fig. 6c). These vesicles, however, are rarely found at the end stage of the erythrocyte enlargement, thus supporting the view that adhesion of the LP-X vesicles is followed by fusion with the erythrocyte membrane.

Recent experiments show that erythrocytes from patients with cholestasis contain in some cases more than twice the number of lecithin molecules per cell as compared to control cells. Treatment of these enlarged cells with phospholipase $A_2$ and sphingomyelinase reveals that roughly 25-40% of the incorporated lecithin cannot be degraded from the outside, indicating that a substantial fraction of LP-X fuses with the inner half of the cell membrane bilayer as well. More exact data cannot be given at present, since these cells often contain some adhering LP-X-vesicles at the outside which are probably degraded by phospholipase action on the cell exterior, thus disturbing the results obtained.

## REFERENCES


Bar, R.S., D.W. Deamer and D.G. Cornwell, Surface area of human erythrocyte lipids: Reinvestigation of experiments on plasma membrane, Science, 153, 1010, 1966.

Bretscher, M.S., Phosphatidylethanolamine: Differential labelling in intact cells and cell ghosts of human erythrocytes by a membrane-impermeable reagent, J. Mol. Biol., 71, 523, 1972.

Cooper, R.A., Lipids of human red cell membrane: Normal composition and variability in disease, Semin. Hematol., 7, 296, 1970.

van Deenen, L.L.M., J. de Gier, Lipids of the red cell membrane, in The Red Blood Cell, edited by D. Surgenor, pp. 147-211, Academic Press, New York, 1974.

Demel, R.A., W.S.M. Geurts van Kessel, R.F.A. Zwaal, B. Roelofsen and L.L.M. van Deenen, Relation between phospholipase action on human red cell membrane and the interfacial phospholipid pressure in monolayers, Biochim. Biophys. Acta, 406, 97, 1975.

Gordesky, S.E. and G.V. Marinetti, The asymmetric arrangement of phospholipids in the human erythrocyte membrane, Biochem. Biophys. Res. Commun., 50, 1027, 1973.

Gorter, E. and F. Grendel, On bimolecular layers of lipids on the chromocytes of the blood, J. Exp. Med., 41, 439, 1925.

Gul, S. and A.D. Smith, Haemolysis of intact human erythrocytes by purified cobra venom phospholipase $A_2$ in the presence of albumin and $Ca^{2+}$, Biochim. Biophys. Acta, 367, 271, 1974.

Hamilton, R.L., R.J. Havel, J.P. Kane, A.E. Blaurock and T. Sata, Cholestasis: Lamellar structure of the abnormal human serum lipoprotein, Science, 172, 475, 1971.

Kahlenberg, A., C. Walker and R. Rohrlick, Evidence for an asymmetric distribution of phospholipids in the human erythrocyte membrane, Can. J. Biochem., 52, 803, 1974.

Renooy, W., L.M.G. van Golde, R.F.A. Zwaal, B. Roelofsen and L.L.M. van Deenen, Preferential incorporation of fatty acids at the inside of human erythrocyte membranes, Biochim. Biophys. Acta, 363, 287, 1974.

Renooy, W., L.M.G. van Golde, R.F.A. Zwaal and L.L.M. van Deenen, Topological asymmetry of phospholipid metabolism in rat erythrocyte membranes, Eur. J. Biochem., 61, 53, 1976.

Seidel, D., P. Aloupovic and R.H. Furman, A lipoprotein characterizing obstructive jaundice. II. Isolation and partial characterization of the protein moieties of low density lipoproteins, J. Clin. Invest., 49, 2396, 1970.

Verkleij, A.J., R.F.A. Zwaal, B. Roelofsen, P. Comfurius, D. Kastelijn and L.L.M. van Deenen, The asymmetric distribution of phospholipids in the human red cell membrane. A combined study using phospholipases and freeze-etch electron microscopy, Biochim. Biophys. Acta, 323, 178, 1973.

Verkleij, A.J., I.L.D. Nauta, J.M. Werre, J.G. Mandersloot, B. Reinders, P.H.J.Th. Ververgaert and J. de Gier, The fusion of abnormal plasma lipoprotein (LP-X) and the erythrocyte membrane in patients with cholestasis studied by electron microscopy, Biochim. Biophys. Acta, 1976, in the press.

Zwaal, R.F.A., B. Roelofsen and C.M. Colley, Localization of red cell membrane constituents, Biochim. Biophys. Acta, 300, 159, 1973.

Zwaal, R.F.A., B. Roelofsen, P. Comfurius and L.L.M. van Deenen, Organisation of phospholipids in human red cell membranes as detected by the action of various purified phospholipases, Biochim. Biophys. Acta, 406, 83, 1975.

THE MOLECULAR MOTION OF SPIN LABELED AMPHIPHILIC MOLECULES IN MODEL MEMBRANES

Anders Ehrenberg, Yuhei Shimoyama and
L. E. Göran Eriksson

Department of Biophysics, Stockholm University
Arrhenius Laboratory, Fack
S-104 05 Stockholm, Sweden

## ABSTRACT

The theoretical background of a method to evaluate EPR spectra of spin labeled amphiphilic molecules in lipid bilayers is outlined. A model of restricted motion is applied including restricted rotation around the molecular long axis and rapid tumbling of this axis within the confines of a cone. The validity of the model is critically tested by application to oriented lipid multibilayer samples measured at three microwave frequencies. It is shown that detailed information about the mode of ordering and motion of the spin label molecule may be obtained.

## INTRODUCTION

Nitroxide spin labeled amphiphilic molecules in lipid bilayers of membranes and membrane models may yield information about the state of molecular organization and molecular motion in the lipid bilayers (Gaffney and McConnell, 1974; Jost et al., 1971; Berliner, 1976). The estimate of degree of order from the spin label EPR spectra is fairly straight forward. The determination of motional parameters as diffusion coefficients or correlation times is more difficult and may be dependent on the model for the molecular motion used for the simulation of the EPR spectra. In each case it must be shown that the model applied is physically realistic and compatible with the system studied.

We have recently (Israelachvili et al., 1974, 1975) developed a model for the calculation of the EPR spectral line shapes for a spin labeled molecule undergoing rapid restricted motion in a lipid bilayer. This restricted motion is composed of twisting or rotation around the long molecular axis which simultaneously is tumbling within the confines of a cone.

## THEORETICAL BASIS OF THE MODEL

The model was originally developed for application to unoriented systems (Israelachvili et al., 1974, 1975). Full details of the theoretical basis for such systems are given in these references. Here the general procedure of treatment and main results will be outlined. As usual the starting point is the spin-Hamiltonian of a nitroxide radical in the laboratory frame (XYZ):

$$\mathcal{H} = \beta_e \, \underline{H} \cdot \underset{\sim}{g} \cdot \underline{S} + \underline{S} \cdot \underset{\sim}{T} \cdot \underline{I} \qquad (1)$$

where $\beta_e$ is the Bohr magneton and the laboratory coordinate system is defined by the direction of the magnetic field ($\underline{H} = \underline{H}_Z$). For simplicity it has been assumed that the molecular axes are co-parallel with the nitroxide radical axes.

The eigenenergies of the three EPR lines of the nitroxide of the tumbling molecule are given by

$$\begin{aligned} E_o &= \beta_e \, \underline{H}_Z \langle g_{ZZ} \rangle = \beta_e \, \underline{H}_Z \, g' \\ E_\pm &= E_o \pm \left( \langle T_{XZ} \rangle^2 + \langle T_{YZ} \rangle^2 + \langle T_{ZZ} \rangle^2 \right)^{1/2} = E_o \pm T' \end{aligned} \qquad (2)$$

Here $g_{ZZ}$, $T_{XZ}$, $T_{YZ}$ and $T_{ZZ}$ are functions of the principal values along the molecular axes (xyz) of the diagonal g- and T-tensors and the Eulerian angles ($\theta$, $\phi$ and $\psi$) relating the molecular and laboratory frames with each other. The brackets < > show that suitable averaging has to be performed over these angles, which gives g' and T' as the effective values of g and T.

In our model the molecular motion is determined by two independent variables: (I) the angular amplitude $\pm\phi_o$ for the rapid rotation or twisting of the molecule about its long axis, and (II) the semi-cone angle $\beta_o$ defining the cone, within the confines of which the long axis of the molecule (z-axis) is tumbling rapidly.

These two angles, $\phi_o$ and $\beta_o$, are independent variables. The two modes of motion are, however, related with each other through the following model (see Fig. 1): The angle $\chi$ between the plane containing the instantaneous molecular z-axis and the cone axis, and that containing the z-axis and the mean x-axis is independent of $\beta$.

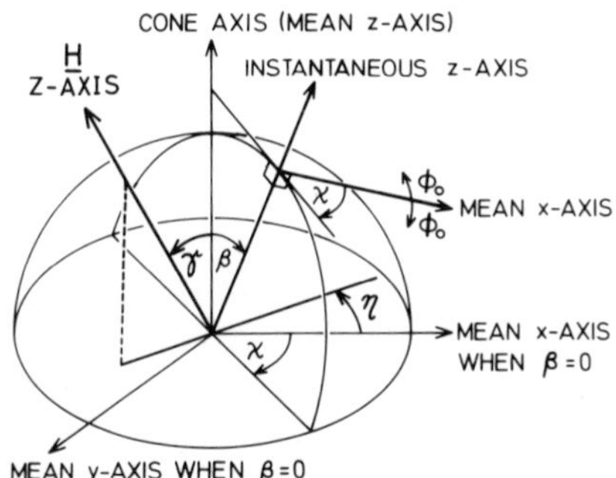


Fig. 1. The present model assumes that the molecule rotates or twists about its long molecular axis with angular amplitude $\pm\phi_o$, simultaneously tumbling rapidly within a cone of semi-cone angle $\beta_o$. The figure shows the instantaneous orientation of the molecular axis relative to the cone axis, where it is assumed that $\beta$ and $\chi$ are independent. The angle $\gamma$ is between the cone axis and the external field direction. The angle $\eta$ is (when $\beta = 0$) between the plane containing the field axis and the molecular z-axis and the plane containing the z-axis and the mean x-axis.

Let us introduce $\gamma$ as the angle between the fixed cone axis and the external field direction, and $\eta$ as the angle (when $\beta = 0$) between the plane containing the field axis Z and the molecular long axis z (which coincides with the cone axis, since $\beta = 0$) and the plane containing the z-axis and the mean x-axis. It is now possible to evaluate $\langle g_{ZZ} \rangle$ for our cone model, which gives

$$g'(\gamma,\beta_o,\eta,\phi_o) = \bar{g}_{xx}\sin^2\gamma\cos^2\eta + \bar{g}_{yy}\sin^2\gamma\sin^2\eta + \bar{g}_{zz}\cos^2\gamma \quad (3)$$

where $\bar{g}_{xx}$, $\bar{g}_{yy}$ and $\bar{g}_{zz}$ are functions of $\phi_o$ and $\beta_o$ and the principal values of the g-tensor ($g_{xx}$, $g_{yy}$, $g_{zz}$).

Similarly $\langle T_{ZZ} \rangle$ is evaluated. However, to obtain the complete expression for T' several pseudosecular terms have to be evaluated which would be very tedious. The following reasonable form of T' was therefore assumed:

$$T'(\gamma,\beta_o,\eta,\phi_o) = (\bar{T}^2_{xx}\sin^2\gamma\cos^2\eta + \bar{T}^2_{yy}\sin^2\gamma\sin^2\eta + \bar{T}^2_{zz}\cos^2\gamma)^{1/2} \quad (4)$$

Here $\bar{T}_{xx}$, $\bar{T}_{yy}$ and $\bar{T}_{zz}$ are functions of $\phi_o$, $\beta_o$ and the principal values of the T-tensor ($T_{xx}$, $T_{yy}$, $T_{zz}$). These functions are analogous to the corresponding expressions for $\bar{g}_{xx}$ etc. and may be obtained by substituting $T_{xx}$ for $g_{xx}$ etc.

In all the limiting cases where rigorous expressions exist it could be shown that this equation gives the correct result.

We have assumed that the total linewidth may be expressed in the form (Fraenkel, 1967)

$$\Gamma_m = \Gamma_r + <(H-<H>)^2> \gamma_e \tau \quad (5)$$

where $\Gamma_m$ is the half-width at half-height of the absorption line; m denotes that the width depends on the nuclear magnetic quantum number; $\Gamma_r$ is the residual linewidth which is independent of m but may include unresolved hyperfine splittings; H is the instantaneous resonance field at the fixed microwave frequency $\nu$; $\gamma_e$ is the free electron magnetogyric ratio; and $\tau$ is the correlation time of the motion. In equation (5) the time dependent fluctuation $\delta H = H - <H>$ must obey the condition of motional narrowing (Freed and Fraenkel, 1963):

$$<\delta H^2> \gamma_e^2 \tau^2 < 1 \quad (6)$$

For isotropic motion of a nitroxide radical the theory is thus applicable for $\tau < 2 \cdot 10^{-9}$ s. For restricted anisotropic motion (small $\beta_o$ and $\phi_o$) the fluctuation $\delta H$ decreases and hence correspondingly larger $\tau$-values will still be compatible with the theory.

Neglecting non-secular terms, i.e. assuming $h\nu = \beta_e g_{zz} H + mT_{zz}$, a general expression of $\Gamma_m$ of the following form could be deduced

$$\Gamma_m = \Gamma_r + f(\nu, g, T, m)\ \gamma_e \tau \quad (7)$$

Here f is a function of the microwave frequency $\nu$, the principle values $g_{xx}$, $g_{yy}$, $g_{zz}$, $T_{xx}$, $T_{yy}$ and $T_{zz}$ of the g- and T-tensors, the nuclear magnetic quantum number m, and suitable averages over the Eulerian angles $\theta$ and $\phi$. This general expression is very complicated.

The averages are nevertheless straight forward to evaluate in two important limiting cases: (a) axial rotation only ($\beta_o = 0$) and (b) complete axial rotations ($\phi_o = 90^o$), plus motion within a cone ( $\beta_o > 0$). For an elongated molecule like a steroid spin label intercalated into a lipid bilayer it is reasonable to assume that there is no appreciable tumbling within a cone until the axial rotations are complete, which also is compatible with experimental data. Hence it should be sufficient to consider these two limiting cases only.

For case (a), axial rotations only, we obtain

$$\Gamma_m = \Gamma_r + \left(\frac{h\nu}{\beta_e g_o^2}\ \delta g + m\Delta T\right)^2 f_a(\gamma, \eta, \phi_o) \gamma_e \tau_a \quad (7a)$$

where $\tau_a$ is the correlation time for the axial rotation of the molecule.

For case (b), complete axial rotation plus motion within a cone, we have

$$\Gamma_m = \Gamma_r + \left(\frac{h\nu}{\beta_e g_o^2}\,\Delta g + m\Delta T\right)^2 f_{bt}(\gamma,\beta_o)\gamma_e\tau_t$$

$$+ \left(\frac{h\nu}{\beta_e g_o^2}\,\delta g + m\delta T\right)^2 f_{ba}(\gamma,\beta_o)\gamma_e\tau_{at} \qquad (7b)$$

Here $\tau_t$ is the tumbling correlation time. The third term arises from both axial rotations and tumbling, so that $\tau_{at}$ represents a mixed correlation time intermediate between $\tau_a$ and $\tau_t$.

In equations (7a) and (7b) $\delta g = g_{xx} - g_{yy}$, $\delta T = T_{xx} - T_{yy}$, $\Delta g = g_{zz} - \frac{1}{2}(g_{xx} + g_{yy})$ and $\Delta T = T_{zz} - \frac{1}{2}(T_{xx} + T_{yy})$. It should be noted that the z-axis is the long molecular axis and that of molecular rotation at the same time.

## APPLICATION TO UNORIENTED SAMPLES

The present model was first used (Israelachvili et al., 1974, 1975) to evaluate the motion of 3-doxyl-cholestane spin label molecules (Fig. 2) in vesicles of egg yolk lecithin (EYL) over a wide range of temperatures. The vesicles were prepared by sonication. Using the molecular axes indicated in Fig. 2 we note that for this radical there is a large difference between $T_{xx}$ and $T_{yy}$, so that the spectra will be sensitive to changes in both $\phi_o$ and $\beta_o$.

Based on equations (2)-(4) and (7a) or (7b) a program was developed for calculation of simulated spectra on an IBM 360/75 computer. Each spectrum was simulated by numerical integration of $\eta$ and $\gamma$ at $2.5^o$ intervals, with $\sin\gamma d\gamma$ as weighting function because of the random distribution of molecular orientations. The following main input parameters were repeatedly adjusted to obtain best-fit simulations, in case (a): $\phi_o$, $\tau_a$ and $\Gamma_r$; and in case (b): $\beta_o$, $\tau_t$, $\tau_{at}$ and $\Gamma_r$. It was also found necessary to in-

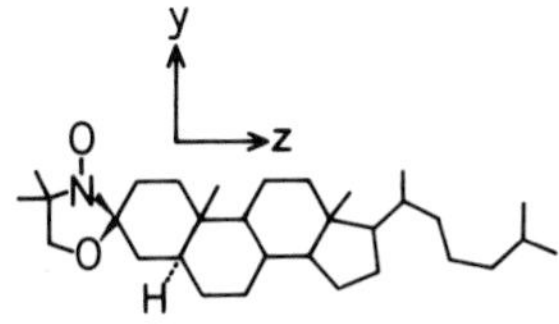


Fig. 2. Formula and principal axes of 3-doxyl-cholestane spin label.

Table I. The principal values of g- and T-tensors used for the simulation of EPR spectra of cholestane spin label in lecithin vesicles.

| | |
|---|---|
| $g_{xx}$ = 2.0083 (best-fit) | $T_{xx}$ = 0.56 mT (best-fit) |
| $g_{yy}$ = 2.0021 (measured) | $T_{yy}$ = 3.42 mT (measured) |
| $g_{zz}$ = 2.0064 (best-fit) | $T_{zz}$ = 0.56 mT (best-fit) |
| $g_o$ = 2.0056 | $T_o$ = 1.51 mT |

clude a possibility to vary the lineshape function. Also minor adjustments had to be made of the g-components taken from literature in order to keep the central zero-crossing at the correct position within the spectrum. The final principal values of the g- and T-tensors used are shown in Table I.

The simulated EPR spectra for several temperatures are shown in Fig. 3 where comparison also is made with the experimental recordings. The agreement is in general very good. Only the spectrum at -10$^{o}$C shows an inferior fit, which it has not been possible to improve. One possibility would be that at this temperature there are two phases present but we have no further support for such a conclusion.

The values of $\phi_o$, $\beta_o$ and $W_r$ used in the final best-fit simulations of Fig. 3 are given in Table II which shows how these parameters vary with temperature. The increase with temperature of $\phi_o$, and $\beta_o$ are as expected. The monotonic decrease of $W_r$ when temperature increases is in accord with the decreasing contribution of anisotropy in any unresolved hyperfine coupling as motion increases.

Each simulated EPR spectrum also gave a best-fit correlation time $\tau_a$, or correlation times $\tau_t$ and $\tau_{at}$. At each temperature an effective correlation time could be calculated from

$$\tau_{eff}^{-1} = \begin{cases} \tau_a^{-1} & \text{if } \phi_o < 90^o,\ \beta_o = 0 \\ \tau_{at}^{-1} + \tau_t^{-1} & \text{if } \phi_o = 90^o,\ \beta_o > 0 \end{cases} \quad (8)$$

The Arrhenius graph of log $\tau_{eff}^{-1}$ vs $T^{-1}$ (Fig. 4) indicates a linear dependence over the whole temperature range and one single activation energy of 31.8 kJ·mole$^{-1}$ (= 7.6 kcal mole$^{-1}$), which is a reasonable value for a lipid system. (Note that the figures earlier given by Israelachvili et al. (1975) were not correct).

Even at the lowest temperatures employed our results fulfill the condition for motional narrowing, Eqn 6.

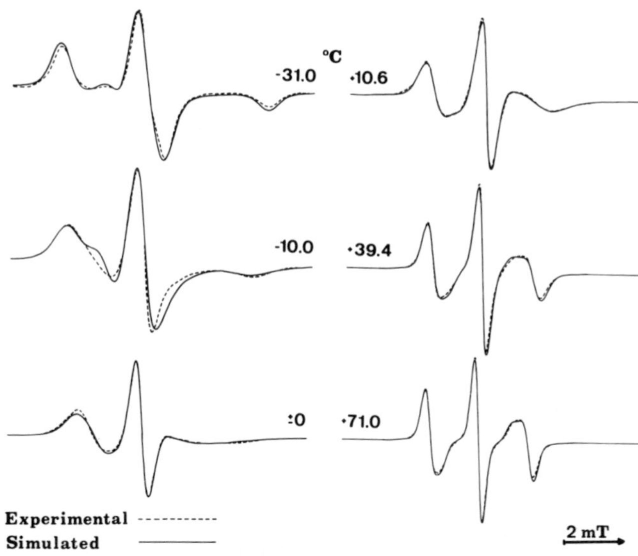


Fig. 3. Comparison between experimental and simulated EPR spectra at several temperatures for the 3-doxyl-cholestane spin label, 1 mole% in sonicated vesicles of egg yolk lecithin, 15 mg/ml, in aqueous buffer, 0.1 M NaCl, 50 mM Tris, pH 7.2. Simulated spectra were scaled to the same amplitude of the central peak as of the corresponding experimental ones. The lineshape function has equal weights of Gaussian and Lorentzian contributions. From Israelachvili et al. (1975).

Table II. Ordering and motional parameters obtained by simulation of EPR spectra of 3-doxyl-cholestane spin label in egg yolk lecithin vesicles as shown in Fig. 3.

| Temp. | Rot. angle | Cone angle | Order para-meter | Res. line-width* | Correlation times | | |
|---|---|---|---|---|---|---|---|
| | $\phi_o$ | $\beta_o$ | $S_{zz}$ | $W_r$ | $\tau_a$ | $\tau_t$ | $\tau_{at}$ |
| °C | deg. | deg. | | mT | ns | ns | ns |
| -31.8 | 6.0 | 0 | 1.00 | 0.62 | 10.0 | - | - |
| -10.0 | 43.0 | 0 | 1.00 | 0.60 | 5.0 | - | - |
| ± 0 | 66.0 | 0 | 1.00 | 0.42 | 3.0 | - | - |
| +10.6 | 90.0 | 25.0 | 0.86 | 0.42 | - | 3.0 | 1.5 |
| +39.4 | 90.0 | 52.0 | 0.50 | 0.35 | - | 2.0 | 0.3 |
| +71.0 | 90.0 | 59.0 | 0.39 | 0.30 | - | 0.5 | 0.1 |

* A superposition of Gaussian and Lorentzian functions in an 1:1 ratio both with the indicated linewidths was employed. When applying equations (7a) and (7b) to calculate the derivative linewidths, $W_m$, proper factors had to be used for the two lineshapes: $W_m(\text{Gaussian}) = (2/\ln 2)^{1/2}\,\Gamma_m$ and $W_m(\text{Lorentzian}) = (2/\sqrt{3})\,\Gamma_m$.

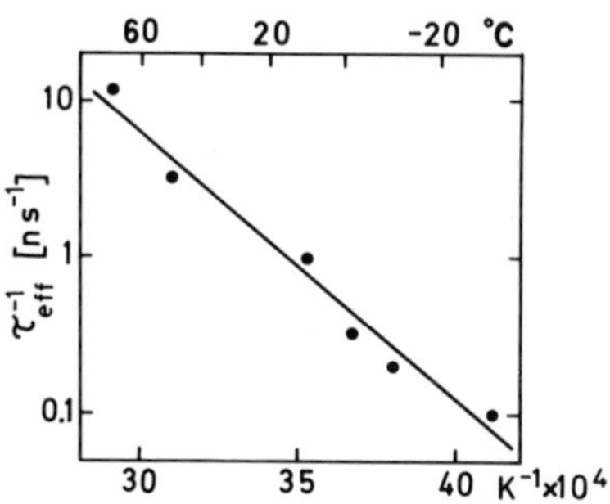


Fig. 4. Arrhenius graph of the correlation times determined by spectral simulation vs. reciprocal temperature for 3-dyxol-cholestane spin label in aqueous egg lecithin dispersion. From Israelachvili et al. (1975).

Before concluding this section we would like to emphasize the importance of evaluating the motional parameters of a molecule with restricted motion in an anisotropic system by means of a model that accounts for these properties. In literature the theory developed for rapid isotropic motion is much too often used to estimate correlation times of such systems from amplitude and/or linewidth ratios. This is bound to lead to erroneous results even when the tendencies to anisotropic motion are rather small. This is obvious from a comparison of such calculations with the present results at the higher temperatures where the motions most closely would approximate the isotropic case. Thus the experimental spectra of Fig. 3 at $71^{\circ}$ and $39^{\circ}$ evaluated with the theory for isotropic motion according to Wilson and Kivelson (1966) give $\tau = 1.8$ and $2.7$ ns, respectively, compared with $\tau_{eff} = 0.08$ ns and $0.26$ ns, using our model for restricted anisotropic motion.

## APPLICATION TO ORIENTED SAMPLES

### Validity Test of Model

In the previous section we have shown that our model for restricted molecular motion successfully accounts for the EPR spectra of 3-doxyl-cholestane spin label molecules in randomly oriented lipid bilayers. One might still ask how well this model reflects physical reality, since, as we pointed out in the introduction, the motional parameters derived are to some extent model dependent. A test of this point would be to evaluate EPR spectra of the spin label in oriented multibilayer samples for more than one direction. This test was made even more critical by measuring the same oriented sample at more than one microwave frequency and comparing the sets of parameters defining state of orientation and motion obtained by the simulation procedure at each of these frequencies.

The oriented samples were prepared on thin glass plates cut from cover glasses for microscopy. The lipid mixture in an organic solvent was deposited and dried on this plate which was mounted in an EPR sample tube and exposed over a suitably saturated salt solution to maintain the desired constant humidity. Temperature was controlled with a gas flow and a heater-sensor system constructed in the laboratory.

Well oriented lipid multibilayers were formed on the plates as revealed by the EPR spectra. The spin label used was the same as before (Fig. 2). EPR spectra were routinely recorded in two directions: with the magnetic field parallel to the plate where the line separation has a maximum and perpendicular to the plate where the line separation has a minimum.

It was necessary to introduce some changes in the simulation algorithm. For a randomly oriented system integration was carried out over the whole sphere. Instead, we must now account for the distribution of the molecules with respect to the direction parallel and perpendicular to the plate. The approach used by Schindler and Seelig (1973) and McFarland and McConnell (1971) was employed. We introduced a Gaussian distribution function of the direction of the cone axes by means of the formula

$$P(\vartheta) = \sin\vartheta \exp\left(-\frac{(\vartheta-\bar{\vartheta})^2}{2\vartheta_o^2}\right) \tag{9}$$

Here $\vartheta_o$ is the spread angle defining the width of the distribution and $\bar{\vartheta}$ is a tilt angle defining the mean angle between the center of the distribution and the direction perpendicular to the sample film. The weighting function $\sin\vartheta$ accounts for the constant solid angle.

A variation of the residual linewidth was also incorporated in the program using the relationship applied by Goldman et al. (1972) and Schindler and Seelig (1973)

$$\Gamma_r = X_o + X_2 \cos^2\theta \tag{10}$$

With these changes in the computer program the simulation procedure is as follows:

From the spectra at X-band microwave frequency in the two directions parallel and perpendicular to the plate it is easy to derive preliminary parameters for the starting simulations. The distance between the zero crossings of the high and low field lines are $2T'_{\parallel}$ and $2T'_{\perp}$, respectively.

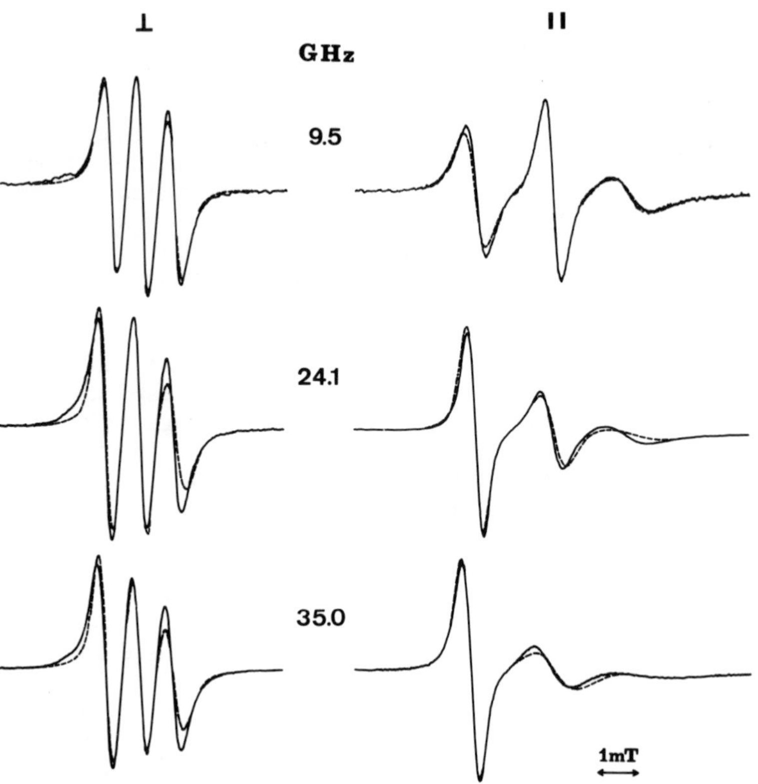


Fig. 5. Comparison of experimental (solid line) and simulated (broken line) EPR spectra of 3-doxyl-cholestane spin label (1 mole%) incorporated in multibilayers of egg yolk lecithin. The same specimen (plate for Q-band cavity size) was used in a course of experiments at three different frequencies (X-, K-, Q-band). Spectra were recorded in the magnetic field direction parallel (∥) and perpendicular (⊥) to the plate, at 23°C and under controlled relative humidity of 90%. Parameters used in the simulated spectra are as shown in Table III.

If $T'_{\parallel} > 1/2(T_{xx} + T_{yy})$ we have the order parameter $S_{zz} = 1$ and $\beta'_o = 0$ and estimate a preliminary $\phi'_o$ from $T'_{\parallel}$ using the formula

$$\sigma = \frac{\sin 2\phi'_o}{2\phi'_o} = \frac{2T'_{\parallel} - (T_{xx} + T_{yy})}{T_{yy} - T_{xx}} \tag{11}$$

If $T'_{\parallel} < 1/2(T_{xx} + T_{yy})$ we have $\phi'_o = 90^o$ and calculate

$$S'_{zz} = \frac{T'_{\parallel} - T'_{\perp}}{\frac{1}{2}(T'_{xx} + T'_{yy}) - T_{zz}} = \frac{1}{2}\cos\beta'_o(1 + \cos\beta'_o) \tag{12}$$

Table III. Ordering and motional parameters obtained by simulation of pairs of EPR spectra of oriented samples of egg yolk lecithin (without and with added cholesterol) containing 1 mole% of 3-doxyl-cholestane spin label. Spectra were taken at three microwave frequencies. In each case spectra were taken in the direction parallel and perpendicular to the plate. In all these simulations $\phi_o = 90^o$, $\vartheta_o = 5.0^o$ and $\bar{\vartheta} = 0^o$ was used. For experimental conditions see Fig. 5.

| Res. freq. | Choles-terol conc. | Cone angle | Order para-meter | Residual linewidth parameters * | | Correlation times | |
|---|---|---|---|---|---|---|---|
| | | $\beta_o$ | $S_{zz}$ | $X_o$ | $X_2$ | $\tau_t$ | $\tau_{at}$ |
| GHz | mole% | deg. | | mT | mT | ns | ns |
| 9.5 | 0 | 30.0 | 0.81 | 0.37 | -0.04 | 8.0 | 1.5 |
| | 30 | 23.0 | 0.86 | 0.40 | -0.05 | 8.0 | 1.5 |
| 24.1 | 0 | 30.0 | 0.81 | 0.35 | -0.03 | 8.0 | 1.5 |
| | 30 | 23.0 | 0.86 | 0.37 | -0.05 | 8.0 | 1.5 |
| 35.0 | 0 | 30.0 | 0.81 | 0.35 | -0.03 | 8.0 | 1.5 |
| | 30 | 23.0 | 0.86 | 0.37 | -0.05 | 8.0 | 1.5 |

*A superposition of Gaussian and Lorentzian functions in an 1:4 ratio both with the indicated linewidth was employed. See text about $X_o$ and $X_2$. Cf also footnote of Table II.

The first pair of simulations (in the two directions) is now made with these values of $\beta_o'$ and $\phi_o'$, having $\bar{\vartheta}$ and $\vartheta_o = 0$, and using reasonable guesses for the correlation time(s). Starting values of the linewidth parameters $X_o$ and $X_2$ were obtained from the linewidths of the centre line, $m = 0$, for both directions and by means of Eqn 10. By trial and error, using visual inspection to judge quality of fitting, the whole set of parameters is adjusted for best-fit to the pair of spectra. When spectra had been taken also at higher microwave frequencies the same input parameters as for X-band spectra were used for the simulations also of these other pairs of spectra.

In the first series of critical test experiments we have used the same material, egg yolk lecithin (EYL) as was used in the experiments on vesicle samples. The spin label was also the same, 3-doxyl-cholestane. The samples were of two kinds: only EYL and EYL plus cholesterol in molar ratio 7:3. EPR spectra were measured with three microwave frequencies at 23°C.

Experimental and simulated spectra for the sample containing no cholesterol are shown in Fig. 5, and the ordering and motional parameters obtained for both samples are given in Table III.

At all three microwave frequencies best-fit simulations for each sample led to identical sets of simulation parameters except for the residual linewidth. The differences in the linewidth parameters $X_o$ and $X_2$ are of such minor magnitude that adjustment for each sample to equal values would not affect the quality of the spectral fit appreciably. The overall fit between simulated and experimental spectra is quite satisfactory and assures that the model developed in a reasonable way reflects the physical situation of the probe molecule in its environment.

From Table III we also note that addition of cholesterol increases the ordering parameter and decreases the amplitude of the tumbling motion, whereas the correlation times, most significantly $\tau_{at}$, remain unchanged. This suggests that cholesterol stiffens the membrane by reducing the cooperative or collective motion whereas the area available to the spin labeled steroid molecule is not changed. We calculate $\tau_{eff}$ = 1.26 ns for the multibilayer and $\tau_{eff}$ = 0.61 ns for the vesicles in aqueous suspension at 23°C. The longer correlation time in the multibilayer is not unreasonable in view of the limited water content of the multibilayer, ca 20 wt%, and the different microscopic geometry of the two types of sample.

The simulation procedure also reveals that a change in $\tau_{at}$ by more than 15% or a change of $\beta_o$ by more than 2 to 5° (depending on the value of $\beta_o$) led to inferior fit.

## Further Applications

The model has now also been applied to evaluate EPR spectra of 3-doxyl-cholestane spin label interchalated in multibilayers of several lipids. We have been particularly interested to follow the course of change with temperature of motional and ordering parameters, and wish to report some of the preliminary results here.

For EYL $\tau_{eff}$-values within the temperature range 20 to 60°C give straight lines in the Arrhenius graph. From these we calculate an activation energy of 42.0 kJ mole$^{-1}$ using data from two microwave frequencies, 9 and 35 GHz. With added 30 mole% of cholesterol a slightly smaller activation energy was obtained, 39.8 kJ mole$^{-1}$. These values should be compared with the value of the corresponding activation energy in EYL vesicles, 31.8 KJ mole$^{-1}$, cf Fig. 4.

Dipalmitoyl lecithin (DPL) with 30 mole% of cholesterol added was studied over the temperature range -5 to +71°C. Fig. 6 shows the temperature variation of the hyperfine splitting of the spin label molecule in both parallel ($T'_{\parallel}$) and perpendicular ($T'_{\perp}$) direc-

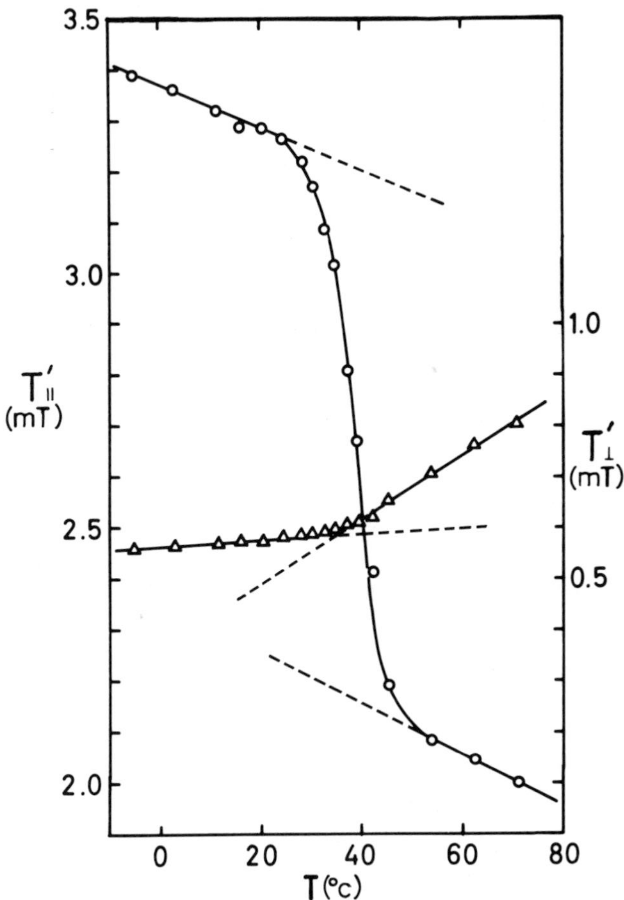


Fig. 6. Temperature variation of apparent hyperfine splittings $T'_{\parallel}$ (-o-o-) and $T'_{\perp}$ (-Δ-Δ-) measured from the EPR spectra of 3-doxyl-cholestane spin label (1 mole%) in oriented multibilayers of dipalmitoyl lecithin with added cholesterol (30 mole%). $T'_{\parallel}$ is half the distance between the derivative extrema at high and low field of spectra taken in the parallel direction. $T'_{\perp}$ is half the distance between the high and low field zero crossings of spectra taken in the perpendicular direction.

tion. $T'_{\parallel}$ shows a gradual change at the lower temperature region, -5 to +30°C, and also at the higher region, 45 to 70°C, but at different levels. In the intermediate region, 30 to 45°C, there is a steep change between these levels with a midpoint at 39°C. The variation of $T'_{\perp}$ is different but shows a breaking point at the same temperature. These curves indicate a phase transition between two states with different modes of molecular motion, the gel and liquid crystalline states, respectively. A similar behaviour was observed for pure DPL but with a slightly lower transition point at 36°C.

For DPL with 30 mole% cholesterol the Arrhenius graph of $\tau_{eff}$ shows a breaking point at about 35°C. Above and below this temperature the activation energy of the motion of the steroid spin label is found to be 65 and 19 kJ mole$^{-1}$, respectively. These values indicate that below the transition temperature there is a fairly well defined volume for the steroid spin label to move within, whereas above the transition point interaction with the surrounding molecules is greatly increased so that the motion is more collective in character.

For DPL without cholesterol EPR spectra possible to analyze by the simulation procedure could so far be obtained only in the high temperature range, 45 to 70°C. For this range we obtain an activation energy of 25 kJ mole$^{-1}$, which is only slightly higher than the value in the lower temperature range with cholesterol.

This observation is in contrast to the EYL case where the activation energy was found to be insensitive to the presence of cholesterol. This is in accord with EYL being a more heterogeneous material as compared to DPL.

The spectra with DPL without added cholesterol in the lower temperature range could not be analyzed with the present model alone. The spectra are more complicated and suggest such possibilities as a distribution of spin label molecules between two or more phases or aggregation of spin label molecules so that magnetic dipolar interaction between them becomes important. Electron microscopy studies (Verkleij et al., 1974) have revealed that the supramolecular structure of pure DPL might be different to that of a mixture of DPL and cholesterol.

The observed transition temperature for the motion of the steroid spin label molecule in DPL with 30 mole% cholesterol is of particular interest. From thermal analysis (Ladbrooke et al., 1968) as well as other methods (Galla and Sackmann, 1974) it is known that pure DPL has a transition from gel state to liquid crystalline state at ca 42°C with a pretransition at ca 35°C (Rand et al., 1975). It is commonly believed that mixing with cholesterol washes out such transitions making the liquid crystalline state more gel-like and vice versa. Our experimental results with the steroid spin label indicate that the position and sharpness of the transition is hardly influenced by cholesterol and reflects the chain melting point of DPL. On the other hand we find that above the transition point cholesterol increases the activation energy for the probe. This is in agreement with the experience that cholesterol makes DPL more rigid in the liquid crystalline state. The mode of motion of spin label molecules added in small proportion might be sensitive to quite small changes in the host system dependent on the particular way these added molecules fit into the system.

## ACKNOWLEDGMENTS

The first part of this work was carried out together with Jacob Israelachvili, Jan Sjösten, Magdalena Ehrström and Astrid Gräslund. The stimulating collaboration with them is vividly and thankfully acknowledged. Skilled technical assistance has been given by Lars Mittermaier. We also wish to thank Dr. Anders Lund, The Swedish Research Councils' Laboratory, Studsvik, Nyköping, for permission to use the Q-band EPR spectrometer. Yuhei Shimoyama received a fellowship from the Rotary Foundation. Financial support was further received from the Swedish Natural Science and Medical Research Councils.

## REFERENCES

Berliner, L. J., (ed.) "Spin labeling. Theory and applications", Academic Press, New York, 1976.

Fraenkel, G. K., Linewidth and frequency shifts in electron spin resonance spectra, J. Phys. Chem. 71, 139, 1967.

Freed, J. H. and G. K. Fraenkel, Theory of linewidth in electron spin resonance spectra, J. Chem. Phys. 39, 326, 1963.

Gaffney, B. J. and H. M. McConnell, The paramagnetic resonance spectra of spin labels in phospholipid membranes, J. Magn. Resonance 16, 1, 1974.

Galla, H.-J. and E. Sackmann, Lateral diffusion in the hydrophobic region of membranes: Use of pyrene excimers as optical probes, Biochim. Biophys. Acta 339, 103, 1974.

Goldman, S. A., G. V. Bruno, C. F. Polnaszek and J. H. Freed, An ESR study of anisotropic rotational reorientation and slow tumbling in liquid and frozen media, J. Chem. Phys. 56, 716, 1972.

Israelachvili, J., J. Sjösten, L. E. G. Eriksson, M. Ehrström, A. Gräslund and A. Ehrenberg, Theoretical analysis of the molecular motion of spin labels in membranes. ESR spectra of labeled Bacillus subtilis membranes, Biophys. Biochim. Acta 339, 164, 1974.

Israelachvili, J., J. Sjösten, L. E. G. Eriksson, M. Ehrström, A. Gräslund and A. Ehrenberg, ESR spectral analysis of the molecular motion of spin label in lipid bilayers and membranes based on a model in terms of two angular motional parameters and rotational correlation times, Biochim. Biophys. Acta 382, 125, 1975.

Jost, P., L. J. Libertini, V. C. Hebert and O. H. Griffith, Lipid spin labels in lecithin multilayers. A study of motion along fatty acid chains, J. Mol. Biol. 59, 77, 1971.

Ladbrooke, B. D., R. M. Williams and D. Chapman, Studies on lecithin-cholesterol-water interactions by differential scanning calorimetry and X-ray diffraction, Biochim. Biophys. Acta 150, 333, 1968.

McFarland, B. G. and H. M. McConnell, Bent fatty acid chains in lecithin bilayers, Proc. Nat. Acad. Sci. USA 68, 1274, 1971.

Rand, R. P., D. Chapman and K. Larsson, Tilted hydrocarbon chains of dipalmitoyl lecithin become perpendicular to the bilayer before melting, Biophys. J. 15, 1117, 1975.

Schindler, H. and J. Seelig, ESR spectra of spin labels in lipid bilayers, J. Chem. Phys. 59, 1841, 1973.

Schindler, H. and J. Seelig, ESR spectra of spin labels in lipid bilayers. II. Rotation of steroid spin probes, J. Chem. Phys. 61, 2946, 1974.

Verkleij, A. J., P. H. J. Ververgaert, B. de Kruyff and L. L. M. van Deenen, The distribution of cholesterol in bilayers of phosphatidylcholines as visualized by freeze fracturing, Biochim. Biophys. Acta 373, 495, 1974.

Wilson, R. and D. Kivelson, ESR linewidths in solution. I. Experiments on anisotropic and spin-rotational effects, J. Chem. Phys. 44, 154, 1966.

# DYNAMICS OF MEMBRANE-ASSOCIATED ENERGY-TRANSDUCING CATALYSTS. A STUDY WITH MITOCHONDRIAL ADENOSINE TRIPHOSPHATASE INHIBITOR

L. Ernster, K. Asami*, K. Juntti, J. Coleman**
and K. Nordenbrand

Department of Biochemistry, Arrhenius Laboratory
University of Stockholm, Stockholm, Sweden

## 1. INTRODUCTION

In recent years evidence has accumulated which strongly supports the concept that energy transfer between various membrane-associated energy-transducing units - electron-transport catalysts, ATP-synthesizing enzymes and ion-translocators of mitochondria, chloroplasts and prokaryotes - can take place via a transmembrane proton gradient and/or a membrane potential, in accordance with the chemiosmotic hypothesis of Mitchell (1961, 1966). The mechanisms by which such an electrochemical gradient is formed and utilized are not known but presumably involve the operation of proton pumps that are components - probably subunits - of the various energy-transducing units (cf. Ernster (1975) for review). It is also not known whether the formation of an electrochemical gradient by a given energy-transducing unit is the primary event of energy conservation, or whether it is preceded by a chemical - e.g. conformational - change that in turn gives rise to an electrochemical

---

*Present address: National Institute of Radiological Sciences, Chiba-shi, Japan.

**Fellow of European Molecular Biology Organization.
Present address: Department of Biology, University of the West Indies, Bridgetown, Barbados.

<u>Abbreviations</u>: AI, ATPase inhibitor; ANS, 8-anilino-naphthalene-1-sulfonate; FCCP, carbonyl cyanide p-trifluoromethoxyphenylhydrazone; LDH, lactate dehydrogenase; PEP, phospho(enol)pyruvate; PK, pyruvate kinase.

gradient. Furthermore, the question arises as to whether transmembrane electrochemical gradients are the sole way of interaction between energy-transducing units located in the same membrane, or whether there may also occur more direct interactions between such units. The purpose of this paper is to summarize experimental data obtained in our laboratory over the last few years in studies of the ATPase inhibitor of Pullman and Monroy (1963) that may be relevant to some of these problems.

In 1963 Pullman and Monroy described the isolation of an ATPase inhibitor from beef-heart mitochondria. The inhibitor was a protein of a molecular weight of about 10,000. It inhibited the ATPase activity of both submitochondrial particles and of the soluble mitochondrial ATPase "coupling factor 1" ($F_1$), but did not inhibit oxidative phosphorylation. They suggested that the inhibitor may be involved in mitochondrial respiratory control.

In 1970 it was found in our laboratory (Asami et al., 1970) that the ATPase inhibitor also inhibited various ATP-dependent energy-linked reactions catalyzed by submitochondrial particles, such as the ATP-driven reductions of $NAD^+$ by succinate and of $NADP^+$ by NADH (i.e., the nicotinamide nucleotide transhydrogenase reaction) as well as the ATP-dependent ANS-fluorescence enhancement. The same reactions when driven by energy derived from the respiratory chain were unaffected by the ATPase inhibitor. The inhibitor also did not affect oxidative phosphorylation, in accordance with the findings of Pullman and Monroy (1963), nor did it influence the oligomycin-induced respiratory control exhibited by the particles. It was suggested (Asami et al., 1970) that the ATPase inhibitor may act as a directional regulator of oxidative phosphorylation, inhibiting the back-flow of energy from ATP to the electron-transport system.

The same year Horstman and Racker (1970) reported that the inhibition of the ATPase activity of both submitochondrial particles and purified $F_1$ by the isolated ATPase inhibitor was enhanced if the inhibitor was preincubated with the ATPase in the presence of ATP and $Mg^{++}$. This finding was confirmed in our laboratory (Ernster et al., 1973) and it was found, moreover, that subsequent incubation of inhibitor-, ATP- and $Mg^{++}$-pretreated particles under conditions of energization from the respiratory chain led to a release of the inhibitor. The release was abolished by uncouplers and oligomycin, and enhanced by $P_i$ and aurovertin. Similar observations were made by Van de Stadt et al. (1973), who found that both respiratory energy and ADP promote the dissociation of the ATPase inhibitor from submitochondrial particles. They pointed out that the ATP/ADP ratio may regulate the sensitivity of $F_1$ to the inhibitor, and concluded that their results supported the earlier proposal (Asami et al., 1970) concerning the role of the ATPase inhibitor as a directional regulator of respiratory chain-linked energy transfer.

Another interesting feature of the ATPase inhibitor that has emerged from recent studies concerns some quantitative aspects of its effects on various ATP-dependent reactions of submitochondrial particles. It was found (Ernster et al., 1973) that the ATP-driven succinate-linked $NAD^+$ reduction and nicotinamide nucleotide transhydrogenase reaction were equally sensitive to the ATPase inhibitor. This was in contrast to the effects of FCCP, oligomycin, or replacement of ATP by ITP, all of which inhibited the succinate-linked $NAD^+$ reduction more strongly than the transhydrogenase reaction. Moreover, the ATPase-inhibitor "titers" of the two ATP-driven electron-transport reactions were equal to that of the ATPase reaction catalyzed by the same particles when the latter was measured under conditions of maximal activity, i.e., in the presence of FCCP. These findings were interpreted as evidence for an "assembly-like" interaction of the ATPase and the ATP-driven electron-transport systems. However, the validity of these results was questioned by Lang and Racker (1974), who found no inhibition of the ATP-driven electron-transport reactions by the ATPase inhibitor.

The sections that follow summarize our current state of knowledge of the various aspects of ATPase-inhibitor action outlined above. First, some kinetic data on the effect of ATPase inhibitor on the ATPase activity of submitochondrial particles will be presented and discussed in relation to the possible mode of action of the inhibitor. Second, various factors and conditions influencing the interaction of the inhibitor with the membrane-bound ATPase will be considered. These results are consistent with the conclusion that the ATPase occurs in two conformational states, one predominant in the presence of ATP and possessing a high affinity for the inhibitor, and another, predominant in the presence of respiratory energy, $P_i$ and ATP, and possessing a low affinity for the inhibitor. Finally, some newer data concerning the effects of the ATPase inhibitor on ATP-dependent electron-transport reactions of submitochondrial particles will be presented, which eliminate the criticism raised by Lang and Racker (1974) and which support the earlier conclusion (Ernster et al., 1973) of an assembly-like interaction between mitochondrial ATPase and transhydrogenase. This conclusion is further supported by data obtained with a reconstituted system consisting of $F_1$-depleted submitochondrial particles and purified $F_1$ ATPase. A brief account of these results has already been published recently (Ernster, 1975).

## 2. KINETICS OF INHIBITION OF THE ATPase ACTIVITY OF SUBMITOCHONDRIAL PARTICLES BY ATPase INHIBITOR

As already reported briefly (Juntti et al., 1971) and subsequently confirmed by Van de Stadt et al. (1973), the ATPase inhibitor acts on the ATPase in a noncompetitive manner. This is illustrated in Fig. 1A, using beef-heart EDTA particles as the source of ATPase. The particles were preincubated with varying amounts of the purified inhibitor in the presence of ATP and $Mg^{++}$ as described by Horstman and Racker (1970). ATPase activity was determined according to Nishimura et al. (1962) by measuring the initial rate of increase in proton concentration with a sensitive pH-meter.

The $K_m$ value for ATP that can be deduced from the data in Fig. 1A is 0.13 mM. This value is somewhat lower than those reported in the literature with both particle-bound and soluble beef-heart $F_1$ (cf. Senior (1973) and Pedersen (1975) for reviews). Occasionally we also obtained higher $K_m$ values, but this could be ascribed to the presence in the assay system (usually in the ATP solution) of small amounts of ADP, which is a strong inhibitor of the ATPase (see below). Our $K_m$ value for the particle-bound beef-heart ATPase agrees well with that reported (Mitchell and Moyle, 1970) for the ATPase of rat-liver submitochondrial particles (0.106 mM).

As expected, a noncompetitive inhibition with the purified ATPase inhibitor was also obtained when ITP rather than ATP was used as the substrate for the ATPase of beef-heart EDTA particles (Fig. 1B). In accordance with earlier reports (cf. Pedersen, 1975), ITP was a relatively poor substrate for the ATPase; from a comparison of the data in Figs. 1A and B it appears that the difference in efficiency between the two substrates is due primarily to their different $K_m$ values (3.4 mM for ITP vs 0.13 mM for ATP).

Also in agreement with previous reports (Ernster et al., 1973), the ATPase inhibitor did not influence the competitive inhibition of the ATPase by ADP (Fig. 2). The $K_i$ value of ADP, approximately 10 μM, again was lower than those earlier reported for both the particle-bound and soluble beef-heart ATPases (cf. Senior, 1973) but was similar to that reported (Mitchell and Moyle, 1970) for the ATPase of rat-liver submitochondrial particles (8.6 μM).

It has been briefly reported (Ernster et al., 1973) that the ATPase inhibitor does not alter the oligomycin sensitivity of the ATPase activity of EDTA particles, and, conversely, that oligomycin does not alter the sensitivity of ATPase to the ATPase inhibitor. Fig. 3 shows data supporting these conclusions and demonstrates, furthermore, that the presence of aurovertin did not appreciably alter either the oligomycin or the ATPase-inhibitor titer

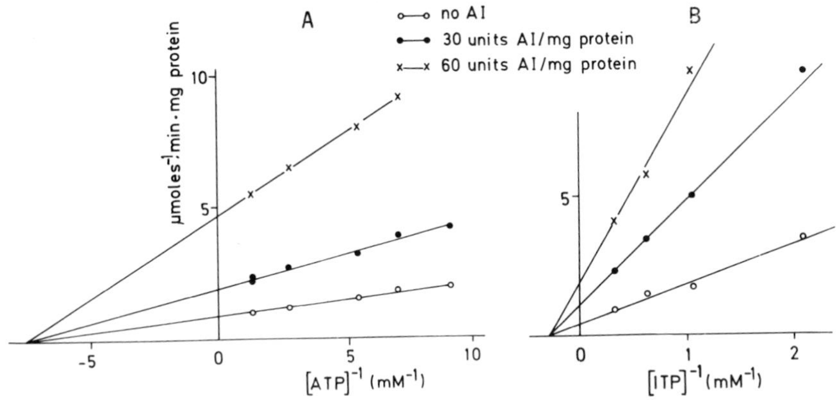


Figure 1 Effect of purified ATPase inhibitor on the ATPase and ITPase activities of EDTA particles. EDTA particles were prepared from beef-heart mitochondria according to Lee and Ernster (1967). ATPase inhibitor (AI) was purified according to Horstman and Racker (1970) and preincubated with the particles at 30°C for 10 min. in a medium containing 0.25 M sucrose, 10 mM Tris-TES buffer (pH 6.5), 0.45 mM ATP and 0.45 mM $MgSO_4$. Unit of AI is defined according to Pullman and Monroy (1963). ATPase activity was determined by measuring the change in pH of the reaction medium as described by Nishimura et al. (1962). The assay system contained in a final volume of 4 ml: 150 mM KCl, 3.3 mM glycylglycine buffer (pH 7.5), 5 mM $MgCl_2$ and 0.2 mg particle protein. The reaction was started by the addition of the indicated amounts of ATP or ITP from a solution containing equimolar concentrations of ATP or ITP and of $MgCl_2$ at pH 7.4. After a few minutes the reaction was terminated by the addition of 0.8 μg oligomycin and the change in pH was determined by back-titration with a standard solution of NaOH. Temperature, 30°C.

of the particle-bound ATPase.

It may be concluded from the data presented above that the ATPase inhibitor acts as a noncompetitive inhibitor of mitochondrial ATPase, at a site which is not affected by oligomycin or aurovertin. While both the ATPase inhibitor and aurovertin act on the $F_1$ moiety of the ATPase complex, the effect of aurovertin clearly differs from that of the ATPase inhibitor in being uncompetitive with respect to ATP (Chang and Penefsky, 1973) and resulting in a lowering of the $K_i$ value for ADP (Mitchell and Moyle, 1970). Furthermore, aurovertin inhibits ATP synthesis more strongly than its reversal (Lee and Ernster, 1968; Mitchell and Moyle, 1970), whereas the converse is true for the ATPase inhibitor

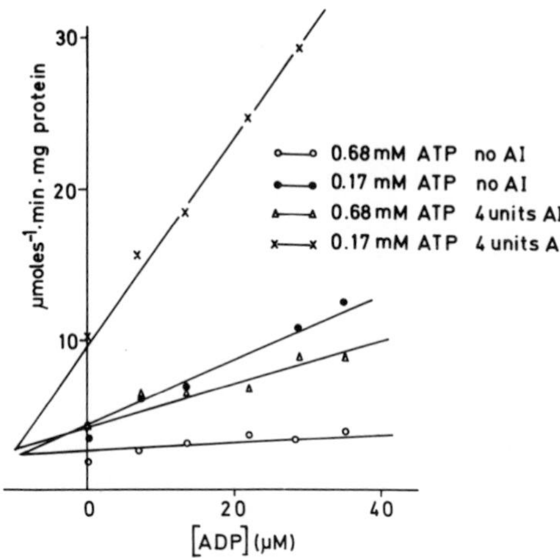


Figure 2 Lack of effect of ATPase inhibitor (AI) on the inhibition of the ATPase activity of EDTA particles by ADP. Conditions were as in Fig. 1. ADP in the amounts indicated was added together with equimolar concentrations of $MgCl_2$.

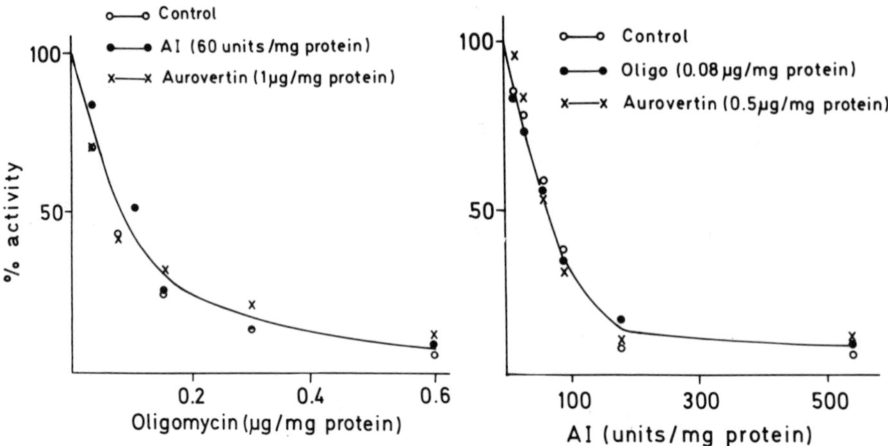


Figure 3 Combined effects of ATPase inhibitor (AI), aurovertin and oligomycin on the ATPase activity of EDTA particles. ATPase activity was assayed spectrophotometrically according to Pullman et al. (1960). The reaction medium contained in a final volume of 3 ml: 30 mM KCl, 5 mM Tris-acetate (pH 7.5), 3 mM $MgCl_2$, 3 mM ATP, 0.2 mM NADH, 0.75 mM PEP, 45 μg LDH, 60 μg PK, 1 μM FCCP, 1 μM rotenone and the indicated amounts of AI, aurovertin and oligomycin. 100 % ATPase activities (μmoles/min/mg protein) were: control, 3.26; AI, 1.51; oligomycin, 1.23; aurovertin (0.5 μg/mg protein), 2.32; aurovertin (1 μg/mg protein), 2.16. Treatment of particles with AI was done as in Fig. 1.

(Pullman and Monroy, 1963). In this respect, the effect of the ATPase inhibitor is similar to that of imido-ATP, which, however, inhibits ATPase competitively with respect to ATP (Penefsky, 1974; Pedersen et al., 1974). The ATPase-inhibitor effect also differs from the recently-described inhibition of soluble $F_1$ ATPase by certain metal-ion trichelates of bathophenanthroline which is relieved by uncouplers (Phelps et al., 1975a,b) and which interferes with the interaction of $F_1$ with aurovertin (Ernster et al., 1976). On the other hand, both the ATPase inhibitor (Pullman and Monroy, 1963) and the bathophenanthroline trichelates (Ernster et al., 1976) have been shown to stabilize soluble $F_1$ against cold-inactivation.

## 3. FACTORS INFLUENCING THE BINDING AND RELEASE OF ATPase INHIBITOR

Horstman and Racker reported in 1970 that incubation of the ATPase inhibitor together with the ATPase in the presence of ATP and $Mg^{++}$ was necessary to obtain inhibition of the ATPase activity. Data in Fig. 4 confirm this finding with purified ATPase inhibitor and EDTA particles as the source of ATPase. The small degree of inhibition obtained in the absence of ATP and $Mg^{++}$ during preincubation can be attributed to an effect of the ATP and $Mg^{++}$ present in the assay system used for the measurement of ATPase activity, as indicated by the fact that the inhibition in that case developed gradually during the assay.

We have briefly reported (Ernster et al., 1973) that incubation of inhibitor-treated EDTA particles under conditions of energy-generation from the respiratory chain led to a relief of the inhibition of ATPase activity. Table I summarizes the conditions used in these experiments. First, EDTA particles were incubated at 30° for 10 min with purified ATPase inhibitor in the presence of ATP and $Mg^{++}$ so as to obtain an inhibition of the ATPase activity by 80-85 %. Then, a 25-μl aliquot of the incubation mixture, containing 0.25 mg inhibitor-treated particles, was transferred to 1 ml of a reaction medium containing succinate, $P_i$ and a low concentration of oligomycin (the latter in order to ensure a maximal phosphorylating capacity of the EDTA particles) and reincubated at 30° for varying lengths of time. Finally, a 0.1 ml aliquot of this mixture was used for the measurement of ATPase activity at 30° spectrophotometrically by linking the ATPase reaction to the pyruvate kinase and lactate dehydrogenase systems.

Fig. 5 shows the effects of preincubation under conditions of energy-generation from the respiratory chain, as specified in Table I, on the ATPase activity of inhibitor-treated EDTA particles. It may be seen that the preincubation led to a relief of the inhibition of ATPase activity which tended to reach maximum after about

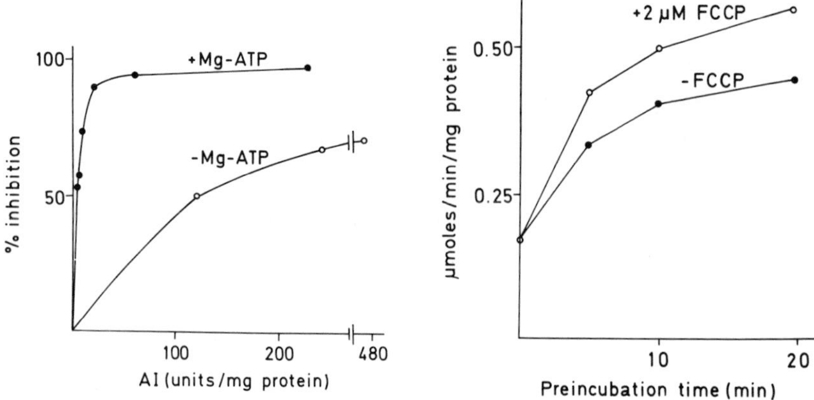


Figure 4 (left) Effect of ATP and $Mg^{++}$ on the inhibition of the ATPase activity of EDTA particles by purified ATPase inhibitor (AI). Treatment with AI was done as described in Table I. When indicated, ATP and $Mg^{++}$ were omitted. ATPase activity was assayed spectrophotometrically as described in Fig. 3.

Figure 5 (right) Reactivation of ATPase in ATP-inhibitor treated EDTA particles upon preincubation under conditions of energy-generation from the respiratory chain. For conditions, see Table I.

Table I

Conditions used for studying the relief of inhibition of ATPase activity from ATPase-inhibitor (AI) treated EDTA particles (ESP)

| AI treatment (10 min, 30°) | Preincubation (30°) | ATPase assay (30°, 340 nm) |
|---|---|---|
| 2 mg ESP | 0.25 mg AI-treated ESP | 25 μg preincub. ESP |
| 14 μg AI | 0.19 M sucrose | 0.23 M sucrose |
| 15 mM Tris-TES, pH 6.7 | 10 mM Tris-Ac, pH 7.4 | 50 mM Tris-Ac, pH 7.4 |
| 1.5 mM ATP | 4 mM succinate | 30 mM KCl |
| 1.5 mM $MgSO_4$ | 3 mM $P_i$ | 5 mM $MgSO_4$ |
| Final vol., 0.2 ml | 5 mM $MgSO_4$ | 3 mM ATP |
| | 0.02 μg oligomycin | 0.75 mM PEP |
| | Final vol., 1 ml | 0.1 mg PK |
| | | 20 μM NADH |
| | | 15 μg LDH |
| | | 2 μM FCCP |
| | | 1.5 μM rotenone |
| | | Final vol., 1 ml |

Table II

Factors influencing reactivation of ATPase of AI-treated EDTA particles during preincubation

| Additions or omissions during preincubation | ATPase activity (μmoles/min/mg ESP) | | | | |
|---|---|---|---|---|---|
| | Exp. 1* | Exp. 2 | Exp. 3** | Exp. 4 | Exp. 5 |
| Complete system | 0.79 | 0.75 | 0.38 | 0.69 | 0.64 |
| - succ | 0.23 | | | | |
| - oligomycin | 0.32 | 0.46 | 0.36 | | |
| - succ, - oligomycin | | 0.26 | 0.26 | | |
| - $P_i$ | | | | 0.49 | |
| + antimycin (1 μg/mg protein) | 0.17 | | 0.19 | | |
| + KCN (2 mM) | | 0.26 | 0.24 | | |
| + FCCP (2 μM) | | 0.17 | | 0.21 | 0.19 |
| + ATP (3 mM) | | | | | 0.58 |
| + ATP + PEP + PK | | | | | 0.17 |
| + aurovertin (0.9 μg/mg protein) | | 1.25 | 0.75 | 0.89 | |

*3 mM $P_i$ in assay medium.

**Same as Exp. 2 except that no FCCP was added to assay medium.

20 minutes. Furthermore, the uncoupler FCCP, which had no effect on the inhibited system, was required for maximal activity when the inhibition was relieved.

Table II summarizes the results of experiments concerning various factors influencing the reactivation of the ATPase of inhibitor-treated EDTA particles. Omission of succinate or addition of antimycin or cyanide abolished, as expected, the relief of inhibition of the ATPase. Omission of the small amount of oligomycin, or, alternatively (not shown), increase of the oligomycin concentration to that giving an inhibition of oxidative phosphorylation, also counteracted the reactivation of the ATPase. A similar effect was obtained by including ATP in the preincubation medium, provided phospho(enol)pyruvate and pyruvate kinase were also added, to maintain the concentration of ATP. Significantly, the omission of $P_i$ from the preincubation medium diminished the extent of reactivation (Exp. 4), and, interestingly, addition of aurovertin quite substantially enhanced it (Expts. 2-4). This increase, which was observed also when the ATPase activity was tested in the absence of FCCP (Expt. 3), was not simply due to a replacement of the low concent-

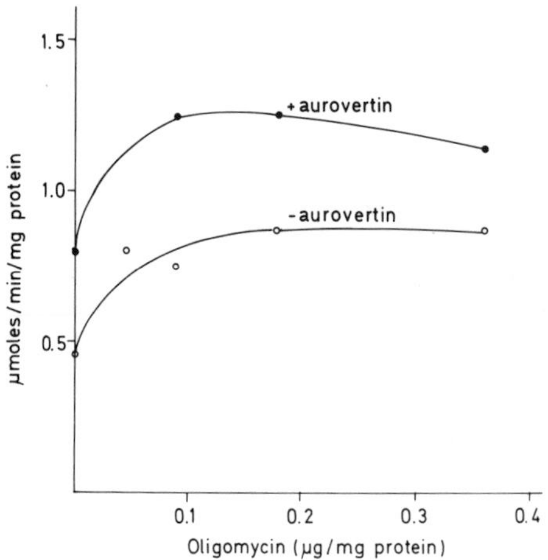


Figure 6. Effect of aurovertin on the reactivation of ATPase in ATPase-inhibitor treated EDTA particles upon preincubation under conditions of energy-generation from the respiratory chain. For conditions, see Table I. When indicated, aurovertin (0.9 µg/mg protein) was added to the preincubation medium.

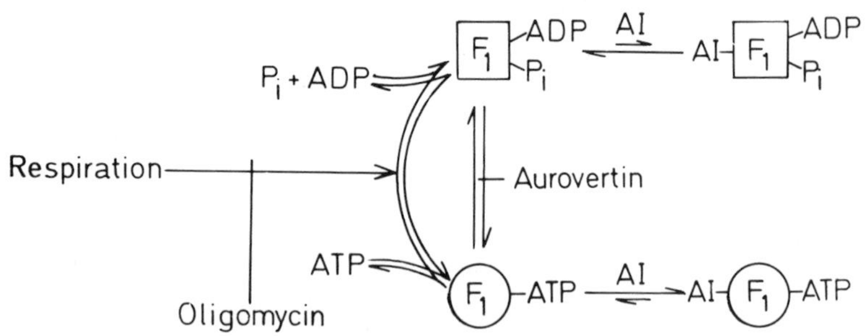


Figure 7 Proposed mechanism of interaction of ATPase inhibitor (AI) with the mitochondrial ATPase $F_1$. Squares and circles symbolize two different conformational states of $F_1$.

rations of oligomycin, and occurred in the absence of the latter (Fig. 6).

Mg-ATP particles contain relatively much ATPase inhibitor (Ernster et al., 1973) which can be released by treatment of the particles with Sephadex (Racker and Horstman, 1967). As shown by Van de Stadt et al. (1973), the inhibitor can also be released by incubating the particles under conditions of low ATP/ADP ratio or energy-generation from the respiratory chain. We have confirmed the latter results (cf. Table IV) and found that in this system, just as in the case of the inhibitor-treated EDTA particles, the relief of the inhibition is prevented by oligomycin and promoted by aurovertin.

The available information is consistent with the mechanism schematically depicted in Fig. 7. It is visualized that $F_1$ occurs in two conformational states, one predominant in the presence of ATP and having a high affinity for the ATPase inhibitor, and another, predominant in the presence of $P_i$, ADP and of energy derived from the respiratory chain, and having a low affinity for the ATPase inhibitor. Respiratory energy may be required either for enabling the enzyme to bind $P_i$ and ADP, or for the release of ATP, in accordance with recent proposals by Boyer (1974) and by Slater et al. (1974). Oligomycin inhibits the respiration-induced energization of $F_1$. Aurovertin prevents the conversion of the (ADP + $P_i$)-binding form of the enzyme into the ATP-binding form, and thereby promotes the release of the ATPase inhibitor.

The data further substantiate the concept (Asami et al., 1970) that the ATPase inhibitor acts as a regulator of the ATP-synthesizing system, controlling its reversibility according to the prevailing "phosphate potential". A somewhat similar control of the transhydrogenase system by means of the prevailing "nicotinamide nucleotide potential" has earlier been suggested on the basis of kinetic data (Rydström et al., 1970, 1972), although, in that case, no special inhibitory subunit of the enzyme has so far been identified. A conformational change of cytochrome c oxidase under the influence of the prevailing membrane potential has been described by Wikström (1975) and discussed in relation to its possible role in energy-linked respiratory control; again, information is still lacking regarding the subunit(s) involved in this effect. The precise relationship of these energy-linked conformational changes of various mitochondrial energy-transducing units to the mechanism and control of energy conservation, and, in particular, to the generation of transmembrane electrochemical gradients, remains to be established.

## 4. "TITRATION" OF ATP-DEPENDENT ACTIVITIES OF SUBMITOCHONDRIAL PARTICLES WITH ATPase INHIBITOR

We have earlier compared the effects of ATPase inhibitor with those of oligomycin and FCCP on the ATP-driven succinate-linked $NAD^+$ reduction and nicotinamide nucleotide transhydrogenase activities of EDTA particles (Ernster et al., 1973). It was found that the two reactions were equally sensitive to the ATPase inhibitor, and their ATPase inhibitor "titers" were virtually identical to that of the ATPase reaction catalyzed by the same particles (Fig. 8C). In contrast, oligomycin and FCCP inhibited much more efficiently the succinate-linked $NAD^+$ reduction than the transhydrogenase reaction (Figs. 8A and B). The same was true when ATP was replaced by ITP as the energy source for these reactions (Table III).

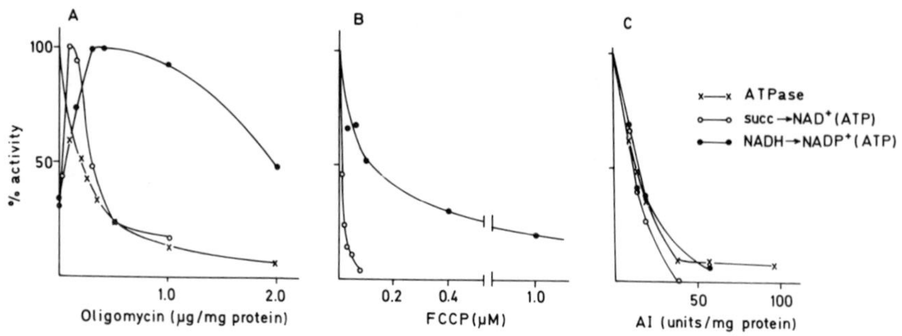


Figure 8 Effects of oligomycin, FCCP and ATPase inhibitor (AI) on ATP-dependent reactions of EDTA particles. From Ernster et al. (1973).

Table III

Comparison of the Efficiencies of ATP and ITP as Substrates for Nucleoside Triphosphatase and Nucleoside Triphosphate--Driven Energy-Linked Transhydrogenase and Succinate-Linked $NAD^+$ Reduction in EDTA Particles from Beef Heart

| NTP* | NTPase | | NADH → $NADP^+$ (NTP)** | | Succ → $NAD^+$ (NTP)** | |
|---|---|---|---|---|---|---|
| | nmoles /min/mg protein | rel. act. | nmoles /min/mg protein | rel. act. | nmoles /min/mg protein | rel. act. |
| ATP | 2520 | 100 | 59.5 | 100 | 32.1 | 100 |
| ITP | 1030 | 41 | 26.5 | 45 | 3.2 | 10 |

*2.5 mM ATP and 2.9 mM ITP were used.

**Increase in activity due to addition of NTP. Oligomycin (0.1 µg/mg protein) was present.

A difficulty in interpreting these results arose from the fact that oligomycin in low concentrations stimulated the ATP-dependent succinate-linked $NAD^+$ reduction and nicotinamide nucleotide transhydrogenase activities of these particles (cf. Fig. 8A), and that the effects of FCCP and ATPase inhibitor, shown in Figs. 8B and C,

Table IV

Effect of removal of ATPase inhibitor on various activities of $Mg^{++}$-ATP particles

$Mg^{++}$-ATP particles were prepared as described by Lee and Ernster (1967). ATPase inhibitor was removed by preincubation of the particles aerobically in the presence of succinate, $P_i$, ADP, glucose and hexokinase according to Van de Stadt et al. (1973). Activities were assayed as described by Asami et al. (1970) except that no oligomycin was added, and that ATP, when present, was added to the assay system prior to the particles. Activities are expressed in nanomoles/min/mg protein.

| Activity | Before pretreatment | After pretreatment | Times increase |
|---|---|---|---|
| ATPase (+ FCCP) | 178 | 1570 | 8.8 |
| Succ → $NAD^+$ (ATP) | 12 | 115 | 9.6 |
| NADH → $NADP^+$ | 58 | 63 | 1.1 |
| NADH → $NADP^+$ (ATP)* | 6 | 56 | 9.3 |
| NADH → $NADP^+$ (succ)* | 113 | 95 | 0.8 |

*Increase in activity due to ATP or succinate.

were tested in the presence of maximally stimulating concentrations of oligomycin. Although it had been found (Ernster et al., 1973; cf. also Fig. 3 of this paper) that oligomycin and the ATPase-inhibitor did not alter their mutual titers of the ATPase activity of EDTA particles, the possibility existed that such a change in titers did occur in the case of the ATP-driven electron-transport reactions. Indeed, it has been proposed by Lang and Racker (1974) that the inhibition of the ATP-dependent reduction of $NAD^+$ by succinate, observed with ATPase inhibitor in "A particles" in the presence of low concentrations of rutamycin, may be due to a shift in the susceptibility of the ATP-dependent succinate-linked $NAD^+$ reduction to rutamycin.

To clarify this point, it was desirable to find conditions which allowed a titration of the ATP-dependent electron-transport reactions of submitochondrial particles in the absence of oligomycin. $Mg^{++}$-ATP particles, which do not require oligomycin to exhibit maximal ATP-dependent succinate-linked $NAD^+$ reduction and transhydrogenase activities, were not suitable for this purpose, since these particles contain large amounts of ATPase inhibitor, as indicated by their relatively low ATPase activity which is virtually unaffected by added inhibitor (Ernster et al., 1973). Removal of the endogenous ATPase inhibitor by treatment of the $Mg^{++}$-ATP particles with Sephadex G-50, as described by Racker and Horstman (1967), also proved unsuitable, since this treatment rendered these particles similar to EDTA particles in that they require low concentrations of oligomycin for maximal ATP-dependent electron-transport activities (Ernster et al., 1973).

It was found, however, that if, instead of Sephadex treatment, the ATPase inhibitor was removed by preincubating the $Mg^{++}$-ATP particles aerobically in the presence of succinate, $P_i$, ADP, and an ADP-regenerating system (hexokinase + glucose), using conditions similar to those described by Van de Stadt et al. (1973), the resulting particles remained tightly-coupled and required no oligomycin for maximal ATP-driven succinate-linked $NAD^+$ reduction and transhydrogenase activities. Table IV compares various activities of $Mg^{++}$-ATP particles before and after such a pretreatment. It may be seen that the pretreatment resulted in an 8- to 10-fold increase in both the ATPase and the ATP-driven succinate-linked $NAD^+$ reduction and transhydrogenase activities of the particles. There was virtually no change in the activities of the nonenergy-linked and the succinate-driven transhydrogenase reactions. The pretreatment also resulted in an increase of the ATP-induced, but not of the succinate-induced, enhancement of ANS-fluorescence of the particles. These data are thus consistent with the conclusion that the pretreatment affects the ATP-dependent reactions of the particles in a specific and parallel fashion.

Fig. 9 shows titration of the ATP-dependent activities of the pretreated ATPase inhibitor-depleted $Mg^{++}$-ATP particles with oligomycin, FCCP and purified ATPase inhibitor; for testing the effect of the ATPase inhibitor the depleted particles were incubated with the inhibitor in the presence of ATP and $Mg^{++}$ (cf. Horstman and Racker, 1970; see also Table I, first column) prior to assay. It is evident that, in contrast to the EDTA-particles used in earlier studies (cf. Fig. 8) these particles did not require oligomycin to exhibit maximal ATP-dependent succinate-linked $NAD^+$ reductase and transhydrogenase activities. Consequently, the effects of FCCP and ATPase inhibitor were tested in the absence of oligomycin. It may be seen that both oligomycin (Fig. 9A) and FCCP (Fig. 9B) inhibited the ATP-driven transhydrogenase reaction more efficiently than the ATP-driven succinate-linked $NAD^+$ reduction. The sensitivity of the succinate-linked $NAD^+$ reduction to oligomycin was similar to that of the ATPase reaction (Fig. 9A). On the other hand, purified ATPase inhibitor inhibited the ATPase and the ATP-dependent succinate-linked $NAD^+$ reduction and transhydrogenase reactions with virtually equal titers (Fig. 9C). It is thus clear that the ATPase

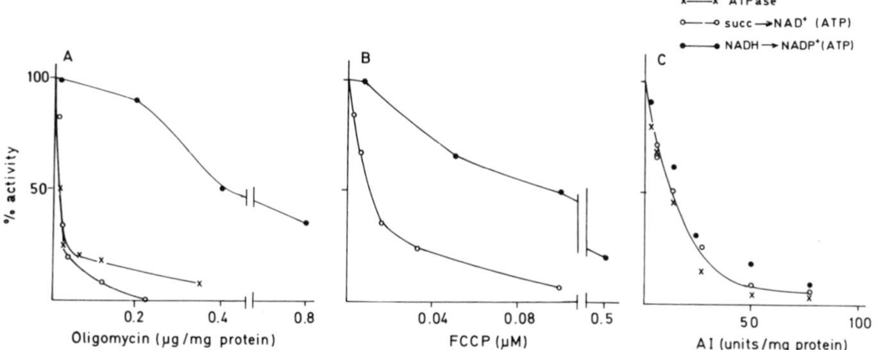


Figure 9 Effects of oligomycin, FCCP and ATPase inhibitor (AI) on ATP-dependent activities of AI-depleted $Mg^{++}$-ATP particles. AI-depleted particles were prepared and assayed as described in Table IV. When indicated, the particles were treated with purified AI as described in Table I. 100 % activities (µmoles/min/mg protein) were: ATPase, 1.53; ATP-driven succinate-linked $NAD^+$ reduction, 0.19; ATP-driven transhydrogenase, 0.16.

inhibitor does inhibit the two ATP-driven electron-transport reactions even in the absence of oligomycin and, hence, that the inhibition cannot be due to an increased susceptibility of these reactions to oligomycin as suggested by Lang and Racker (1974). From a comparison of the effect of oligomycin on the three ATP-dependent reactions (Fig. 9A) it is also evident that the ATPase activity may be rate-limiting for the succinate-linked $NAD^+$ reduction but is not rate-limiting for the transhydrogenase. The equal ATPase-inhibitor titers of the ATPase and ATP-driven transhydrogenase reactions strongly indicate, therefore, that the ATPase and transhydrogenase interact in an assembly-like fashion.

While these experiments eliminated the criticism raised by Lang and Racker (1974), there remained another point of uncertainty. It was assumed in the experiments described above that, under the conditions employed, the ATPase inhibitor was bound to the ATPase in a practically irreversible fashion. This assumption was based on the findings that ATP (+ $Mg^{++}$) promoted the binding of the inhibitor to the enzyme and that there occurred no release of the inhibitor from the enzyme unless energy was supplied from the respiratory chain. Moreover, it was ascertained that there occurred no release of the inhibitor when both ATP (+ $Mg^{++}$) and an oxidizable substrate (succinate, NADH or NADPH, alone or in combination) were present, i.e., under the conditions prevailing during the measurements of the ATP-driven succinate-linked $NAD^+$ reduction and transhydrogenase activities. In fact, in the experiments shown in Table IV and Fig. 9, ATP (+ $Mg^{++}$) was consistently added to the particles prior to the addition of oxidizable substrates, in order to prevent release of the inhibitor. It is not clear whether Lang and Racker (1974) have taken similar precautions, lack of which would explain their failure to obtain an inhibition of the ATP-driven succinate-linked $NAD^+$ reduction by ATPase inhibitor.

## 5. RECONSTITUTION OF ATP-DRIVEN TRANSHYDROGENASE IN ATPase-INHIBITOR- AND $F_1$-DEPLETED SUBMITOCHONDRIAL PARTICLES

Further evidence for an assembly-like interaction between ATPase and transhydrogenase was obtained in experiments with $F_1$-depleted particles after reconstitution with purified $F_1$. EDTA particles were depleted of ATPase inhibitor and $F_1$ by treatment with Sephadex and urea according to Racker and Horstman (1967; cf. also Ernster et al., 1974; Ernster, 1975). To these particles increasing amounts of purified $F_1$ were added and the oligomycin-sensitive ATPase and ATP-driven transhydrogenase activities were measured; in addition, as a control, the respiration-driven transhydrogenase activity was also followed, using succinate as substrate. As shown in Fig. 10, the ATP-driven transhydrogenase activity increased in a fashion parallel to that of the oligomycin-sensitive ATPase upon the addition of increasing amounts of $F_1$, the two activities reaching maximum at the same, "saturating" level of $F_1$, in spite of the fact that the latter was over 50 times higher than the former. The respiration-driven transhydrogenase activity was, as expected, unaffected by $F_1$.

Although these results thus seem to lend independent support to the concept of an assembly-like interaction between mitochondrial ATPase and transhydrogenase, it should be pointed out that they are still to be regarded as preliminary and in need of further controls. For example, the ATP-driven transhydrogenase activity at various levels of added $F_1$ was measured in the presence of a fixed, low concentration of oligomycin. Probably, the optimal oligomycin concentration varies with the amount of $F_1$ present (cf. Ernster et al., 1974). Indeed, the final decision as to whether there is a direct, molecular interaction between transhydrogenase and ATPase, and between membrane-associated energy-transducing units in general, will probably have to await the results of experiments with isolated, purified systems, preferably in the absence of a vesicular membrane structure.

On the other hand it should be pointed out that the present data do not constitute the only evidence in favor of the existence of discrete energy-transfer assemblies or domains in the mitochondrial inner membrane. Lee et al. (1969) have reported that the oligomycin-induced respiratory control of submitochondrial particles is unaffected by partially inhibitory concentrations of various electron-transport inhibitors such as rotenone, antimycin, cyanide

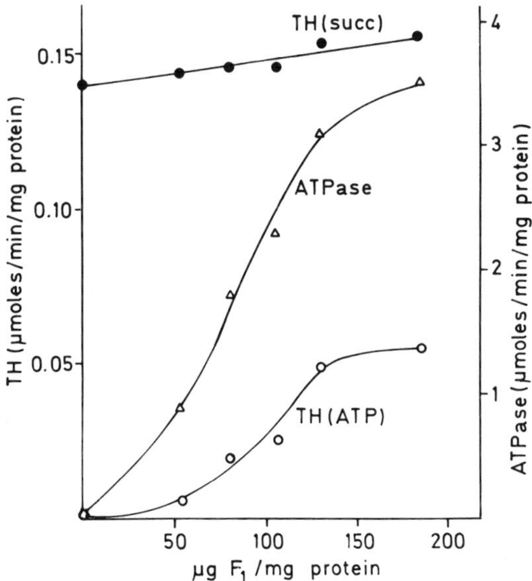


Figure 10 Effects of $F_1$ on the oligomycin-sensitive ATPase and the ATP- and succinate-driven transhydrogenase (TH) activities of ATP-ase-inhibitor- and $F_1$-depleted submitochondrial particles. ATPase-inhibitor- and $F_1$-depleted particles were prepared by treatment with Sephadex G-50 and urea as described by Ernster et al. (1974), using EDTA particles as the starting material. Beef-heart $F_1$ was purified according to Horstman and Racker (1970). ATPase and transhydrogenase activities were assayed as described by Asami et al. (1970).

or azide. Similar indications were subsequently obtained by Baum et al. (1971) from measurements of the effects of rotenone and oligomycin on the rate of ATP-driven succinate-linked $NAD^+$ reduction. Although the findings of Lee et al. (1969) were challenged by Hinkle et al. (1975), recent experiments performed in Dr. Lee's laboratory eliminate this criticism (C.P. Lee, personal communication). Studies of the quantitative relationship between oligomycin-induced respiratory control and energy-linked ANS-fluorescence enhancement in submitochondrial particles (Nordenbrand and Ernster, 1971; Ernster et al., 1971) have given results consistent with the conclusion that the three energy-coupling sites of the respiratory chain are located in different environments of the membrane as revealed by different intensities of the ANS probe. Strong evidence that the ANS probe primarily measures localized events in the membrane, rather than a bulk membrane potential, has recently been obtained by Ferguson et al. (1975) in studies of the effects of the ATPase inhibitor 4-chloro-7-nitrobenzofurazan on the ATP-induced ANS response of submitochondrial particles. Just as the present

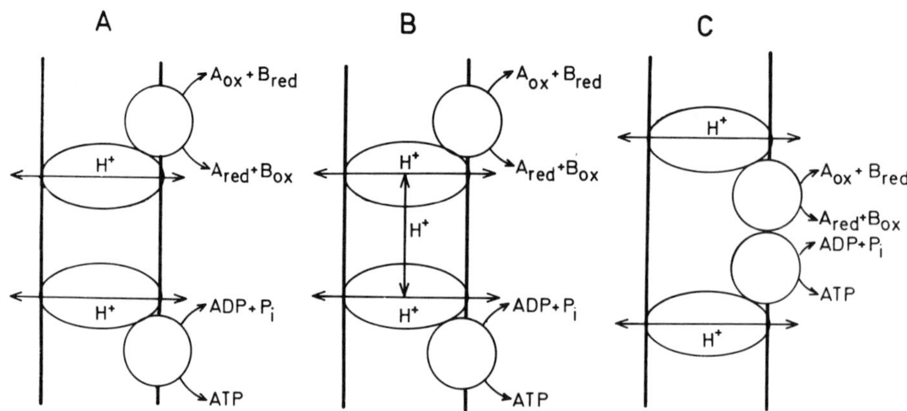


Figure 11 Possible interactions of membrane-associated energy-transducing catalysts.
A: by transmembrane $H^+$ flux only (Mitchell, 1961)
B: also by intramembrane $H^+$ flux (Williams, 1961)
C: also by direct conformational interaction (Boyer, 1965)

data, these results are consistent with the concept that each ATP-ase molecule can act only within a limited segment or domain of the membrane.

Taken together, the available evidence seems to provide strong indications for a dynamic state of membrane-associated energy-transducing catalysts, in which energy conservation and energy transfer may well occur by way of conformational interactions directly (Boyer, 1965) or through intramembrane proton fluxes (Williams, 1961) in addition to bulk transmembrane electrochemical potentials (Mitchell, 1961) (Fig. 11).

## 6. SUMMARY

The effects of a mitochondrial ATPase inhibitor protein (Pullman and Monroy, 1963) on ATP-dependent reactions of submitochondrial particles have been investigated. In confirmation of earlier results (Asami et al., 1970) the purified protein inhibits ATP-dependent succinate-linked $NAD^+$ reduction and nicotinamide nucleotide transhydrogenase activities of submitochondrial particles. The inhibition can also be demonstrated in the absence of oligomycin, using tightly coupled, inhibitor-depleted particles prepared according to Van de Stadt et al. (1973), thus eliminating criticism raised by Lang and Racker (1974). The concept that the inhibitor may act as a regulator of energy transfer between the mitochondrial electron-transport and ATP-synthesizing systems (Asami et al., 1970) is thus further substantiated. The results also allow an extension of the concepts that mitochondrial ATPase occurs in two conformational states, with different affinities for the ATPase inhibitor (Ernster et al., 1973; Van de Stadt et al., 1973) and that the mitochondrial ATPase and electron-transport systems may transfer energy by reactions taking place within the membrane, through direct conformational interactions (Boyer, 1965) or intramembrane proton fluxes (Williams, 1961), in addition to an interaction by way of transmembrane electrochemical gradients (Mitchell, 1961).

---

This work has been supported by a grant from the Swedish Natural-Science Research Council.

---

## REFERENCES

Asami, K., K. Juntti and L. Ernster, Possible regulatory function of a mitochondrial ATPase inhibitor in respiratory chain-linked energy transfer, Biochim. Biophys. Acta **205**, 307-311, 1970.

Baum, H., G.S. Hall, J. Nalder and R.B. Beechey, On the mechanism of oxidative phosphorylation in submitochondrial particles, in _Energy Transduction in Respiration and Photosynthesis_, edited by E. Quagliariello, S. Papa and C.S. Rossi, pp. 747-755, Adriatica Editrice, 1971.

Boyer, P.D., Carboxyl activation as a possible common reaction in substrate-level and oxidative phosphorylation and in muscle contraction, in _Oxidases and Related Redox Systems_, edited by T.E. King, H.S. Mason and M. Morrison, pp. 994-1017, John Wiley & Sons, New York, 1965.

Boyer, P.D., Conformational coupling in biological energy transductions, in _BBA Library_ 13, 289-301, 1974.

Chang, T.M. and H.S. Penefsky, Aurovertin, a fluorescent probe of conformational change in beef heart mitochondrial adenosine triphosphatase, J. Biol. Chem. **248**, 2746-2754, 1973.

Ernster, L., Chemical and chemiosmotic aspects of mitochondrial energy transduction, in _Proc. 10th FEBS Meeting. Fed. Europ. Biochem. Soc. Symp._ Vol. 40, edited by P. Desnuelle and A.M. Michelson, pp. 253-276, North-Holland, Amsterdam, 1975.

Ernster, L., K. Nordenbrand, C.P. Lee, Y. Avi-Dor and T. Hundal, Qualitative and quantitative aspects of the energized state of mitochondria. Studies with the fluorescence probe 8-anilino-1-naphthalene sulphonic acid, in Energy Transduction in Respiration and Photosynthesis, edited by E. Quagliariello, S. Papa, C.S. Rossi, pp. 57-87, Adriatica Editrice, Bari, 1971.

Ernster, L., K. Juntti and K. Asami, Mechanisms of energy conservation in the mitochondrial membrane, J. Bioenergetics 4, 149-159, 1973.

Ernster, L., K. Nordenbrand, O. Chude and K. Juntti, Relationship of components of the ATPase system to the oligomycin-induced respiratory control of submitochondrial particles, in Membrane Proteins in Transport and Phosphorylation, edited by G.F. Azzone, M.E. Klingenberg, E. Quagliariello and N. Siliprandi, pp. 29-41, North-Holland, Amsterdam, 1974.

Ernster, L., K. Nordenbrand, T. Hundal and C. Carlsson, Studies of the reaction mechanism of mitochondrial adenosine triphosphatase by means of inhibitors, 10th International Congress of Biochemistry, Hamburg, Abstract, 1976.

Ferguson, S.J., W.J. Lloyd and G.K. Radda, Properties of mitochondrial and bacterial ATPases, in Electron Transfer Chains and Oxidative Phosphorylation, edited by E. Quagliariello, S. Papa, F. Palmieri, E.C. Slater and N. Siliprandi, pp. 161-166, North-Holland, Amsterdam, 1975.

Hinkle, P.C., Y.-S.L. Tu and J.J. Kim, Studies of respiratory control in submitochondrial particles and reconstituted systems, in Molecular Aspects of Membrane Phenomena, edited by M.R. Kaback, M. Neurath, G.K. Radda, R. Schwyzer and W.R. Wiley, pp. 222-232, Springer-Verlag, New York, 1975.

Horstman, L.L. and E. Racker, Partial resolution of the enzymes catalyzing oxidative phosphorylation. XXII. Interaction between mitochondrial adenosine triphosphatase inhibitor and mitochondrial adenosine triphosphatase. J. Biol. Chem. 245, 1336-1344, 1970.

Juntti, K., K. Asami and L. Ernster, Studies on the mode of action of mitochondrial ATPase inhibitor, 7th FEBS Meeting, Varna, Abstract No. 660, 1971.

Lang, D.R. and E. Racker, Effects of quercetin and $F_1$ inhibitor on mitochondrial ATPase and energy-linked reactions in submitochondrial particles. Biochim. Biophys. Acta 333, 180-186, 1974.

Lee, C.P. and L. Ernster, Energy-coupling in nonphosphorylating submitochondrial particles, Meth. Enzymol. 10, 543-548, 1967.

Lee, C.P. and L. Ernster, Studies of the energy-transfer system of submitochondrial particles. 2. Effects of oligomycin and aurovertin, Eur. J. Biochem. 3, 391-400, 1968.

Lee, C.P., L. Ernster and B. Chance, Studies of the energy-transfer system of submitochondrial particles. Kinetic studies of the effect of oligomycin on the respiratory chain of EDTA particles. Eur. J. Biochem. 8, 153-163, 1969.

Mitchell, P., Coupling of phosphorylation to electron and hydrogen transfer by a chemi-osmotic type of mechanism, Nature 191, 144-148, 1961.

Mitchell, P., Chemiosmotic coupling in oxidative and photosynthetic phosphorylation, Glynn Research, Bodmin, Cornwall, England, 1966.

Mitchell, P. and J. Moyle, Influence of aurovertin on affinity of mitochondrial adenosine triphosphatase for ATP and ADP, FEBS Letters 6, 309-311, 1970.

Nishimura, M., T. Ito and B. Chance, Studies on bacterial photophosphorylation. III. A sensitive and rapid method of determination of photophosphorylation, Biochim. Biophys. Acta 59, 177-182, 1962.

Nordenbrand, K. and L. Ernster, Studies of the energy-transfer system of submitochondrial particles. Fluorochrome response as a measure of the energized state. Eur. J. Biochem. 18, 258-273, 1971.

Pedersen, P.L., Mitochondrial adenosine triphosphatase, J. Bioenergetics 6, 143-175, 1975.

Pedersen, P.L., H. Le Vine and N. Cintrón, Activation and inhibition of mitochondrial ATPase of rat liver mitochondria, in Membrane Proteins in Transport and Phosphorylation, edited by G.F. Azzone, M.E. Klingenberg, E. Quagliariello and N. Siliprandi, pp. 43-54, North-Holland, Amsterdam, 1974.

Penefsky, H.S., Differential effects of adenyl imidodiphosphate on adenosintriphosphate synthesis and the partial reactions on oxidative phosphorylation, J. Biol. Chem. 249, 3579-3585, 1974.

Phelps, D.C., K. Nordenbrand, B.D. Nelson and L. Ernster, Inhibition of purified mitochondrial ATPase ($F_1$) by bathophenanthroline and relief of the inhibition by uncouplers, Biochem. Biophys. Res. Commun. 63, 1005-1012, 1975a.

Phelps, D.C., K. Nordenbrand, T. Hundal, C. Carlsson, B.D. Nelson and L. Ernster, Some recent aspects of mitochondrial ATPase action, in Electron Transfer Chains and Oxidative Phosphorylation, edited by E. Quagliariello, S. Papa, F. Palmieri, E.C. Slater and N. Siliprandi, pp. 385-400, North-Holland, Amsterdam, 1975b.

Pullman, M.E., H.S. Penefsky, A. Datta and E. Racker, Partial resolution of the enzymes catalyzing oxidative phosphorylation. 1. Purification and properties of soluble, dinitrophenol-stimulated adenosine triphosphatase, J. Biol. Chem. 235, 3322-3329, 1960.

Pullman, M.E. and G.C. Monroy, A naturally occurring inhibitor of mitochondrial adenosine triphosphatase, J. Biol. Chem. 238, 3762-3769, 1963.

Racker, E. and L.L. Horstman, Partial resolution of the enzymes catalyzing oxidative phosphorylation. XIII. Structure and function of submitochondrial particles completely resolved with respects to coupling factor 1. J. Biol. Chem. 242, 2547-2551, 1967.

Rydström, J., A. Teixeira da Cruz and L. Ernster, Factors governing the kinetics and steady state of the mitochondrial nicotinamide nucleotide transhydrogenase system, Eur. J. Biochem. 17, 56-62, 1970.

Rydström, J., A. Teixeira da Cruz and L. Ernster, Reaction mechanism of mitochondrial nicotinamide nucleotide transhydrogenase, in Biochemistry and Biophysics of Mitochondrial Membranes, edited by G.F. Azzone, E. Carafoli, A.L. Lehninger, E. Quagliariello and N. Siliprandi, pp. 177-200, Academic Pres s, New York, 1972.

Senior, A.L., The structure of mitochondrial ATPase, Biochim. Biophys. Acta 301, 249-277, 1973.

Slater, E.C., J. Rosing, D.A. Harris, R.J. Van de Stadt and A. Kemp, Jr., The identification of functional ATPase in energy-transducing membranes, in Membrane Proteins in Transport and Phosphorylation, edited by G.F. Azzone, M.E. Klingenberg, E. Quagliariello and N. Siliprandi, pp. 137-147, North-Holland, Amsterdam, 1974.

Van de Stadt, R.J., B.L. de Boer and K. van Dam, The interaction between the mitochondrial ATPase ($F_1$) and the ATPase inhibitor, Biochim. Biophys. Acta 292, 338-349, 1973.

Wikström, M.K.F., Energy-linked conformational changes in cytochrome c oxidase of the mitochondrial membrane, in Electron Transfer Chains and Oxidative Phosphorylation, edited by E. Quagliariello, S. Papa, F. Palmieri, E.C. Slater and N. Siliprandi, pp. 97-103, North-Holland, Amsterdam, 1975.

Williams, R.J.P., Possible functions of chains of catalysts, J. Theoret. Biol. 1, 1-17, 1961.

# THE MYELIN MEMBRANE AND ITS BASIC PROTEINS

E. H. Eylar

Playfair Neuroscience Institute & Dept. of Biochemistry

University of Toronto, Toronto, Canada M5S 2J5

## ABSTRACT

Peripheral nerve myelin contains a wider variety of protein components than central nervous system myelin. Common features are the presence of the basic A1 protein, as shown by amino acid sequence, and a high proportion of a hydrophobic protein such as the PO protein (PNS) and the FL proteolipid (CNS) which are likely amphipathic components. It was proposed that the A1 protein only partially penetrates into the lipid layer and thus is externally situated such that it can provide a structural security in the lateral dimension. The hydrophobic proteins may provide a vertical (radial) stabilization to myelin via linear polymerization across the lammelae layers. The PNS and CNS myelin also differ in immunologic properties since the A1 protein, which induces allergic emcephalomyelitis in animals when given alone or as part of CNS myelin, is masked in PNS myelin. By contrast, PNS myelin induces allergic neuritis in monkeys where the responsible antigen appears to be the P2 protein, a basic protein absent in CNS myelin. The rabbit P2 protein (12,000 MW) is a more compact, highly structured molecule in contrast to the open, extended double-chain conformation of the A1 protein (18,000 MW).

The role of these proteins in human demyelination diseases appears crucial. Guillian-Barre patients show a cellular hypersensitivity to the P2 protein; multiple sclerosis (MS) subjects show a hypersensitivity to the A1 protein prior to and during exacerbation. These data suggest that allergic encephalomyelitis and neuritis are relevant animal models for early phases of the respective human disease, and provide a rationale for consideration of clinical studies aimed to suppress MS; in monkeys with severe clinical signs of allergic encephalomyelitis, the A1 protein was shown to suppress and reverse the course of the disease.

## INTRODUCTION

Myelin is the membraneous material that surrounds the nerve axon, presumably as an insulation which aids in the conductivity of the nerve impulse. Myelin is considered a classical example of a lipid bilayer, actually a double bilayer, being derived presumably from a double layer wrapping of the plasma membrane of the Schwann cell (PNS)* or oligodendroglial cell (CNS). Our present concepts of myelin structure derive primarily from the pioneering work of Schmitt and Baer (1), and Finean (2), who showed by X-ray and electron microscopic studies that myelin is constructed from layers, or lamellae, of 150 - 170Å repeating distance between the double bilayers. However accurate as this picture may be, the notion of continuity between the plasma membrane and myelin may be outmoded since myelin differs markedly in many respects from conventional plasma membranes (3, 4). Myelin has a low protein content (10 - 20%), little glycolipid (5) and glycoprotein (6), very low enzymatic activities, and contains a relatively high content of cerebroside and rarer lipids such as cerebroside sulfate and triphosphoinositide (3). Myelin particularly has qualitatively few proteins and these have relatively low molecular weight (7).

The purpose of this report is to discuss myelin structure in terms of present-knowledge of its components, particularly the basic proteins referred to as the A1* and P2 proteins (4). Myelin may be far more stable and rigid than the fluid mosaic model (8) characterizing many biological membranes since it has some unique proteins (A1 and P2 proteins), stable lipid elements (9) and a high cholesterol content. For our understanding of myelin structure and its immunologic properties, the study of myelin proteins have proven invaluable (4). Myelin is also an important subject for study because many serious nervous system diseases in humans are demyelinating diseases such as multiple sclerosis (MS) and Guillian-Barre (GB) syndrome.

\- - - - - -

| | | | |
|---|---|---|---|
| PNS | peripheral nervous system | EAE | allergic encephalomyelitis |
| CNS | central nervous system | EAN | allergic neuritis |
| GB | Guillian-Barre | PAGE | polyacrylamide gel electrophoresis |
| MS | multiple sclerosis | FL | Folch-Lees |

*The term A1 protein, is often referred to as the myelin basic protein, but the use of A1 is preferred for this report to distinguish it from other basic proteins of myelin such as the P2 protein and the small basic protein of rat and mouse myelin.

COMPARISON OF CNS AND PNS MYELIN PROTEINS. The preparation of highly purified CNS myelin (3) is easily accomplished using sucrose or cesium chloride gradients (7) although traces of other membranes are likely contaminates. PNS myelin is far more difficult to isolate, however, because of difficulties with connective tissue. We have now worked out procedures for large scale preparation of highly purified PNS myelin on sucrose gradients (7, 11) in the case of rabbit sciatic nerve. Larger peripheral nerves (from human, horse, bovine, etc.) still provide difficulties, but myelin has been prepared successfully from bovine intradural roots (3, 9). Purity is established by: electron microscopy; the level of 2', 3' - cyclicnucleotide - 3' phosphohydrolase (a myelin enzyme); the absence of enzymes such as 5' -nucleotidase, glycosyl transferases, ATP ase, and mitochondrial enzymes; and the protein and lipid profiles. The protein profile of rabbit spinal cord and sciatic nerve myelin is shown in Fig. 1 along with purified A1 and P2 proteins and standards. The myelin samples differ markedly since only 2 major bands are readily discernable, the A1 protein and FL proteolipid, in the CNS myelin, whereas 4 major bands are seen in the PNS myelin.

The four bands at 30K, 23K, 18K, and 12K in the PNS myelin represent the P0, P4, A1 and P2 proteins respectively. In Fig. 2, where better resolution has been achieved, a band between A1 and P2 is clearly observed (P5 band). In addition to the five bands shown in Fig. 2, another band is resolved, as shown previously (7) using a discontinuous PAGE system in SDS, and referred to now as P3 (previously called Y). It migrates just slightly faster than P0 but is obscured in the 10% polyacrylamide gels shown here because of the relatively large quantity of P0 protein.

The P4 band (previously X bands) of PNS myelin moves at the same position as the FL proteolipid (23K) of CNS myelin. Fig. 1. There is another proteolipid (20K) in CNS myelin, not seen in Fig. 1, which migrates between the FL proteolipid (23K) and A1 protein (18K) and referred to as the DM-20 protein (12). The P4 protein of PNS myelin does not appear to be the same as either of the CNS proteolipids, however, since it appears to be a glycoprotein and is not soluble in chloroform-methanol. Small molecular weight proteins are often seen in PNS myelin migrating faster than the P2 band. Whether these are artifacts absorbed to lipids or represent true small molecular weight components is unknown.

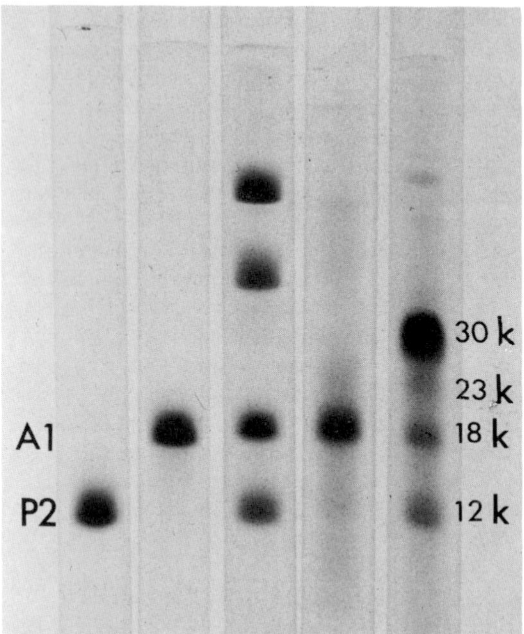


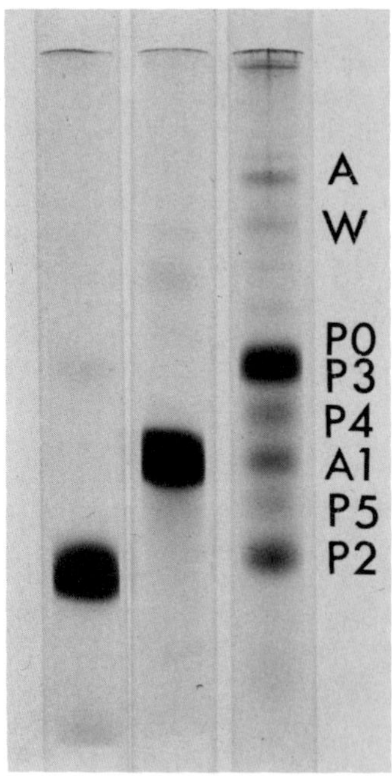


Figure 1 (left) Polyacrylamide (10%) gel electrophoresis in SDS (left to right): purified rabbit P2 protein; purified A1 protein; standard proteins albumin, ovalbumin, A1 and P2 proteins (25 μg each); rabbit spinal cord myelin; rabbit sciatic nerve myelin. Myelin samples were defatted with ethanol-ether (1:3); 100 μg was then used. The F1 proteolipid band appears light because some was extracted by the lipid solvents.

Figure 2 (right) PAGE (10%) in SDS. From left to right: P2 protein (50 μg); A1 protein (50 μg); and rabbit sciatic nerve myelin (100 g defatted material). Proteins are designated A for albumin and W for Wolfgram protein. The myelin proteins are designated according to observed bands. The P3 band migrates together with P0 band in this system.

One of the striking features of the distribution of myelin protein in both PNS and CNS myelin is that nearly all the proteins have a molecular weight below 30,000 whereas erythrocyte stroma (13) for example, has nearly all proteins greater than 30,000 MW. This comparison further emphasizes the difference between plasma membrane and myelin proteins. Of the protein bands in myelin above 30,000 MW, the Wolfgram protein, which occurs in both PNS and CNS myelin, usually as a faint doublet, is the most conspicuous. The question arises however whether this is a true myelin protein, which has never been rigorously established, or a contaminant. Undoubtedly some of the faint, high MW bands of myelin arise from trace contamination by other membranes or soluble proteins, even though myelin itself can be highly purified by density gradient sedimentation (7); the glycoprotein (6) and the cyclic nucleotide phosphohydrolase of CNS myelin represent high molecular weight proteins that are probably intrinsic components. However, as a clear example of an absorbed contaminant (Fig. 2), serum albumin binds avidly to PNS myelin. Although the PNS myelin can be repurified extensively, traces of albumin remain. We recently isolated the absorbed albumin and showed it to be identical to serum albumin in both amino acid composition and immunologic properties (11).

In order to establish an accurate profile of myelin proteins, two points are essential: that the isolated protein be homogeneous as shown by PAGE and immunologic procedures, and that its membrane localization be established by use of antibody containing fluorescent or dense (ferritin, peroxidase)markers. These criteria are especially important in dealing with lesser components, and we hope to utilize this approach in our study of the many PNS myelin proteins. For the major components, preparation of homogeneous protein is adequate since myelin can be highly purified. It is by this approach that we establish (14) that the A1 protein exists in both CNS and PNS myelin. We prepared A1 protein from highly purified rabbit spinal cord and sciatic nerve myelin and compared their amino acid sequence. They were identical over the entire 168 residues of the sequence shown in Table 1. Thus we concluded that the A1 protein is a component of both types of myelin, although it comprises only 14% of rabbit sciatic nerve myelin protein (7), compared to 30% in CNS myelin (15).

Table 1: The Complete Amino Acid Sequence of the Rabbit A1 Protein Derived from Sciatic Nerve and Spinal Cord

N-Ac-Ala-Ser-Gln-Lys- Arg-Pro-Ser-Gln-Arg-His-Gly-Ser-Lys-Tyr-Leu-Ala-
20 30
Thr-Ala-Ser-Thr-Met-Asp-His-Ala-Arg-His-Gly-Phe-Leu-Pro-Arg-His-
40
Arg-Asp-Thr-Gly-Ile-Leu-Asp-Ser-Ile-Gly-Arg-Phe-Phe-Ser-Ser-Asp-
50 60
Arg-Gly-Ala-Pro-Lys-Arg-Gly-Ser-Gly-Lys-Asp-His-Ala-Ala-Arg-Thr-Thr-
80
His-Tyr-Gly-Ser-Leu-Pro-Gln-Lys-Ser-Gly-His-Arg-Pro-Gln-Asp-Glu-
90
Asn-Pro-Val-Val-His-Phe-Phe-Lys-Asn-Ile-Val-Thr-Pro-Arg-Thr-Pro-Pro-
100
Pro-Ser-Gln-Gly-Lys-Gly-Arg-Gly-Leu-Ser-Val-Thr-Arg-Phe-Ser-Trp-Gly-
120 130
Ala-Glu-Gly-Gln-Lys-Pro-Gly-Phe-Gly-Tyr-Gly-Gly-Arg-Ala-Ala-Asp-Tyr-
140
Lys-Ser-Ala-His-Lys-Gly-Leu-Lys-Gly-Ala-Asp-Ala-Gln-Gly-Thr-Leu-Ser-
150 160
Arg-Leu-Phe-Lys-Leu-Gly-Gly-Arg-Asp-Ser-Arg-Ser-Gly-Ser-Pro-Met-Ala-
168
Arg-Arg-COOH.

MYELIN PROTEINS. The basic proteins (A1 and P2) and the FL proteolipid are the only myelin proteins which have been isolated. The FL proteolipid (16) is soluble in chloroform-methanol, binds lipids and contains two moles of fatty acid convalently-bound. Although it tends to aggregate, the apoprotein may become soluble in aqueous systems and assume β-structure or α-helical forms under appropriate conditions (17). It shows a MW of 23,000 in SDS-PAGE (Fig. 1). Like the proteolipid, the PO protein of PNS myelin is very hydrophobic and quite insoluble in aqueous solution, but is not soluble in chloroform-methanol. Both proteins comprise approximately 50% of the total myelin proteins (3, 7) in CNS and PNS myelin respectively, and therefore must play a major structural role. The PO protein may be a glycoprotein (18, 19) since it appears to stain by the periodic acid-Schiff technique. Since the FL proteolipid and PO protein are insoluble and very resistant to proteolytic enzymes, their characterization has proceeded slowly.

Since the A1 protein was one of the first membrane proteins to be well characterized, it has been studied for its unusual properties, and for its ability to elicit EAE in animals (see ref 4 and 10 for reviews). For our understanding of its role in myelin structure, it will be instructive to compare the A1 with the P2 protein, the latter being found only in PNS myelin. The P2 protein is smaller, and distinguished from the A1 protein by its large proportion of secondary structure (20, 21) in which β-structure predominates (Table II). Quite probably therefore the P2 protein assumes a compact conformation in contrast to the A1 protein which is highly extended (4), in part because of restrictions imposed by its proline-rich region (Table I), Pro-Arg-Thr-Pro-Pro-Pro (residues 94-99), which induces a bend. Thus the most reasonable conformation for the A1 molecule is not a random coil but an open, double chain structure (22). The A1 molecule further unfolds in 8M urea since the intrinsic viscosity doubles (23). The P2 protein unfolds as well in urea since the secondary structure is lost. In neither protein does sulfhydryl cross-linking play a role.

Table II: Physico-Chemical Properties of the Rabbit A1 and P2 Proteins

| Property | A1 Protein | P2 Protein |
|---|---|---|
| Molecular weight | 18,100 (sequence) | 12,000 (SDS-PAGE) |
| Secondary structure<br>(CD analysis) in saline<br>In 8M urea | <br>none<br>none | (29% β-structure<br>(10% α-helix<br>94% random |
| $NH_2$ -terminal residue | acetylated alanine | acetylated |
| Axial ratio | 10-14 (viscosity) | - |
| Conformation | Highly extended | Probably compact |
| Isoelectric point | 12 or more | 9-10 |

The amino acid composition of the two proteins markedly contrast (Table III). Not only is the P2 protein less basic, having no histidine, but it has much less proline, and relatively more non-polar residues, which may account in part for its ability to acquire secondary structure. Although the number of basic residues in the P2 protein approximately equals the acid residues, most of the latter are amidated (20). Thus the calculated isoelectric point falls between pH 9-10. Our preliminary sequence studies on the P2 protein shows no obvious similarity with the A1 protein (11). Although the $NH_2$ -terminus is blocked, a methionine residue is located nearby in both proteins and use of CNBr provides a large peptide which can be sequenced in the automatic Edman sequenator.

Table III: Amino Acid Composition of Rabbit Myelin Basic Proteins

| | Residues/mole | |
|---|---|---|
| Amino Acid | A1 Protein | P2 Protein |
| Lysine | 12 | 13 |
| Histidine | 9 | 0 |
| Arginine | 19 | 5 |
| Aspartic acid | 11 | 10 |
| Threonine | 9 | 9 |
| Serine | 18 | 6 |
| Glutamic acid | 9 | 11 |
| Proline | 12 | 2 |
| Glycine | 24 | 8 |
| Alanine | 14 | 5 |
| Half cystine | 0 | 0 |
| Valine | 4 | 7 |
| Methionine | 2 | 2 |
| Isoleucine | 3 | 6 |
| Leucine | 9 | 9 |
| Tyrosine | 4 | 2 |
| Phenylalanine | 8 | 5 |
| Tryptophan | 1 | 1 |
| TOTAL | 168 | 101 |

COMPARISON OF CNS AND PNS MYELIN. Questions concerning myelin structure must eventually account for the difference in protein components between PNS and CNS myelin, but at this point, we can benefit by considering the similarities. The most obvious similarity between the two myelin forms is the presence of A1 protein, but to a lower extent in rabbit PNS myelin (14% compared to 30% of the total protein). The PNS myelin has several other protein components (Fig. 2) and its protein to lipid ratio is considerably smaller than for CNS myelin. Apparently the differences in protein composition and content do not influence the myelin structure grossly since electron microscopic and X-ray studies (24) show a repeat distance between the major period line of only 24Å more for PNS over CNS myelin.

Although we do not know for certain where the proteins are positioned within the myelin framework, we can infer that the A1 protein occupies an external position as shown by consideration of Table IV. The A1 protein is know to interact strongly with acidic lipids such as sulfatide (25) and triphosphoinositide (26a) in predominantly electrostatic linkage. In this regard, we found that the A1 protein would bind and partially penetrate into the surface of negatively charged lipids in liposomes.(26)

Table IV: Evidence for an Extrinsic Localization of the A1 Protein in Myelin.

1. X-ray studies suggest a considerable proportion of myelin protein positioned at the surface of the lipid bilayer (Casper, Kirschner, Nature New Biol. 231, 45, 1971).

2. The interaction of A1 protein with liposomes shows a surface localization along with some penetration or distortion of the lipid bilayer (Papahadjopoulos et al, BBA, 401, 317 (1975).

3. Interaction between A1 protein and lipid films produces a pronounced increase in surface pressure when sulfatide is used. In the A1 protein-sulfatide complex, portions of the N-terminal region up to 100 residues are masked by the lipid, but that of the C-terminal portion is accessible to trypsin (London et al, BBA, 113, 520, 1973).

4. The highly basic character of the A1 protein suggest a predominantly electrostatic interaction with negatively charged lipids. However, some regions are nonpolar and could partially penetrate into the lipid matrix (Eylar et al, J. Biol. Chem. 246, 5770, 1971).

Several regions of the A1 molecule of 9-10 residues are hydrophobic (lacking basic residues), and might be expected to penetrate the lipid layer (27). The ingenious experiments of van Deenen, London, Demel and co-workers (25) suggest that the $NH_2$-terminal region of the A1 molecule, where most of the nonpolar regions exist, is protected by interaction with lipids in films, but that the C-terminal region is more accessible.

All of the data (Table IV) suggest an extrinsic position of the A1 protein, as diagramed schematically in Fig. 3, with the $NH_2$-terminal region probably penetrating more into the lipid layer where it can partake in hydrophobic bonding. Controversy exists on whether the A1 protein is localized in the major or minor period line of myelin (28). Although this question has yet to be resolved, it appears that the A1 protein is asymmetrically in either one or the other of these positions. It should be noted that these lines are generally equated with the outer surface (minor) or inner surface (major) of the plasma membrane of the oligodendroglial or Schwann cell. This concept may be outmoded because of the great disparity discussed earlier between myelin and plasma membranes; a sharp chemical discontinuity exists between the plasma membrane and myelin, not observable in the electron microscope. In any case, it seems most reasonable for the A1 protein to be accessible and thus an extrinsic protein in the classification of Vanderkooi (29) or Singer (8). This position would also be consistent with its ability to react with sensitized lymphocytes which mediate immunopathologic events in EAE (10).

By this reasoning the P2 protein would also classify as an external protein if it is the antigen responsible for EAN. Like the A1 protein, it will penetrate films of acidic lipids and increase the surface pressure (25). Thus it probably interacts predominantly with negatively charged lipid groupings electrostatically but may penetrate somewhat and engage in hydrophobic bonding. It is probably not an amphipathic protein since it has much more hydrophilic than hydrophobic character, and is very soluble in aqueous media unlike the FL proteolipid or PO protein.

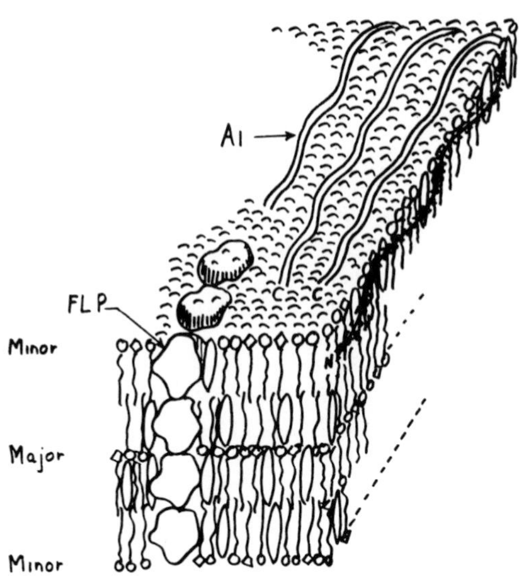


Figure 3 Theoretical model proposed for a bilayer of CNS myelin membrane. The A1 protein is shown in its open, double chain conformation as it may interact with the charged phosphate and sulfate groups of the lipid layer; certain regions of the A1 molecule may partially penetrate into the nonpolar lipid matrix. The proteolipid is shown in a globular conformation penetrating deeply into the lipid layer and interacting strongly with cholesterol and phospholipids; it is partly exposed at the surface, and is pictured in a polymer form extending through the double bilayer as it may extend through many bilayers.

MYELIN STRUCTURE. In this regard it is tempting to speculate on the biological role of these structural proteins. It is possible that for both PNS and CNS myelin, the genome for the A1 protein is derepressed by a signal from the axon or upon interaction of the glial (or Schwann) cell surface with the axonal membrane. Such a basic, open protein would provide an ideal molecule with which to interact with negatively charged lipids such as sulfatide and triphosphoinositide and help to focus the organization of the myelin membrane. In this regard it should be noted that the A1 protein, when combined with acidic lipids, formed a double leaflet having remarkable similarities to myelin (3). Thus the A1 protein could provide the direction for spontaneous aggregation into the classical double leaflet form of myelin. In this picture, the A1 protein would also provide a type of horizontal stabilization. In other words, the presence of a positively charged network of basic A1 protein over the lipid leaflet, would lend rigidity to the lipid surface and interfere with fluid movement. Such a role for the A1 protein would surely subserve the role for myelin as a relatively inert insulator of the nerve axon. It is of interest that basic proteins like the A1 protein are not found in membranes such as plasma membranes where fluidity is a feature. Whereas the A1 protein may provide lateral stabilization, the FL proteolipid may provide a radial stabilization. Recent freeze fraction studies by Pinto Da Silva and Miller (31) suggest protein particles arrayed through the bilayer and across the lammelae, a function possibly performed by the FL proteolipid. This apoprotein in fact aggregates markedly in aqueous systems (16, 17). The FL proteolipid is well suited for this role since it is a major protein of myelin, strongly interacts with lipids, and has a high proportion of nonpolar residues. This hydrophobic protein is highly compact with a diameter of about 30-35Å (23,000 MW) or half a bilayer; probably it is a true amphipathic protein submerged into half the bilayer with a hydrophilic portion extending into the aqueous phase where it interacts with specific labelling agents (32). Thus the particles seen in freeze fracture would not be large. Supporting the role of the FL proteolipid as a radially directed linear aggregate is its ability to penetrate the lipid matrix of liposomes (26). In PNS myelin, an analagous role could be assumed by the PO protein, the major protein, which is highly hydrophobic and which could easily penetrate the lipid layer.

What about other protein components of PNS myelin? We know very little about most of these proteins except for the P2 protein. While the P2 protein, like the A1 protein, also shows a strong interaction with acidic lipids in surface film studies (33), it is probable that this protein likewise is externally located, because of its basic character, and not deeply submerged into the lipid matrix. It was proposed (30) that it is this protein which provides the difference in spacing of 20Å between repeating lines in CNS and PNS myelin, a suggestion based on the finding that the P2 protein formed a double leaflet membrane with acidic lipids which exceeded that formed by A1 protein by 20Å. By comparison with analagous basic proteins such as cytochrome C, ribonclease, and lysozyme, it is reasonable to suppose that the P2 protein assumes a relatively compact structure because of its high percentage of secondary structur Its size and shape could ideally account for the greater repeat distance. Moreover, at least 30Å or so of the repeat distance in myelin must be accounted for by protein since the double leaflet of lipid would only be 120-130Å, far short of the 150-180Å thickness.

Thus it seems logical to propose that the role of the major myelin proteins are primarily for biogenesis and structural stability; the basic A1 protein for horizontal rigidity, and the nonpolar proteins (FL proteolipid and PO protein) for vertical rigidity across the many lammelae. The P2 protein is probably in an external location, like the A1 protein, where its main interaction is electrostatic although it may too penetrate partially into the lipid layer. By comparison with artificial membranes it may account for the greater repeat distance in PNS myelin. The P2 protein is not essential to the structural integrity of PNS myelin, however, since it is absent or very low in guinea pig myelin (7), and varies widely with species.

IMMUNOLOGIC PROPERTIES. The immunologic properties of myelin are of special concern because it appears that human demyelinating diseases such as multiple sclerosis and Guillian-Barre syndrome may involve immunologic factors. It was established by Laatch et al (34) that CNS myelin elicited allergic encephalomyelitis in guinea pigs, whereas we (35) have shown that rabbit PNS myelin induced classical allergic neuritis in monkeys. It is generally accepted that EAN is a reasonable if not an accurate model of the human disease (GB syndrome). Both EAN and the BG syndrome appear to be mediated by lymphocytes (14); the clinical course and the hostologic lesions appear similar in both cases (36), and demyelination of PNS myelin by macrophages can be seen in both instances (35). Since EAN involves only the PNS and not the CNS, it is logical to assume that the responsible antigen is present in PNS myelin only. We believe this antigen to be the P2 protein since in some monkeys and rabbits, but not all, it induced disease. The failure of the purified P2 protein to induce EAN in 100% of the animals, as does PNS myelin, is likely due to its conformation. Because of its high degree of secondary structure, the conformation of the isolated P2 protein may differ from its conformation _in situ_. Thus an immunologic response to the isolated P2 protein may not recognize the P2 protein in its myelin environment, and thus fail to induce EAN.

It is remarkable that the CNS and PNS myelin behave differently immunologically. The A1 protein is common to both these membranes, yet it appears that it is masked in PNS myelin. For example, when CNS myelin is injected into rabbits, it induces EAE and concommitantly, lymphocytes sensitized to this protein (4, 10). In contrast, when PNS myelin is injected into monkeys, EAE is _not_ induced nor are lymphocytes detected which respond to the A1 protein (35). Injection of PNS myelin does induce a lymphocytic response to the P2 protein, however, as shown by the occurrence of EAN in which demyelination of peripheral, but not central myelin, occurs. Since lymph nodes absorb P2 protein but not A1 protein in rabbits sensitized to PNS myelin (37), it is clear that the A1 protein is masked.

Thus it is apparent that the A1 protein behaves differently in CNS and PNS myelin. These results are consistent with the position of these proteins in the myelin ultrastructure proposed earlier. The P2 protein may mask expression of the A1 protein in PNS myelin by virtue of a more external projection from the membrane due to its globular conformation. Moreover, the A1 protein may be partially buried in the lipid layer since only discrete sites appear accessible for interaction with lymphocytes in various animals in the course of EAE. Bergstrand (38) has shown by the MIF test that cellular sensitivity occurs to many regions of the A1 molecule; yet in guinea pigs, the disease inducing site is located in a very discrete segment of only 9 residues (112-120). The synthetic nonapeptide will itself elicit EAE (4). In monkeys, and perhaps also in humans, the disease-inducing site occupies approximately 20-30 residues of the COOH-terminal region (39). Pepsin cleavage of the Leu-Phe bond, res. 151-152, gives a peptide of 17 residues which is active in monkeys (see Table III).

RELEVANCE TO MS AND GB SYNDROME. One of the major questions confronting researchers on demyelinating diseases is how accurate is the EAE model for multiple sclerosis. While there is quite general acceptance of EAN as a model for GB syndrome, EAE and MS differ considerably in the time-course of the disease and the histologic appearance of the lesion. EAE appears to result from a cellular hypersensitivity response to the injected A1 protein, while MS likely originates from a viral infection (40). Recently, however, we have presented evidence linking immunologic events in MS with those occurring in EAE. That is, that the crucial early events in MS may be mediated by lymphocytes sensitized to the A1 protein just as in EAE. We can summarize our findings as follows:

(1) In studies (41, 42) of over 250 patients (over 100 with MS), it was found that the level of sensitized lymphocytes to the A1 protein circulating in the blood correlated with the severity of acute attack as shown by the macrophage migration inhibition (MIF) test. As shown in Fig. 4, the MIF response was greatest in patients within 3 weeks of exacerbation and progressively less in convalescing and chronic patients. This study demonstrated for the first time the importance of immunologic studies during the proper time course of the disease and reconciles other studies which failed to detect a response (42).

(2) We found (43) that in several patients fortuitously examined prior to a severe exacerbation, sensitization to the A1 protein had developed 1 - 3 weeks prior to attack. This exciting finding suggests that hypersensitization to the A1 protein may be an early instigating event and not simply an epiphenomenon of the disease process. Although sensitized lymphocytes may occur in some subjects suffering from strokes it occurs at least two weeks following an insult, unlike MS (42). Moreover, the degree of the MIF response is much less than in MS (42).

(3) The response of lymphocytes from MS subjects to the COOH-terminal peptide T (a 54 residue peptide (4) obtained by cleavage of the tryptophanyl-COOH peptide linkage with BNPS-skatole) shows the same variation with severity of the disease as to the A1 protein (44). This finding is of significance because peptide T contains the region which induces disease in monkeys (39). The other 116 residue segment, Peptide L, is not particularly active in monkeys or MS lymphocytes as shown by the MIF test (44).

Not only in MS but in GB syndrome, cellular hypersensitivity was shown (45) by response of peripheral blood lymphocytes in the MIF test in the early severe course of the disease (Fig. 5). It is of interest that the MIF response is directed to the P2 protein, and not the A1 protein. These data would also suggest that the A1 protein in human peripheral nerve is masked as it is in rabbit sciatic nerve. This result is fully compatible with our studies in EAN in monkeys in which the responsible antigen appears to be the P2 protein (15). In both EAN and GB syndrome, certain areas of the PNS are especially vulnerable, and the CNS is rarely involved.

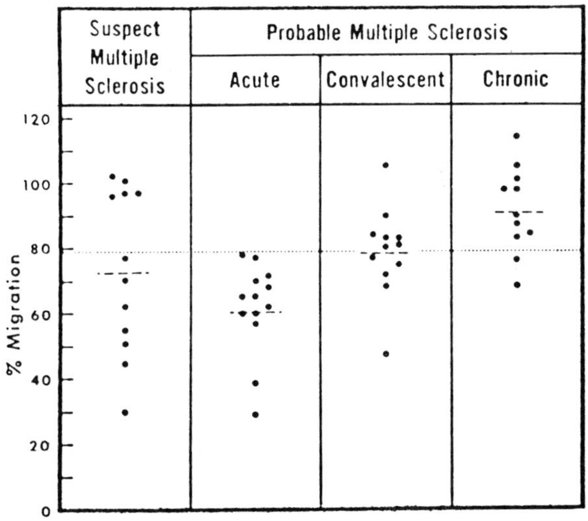

Figure 4: The production of MIF in MS patients with respect to their last exacerbation: acute (3 weeks); convalescent (4-12 weeks); and chronic (after 6 months). They are compared with patients suspected of having MS. Normal controls gave an average MIF value of 97. The line drawn at 79 represents 2 standard deviations below this value.

CONCLUSION. We interpret these data to suggest that in the early stages of MS the etiology may be similar to EAE. That is to say, that periodically cellular hypersensitivity occurs in the COOH region of the A1 protein, the very region active in monkeys (39). The resulting effector T cells then migrate to the CNS, encounter the myelin, activate macrophages which directly perform the demyelinating process and thus elicit the characteristics of acute exacerbation found in MS. Monkeys with EAE generally show clinical and histologic signs similar to early MS subjects. Obviously the induction of pathologic events in MS must differ from those in EAE where myelin or A1 protein is administered directly in Freund's complete adjuvant. It is likely that the agent responsible for MS is a virus, such as the measles virus, and possibly acts as a carrier for the A1 protein. In GB syndrome the

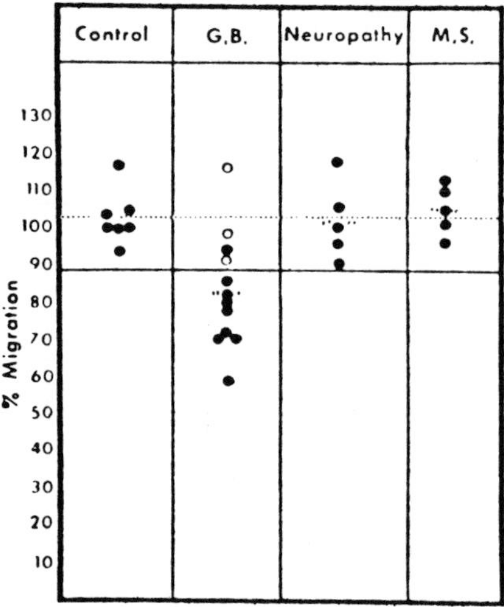


Figure 5: MIF production is shown in GB syndrome and other PNS diseases. Mean migration in each column is indicated by the dashed line. The antigens used were either purified rabbit sciatic nerve P2 protein or homogenized human sciatic nerve. Values less than 2 standard deviations in controls, 86, are considered significant

data are more compelling since this disease is known to follow, after a few weeks,viral infection in 60-70% of the cases. Yet the cellular immune response is directed primarily to the P2 protein, rather than the virus. These data suggest that in GB syndrome, and possibly in MS, the virus initiates the infection and cellular immune response that eventually leads to the autoimmune disease.

Only recently have the immunologic parameters begun to be elucidated in MS. In addition to the cellular hypersensitivity to A1 protein discussed above, there appears to be an immune deficiency in the cellular response to measles and other viruses (46). How a cellular hyper- and hyposensitivity may co-exist is not known, but it may relate to the finding that a very high percentage of MS subjects have the HL-7a and 3a histocompatibility genes (47) compared to normals. Thus the immune parameters in MS may be partially regulated by immune response genes that limit cellular sensitivity to certain antigens (viruses), yet permit a cell-mediated hypersensitive response to the Al protein as part of a viral-Al protein complex. Such events may also occur in GB syndrome. In addition to genetic factors, the main differences between the human and experimental diseases may be due to differences in the mode of induction and the site of the pathology, and not in the immunologic events.

SUPPRESSION OF EAE AND CLINICAL APPLICATION. We have found that in monkeys with unmistakable clinical signs of EAE, the course of the disease could be reversed and suppressed by daily administration of Al protein in oil (48). This result was unexpected in view of previous unsuccessful attempts to reverse the course of severe EAE with immunosuppressive drugs and antilymphocytic serum (10). As shown in Table V, many combinations of Al protein from various species were effective. EAE induced with human Al protein could be suppressed with bovine Al protein or that from other species. It is noteworthy that peptide T (the COOH terminal peptide) was also able to suppress the disease.

The most important factor in the suppressive treatment is the critical period over which the protein must be given. For monkeys it varied from 12-17 days. For example, if the suppressive treatment was stopped at day 10 or so of the critical period, the clinical state of the monkey, although nearly normal, would retrogress and full clinical signs leading to death would appear over a 2 - 5 day period. Once the critical period was exceeded, however, the monkeys showed no signs of EAE, even after several years, and thus appear permanently suppressed (*unless rechallenged with Al protein in appropriate adjuvant).

These results provide a sound basis on which to devise a protocol which could be used for clinical studies with MS subjects. It seems reasonable that if an approach to MS can be made which has the appropriate protocol and rationale and a low risk, then it should be attempted on a modest scale. The rationale is provided by the studies described above showing the cellular hypersensitivity to the same region of the Al molecule active in monkeys, occurring in MS subjects prior to and during acute attack (42, 43). Thus a chemical and immunologic link to the EAE model exists in MS. We believe it is essential to choose MS subjects for study who are in relatively early stages of the disease since monkeys with EAE responded better if treatment was initiated as early as possible after clinical signs were detected.

Table V: Suppression of EAE in Rhesus Monkeys

| Inducing Protein | No. of Animals | Suppressing Protein | No. Dead | No. Recovery |
|---|---|---|---|---|
| Human A1 | 20 | Human A1 | 6 | 14 |
| Human A1 | 6 | Bovine A1 | 1 | 5 |
| Human HNB A1 | 4 | Human HNB A1 | 0 | 4 |
| Monkey A1 | 7 | Monkey A1 | 2 | 5 |
| Monkey A1 | 4 | Human A1 | 1 | 3 |
| Peptide T | 2 | Peptide T | 0 | 2 |
| Monkey A1 | 3 | Peptide T | 1 | 2 |
| TOTAL | 51 | | 12 | 39 |

In the monkeys studied, EAE was induced with 5 mg A1 protein (emulsified in Difco H37 Ra Freund's adjuvant) given in two 0.1 ml injections in the footpad. Peptide T (2 mg) was also used. Suppressive treatment began with i.m. injection of 10 mg A1 protein in saline emulsified in mineral oil (1:1) when the animal first showed clear clinical signs of disease such as limb weakness, paralysis, ataxia, etc. Subsequently, one injection per day of 2 mg A1 protein was given for 16 days. Also penicillin G (100,000 units/day) was given, and in most cases, animals were fed twice daily with 20 cc of AB-dextrose solution.

A high safety factor appears to exist in the use of the highly purified A1 protein since it is only encephalitogenic when given in combination with Freund's adjuvant or used in whole brain homogenat Moreover, a factor (49) appears to exist in serum which inactivates the encephalitogenic properties of the A1 protein. Antibody to the A1 protein does not mediate events in EAE, nor does it demyelinate nerve cultures (50). Antibody to A1 protein may even prove beneficial since animals preimmunized with basic protein are protected from EAE (51).

Whether such a clinical trial as proposed here would be successfu or not cannot be predicted, but it is of interest that a basic biochemical consideration of the membrane proteins of myelin has led to such clinical possibilities in this futile and devastating disease. If a tissue antigen can be shown to alter the course of a human disease therapeutically, then there are other pharmacological applications where this approach might be feasible.

ACKNOWLEDGMENT. Our recent work was supported by the MRC of Canada and was carried out in collaboration with Drs. J. Jackson, S. Brostoff, W. Sheremata, H. Wisniewski, A. Ishaque, W. Roomi.

## REFERENCES


1. Schmitt, F.O., Baer, R., and Palmer, K., J. Cellular Comp. Physiol., 18, 31, 1941.
2. Finear, J.B., Ann. N.Y. Acad. Sci., 122, 51, 1965.
3. Davison, A.N. in Myelination (Davison, A., and Peters, A.), p. 90, Charles Thomas, Springfield, Ill., 1970.
4. Eylar, E.H. in Functional and Structural Proteins of the Nervous System (Davison, A., Mandel, P., and Morgan, I., eds.) p. 215-239, Plenum Press, New York, 1972.
5. Suzuki, K., Poduslo, S., and Norton, W.T., Biochim. Biophys. Acta, 144, 375, 1967.
6. Quarles, R., Everly, J. and Brady, R., J. Neurochem, 21, 1177, 1973.
7. Greenfield, S., Brostoff, S., Eylar, E.H., and Morell, P., J. Neurochem, 20, 1207, 1973.
8. Singer, S.J., Ann. N.Y. Sci. 195, 16, 1972.
9. O'Brien, H., Science, 147, 1099, 1965.
10. Eylar, E.H. in Multiple Sclerosis (Wolfgram, F., Ellison, G., Stevens, J., and Andrews, J., eds), p.449-481, Acad. Press, New York, 1972).
11. Ishaque, A., Roomi, W., and Eylar, E.H., to be published.
12. Agrawal, H.C., Burton, R., Fishman, M., Mitchell, R., and Prensky, A., J. Neurochem, 19, 2083, 1972.
13. Steck, T., J. Cell. Biol. 62, 1, 1974.
14. Brostoff, S., and Eylar, E.H., Arch. Biochem. Biophys, 153, 590, 1972.
15. Eylar, E.H., Salk, J., Beveridge, G., and Brown, L., Arch. Biochem. Biophys, 132, 34, 1969.
16. Folch-Pi, J., and Stoffyn, P., Ann. N.Y. Acad. Sci. 195, 86, 1972.
17. Moscarello, M., Gagnon, J., Wood, D., Anthony, J., and Epand, R., Biochemistry, 12, 3402, 1973.
18. Everly, J., Brady, R.O. and Quarles, R., J. Neurochem, 21, 329, 1973.
19. Wood, J., and Dawson, R.M.C., J. Neurochem., 21, 717, 1973.
20. Brostoff, S., Burnett, P., Lampert, P., and Eylar, E.H., Nature new Biol, 235, 210, 1972.
21. Brostoff, S., Karkhanis, W., Carlo, D., Reuter, W., and Eylar, E.H., Brain Res., 86, 449, 1975.
22. Brostoff, S., and Eylar, E.H., Proc. Nat. Acad. Sci. U.S.A., 68, 765, 1971.
23. Epand, R., Moscarello, M., Zierenberg, B., and Vail, W., Biochemistry, 13, 1264, 1974.
24. Casper, D., and Kirschner, D., Nature new Biol., 231, 46, 1971.
25. London, Y., Demel, R, R., Kessel, G., Vossenberg, F., and van Deenen, L., Biochim. Biophys. Acta, 311, 520, 1973.
26. Papahadjopoulos, D., Moscarello, M., Eylar, E.H., and Isac, T., Biochim. Biophys. Acta, 401, 317, 1975.

26a. Palmer, F., and Dawson, R., Biochem. J., 111, 629, 1969.
27. Eylar, E.H., Brostoff, S., Hashim, G., Caccam, J., and Burnett, P., J. Biol. Chem., 246, 5770, 1971.
28. Rauch, H., and Einstein, E.R., Rev. Neuroscience, I, 283, 1974.
29. Capaldi, R., and Vanderkooi, G., Proc. Nat. Acad. Sci. 69, 930, 1972.
30. Mateu, L., Luzzati, W., London, W., Gould, R., Vossenberg, F.G.A., and Olive, J., J. Mol. Biol., 75, 697, 1973.
31. Da Silva, P., and Miller, R., Proc. Nat. Acad. Sci. U.S.A., 72, 4046, 1976.
32. Wood. D., Vail, W., amd Moscarello, M., Brain Res. 93, 463, 1975.
33. Demel, R., London Y., van Kessel, G., Vossenberg, F., and van Deenen, L., Biochim. Biophys. Acta, 311, 507, 1973.
34. Laatsch, R., Kies, M., Gordon, S., and Alvord, E.C., J. Exp. Med., 115, 778, 1962.
35. Wisniewski, H., Brostoff, S., Carter, H., and Eylar, E.H., Arch. Neurol., 30, 347, 1974.
36. Arnason, B., Asbury, A., Astrom, K., and Adams, R., Trans. Am. Neurol. Assoc. 93, 133, 1968.
37. Brostoff, S., Sacks, H., Del Canto, M., Johnson, A., Raine, C., and Wisniewski, H., J. Neurochem, 23, 1037, 1974.
38. Bergstrand, H., Eur. J. Immunol., 2, 266, 1972.
39. Karkhanis, Y., Carlo, D., Brostoff, S., and Eylar, E.H., J. Biol. Chem., 250, 1718, 1975.
40. News and Views, Nature, 260, 190, 1976.
41. Sheremata, W., Cosgrove, J., and Eylar, E.H., N. Eng. J. Med., 291, 1417, 1974.
42. Sheremata, W., Cosgrove, J., and Eylar, E.H., J. Neuro. Sc., in press.
43. Sheremata, W., Cosgrove, J., and Eylar, E.H., Trans. Am. Neurol. Assoc. 99, 49, 1974.
44. Sheremata, W., and Eylar, E.H., 5th Int. Cong. Neurochem, Barcelona, 1975.
45. Sheremata, W., Colby, S., Karkhanis, Y., and Eylar, E.H., Can. J. Neurol. Sci., 2, 87, 1975.
46. Utermohlen, W., and Zabriskie, Lancet, 2, 1147, 1973.
47. Jersild, C., Dupont, B., Fog, T., Hansen, G., Neilsen, L., Thomsen, M., and Svejgaard, A., Transplant. Proc. V, 1791, 1973.
48. Eylar, E.H., Jackson, J., Rothenberg, B., and Brostoff, S., Nature, 236, 74, 1972.
49. Bernard, C., and Lamoureux, G., Cell Immunol, 16, 182, 1975.
50. Seil, F., Rauch, H., Einstein, E.R., and Hamilton, A., J. Immunol., 111, 96, 1973.
51. Alvord, E., Shaw, C., Hruby, S., and Kies, M., Ann. N.Y. Acad. Sci. 122, 333, 1965.

# REGULATION OF PANCREATIC PHOSPHOLIPASE $A_2$ ACTIVITY BY DIFFERENT LIPID-WATER INTERFACES

M.C.E. van Dam-Mieras, A.J. Slotboom[+], H.M. Verheij, R. Verger[*] and G.H. de Haas

Laboratory of Biochemistry
State University of Utrecht, Transitorium 3
University Centre "De Uithof"
Padualaan 8
Utrecht
The Netherlands

Summary

Pancreatic phospholipase $A_2$ interacts with lipid-water interfaces by means of a specific region, the Interface Recognition Site (IRS), which most probably penetrates to a certain extent into the hydrophobic interior of the lipid phase. This process causes a dramatic increase in the rate of hydrolysis. The IRS embraces at least the rather apolar N-terminal sequence of the polypeptide chain: Ala. Leu. Trp. Gln. Phe. Arg. Ser. Met and its most effective configuration seems to be stabilised by an ionpair between the $\alpha$-$\overset{+}{N}H_3$ group of the N-terminal amino acid Ala and a buried carboxylate function.

[*] Centre de Biochimie et de Biologie Moleculaire, 31 Chemin J. Aiguier, 13009 Marseille, Cédex 2, France.

[+] To whom correspondence should be addressed.

Using a series of specifically modified phospholipases, in which the native N-terminal amino acid L-Ala has been deleted or substituted by other amino acids, the properties of the IRS were compared by spectroscopic techniques and monolayer kinetics.
Substitution of L-Ala by Gly or β-Ala does not seriously impede the penetrating properties of the enzyme. Chain elongation, chain shortening or even replacement of L-Ala by D-Ala, however, weakens the IRS in such a manner, that penetration of the relatively close-packed micelles becomes impossible.
These enzymes still interact with monomolecular surface films up to well-defined surface pressures.

## Introduction

Certain proteins display enzymatic activity only if they are anchored in well-defined lipid-water interfaces. Well-known examples are the membrane-bound enzymes such as the sarcoplasmic reticulum ATPase and the mitochondrial β-hydroxy butyrate dehydrogenase. In these cases it can be expected, that small structural changes in the lipid molecule will modify the physicochemical parameters of the lipid-water interface. If such changes are transmitted to the anchored enzyme molecule and influence the kinetic properties of the latter, one can state, that membrane-bound enzymes are under allosteric control by certain physico-chemical parameters of the interface. Up to now the study of these interesting systems is seriously impeded by difficulties in the purification. Very few techniques are available to isolate intact lipoprotein complexes in a pure state. Delipidation followed by purification of the apoprotein in the presence of detergents, is often accompanied by irreversible loss of enzyme activity and polymerisation of the protein upon removal of the detergent. Therefore model systems are required and in particular studies on immobilized enzymes and lipolytic enzymes appear to be promising. In this paper we will discuss how studies of lipolytic enzymes, acting on water-insoluble substrates, may contribute to our knowledge of lipid-protein interaction. Although these water-soluble enzymes do hydrolyze monomeric substrates in a homogeneous system, it is known, that their catalytic activity dramatically increases upon interaction with certain lipid-water interfaces.

Most probably part of the enzyme molecule will penetrate into the lipid-water interface and intriguing questions are:

- Which part of the protein is involved?
- How far does the enzyme penetrate into the hydrophobic core?
- What is the three-dimensional structure of the penetrating moiety?
- How does the architecture of the lipid packing influence penetration and enzyme activity?

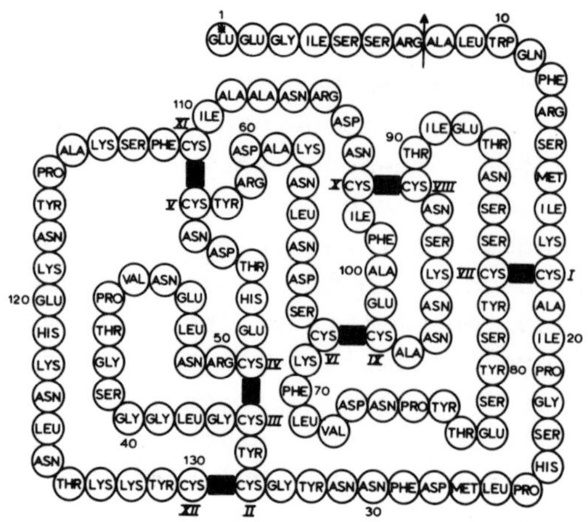


FIGURE 1. Primary structure of porcine pancreatic prophospholipase $A_2$. *stands for pyroglutamic acid. The sequence $His^{53}.Thr^{54}$ should read $Thr^{53}.His^{54}$ (Verheij et al., to be published). AMPREC refers to fully $\varepsilon$-amidinated prophospholipase $A_2$, AMPA refers to $\varepsilon$-amidinated phospholipase $A_2$.

The present study is based on porcine pancreatic phospholipase $A_2$ (EC 3.1.1.4), a single-chain compact molecule consisting of 123 amino acids, cross-linked by disulfide bridges(1,2). The protein catalyzes the specific hydrolysis of fatty acid ester bonds at the 2-position of 3-sn-phosphoglycerides(3).
The enzyme is produced by the pancreas as a zymogen which differs from the active phospholipase by the presence of an additional heptapeptide covalently linked to the N-terminal amino acid $Ala^8$ (Cf. Figure 1). The X-ray structure (resolution 3 Å) of the zymogen has been elucidated recently by Drenth and colleagues (to be published).

Both the zymogen and active phospholipase A are able to hydrolyze monomeric substrates with a low, but similar efficiency, indicating that the active site of the enzyme preexists in the zymogen. The fact that suitable lipid-water aggregates are degraded by the active phospholipase up to 3-4 orders of magnitude more rapidly than monomers, whereas the

zymogen is completely unable to attack such organised structures, has been interpreted by the presence of a so-called Interface Recognition Site (IRS) (4,5).
This is a presumably three-dimensionally structured region on the surface of the active phospholipase which penetrates into the lipid-water interface and which is not present in the zymogen . The presumably hydrophobic region is thought to interact specifically with certain lipid-water interfaces with a concomitant optimization of the active site architecture.*) It has been shown by different techniques (5,6) that this site is not only functionally, but also topographically distinct from the classical active site, which is present in both proteins.As shown before(7), during the limited tryptic hydrolysis of the $Arg^7$-$Ala^8$ bond which transforms the zymogen into the active enzyme, a conformational change takes place in which the newly formed protonated α-aminogroup of the N-terminal alanine forms a salt bridge with a buried negatively charged side chain. The essential role of this salt bridge in stabilising the structure of a functionally active IRS has been discussed(8). Recent spectroscopic studies provided evidence for a direct involvement of the N-terminal sequence Ala.Leu.Trp.Gln.Phe.Arg.Ser.Met in the IRS(8,10).Furthermore it was shown that $Ca^{2+}$ ions are not only needed for catalysis but also to assist the enzyme in penetrating the organised lipid-water interface at alkaline pH. Most probably even at alkaline pH where the $\alpha$-$\overset{+}{N}H_3$ group becomes deprotonated, a functionally active IRS may still be stabilised by the binding of additional $Ca^{2+}$ ions.
In order to delineate further the role of the N-terminal sequence of the enzyme in the recognition process of organized lipid-water interfaces, chemical modifications in this part of the protein molecule have been performed(11) . For this purpose fully $\varepsilon$-amidinated phospholipase $A_2$ (AMPA), which still exhibits 70% of the maximal velocity of the native enzyme,

*) Various other hypotheses to explain the activation of lipolytic enzymes by interfaces have been discussed recently(9).

was used to prepare N-terminally modified enzyme analogs. In the present paper the effects of several chemical modifications which influence only slightly the length of the polypeptide chain ( [Gly$^8$]-AMPA, des-Ala$^8$ -AMPA and [Ala$^7$] -AMPA) or merely change the stereochemical configuration of the N-terminal amino acid ( [D-Ala$^8$] -AMPA), have been studied by spectroscopic techniques and monolayer kinetics. In addition the influence of differently structured lipid-water interfaces on the penetration process will be discussed.

## Results and Discussion

Up to now the important role of the N-terminal amino acid in pancreatic phospholipase $A_2$ in stabilizing the interface recognition site has been related to the presence of a salt bridge between the $\alpha$-amino function and a negatively charged carboxylate group(7,8). The relatively high pK value of the N-terminal L-alanine (8.3 at 25$^{\circ}$C) suggest a rather apolar environment of this saltbridge. This pK value was determined earlier by potentiometric titration (Janssen *et al.* (12). pK determinations of the N-terminally modified phospholipases, however, require other techniques which consume less protein.

Figure 2 shows ultraviolet difference spectra between equimolar solutions of AMPA and AMPREC as a function of pH. Upon lowering the pH of these solutions from 9.0 to 7.0 difference peaks of increasing intensity at 297 and 275 nm and an isosbestic point at 290 nm are observed. Plotting $\Delta$ absorbance as a function of pH (inset Figure 2) gives titration curves for a group with pK 8.5. Taking into account the main molecular difference between AMPA and AMPREC (free $\alpha$-amino group in AMPA and a blocked $\alpha$-amino group in AMPREC), it is concluded that the ultraviolet difference spectrum is caused by protonation of the $\alpha$-amino function in AMPA. Most probably the ion-pair formation which follows, induces a conformational change in the protein perturbing mainly the microenvironment of the unique tryptophan residue.

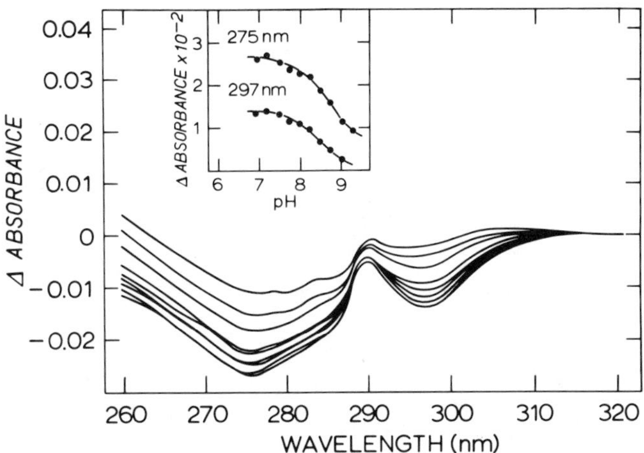


Figure 2: Ultraviolet difference spectra of equimolar solutions of AMPA and AMPREC at pH 9.03, 8.75, 8.50, 8.25, 8.01, 7.75, 7.50, 7.20 and 6.95. Conditions: 10 mM sodium cacodylate, 10 mM borate, 10 mM $CaCl_2$, and 56.6 μM protein, at 25°C. Inset: titration curves obtained by plotting Δ absorbances at 297 and 275 nm obtained from ultraviolet difference spectroscopy versus pH.

As was demonstrated previously(8) also the interaction of phospholipase $A_2$ with micellar lipid-water interfaces can be studied by ultraviolet difference spectroscopy. In Figure 3 the ultraviolet difference spectra obtained upon interaction of various N-terminally modified phospholipases with micelles of the substrate analog n-hexadecylphosphorylcholine are shown. It is clear, that of all modified proteins only the [$Gly^8$]-AMPA analog interacts like AMPA with the lipid-water interface. In both complexes the single tryptophan residue is probably buried in a rather hydrophobic environment(8). Also in a more quantitative way no differences could be detected between both proteins in the interaction process with micelles. Titrating increasing amounts of lipid into solutions of AMPA and [$Gly^8$]-AMPA yields difference spectra with saturation characteristics from which the same apparent dissociation constant $K_D$ = 0.55 mM (pH 6.0 and 25°C) is obtained. These results are in agreement with the previously reported similar kinetic properties of AMPA and [$Gly^8$]-AMPA(11).

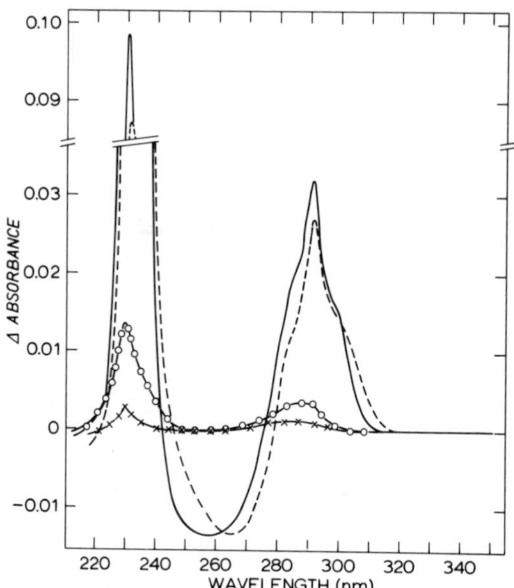


Figure 3: Ultraviolet difference spectra produced by the interaction of equimolar solutions of AMPA (——), [$Gly^8$] -AMPA (– – –), [D-$Ala^8$] -AMPA (o—o—o), des-$Ala^8$-AMPA (x—x—x), [$Ala^7$] -AMPA (x—x—x) and AMPREC(x—x—x) with n-hexadecylphosphorylcholine. Conditions: 0.05 M sodium acetate, 0.1 M NaCl, 0.05 M $CaCl_2$, 15.9 µM protein and 1.54 mM n-hexadecylphosphorylcholine; pH 6.0 and 25°C.

On the other hand, it is clear, that AMPREC and the N-terminally modified proteins, [D-$Ala^8$]- , [$Ala^7$]- and des-$Ala^8$-AMPA, give only negligible difference spectra upon addition of lipid micelles. Apparently these proteins are not able to penetrate into this lipid-water interface which confirms the kinetic experiments demonstrating that they do not display interfacial enzyme activity. It is concluded, that with the exception of the substitution of $Ala^8$ by $Gly^8$ , even minor changes in the N-terminal amino acid of phospholipase $A_2$, which have only a limited influence on the active site (11), eliminate the penetrating power of the protein into certain lipid-water interfaces.

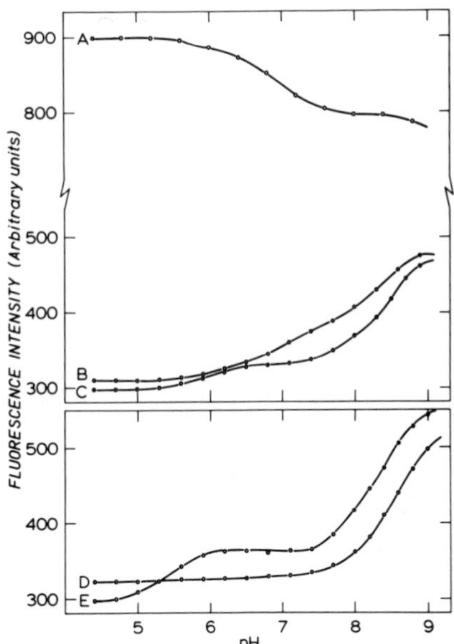


Figure 4: pH dependence of fluorescence intensities of AMPA and AMPREC. A = AMPREC; B = AMPA; C = AMPA in the presence of 10 mM $CaCl_2$; D = [$^{14}$C] -1-bromo-2-octanon inhibited AMPA and E = AMPA in the presence of 10 mM $CaCl_2$ and 0.27 mM n-dodecylphosphorylcholine. Conditions: 1 mM Tris, 0.1 M NaCl, 13.76 μM protein; excitation: 295 nm, intensities were measured at 347 nm for AMPREC and 342 nm for AMPA, (25°C).

The supposed involvement of the single tryptophan residue in the interaction process of phospholipase $A_2$ with micelles makes fluorescence spectroscopy a very attractive technique. Figure 4 shows the fluorescent properties of AMPA and AMPREC as a function of pH in the absence of lipid-water interfaces. In agreement with previous results(7) at slightly acidic or neutral pH the quantum yield of the zymogen (curve A) is considerably higher than that of the active enzyme (curve B) indicating a different microenvironment of tryptophan in both proteins. It is very dangerous, however, to draw any further conclusions about the relative location of tryptophan in both proteins from fluorescence intensity measurements only. The variation of the fluorescence intensity of AMPA as a function of pH (curve B) strongly indicates that deprotonation of one

or more amino acid side chains abolishes or diminishes the quenching of tryptophan fluorescence by neutralizing positive charges in its environment. Moreover the nearly continuous increase of the fluorescence intensity between pH 6 and 9 indicates that at least two residues are involved having pK values of about 7 and 8. Obvious candidates for amino acid side chains possessing such pK values are histidine and the α-amino function, respectively. Because of the fact, that a similar fluorescence-pH profile is found for the horse pancreatic phospholipase $A_2$ (unpublished experiments), an isoenzyme containing only one histidine residue, namely the active site histidine (13), we tentatively attribute the increase in fluorescence intensity in AMPA going from pH 6 to 9 to the successive deprotonation of histidine-54 and of the N-terminal alanine-8 both of which are probably located close to tryptophan-10. The definite identification of the higher pK side chain with the N-terminal amino acid was obtained by specific acetylation of the α-amino group. Such a protein containing a blocked α-amino group does not show an increase in fluorescence intensity between pH 7 and 9. (7).
In order to obtain stronger support for the assignment of the lower pK side chain, the fluorescence-pH course of AMPA was measured also in the presence of $Ca^{2+}$ ions. It is known from previous (14,15) work, that this essential cofactor binds in a 1:1 molar ratio to the enzyme, close to histidine-54 and might therefore change the pK of the histidine side chain. From Figure 4, curve C, it seems evident indeed that the pK of one of the groups whose protonation state influences the fluorescence intensity, decreases from about 6.6 to 6.1. Moreover, upon addition of $Ca^{2+}$ ions and monomeric substrate analog, which is known (14) to bind also in the proximity of histidine-54 the pK of this group descends even to pH 5.4 (curve E). *)

*) The lowering of the pK of histidine-54 from 6.6 to 5.4 in the presence of $Ca^{2+}$ ions and monomeric substrate is compatible with the previously reported $V_{max}$-pH profile(16).

Finally the identification of this group with histidine-54 was obtained after specifically modifying its side chain with the haloketone [$^{14}$C] -1-bromo-2-octanone (unpublished experiments, Cf.Volwerk et al.(14)). The fluorescence-pH profile of this inhibited AMPA is given as curve D (Figure 4) and shows that the fluorescence intensity of the histidine-54 blocked AMPA is determined by the protonation state of only one residue namely the α-amino group of the N-terminus.

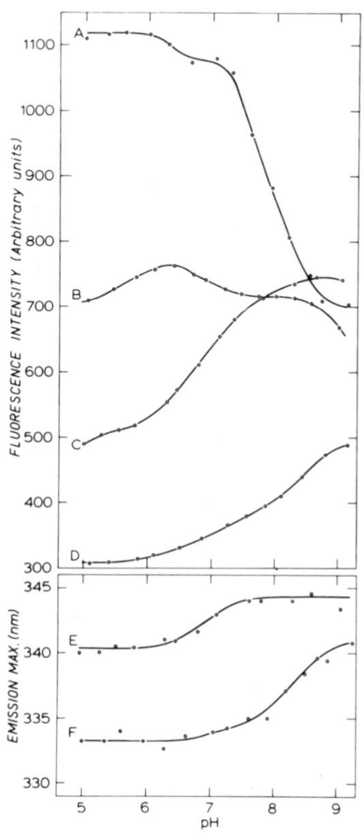


Figure 5: Effect of pH on fluorescence intensities and emission maxima of AMPA and [Gly$^8$] -AMPA in the presence and absence of n-hexadecylphosphorylcholine. A and F=AMPA in the presence of 9.4 mM n-hexadecylphosphorylcholine; B = [Gly$^8$]-AMPA; C and E = [Gly$^8$] -AMPA in the presence of 9.4 mM n-hexadecylphosphorylcholine; D = AMPA. Conditions:0.01 M sodium acetate, 0.01 M Tris, 0.1 M NaCl, 13,76 μM protein; excitation: 295 nm, and intensities were measured at the respective maximal emission wavelengths, (25°C).

Up to now we have not been able to attribute the decrease in fluorescence intensity of the zymogen (curve A) between pH 6 and 8 to the deprotonation of (a) well-defined side chain(s). Figure 5 compares the pH dependence of the fluorescence intensities and emission maxima of AMPA and [Gly$^8$]-AMPA both in the presence and absence of a micellar lipid-water interface. Comparison of curves B and D in Figure 5 shows that in the absence of lipids the [Gly$^8$]-AMPA has a considerable higher quantum yield than AMPA and only at very alkaline pH values the fluorescence intensities become similar. Moreover also from the pH dependence of curve B it is evident that [Gly$^8$]-AMPA resembles very much the behaviour of the zymogen AMPREC (Figure 4, curve A). This indicates a similar environment of tryptophan and consequently of the N-terminus in both proteins and therefore the absence of an ion-pair in [Gly$^8$]-AMPA. How to reconcile this fact with the kinetic (11) and ultraviolet spectroscopic results (Cf. Fig. 3) which clearly showed a very similar behaviour of AMPA and [Gly$^8$]-AMPA?

Comparison of curves A and D (Figure 5) shows that addition of micelles to AMPA at pH 5 gives a 350% increase in fluorescence intensity. At the same time the maximum emission wavelength of the fluorescence drops from about 342 to 333 nm. Neither the increase in quantum yield nor the blue shift of the emission maximum was observed upon addition of micelles to the zymogen AMPREC(Cf. Figure 6). These observations were explained earlier(8) by the presumed interaction of the N-terminal part of phospholipase $A_2$ with the lipid-water interface, a process in which the single tryptophan residue gets shielded from the aqueous environment and becomes buried in the more hydrophobic environment of the lipid-water interface. Upon increase of the pH above 7 the α-amino group of the N-terminus looses its proton and the IRS disappears. The enzyme leaves the lipid-water interface and the quantum yield of the fluorescence approaches that of the lipid free system (curve A, Figure 5). In the absence of $Ca^{2+}$ ions this process is governed by the protonation state of an amino acid residue having a pK of about 8.3, and which is identical to the N-terminal alanine (curve F). If we compare, however, curves B and C, it is clear that addition of micelles to [$Gly^8$]-AMPA at pH 5 results in a 30% decrease of the fluorescence intensity. Concomitantly the emission maximum undergoes a blue shift from about 346 to 340 nm. From the kinetics (11) and the ultraviolet difference spectroscopy (Cf. Figure 3) we know that a productive enzyme-micelle complex is formed and we can understand why an increase in pH up to pH 9.0 finally yields a rather similar quantum yield (curves B and C). Apparently by raising the pH we deprotonate a group essential for the interaction with lipid-water interfaces. Curve E, which depicts the fluorescence emission maximum of the [$Gly^8$]-AMPA as well as curve C point to a pK of about 7. In analogy with AMPA, it seems rather probable that this pK belongs to the α-amino group of the N-terminal glycine-8.

Summarizing we can state that in the absence of micelles the tryptophan residue in [$Gly^8$]-AMPA is located in a similar position as in the zymogen. Therefore it seems very likely that [$Gly^8$]-AMPA, in contrast to AMPA, does not possess a salt bridge between the α-amino group and a buried carboxylate group. Upon addition of micelles, however, [$Gly^8$]-AMPA does not resemble anylonger AMPREC, but much more AMPA as is evident from its ability to interact with lipid-water interfaces and its high enzymatic activity toward micellar substrate. On the other hand it is quite obvious from the differences in the pK values of the α-amino groups, the emission maxima and the intensities of AMPA and [$Gly^8$]-AMPA in the presence of micellar lipid , that the microenvironment of tryptophan and the α-amino group in both enzymes is different. One may wonder whether the difference in the interaction of AMPA and [$Gly^8$]-AMPA with micelles has to be ascribed to the absence of an internal salt bridge in [$Gly^8$]-AMPA.

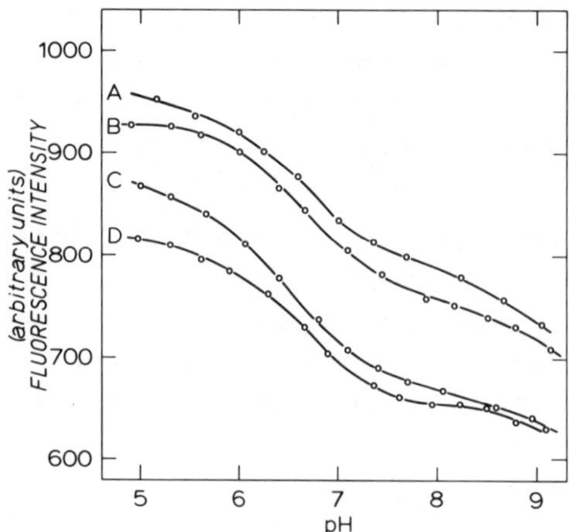


Figure 6: Effect of pH on fluorescence intensities of AMPREC and [D-Ala$^{8}$]-AMPA in the presence and absence of n-hexadecylphosphorylcholine. A = [D-Ala$^{8}$] -AMPA in the presence of 9.4 mM n-hexadecylphosphorylcholine; B = [D-Ala$^{8}$] -AMPA; C = AMPREC in the presence of 9.4 mM n-hexadecylphosphorylcholine; D = AMPREC. Protein: 12.3 μM. Conditions: see Figure 5.

In Figure 6 the fluorescence intensity - pH dependence is shown for AMPREC and [D-Ala$^{8}$]-AMPA, both in the presence and absence of micelles. It is evident that without lipids the D-analog strongly resembles the zymogen. Both proteins are characterized by a rather high quantum yield and a similar emission maximum (346 nm). Addition of lipids gives in both cases only a very small increase in fluorescence intensity and the emission maximum does not shift to lower wavelengths. In agreement with the ultraviolet spectroscopic results (Cf.Figure 3) and the kinetic properties(11) it is concluded, that these proteins are not able to interact with micellar lipid-water interfaces. Apparently the minor change of [L-Ala$^{8}$] into [D-Ala$^{8}$] prevents the salt bridge formation.

Taking together the information obtained by kinetic measurements, ultraviolet difference spectroscopy and fluorimetry, one would be inclined to divide the series of N-terminally modified phospholipases in two classes:

A: proteins with an IRS and showing high interfacial enzyme activity, such as AMPA, [$Gly^8$]-AMPA and [$\beta$-$Ala^8$] -AMPA.
B: proteins lacking an IRS which are unable to penetrate micellar interfaces, such as AMPREC, [D-$Ala^8$] -AMPA, [$Ala^7$]-AMPA, des-$Ala^8$- AMPA and $\alpha$-amino blocked AMPA's .

On the other hand there are indications that such an "all or non" classification is too rigid and dependent on the particular lipid-water interface investigated. For example the class A enzymes [$Gly^8$] -AMPA and [$\beta$-$Ala^8$]-AMPA penetrate, it is true, without difficulties in the loosely packed micelles of short-chain lecithins and no time effects are observed. However, in the egg-yolk assay(17) , these latter enzymes show unusual long lag periods(4) such in contrast to AMPA and full interface activity develops only after several minutes. This might be interpreted by assuming that these proteins do have an IRS, but a less effective one than in AMPA. One may wonder whether the proteins of class B are really devoid of such an IRS or possess a much less effective site allowing them to penetrate in certain lipid-water interfaces and not in others. To test this hypothesis, kinetic experiments were performed with [D-$Ala^8$] -AMPA using different types of micellar lipid-water interfaces. It turned out that neither micelles of 3-sn-dioctanoyl-lecithin nor those of 3-sn-diheptanoyl- and 3-sn-dihexanoyl-lecithins constituted a favourable architecture allowing the protein to penetrate. It is evident, however, that our possibilities to offer different micellar substrates are very limited because the substrate density in the surface is fixed. A much more attractive system which offers the possibility to change continuously the lipid packing, is the monolayer technique(4). Using 3-sn-dinonanoyl lecithin films at the air-water interface at a surface pressure of 8 dynes/cm the following relative *) rates of hydrolysis were obtained at pH 6.0:

*) It should be realized that the velocities measured by monolayer technique are not $V_{max}$ values(9).

| | | | |
|---|---|---|---|
| AMPA | 100% | AMPREC | 0% |
| [D-Ala⁸]-AMPA | 60% | $\alpha$-$NH_2$-blocked AMPA | 0% |
| [L-Ala⁷]-AMPA | 30% | | |
| des-Ala⁸-AMPA | 15% | | |

The most striking fact is the high interface activity of [D-Ala$^8$]-AMPA whereas the zymogen AMPREC and $\alpha$-amino blocked AMPA derivatives, even on monolayers of still lower surface pressure, is completely inactive. Therefore it is evident that part of the modified enzymes classified above in class B, should in fact be grouped in class A. Although they contain no internal salt bridge they do possess an IRS, but this site has a very weak penetrating capacity.

A possibility to measure the "penetrating power" of the IRS is shown in Figure 7. As demonstrated previously(4), the lag times observed in the hydrolysis of lecithin films at the air-water interface by pancreatic phospholipase $A_2$ are related to a slow reversible penetration of the enzyme into the monolayer.

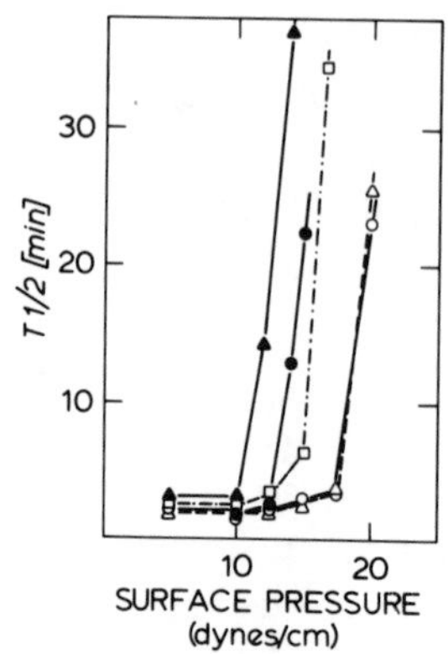


Figure 7: Influence of the surface pressure of a 3-sn-dinonanoyl lecithin on the half-time value of induction (T 1/2).

○—○ = AMPA; △– –△ = [Gly$^8$] -AMPA; □—·—□ = [D-Ala$^8$]-AMPA; ●—● = [Ala $^7$] -AMPA; ▲—▲ = des-Ala$^8$-AMPA.

Conditions: 10 mM Tris-acetate, 0.1 M NaCl, 20 mM $CaCl_2$, pH 6.0, 25°C.

Therefore a plot of the half time of induction (T 1/2) *versus* surface pressure ($\pi$) will be indicative for the penetrating power of the particular enzyme investigated (18). From Figure 7 it is evident that AMPA and [Gly$^8$] -AMPA possess similar penetrating capacities and will be able to hydrolyze lipid-water interfaces having a surface pressure up to 18 dynes/cm. The [D-Ala$^8$] -AMPA remains active up to about 14 dynes/cm, whereas [L-Ala$^7$] -AMPA and des-Ala$^8$-AMPA become already inactive at surface pressures higher than 13 and 11 dynes/cm, respectively.

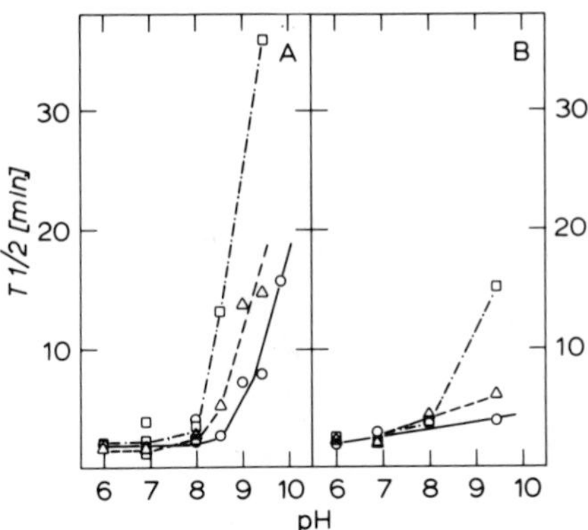


Figure 8: Influence of the pH of the subphase on the half-time value of induction (T 1/2) during the hydrolysis of a 3-*sn*-dinonanoyl lecithin film. ○—○ = AMPA, surface pressure: 12 dynes/cm.; △- -△ = [Gly$^8$] -AMPA, surface pressure: 12 dynes/cm; □-·-□ = [D-Ala$^8$] -AMPA, surface pressure: 10 dynes/cm. Conditions: 10 mM Tris-acetate, 0.1 M NaCl. A: in the presence of 0.5 mM $CaCl_2$; B: in the presence of 20 mM $CaCl_2$. (25°C).

The relation between the "penetrating capacity" and the pK of the $\alpha$-amino group of the N-terminal amino acid is visualized for a few proteins in Figure 8. Whereas AMPA in the presence of low $[Ca^{2+}]$ (Figure 8A) is able to penetrate a certain lipid-water interface ($\pi$ = 10-12 dynes/cm) up until pH 8.5, it is clear that $[Gly^8]$ -AMPA and in particular $[D\text{-}Ala^8]$ -AMPA experience difficulties to interact at this pH and this $Ca^{2+}$ concentration. The latter two enzymes which are presumably devoid of the internal salt bridge loose their proton from the $\alpha$-ammonium group at a lower pH than AMPA (Cf. Figure 5). $Ca^{2+}$ ions are known to increase the apparent pK of the $\alpha$-amino function in the presence of lipid-water interfaces (8). Figure 8B shows that also the modified protein $[Gly^8]$ -AMPA and to a lesser extent $[D\text{-}Ala^8]$ -AMPA are able to interact with lipid-water interfaces up until higher pH values if a high $Ca^{2+}$ concentration is present.

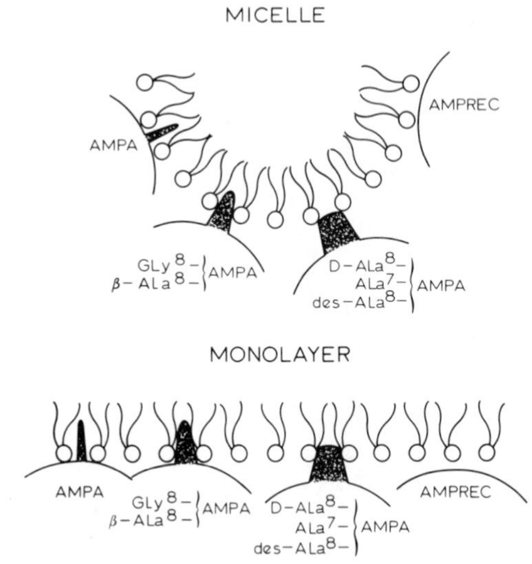


Figure 9: Schematic presentation of the penetration of different lipid-water interfaces by various N-terminally modified pancreatic phospholipases $A_2$.

Figure 9 summarizes in a highly schematical way the above results. Neutral, zwitterionic lipids such as the synthetic short-chain lecithins or the n-alkylphosphorylcholines form micelles in which the "native"-enzyme AMPA and its analogs [Gly$^8$] - and [β-Ala$^8$] -AMPA readily penetrate. The other N-terminally modified phospholipases are completely devoid of an IRS (AMPREC) or possess a recognition site of low penetrating power ([D-Ala$^8$]-; Des-Ala$^8$-; [Ala$^7$] -AMPA) and they are unable to interact with the rather densely packed micelles. Upon increasing the intermolecular distance of the zwitterionic lipid molecules, as can be done with monomolecular surface films, these latter proteins become also able to penetrate the interface. From figure 7 it might be concluded, that the micellar interfaces investigated are characterized by surface pressures between about 14 and 17 dynes/cm.
Up to now the existence of an IRS in pancreatic phospholipase $A_2$ has been connected with an ion-pair between the $\alpha$-$\overset{+}{N}H_3$ group and a negatively charged side chain. Although the presence of such a salt bridge undoubtedly reinforces the penetrating power of the enzyme, the results of the present study demonstrate that also proteins which are probably devoid of a saltbridge ( [Gly$^8$] -AMPA; [D-Ala$^8$] -AMPA) may possess a weak penetrating power. This is in agreement with our previous findings that pancreatic phospholipase A can display high interface activity even at pH values above 9 where presumably the $\alpha$-$\overset{+}{N}H_3$ group is deprotonated (8). However, penetration of micellar lecithins at alkaline pH specifically requires the presence of $Ca^{2+}$ , suggesting that a functionally active IRS may also be induced by certain lipid-water interfaces in the presence of $Ca^{2+}$ ions.

## Acknowledgments

The authors are grateful to Drs. P. Bruijnzeel for his valuable contributions to the fluorimetric studies and to Mrs. P.H.M. Baartmans for her collaboration in the monolayer experiments.
Thanks are due to Mrs. L.C. Mey-Brants and Mr. R. Dijkman for their excellent technical assistance.

## References

1) G.H. de Haas, A.J. Slotboom, P.P.M. Bonsen, L.L.M. van Deenen, S. Maroux, A. Puigserver and P. Desnuelle, Biochim. Biophys. Acta, 221 (1970) 31.

2) G.H. de Haas, A.J. Slotboom, P.P.M. Bonsen, W. Nieuwenhuizen, L.L.M. van Deenen, S. Maroux, V. Dlouha, P. Desnuelle, Biochim. Biophys. Acta, 221 (1970) 54.

3) G.H. de Haas, N. Postema, W. Nieuwenhuizen and L.L.M. van Deenen, Biochim. Biophys. Acta, 159 (1968) 103.

4) R. Verger, M.C.E. Mieras and G.H. de Haas, J. Biol. Chem., 248 (1973) 4023.

5) W.A. Pieterson, J.C. Vidal, J.J. Volwerk and G.H. de Haas Biochemistry, 13 (1974) 1455.

6) R.D. Hershberg, G.H. Reed, A.J. Slotboom and G.H. de Haas, Biochemistry, 15 (1976), accepted for publication.

7) J.P. Abita, M. Lazdunski, P.P.M. Bonsen, W.A. Pieterson and G.H. de Haas, Eur. J. Biochem., 30 (1972) 37.

8) M.C.E. van Dam-Mieras, A.J. Slotboom, W.A. Pieterson and G.H. de Haas, Biochemistry, 14 (1975) 5387.

9) R. Verger and G.H. de Haas, Annual Review of Biophysics and Bioengineering, 5 (1976) 77.

10) F.M. van Wezel, A.J. Slotboom and G.H. de Haas, manuscript in preparation.

11) A.J. Slotboom and G.H. de Haas, Biochemistry, 14 (1975) 5394.

12) L.H.M. Janssen, S.H. de Bruin and G.H. de Haas, Eur. J. Biochem., 28 (1972) 156.

13) A. Evenberg, H. Meyer, H.M. Verheij and G.H. de Haas, Eur. J. Biochem., submitted for publication.

14) J.J. Volwerk, W.A. Pieterson and G.H. de Haas, Biochemistry, 13 (1974) 1446.

15) W.A. Pieterson, J.J. Volwerk and G.H. de Haas, Biochemistry, 13 (1974) 1439.

16) G.H. de Haas, P.P.M. Bonsen, W.A. Pieterson and L.L.M. van Deenen, Biochim. Biophys. Acta, 239 (1971) 252.

17) W. Nieuwenhuizen, H. Kunze and G.H. de Haas, Methods Enzymol. 32B (1974) 147.

18) R. Verger, J. Rietsch, M.C.E. van Dam-Mieras and G.H. de Haas, J. Biol. Chem., 251 (1976), accepted for publication.

# MOLECULAR ORGANIZATION AND THE FLUID NATURE OF THE MITOCHONDRIAL ENERGY TRANSDUCING MEMBRANE

Charles R. Hackenbrock

University of Texas Health Science Center
Southwestern Medical School
Dallas, Texas, U.S.A.

## INTRODUCTION

The inner or energy transducing membrane of the mitochondrion is the site of various metabolic activities, including the sequential transfer of electrons along a chain of respiratory proteins and the coupling of the free energy derived from such transfer to the phosphorylation of ADP. Electron transfer between the heme protein components of the membrane is rapid and can be expected to require protein-protein interactions equal to the half times of their oxidations, for example, as rapid as 2 msec in the case of the oxidation of cytochrome *c* by cytochrome *c* oxidase (Chance *et al.*, 1967). Interactions between other redox components in the membrane, however, may be somewhat slower as indicated by delays in the transfer of reducing equivalents, as for example between the *b* cytochromes and cytochrome $c_1$ and between the flavoproteins and *b* cytochromes. Of interest in this regard are the reported delays in the rate and half time of ATP synthesis coupled to the rapid half time oxidation of cytochrome *c* oxidase (Lemasters & Hackenbrock, 1975; Thayer & Hinkle, 1975). Irrespective of such delays, the sequential and rapid events inherent in electron transfer and energy transduction generally tend to support the inference that the proteins in the energy transducing membrane of the mitochondrion are stabilized in a continuous, rigid protein-protein lattice (Fleisher *et al.*, 1967; Sjöstrand & Barajas, 1970; Capaldi & Green, 1972). In classical agreement, the specific proteins of the respiratory chain have been assumed to be ordered with a recurring lateral intermolecular spacing throughout the plane of the membrane (Klingenberg, 1968; Lehninger, 1970). Further, these notions are supported by the fact that the energy transducing membrane is endowed with an unusually high protein content (75%) compared to various other membranes of eukaryote cells.

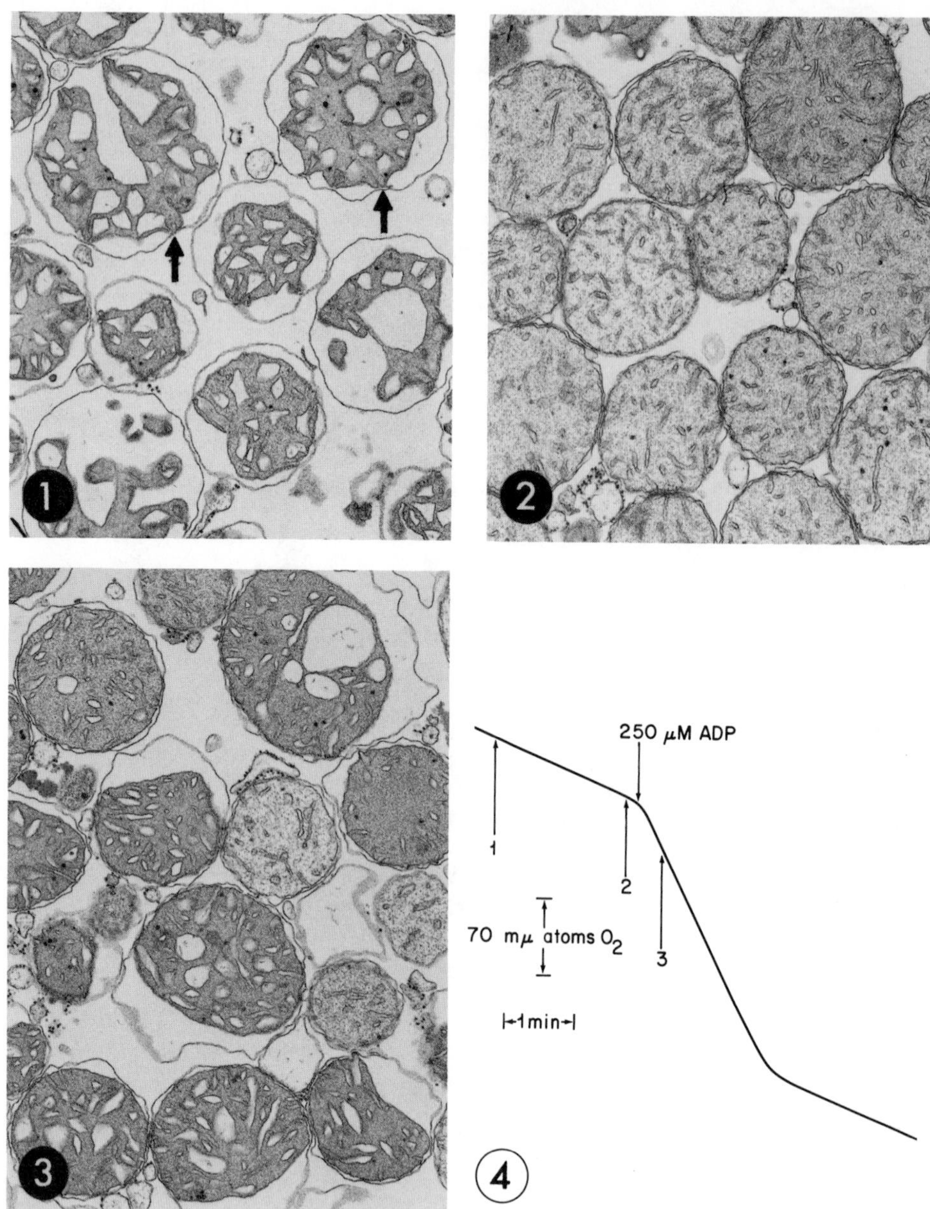


Figs. 1-4. Metabolically-linked structural transformations in isolated rat liver mitochondria: Fig. 1. Freshly isolated mitochondria; Fig. 2. Mitochondria during succinate-supported electron transport; Fig. 3. Mitochondria during oxidative phosphorylation, X 27,000; Fig. 4. Polarographic trace showing ADP-induced oxidative phosphorylation and consecutive times (arrows) at which mitochondria in Figs. 1-3 were fixed.

Other more recent observations which have originated in my laboratory suggest that the energy transducing membrane of the mitochondrion is less rigid, considerably more plastic, and indeed more fluid in its molecular organization than previously recognized. It is these observations, which relate directly to the structure of a natural biological membrane, which I intend to focus on here.

## STRUCTURAL TRANSFORMATIONS LINKED TO ENERGY STATE

Rather dramatic structural transformations in the spatial orientation of the energy transducing membrane parallel changes in the energy state of the mitochondrion (Hackenbrock, 1966; 1968a). Although initially observed in mitochondria isolated from liver, it is now well recognized that such energy-linked structural transformations occur in the energy transducing membrane universally, i.e., irrespective of the cell or tissue type from which the mitochondria are isolated. Of considerable significance in this regard is that identical structural transformations occur in mitochondria in the structurally and functionally intact cell which parallel changes in the energy state of the cell (Hackenbrock, *et al.*, 1971a).

The freshly isolated, metabolically intact mitochondrion consistently reveals the condensed configuration in which the energy transducing membrane is randomly folded without clear distinction between its cristal membrane and inner boundary membrane regions (Fig. 1). Upon energization with respiratory substrate and subsequent initiation of oxidative phosphorylation with ADP, cyclic structural transformations occur in the membrane (Figs. 2-4) which are completely blocked by inhibitors of electron transport and inhibitors and uncouplers of oxidative phosphorylation (Hackenbrock, 1968). Energy-linked structural transformations in the inner mitochondrial membrane are quite complex, the topographical surface details of which can be revealed by scanning electron microscopy using the digitonin-prepared, intact inner membrane-matrix fraction (outer membrane-free mitochondria)(Andrews & Hackenbrock, 1975). A sample of the metabolically intact inner membrane-matrix fraction is shown in thin section in Figure 5 and in whole mount by scanning electron microscopy in Figure 6. The rapid structural transformation in membrane surface reorganization which occurs at the initiation of oxidative phosphorylation is demonstrated in Figures 7 and 8.

I have previously suggested that these metabolically-linked structural transformations in the energy transducing membrane are linked to energy conservation through common conformational changes which occur in the membrane's energy transducing components during steady state transitions in electron transport (Hackenbrock, 1968a, 1972a; Hackenbrock *et al.*, 1971b). This notion, supported by the reality of the dynamic nature of the structure of the energy trans-

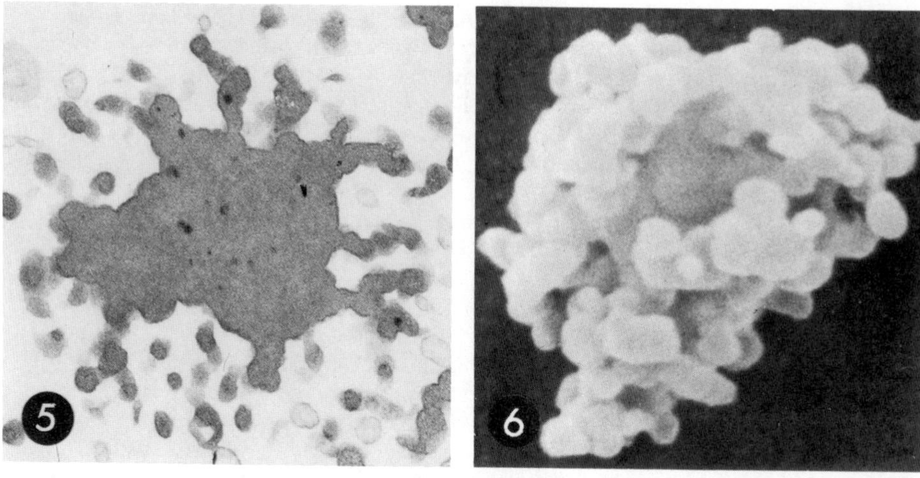

Figs. 5, 6. Surface configuration of the energy transducing membrane of an inner membrane-matrix particle before oxidative phosphorylation; Fig. 5. Transmission image of thin section, X 48,000; Fig. 6. Scanning image of whole mount, X 42,000.

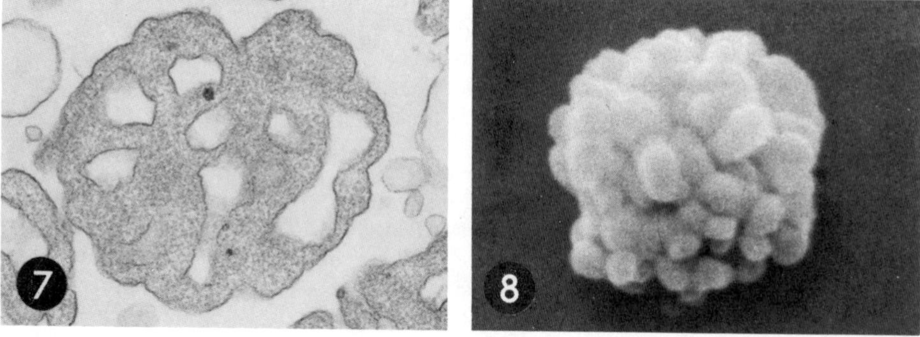

Figs. 7, 8. Surface transformations in the energy transducing membrane of an inner membrane-matrix particle during oxidative phosphorylation: Fig. 7. Transmission image of thin section, X 48,000; Fig. 8. Scanning image of whole mount, X 42,000.

ducing membrane, together with the fact that discrete membrane is a prerequisite for oxidative phosphorylation, focuses attention on a structurally organized membrane as serving to integrate some physical change in its molecular order with energy conservation. These findings and interpretations are consistent with the conformational coupling hypothesis of energy conservation (Boyer, 1965) including a more recent version (Boyer *et al.*, 1975). However, lack of a unifying concept of general acceptance which may serve to resolve the mechanism of energy conservation in the inner mitochondrial membrane persists today and perhaps reveals the inadequacy of our understanding of the molecular organization and its relationship to the rapid metabolic events which occur in this highly complex membrane.

It is not my objective here to present and assess the hypothesis of conformational coupling in the energy transducing membrane, but rather to introduce my findings which initially revealed the plastic nature of the structure of this membrane. Clearly the observations on energy-linked structural transformations in the energy transducing membrane demonstrate that this membrane can undergo rapid geometric surface reorganization which requires a high degree of flexibility, plasticity, and perhaps high degree of motional freedom of its membrane components. It is in this context that I should now like to examine more closely some features of the molecular organization of the energy transducing membrane of the mitochondrion.

## LIPID PHASE TRANSITIONS AND LATERAL TRANSLATIONAL MOBILITY OF INTEGRAL PROTEINS

### Membrane Preparations and Approach

We have combined differential scanning calorimetry (DSC) and freeze fracture electron microscopy to study whole mitochondria (Fig. 1) as well as two types of purified inner membrane-matrix fractions: the structurally condensed inner membrane-matrix fraction in 300 milliosmolar medium (Figs. 5,6) and a structurally spherical inner membrane-matrix fraction in 40 milliosmolar medium (Fig. 9)(Hackenbrock, 1972b). Purified outer mitochondrial membrane (Parsons *et al.*, 1966) was analyzed similarly and is included here for the purpose of comparison. The spherical inner membrane-matrix fraction offers an exceptionally large surface area of the energy transducing membrane for critical analysis by freeze fracture electron microscopy and was prepared especially for this purpose. The purities of the inner and outer membrane preparations are indicated in Table 1. Of significance here is that the inner membrane-matrix preparations were essentially free of monoamine oxidase, an outer membrane marker enzyme, while the outer membrane fraction was enriched 30-fold with respect to monoamine oxidase.

In the DSC studies, the upper transition temperature was determined by the onset temperature on cooling, while the lower transition temperature was determined by the onset temperature on heating. The difference between these two onset temperatures was used to determine the extent of the transition region, as is customary in studies of mixtures of pure lipids (Ladbrooke & Chapman, 1969). It was generally observed that mitochondrial membranes revealed subzero transition temperatures; therefore, membrane

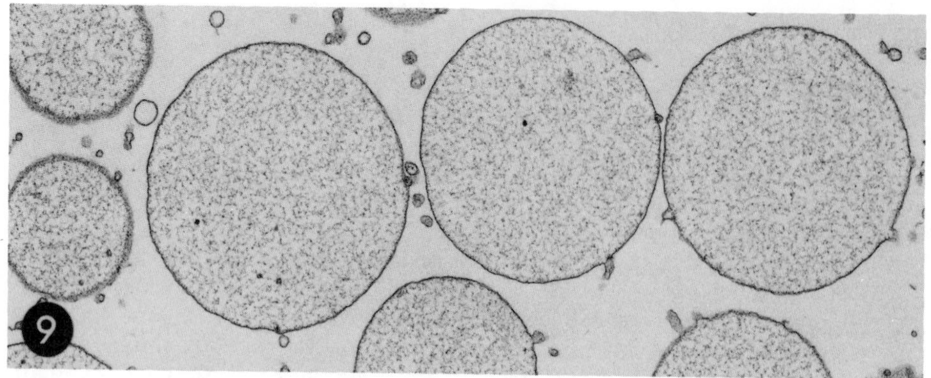

Fig. 9. Purified spherical inner membrane-matrix fraction. X 20,000.

samples were analyzed in the presence of 50% ethylene glycol to adequately depress the water freezing exotherm. Ethylene glycol at 50% was found to depress transition temperatures in mitochondrial membranes (as in bacterial membranes; Steim *et al.*, 1969) by 3°C compared to 25% ethylene glycol, and by 6°C compared to 30% glycerol. This 6°C depression has been corrected for in the DSC traces in order to compare more directly these data with the results of freeze fracture electron microscopy in which 30% glycerol was the membrane cryoprotector. Table 2 summarizes the corrected and uncorrected onset temperature values, as well as the extent of the transition regions for all membrane preparations. Lipid transition exotherms were generally found to be of greater use than endotherms for comparison with ultrastructural observations, since

TABLE 1
Relative Purity of
Fractionated Mitochondrial Membranes

| | Specific Activity* | |
|---|---|---|
| | Monoamine Oxidase | Cytochrome <u>c</u> Oxidase |
| Whole Mitochondria | 12.8 | 720 |
| Condensed Inner Membrane-Matrix Fraction | 0.257 | 1060 |
| Spherical Inner Membrane-Matrix Fraction | 0.014 | 1200 |
| Outer Membrane Fraction | 352.0 | 150 |

*Specific Activity: for monoamine oxidase in nmoles benzylamine oxidized · $min^{-1} \cdot mg^{-1}$ protein; for cytochrome <u>c</u> oxidase in natoms oxygen reduced · $min^{-1} \cdot mg^{-1}$ protein.

TABLE 2
Relative Lipid Phase Transition Temperatures of Whole Mitochondria and Fractionated Membranes[a]

| | Onset Temperature in 50% Ethylene Glycol (%) | | Onset Temperature Corrected (°C)[b] | | Extent of Transition Region (°C) |
|---|---|---|---|---|---|
| | Exothermal | Endothermal | Exothermal | Endothermal | |
| Whole Mitochondria | +3, –5 | –21, –14 | +9, +1 | –15, –8 | 24 |
| Condensed Inner Membrane-Matrix Fraction | –10 | –21, –14 | –4 | –15, –8 | 11 |
| Spherical Inner Membrane-Matrix Fraction | –10 | –21, –14 | –4 | –15, –8 | 11 |
| Outer Membrane Fraction | +3, –5 | –21 | +9, +1 | –15 | 24 |

[a]All temperatures given are averages of 2 or 3 heating or cooling runs and are ± 1°C.
[b]Corrected for 6°C depressions induced by 50% ethylene glycol.

freeze fracturing was performed on membrane samples cooled to various temperatures prior to rapid freezing.

## Whole Mitochondria

An initial but slight onset temperature in the exotherm for whole mitochondria occurred at approximately 9°C at a recording sensitivity of 0.5 mCal•sec$^{-1}$ and was observed to be the result of a transition occurring in the outer membrane (Figs. 10A,B). Cooling runs at a higher sensitivity of 0.2 mCal•sec$^{-1}$ confirmed this finding (Fig. 10C). A second, major onset temperature in the transition exotherm occurred at 1°C in whole mitochondria, as well as in purified outer membrane preparations (Fig. 10).

Membrane fracture faces of whole mitochondria cooled slowly from 30°C to 10°C, i.e., to just above the initial onset temperature of the transition exotherm of whole mitochondria (Fig. 10A), are demonstrated in Figure 11. The exposed fracture faces of the energy transducing membrane contained a strikingly high density of randomly distributed intramembrane particles, which represent integral proteins intercalated into the hydrophobic interior of the bilayer lipid of this complex membrane. Fewer and smaller randomly distributed particles occurred in the outer membrane. We determined previously that the frequency size distribution of the intramembrane particles shows major peaks at 10 and 8 nm in diameter for the inner and outer membranes, respectively (Hackenbrock, 1973). The smallest intramembrane particles in the energy transducing membrane measure between 4 and 5 nm, while approximately 70% of the particles are between 10 and 18 nm in

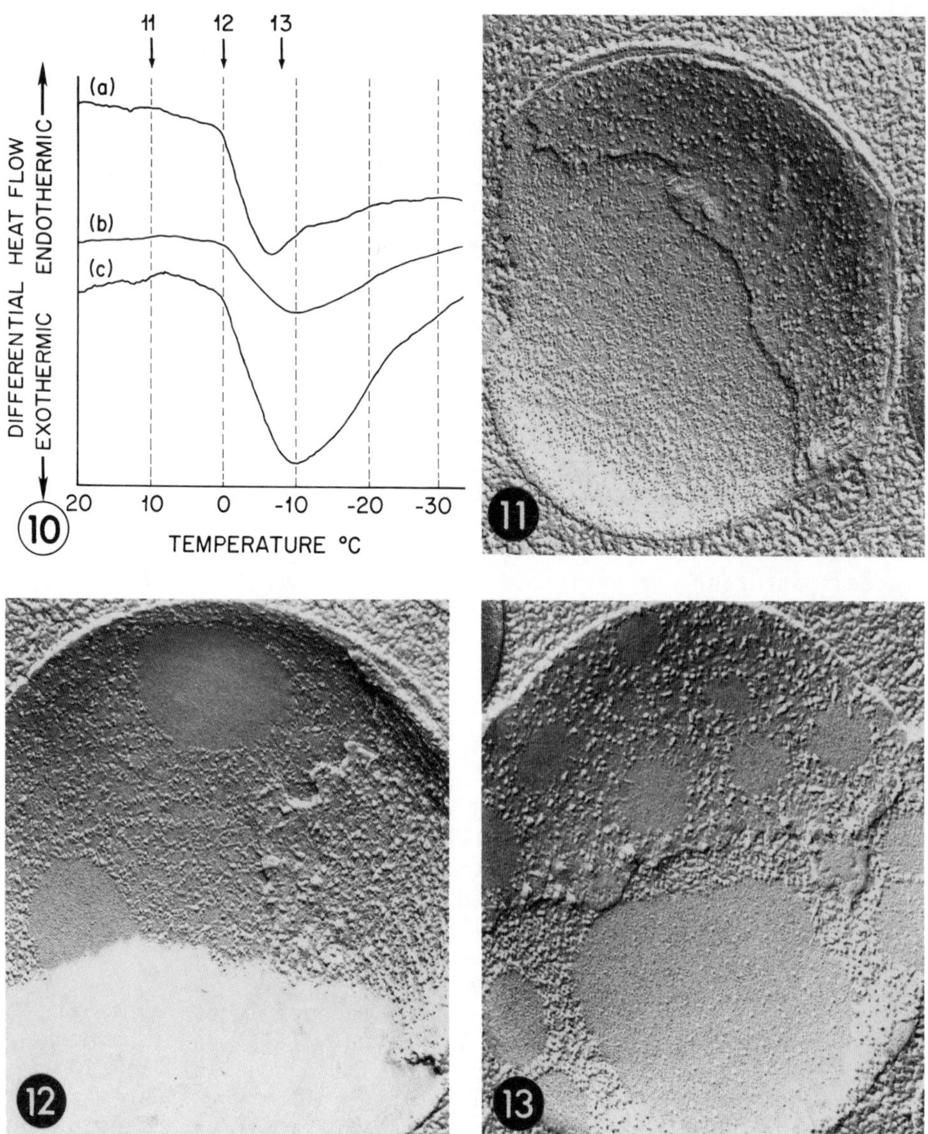


Figs. 10-13. Lipid phase transition and lateral mobility of integral proteins: Fig. 10. DSC cooling runs of whole mitochondria (a) and purified outer membrane (b,c). Recording sensitivity 0.5 mCal•sec$^{-1}$ (a,b), 0.2 mCal•sec$^{-1}$ (c). Arrows identify temperatures on the curve for whole mitochondria which correspond to Figs. 11-13; Fig. 11. Concave fracture faces of both membranes of a mitochondrion frozen from 10°C; Fig. 12. Cooled to 0°C, then frozen; Fig. 13. Cooled to -8°C, then frozen. Note lateral displacement of intramembrane particles. X 67,500.

diameter, which represents an exceptionally high content of relatively large intramembrane particles compared to other membranes of eukaryote cells. This high content of various sizes of intramembrane particles can account for virtually all the integral proteins of this membrane. These consist primarily of the respiratory proteins including the cytochromes (except cytochrome *c*), several membrane-bound dehydrogenases, the non-catalytic subunits of ATPase, and various ion and substrate-translocating integral proteins.

In the results and discussions to follow, I will equate intramembrane particles with integral proteins. That intramembrane particles represent integral proteins has been clearly demonstrated in freeze fracture studies in which purified proteins, such as rhodopsin, glycophorin, microsomal ATPase, and human myelin N-2 protein reconstituted into the bilayer of synthetic phospholipid vesicles, appear as discrete particles in the exposed hydrophobic interior of the vesicle lipid bilayer (Chen & Hubbell, 1973; Grant & McConnell, 1974; Kleeman *et al.*, 1974; Papahadjopoulos *et al.*, 1975). In several of these model systems, integral proteins show thermotrophic lateral translational mobility (Grant & McConnell, 1974; Kleeman *et al.*, 1974). For the energy transducing membrane, it has been estimated that the integral proteins occupy approximately one third of the membrane surface area (Vanderkooi, 1974). Thus, the integral proteins of this membrane are at least afforded the lateral space requirement for translational mobility. The potential for such mobility can be considerable, provided the bilayer lipid is fluid and the proteins are not stabilized by a continuous, rigid protein-protein lattice.

We determined that whole mitochondria cooled slowly to 0°C, i.e., to just below the major onset temperature of their transition exotherm (Fig. 10A), revealed obvious lateral separations between smooth, particle-free regions and intramembrane particle-rich regions in the hydrophobic fracture faces of the outer membrane, whereas no such structural separations occurred in the energy transducing membrane (Fig. 12). However, whole mitochondria cooled slowly to -8°C, i.e., to the peak of their transition exotherm (Fig. 10A), exhibited striking lateral separations between smooth, particle-free regions and particle-rich regions in the energy transducing membrane, as well as in the outer membrane (Fig. 13). Such thermotropic lateral separation leads to aggregation of intramembrane particles and is most likely the result of the growth of protein-excluding regions of a lipid crystalline to gel state phase transition in the bilayer lipid as indicated by our DSC studies. The smooth, particle-free regions in the hydrophobic interior of the membrane represent the low temperature-induced gel state bilayer lipid, while the intramembrane particles represent the redistributed integral membrane proteins.

TABLE 3

ADP:O and Acceptor Control Ratios in Mitochondria Before and After Low Temperature-Induced Lateral Separations in the Inner Membrane

| | Pretreatment with Glycerol Medium[1] | | | |
|---|---|---|---|---|
| Exp. | Temp. in °C | Time in Min. | ADP:O | A.C. |
| 1 | No Pretreatment | | 1.80 | 5.50 |
| 2 | 25 | 2 | 1.48 | 3.38 |
| 3 | 25 | 2 | 1.45 | 3.33 |
| 4 | 0 | 8 | 1.50 | 3.19 |
| 5 | 0 | 8 | 1.59 | 3.05 |
| 6 | −8 | 8 | 1.42 | 3.43 |
| 7 | −8 | 8 | 1.48 | 3.00 |
| 8 | −8 | 15 | 1.56 | 3.05 |

[1]Mitochondria were pretreated in glycerol medium (30% glycerol; 250 mM sucrose; 10 mM Tris, pH 7.4) at temperatures and times indicated prior to polarographic analysis.

The thermotropic lateral separations of membrane components in the hydrophobic interior of the energy transducing membrane, as well as in the outer membrane, are completely and rapidly reversible. Mitochondria cooled to and equilibrated at -8 or -13°C for 15 min. and subsequently warmed to 0 or 30°C for only a few seconds prior to rapid freezing exhibited randomly distributed intramembrane particles. Particle distribution appeared identical to that shown in Figure 11, which is of whole mitochondria not previously cooled to subzero temperatures. Thus, rapid disaggregation, free lateral translational diffusion, and complete randomization of intramembrane particles occurred in the energy transducing membrane as the temperature was raised and is consistent with the melting properties of membrane bilayer lipids. From these observations, I conclude that the integral proteins have a high potential for free lateral translational diffusion in the fluid bilayer lipid of this membrane.

Of considerable significance in these studies was the determination that the thermotropic lateral separation between integral proteins and gel state bilayer lipid, and the reversal of this separation in the energy transducing membrane, were not destructive to electron transport or oxidative phosphorylation. Mitochondria equilibrated at -8°C for 8 or 15 min. and then returned to 30°C were as efficient in oxidative phosphorylation and exhibited the same degree of acceptor control as mitochondria in which membrane components were not induced to undergo thermotropic lateral separations (Table 3).

## Purified Energy Transducing Membrane

The onset temperature in the exotherm transition of both the purified condensed and spherical inner membrane-matrix preparations occurred at -4°C (Fig. 14B,C). This is clearly a lower transition temperature compared to that which occurred in the composite two-membrane system of the whole mitochondrion (Fig. 14A). Reproducibility of the onset temperature was indicated in comparing the exotherms of the two different inner membrane preparations (Figs. 14B,C). In addition, slight but superimposable exothermal deflections occurred above noise level, at 21, 12, and 1°C in both purified inner membrane preparations (Fig. 14B,C). These small deflections may be related to discontinuities in Arrhenius activation energies of several inner membrane enzymes reported from other laboratories over the temperature range of 8 to 27°C (Raison *et al.*, 1971; Lee & Gear, 1974).

In cooling runs, the spherical energy transducing membrane exhibited randomly dispersed intramembrane particles at 30, 10, and 0°C (Fig. 15). However, when cooled to -8°C, i.e., 4°C below the onset temperature of the transition exotherm for the inner membrane (Fig. 14C), extensive lateral separations were revealed between smooth regions and particle-dense regions in the hydrophobic interior of the membrane (Fig. 16). Such separations were indeed impressive at -13°C (Fig. 17), which is well beyond the peak of the transition exotherm and where liquid crystalline to gel state ordering in the bilayer lipid can be expected to be quantitatively maximal (Fig. 14C).

Upon heating from -30°C, DSC revealed a slightly biphasic transition endotherm for the two purified preparations of the energy transducing membrane, as well as for the whole mitochondrion, the first onset temperature of the transition endotherm appearing at -15°C and the second at -8°C (Fig. 18). Table 2 summarizes the onset temperatures of the exotherm and endotherm transitions as well as the extents of the transition regions for all membrane preparations.

That thermotropic lipid phase transitions, lipid-protein lateral separations, and free lateral translational diffusion of integral proteins can occur in the energy transducing membrane, and that such molecular reorganizational events are not destructive to oxidative phosphorylation, reveals the fluid nature of this membrane. The rather low lipid phase transition temperature of the membrane

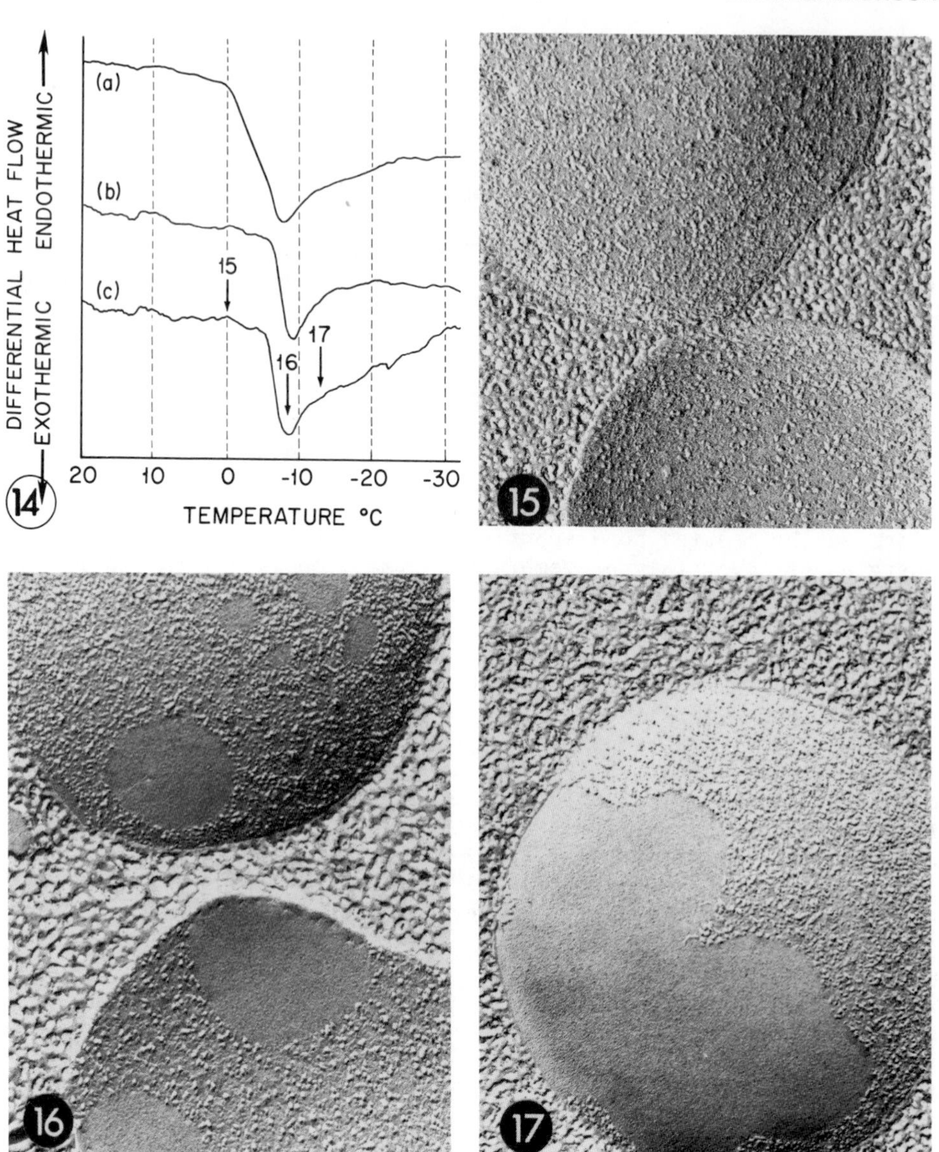


Figs. 14-17. Lipid phase transition and lateral mobility of integral proteins: Fig. 14. DSC cooling runs of whole mitochondria (a), condensed inner membrane-matrix fraction (b), and spherical inner membrane-matrix fraction (c). Arrows identify temperatures on the curve for the spherical inner membrane-matrix fraction which correspond to Figs. 15-17; Fig. 15. Fracture faces of both halves of the spherical energy transducing membrane cooled to 0°C, then frozen; Fig. 16. Cooled to -8°C, then frozen; Fig. 17. Cooled to -13°C, then frozen. Note lateral displacement of intramembrane particles. X 67,500.

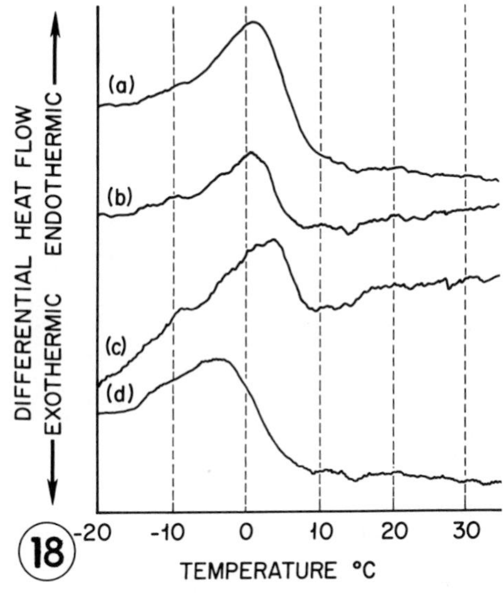


Fig. 18. DSC of heating runs of whole mitochondria (a), condensed inner membrane-matrix fraction (b), spherical inner membrane-matrix fraction (c), and outer mitochondrial membrane fraction (d).

suggests a high degree of fluidity in its bilayer lipid at physiological temperature and is consistent with the known lipid composition of the membrane. Phosphatidylcholine and phosphatidylethanolamine represent approximately 80% of the inner membrane phospholipid and are collectively 53% unsaturated (Colbeau *et al.*, 1971). Cardiolipin, which essentially represents the remainder of the phospholipid, is 90% unsaturated, with oleic (18:1) and linoleic (18:2) acids comprising 80% of the unsaturated fatty acids. The ratio of total saturated to unsaturated membrane phospholipids is 0.65. An absence of cholesterol in the energy transducing membrane, combined with its high content of unsaturated phospholipids, is consistent with the low onset temperature in the transition exotherm, the relatively narrow phase transition region, the high degree of fluidity, and the configurational plasticity of this membrane revealed by our studies. For comparison, it may be noted here that the outer membrane, which exhibited a much higher onset temperature in its transition exotherm and a broader phase transition region, has a saturated to unsaturated phospholipid ratio of 1.75 and a cholesterol to phospholipid-P molar ratio of 0.13.

The energy transducing membrane is composed of 75% protein and approximately 50% of this is integral to the membrane (Capaldi & Tan, 1974; Harmon *et al.*, 1974). The apolar surfaces of the integral proteins, the majority of which are metabolically active and presumably globular proteins, are in hydrophobic association with the acyl chains of the membrane phospholipids and can be

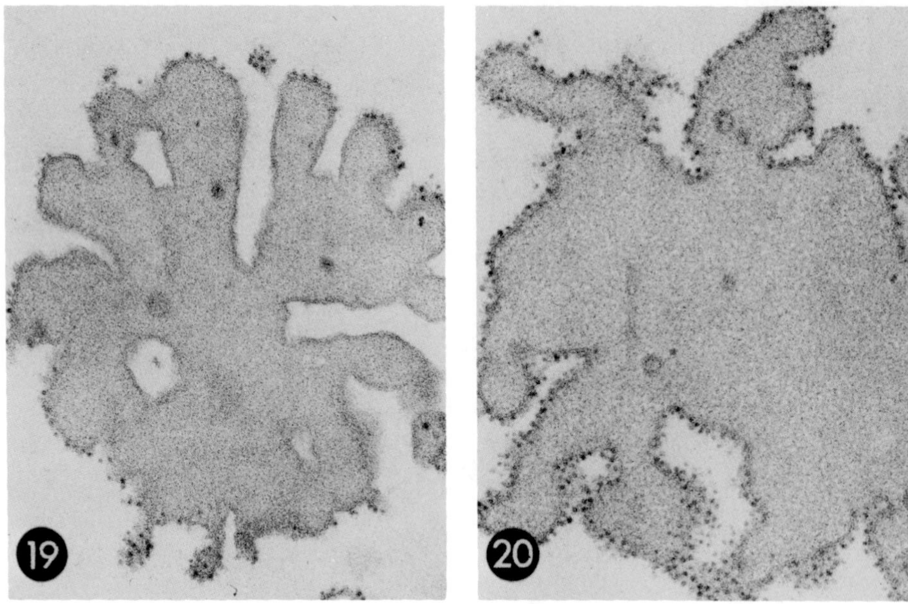


Figs. 19, 20. Location of polycationic ferritin on the energy transducing membrane of an inner membrane-matrix particle after electrostatic binding: Fig. 19. Polycationic ferritin at 30 μg/ml; Fig. 20. Polycationic ferritin at 90 μg/ml. X 48,000.

expected, from studies on other membrane systems, to immobilize anywhere from 15 to 25% of the phospholipid component (McConnell *et al.*, 1972; Trauble & Overath, 1973; Jost *et al.*, 1973). The low lipid phase transition temperature of the energy transducing membrane may therefore be related in part to a low ordering efficiency of the phospholipid acyl chains owing to the high content of integral protein. However, although a high content of integral protein may hinder cooperative interaction of the phospholipid acyl chains, our results revealed a rather narrow lipid phase transition region for this complex membrane. Finally, the large component of peripheral protein may lower the transition temperature in the energy transducing membrane through electrostatic interaction with the polar head groups of the bilayer phospholipids, as has been proposed from studies of protein-lipid electrostatic interactions in model membrane systems (Chapman *et al.*, 1974).

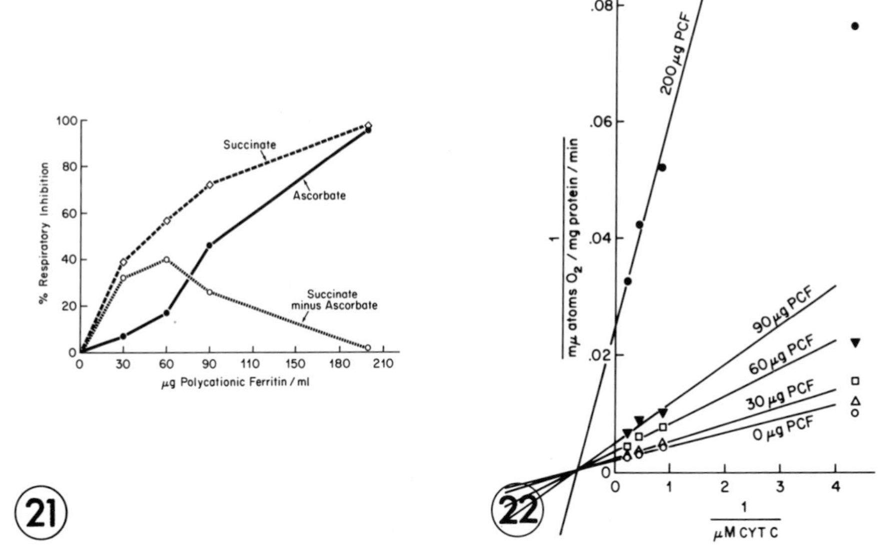


Figs. 21, 22. Metabolic inhibition in the energy transducing membrane of the inner membrane-matrix fraction by electrostatic binding of polycationic ferritin: Fig. 21. Percent inhibition of succinate oxidase and ascorbate-TMPD cytochrome *c* oxidase activity; Fig. 22. Lineweaver-Burk plot of 1/ascorbate-TMPD cytochrome *c* oxidase activity vs. 1/cytochrome *c* concentration at progressively increasing concentrations of polycationic ferritin.

## LATERAL TRANSLATIONAL MOBILITY OF INTEGRAL PROTEINS CONTAINING ANIONIC GROUPS EXPOSED ON THE MEMBRANE SURFACE

### Location and Mobility

We have recently demonstrated that polycationic ferritin, a visually detectable macromolecular ligand, binds electrostatically to the surface of the energy transducing membrane and inhibits electron transport proportional to the degree and location of binding (Hackenbrock, 1975; Hackenbrock & Miller, 1975). At 30 µg/ml, the ligand binds preferentially to high affinity sites on the surface of the inner boundary membrane region of the inner membrane (Fig. 19), and at 90 µg/ml, it binds in addition to lower affinity sites on the surface of the cristal membrane region of the inner membrane (Fig. 20). Ascorbate-TMPD cytochrome *c* oxidase activity and succinate permease activity are inhibited as demonstrated in Figure 21. In the case of ascorbate oxidation, kinetic data show that polycationic ferritin binds to cytochrome *c* oxidase as a non-competitive inhibitor with respect to cytochrome *c* (Fig. 22). Thus polycationic

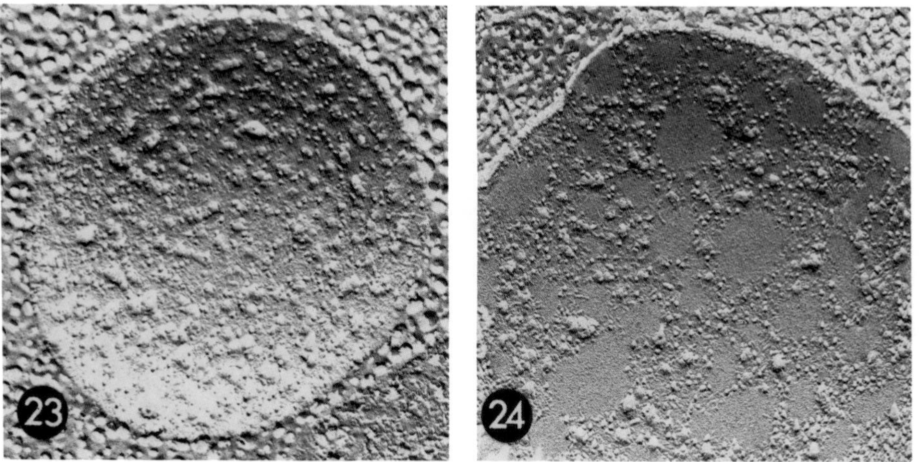

Figs. 23, 24. Concave fracture face of the energy transducing membrane showing ferritin impressions after surface binding of polycationic ferritin at 30 μg/ml: Fig. 23. Membrane frozen from 30°C; Fig. 24. Membrane cooled to -10°C, then frozen, shows ferritin impressions only in regions of intramembrane particles. X 67,500.

ferritin, a large molecule of approximately 11 nm in diameter and containing approximately 65 cationic charges at physiological pH, binds to anionic groups of cytochrome *c* oxidase exposed at the membrane surface thereby dislocating cytochrome *c*, a peripheral membrane protein, from its normal position on the membrane surface.

Polycationic ferritin bound at 25°C of low concentrations (30-60 μg/ml) to the surface of the spherical inner membrane-matrix preparation was found to impress deeply into the highly plastic energy transducing membrane. This permits the ligand's surface-bound distribution to be observed simultaneously with the distribution of intramembrane particles in the hydrophobic fracture face of the freeze fractured membrane (Fig. 23). Impressions of single ferritin molecules measure 21 nm in diameter. With subsequent cooling to -10°C, thermotropic lateral separations occurred between integral proteins and gel state bilayer lipid while the polycationic ferritin comigrated with the integral proteins (Fig. 24).

Thus, polycationic ferritin, bound to anionic groups on the surface of the energy transducing membrane at concentrations which partly inhibit cytochrome *c* oxidase as well as succinate permease, migrates laterally with the integral membrane proteins during liquid crystalline to gel state lipid phase transitions and related lipid-protein structural separations. These results are consistent with our suggestions that the intramembrane particles are the various integral proteins of the energy transducing membrane and that these integral proteins have a high potential for lateral translational mobility in the plane of the membrane.

## Immobilization of Integral Proteins by an Artificial Electrostatic Peripheral Protein Lattice

Polycationic ferritin was bound at 25°C to the spherical energy transducing membrane at high concentration (90 μg/ml), followed by the addition of native (anionic) ferritin. The deep-etched membrane exhibited a complete but not packed ferritin coverage (Fig. 25). Since we found that native ferritin alone did not bind to mitochondrial membranes, and since native ferritin readily precipitated polycationic ferritin from solution, it could be reasonably expected that the native ferritin associated with, and crossbridged the polycationic ferritin which was bound initially to the anionic groups of the various integral proteins exposed on the membrane surface. In this manner, the integral proteins of the energy transducing membrane were essentially immobilized through an electrostatic, relatively continuous, artificial peripheral protein-protein lattice over the surface of the membrane. When cooled to -10°C after such electrostatic latticing, the thermotropic lateral separations between integral proteins and gel state bilayer lipid were significantly reduced (Fig. 26; cf Fig. 24).

The artificial peripheral protein-protein lattice also prevented free lateral translational diffusion of integral membrane proteins. Inhibition of the lateral diffusion of large integral proteins (large intramembrane particles) was more effectively inhibited than the diffusion of small integral proteins (small intramembrane particles). In these experiments, the energy transducing membrane was cooled to -10°C to induce thermotropic lateral separations between integral proteins and gel state bilayer lipid. This was followed by the addition of polycationic ferritin and native ferritin at -10°C to crossbridge and thus immobilize the integral proteins in their low temperature-induced aggregated state. The deep-etched membrane surface exhibited an incomplete coverage of packed ferritin (Fig. 27). Ferritin-free smooth patches of membrane surface presumably represented gel state lipid-rich regions, while the packed ferritin presumably bound to, and crossbridged anionic groups of the aggregated integral proteins exposed at the membrane surface (Fig. 27). After such latticing, the larger integral proteins did not disaggregate through free lateral translational diffusion when the temperature was raised to 0 or 25°C. To be noted was that a very small population of particles of less than average diameter did diffuse back into the smooth, liquid crystalline lipid regions of the hydrophobic interior of the membrane after raising the temperature (Fig. 28).

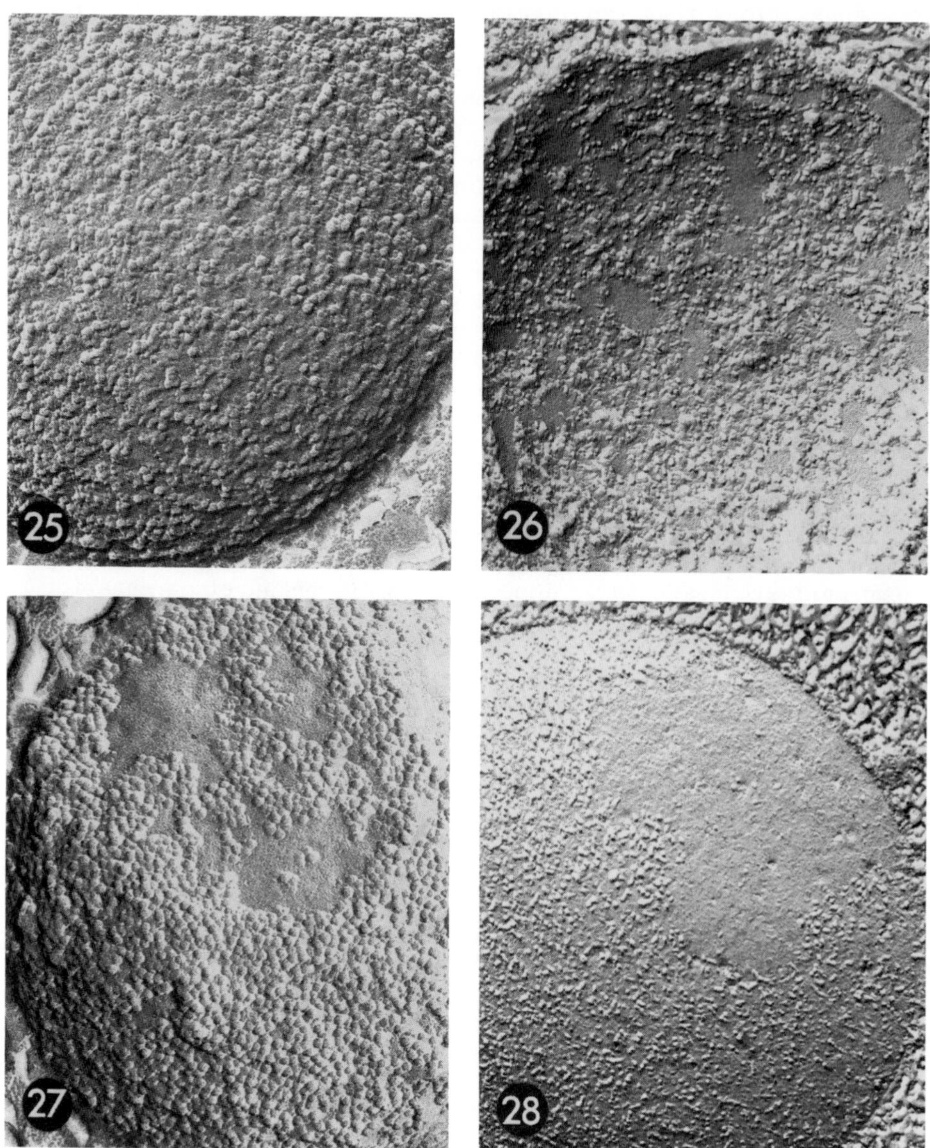


Figs. 25-28. Latticing and immobilization of integral proteins in the energy transducing membrane through polycationic ferritin: Fig. 25. Deep-etched membrane surface after latticing with polycationic followed by native ferritin, then frozen from 25°C; Fig. 26. Concave fracture face after latticing, then cooled to -10°C, then frozen; Fig. 27. Deep-etched membrane surface after cooling to -10°C, then latticed, then frozen; Fig. 28. Convex fracture face after cooling to -10°C, then latticed, then warmed to 25°C, and finally frozen shows large, particle-free regions. Figs. 25, 27, X 115,000; Figs. 26, 28, X 67,000.

Thus, we have determined that metabolically active integral proteins, such as cytochrome *c* oxidase and succinate permease, contain anionic groups which are exposed at the surface of the energy transducing membrane. It would appear that these proteins and most other integral proteins in this membrane can undergo free lateral translational diffusion in the fluid bilayer lipid of the membrane and can be prevented from doing so by immobilization through cross-bridging with an electrostatic, artificial peripheral protein lattice. If a natural electrostatic protein lattice exists in the energy transducing membrane, it is clearly less effective in immobilizing the integral membrane proteins than the artificial electrostatic protein lattice that we have used. I should now like to examine the potential for lateral translational mobility and free diffusion of cytochrome *c* oxidase using a more discriminating approach.

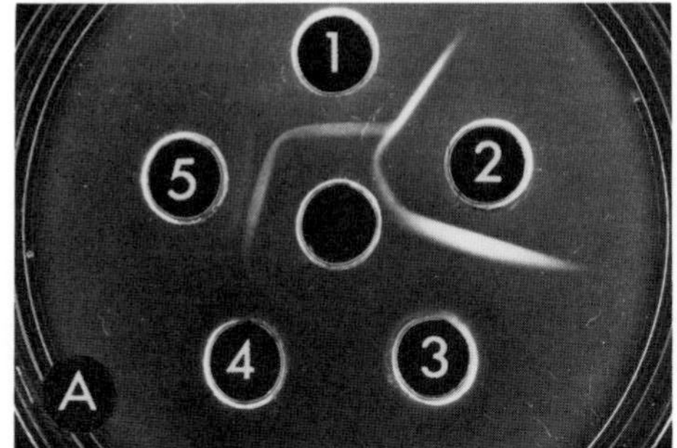


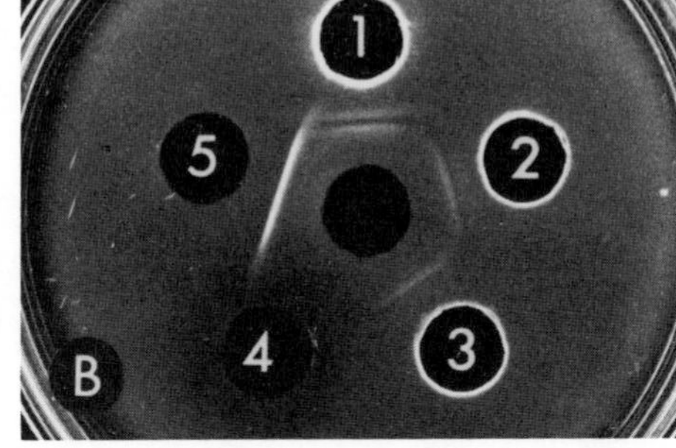


Fig. 29. Double immunodiffusion assay of affinity purified rabbit IgG monospecific for cytochrome *c* oxidase (cyt ox Ab): A. Center well contains purified bovine heart oxidase; well 1 contains cyt ox Ab; well 2 contains affinity purified goat IgG monospecific for rabbit IgG; well 3 contains normal rabbit whole serum; wells 4 and 5 contain crude IgG to the heart oxidase. B. Center well contains cyt ox Ab; well 1 contains 13 mg of Triton X-100-solubilized rat liver inner membrane-matrix fraction; wells 2 and 3 contain lower concentrations of inner membrane-matrix; well 4 contains normal rabbit whole serum; well 5 contains bovine heart cytochrome *c* oxidase.

## LATERAL TRANSLATIONAL MOBILITY AND FREE DIFFUSION OF CYTOCHROME *c* OXIDASE

### Orientation and Site-by-Site Distribution

Before examining the potential for lateral translational mobility and free diffusion of cytochrome *c* oxidase, consideration of the distribution of the oxidase in the energy transducing membrane will be necessary. We have recently prepared several membrane impermeable immunoligands for the purpose of probing the two surfaces of the energy transducing membrane for the spatial, site-by-site distribution of cytochrome *c* oxidase (Hackenbrock & Hammon, 1975). Affinity purified rabbit IgG was prepared using guanidine-treated bovine heart cytochrome *c* oxidase as the immunoabsorbent. For ultrastructural studies, the visual probe used for location of the oxidase was a ferritin-conjugate of the affinity IgG. An IgG monospecific for cytochrome *c* was also prepared. Finally, an affinity purified goat IgG monospecific for rabbit IgG was produced (= goat anti-rabbit IgG).

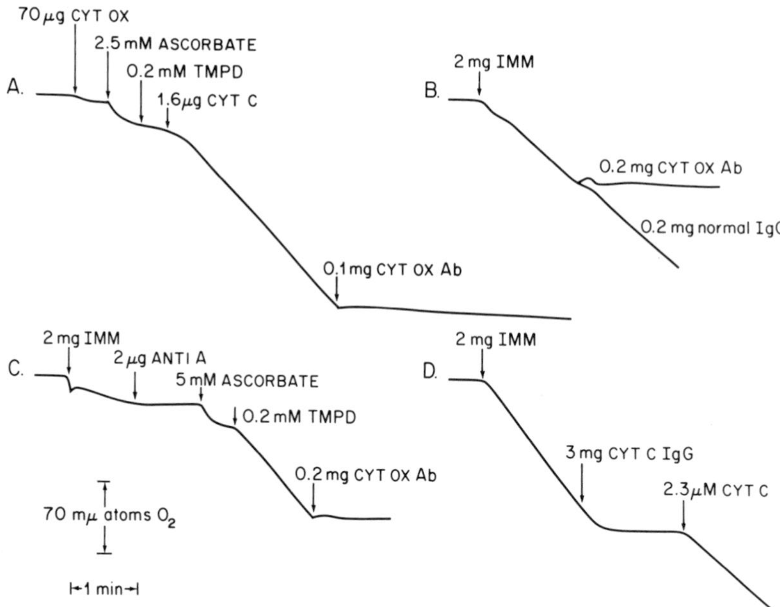


Fig. 30. Polarographic traces of the inhibition of cytochrome *c* oxidase and cytochrome *c* activities by monospecific IgG: A. Complete inhibition of purified bovine heart cytochrome *c* oxidase by cyt ox Ab; B. Complete inhibition of succinate oxidase activity in rat liver inner membrane-matrix fraction by cyt ox Ab; C. Complete inhibition of ascorbate-TMPD cytochrome *c* oxidase activity in rat liver inner membrane-matrix fraction by cyt ox Ab; D. Complete inhibition of succinate oxidase activity in rat liver inner membrane-matrix fraction by cyt *c* IgG.

The affinity purified IgG developed against bovine heart cytochrome *c* oxidase, which I shall call cyt ox Ab, gave a single immunoprecipitin band against the bovine heart oxidase (Fig. 29A; well 1). This band fused completely with a single immunoprecipitin band between cyt ox Ab and goat anti-rabbit IgG (wells 1 and 2). This test proved the specificity of the cyt ox Ab and its identity as a pure IgG. Cyt ox Ab developed against the bovine heart oxidase precipitated only cytochrome *c* oxidase from the detergent solubilized inner membrane-matrix fraction of rat liver mitochondria (Fig. 29B). Although two immunoprecipitin bands occurred with high concentrations of solubilized energy transducing membrane (well 1), both bands showed identify with purified bovine heart cytochrome *c* oxidase (well 5). The two bands of the solubilized membrane may thus be two polymeric forms of the oxidase, or the oxidase may contain two different quantities of bound lipid or detergent.

Affinity purified cyt ox Ab inhibited the activity of purified bovine heart cytochrome *c* oxidase immediately and completely at a ratio of 100 μg cyt ox Ab to 70 μg oxidase (Fig. 30A). Cyt ox Ab inhibited succinate oxidase activity and ascorbate-TMPD cytochrome *c* oxidase activity immediately and completely when it was restricted in its binding to the outer surface of the structurally and functionally intact energy transducing membrane of the condensed inner membrane-matrix fraction (Figs. 30B,C). Similarly, IgG monospecific

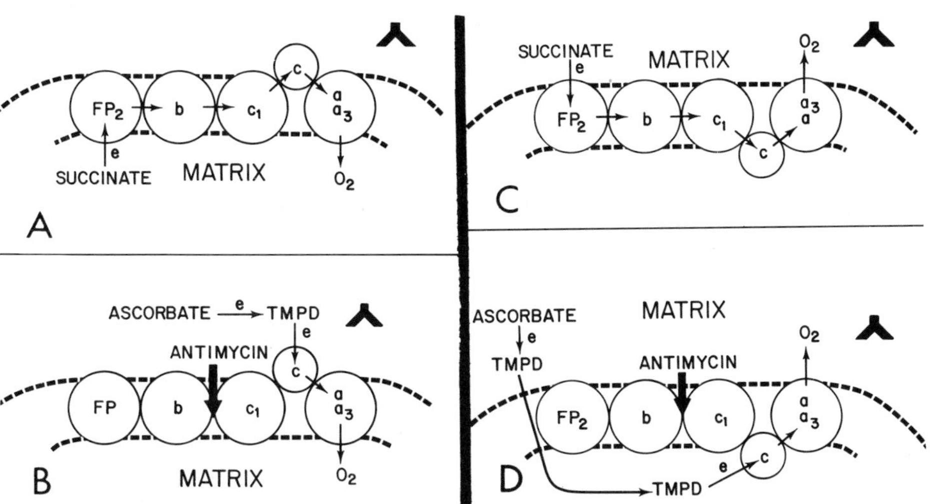


Fig. 31. Pathways of electron flow and accessibility of membrane-bound cytochrome *c* oxidase and cytochrome *c* to monospecific IgG (⅄): A,B. Energy transducing membrane of inner membrane-matrix fraction; C,D. Energy transducing membrane as sonicated inverted vesicles.

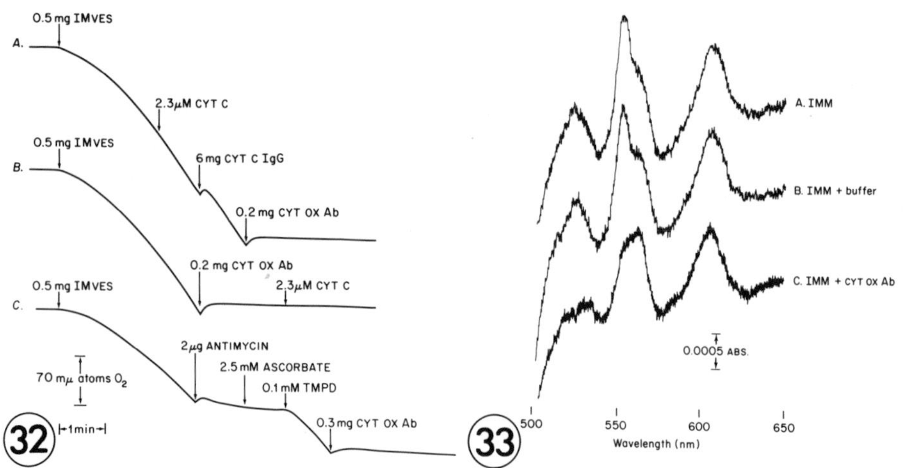


Fig. 32. Polarographic traces of the inhibition of cytochrome *c* oxidase in inverted vesicles of the energy transducing membrane: A,B. Insensitivity of succinate oxidase activity in inverted vesicles to cytochrome *c* or cytochrome *c* IgG and complete inhibition by cyt ox Ab; C. Complete inhibition of ascorbate-TMPD cytochrome *c* oxidase activity by cyt ox Ab.

Fig. 33. Reduced minus oxidized spectra of cytochromes in the energy transducing membrane of the inner membrane-matrix (IMM) fraction: A. Spectra without additions; B. Spectra after addition of 0.1 M sodium phosphate; C. Spectra after addition of 0.1 M sodium phosphate plus cyt ox Ab.

for cytochrome *c* inhibited electron transport immediately and completely when reacted with the outer surface of the membrane (Fig. 30D). These results reveal that *all* functional cytochrome *c* oxidase is exposed on the outer surface of the energy transducing membrane. Figure 31A,B demonstrates graphically the pathways of electron flow utilized in these experiments.

When the spherical inner membrane-matrix preparation is sonicated, completely inverted vesicles of energy transducing membrane are formed (Hackenbrock & Hammon, 1975). Neither cytochrome *c* nor high concentrations of IgG monospecific for cytochrome *c* affect succinate or ascorbate-TMPD supported electron transport in the inverted membrane vesicles (Fig. 32A). However, immediate and complete inhibition of succinate and ascorbate-TMPD oxidation occurs upon addition of low concentrations of cyt ox Ab (Fig. 32B,C). These results reveal that *all* functional cytochrome *c* oxidase is exposed on the inner surface of the energy transducing membrane.

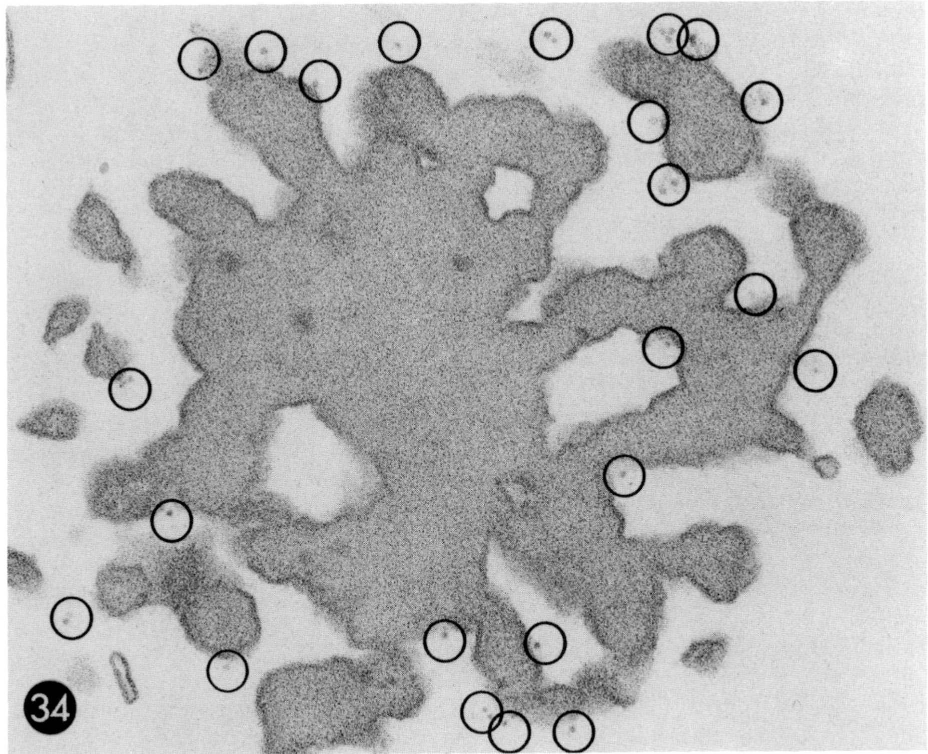


Fig. 34. Site-by-site distribution of cytochrome *c* oxidase on the outer surface of the energy transducing membrane of an inner membrane-matrix particle. Fixed after complete oxidase inhibition by the ferritin conjugate of cyt ox Ab. X 113,000.

Figure 31C,D demonstrates the pathways of electron flow utilized in the studies on the inverted energy transducing membrane.

Collectively then, our data reveal that *all* functional cytochrome *c* oxidase occupies a transmembrane position since it is *totally* accessible on *both* surfaces of the energy transducing membrane to an inhibiting IgG monospecific for the oxidase.

For determining the site-by-site distribution of any membrane protein by immunoelectron microscopy, it is best to verify that the protein is quantitatively unchanged subsequent to the immunoglobulin binding procedures. Cytochrome *c* IgG, for example, removes cytochrome *c* from the membrane surface (Hackenbrock & Hammon, 1975). The reduced minus oxidized difference spectrum (Fig. 33) of the cytochromes of the energy transducing membrane revealed that no loss of cytochrome *c* oxidase (604 nm band) from the membrane or change in the spectrum of the hemes $aa_3$ occurred owing to the binding of cyt ox Ab. No loss occurred in cytochrome *b* (562 nm band) or cytochrome $c_1$ (554 nm band), although most of the cytochrome *c* (550 nm band), a peripheral protein in functional association with cytochrome *c* oxidase, was dislocated from the membrane when cyt ox Ab bound to the outer surface of the intact membrane (Fig. 33C).

The ferritin conjugate of the affinity purified cyt ox Ab inhibited succinate supported electron transport completely and within 10 sec after addition to the condensed inner membrane-matrix fraction. After complete inhibition, microsamples were prepared for electron microscopy. A typical site-by-site distribution of cytochrome *c* oxidase on the outer surface of the intact energy transducing membrane is demonstrated in Figure 34. The oxidase exhibits a relatively disordered lateral distribution in the plane of the membrane with a non-equidistant intermolecular spacing. The shortest intermolecular distance often appeared to be 100 nm, and the oxidase was equally distributed over the cristal and inner boundary membrane regions. In a typical 30 nm thick two dimensional section, approximately 30 cyt ox Ab-ferritin conjugates were bound per energy transducing membrane. This corresponds to a total of approximately 2,000 cytochrome *c* oxidase sites in the energy transducing membrane of the liver mitochondrion. Based on difference spectra, we determined that 1 mg of inner membrane-matrix protein contained 0.35 nmoles of heme $aa_3$ and, assuming a minimum molecular weight of 100,000, we estimated 17,000 monomer cytochrome *c* oxidase molecules per mitochondrion. The difference between our direct analysis of the number of oxidase sites on the surface of the energy transducing membrane by immunoelectron microscopic analysis and our calculated number of oxidase molecules per mitochondrion suggests that the oxidase exists in the functional membrane at least as a hexamer. Thus, in its normal functional orientation in the intact energy transducing membrane, cytochrome *c* oxidase may occur as a relatively large polymeric integral membrane protein.

## Lateral Translational Mobility

We have already seen that the integral proteins of the energy transducing membrane can be observed in the hydrophobic fracture plane as intramembrane particles ranging in size from 4 to 18 nm in diameter. It is also clear that above the onset temperature of the transition exotherm (-4°C), the integral proteins exhibit a completely random distribution (Fig. 15), and at temperatures below the onset temperature of the transition exotherm, structural separations occur laterally between gel state bilayer lipid and the integral proteins (Figs. 16,17). We have taken advantage of thermotropic lateral separations in combination with the use of affinity purified cyt ox Ab to visually follow the lateral movements of cytochrome *c* oxidase in the plane of the energy transducing membrane. In these studies, rabbit cyt ox Ab was added to the spherical inner membrane-matrix preparation at concentrations which completely inhibited succinate supported cytochrome *c* oxidase activity. This was followed by the addition of goat anti-rabbit IgG for the purpose of partially crossbridging the cyt ox Ab which was bound initially to the immunodeterminants of cytochrome *c* oxidase exposed at the membrane surface. Since we determined that above the transition temperature, the minimum intermolecular distance of the oxidase was approximately 100 nm (Fig. 34), we assumed that the oxidase would be only partially immobilized, but through a very specific peripheral immunoglobulin lattice. Without such latticing, it will be recalled that a high degree of thermotropic lateral separation occurs between gel state bilayer lipid and the integral membrane proteins at -13°C, owing to a near maximal liquid crystalline to gel state lipid phase transition which occurs at this temperature (Fig. 17). After immunoglobulin latticing of cytochrome *c* oxidase, however, there was a significant hindrance of the thermotropic lateral movement of integral proteins (Fig. 35). The lateral separation between the integral proteins and gel state bilayer lipid which occurred at -13°C appeared quite distorted, with the smooth, gel state bilayer lipid regions in the hydrophobic fracture faces of the membrane considerably restricted in size (Fig. 35; cf, Fig. 17).

After such limited lateral separation, the immunoglobulin lattice could be observed on the surface of the energy transducing membrane immediately over membrane regions rich in aggregated integral proteins, while no immunoglobulin could be resolved over membrane regions rich in gel state bilayer lipid (Fig. 36). Thus, the immunoglobulin lattice, monospecific for cytochrome *c* oxidase, identifies the thermotropic lateral translational movement of the oxidase with the movement of intramembrane particles in the intact energy transducing membrane. These experiments, however, do not identify specific intramembrane particles with the oxidase in this complex membrane.

## Free Lateral Translational Diffusion of Cytochrome *c* Oxidase and its Inhibition by an Immunoglobulin Lattice

The lateral separation between gel state bilayer lipid and integral proteins that occurs below the onset temperature of the transition exotherm (Figs. 16,17) is completely reversible in the intact energy transducing membrane (Fig. 37). Such reversibility occurs above 0°C with the intramembrane particles randomizing during the gel state to liquid crystalline phase transition in the lipid bilayer and indicates a high degree of free lateral translational diffusion for most, if not all, of the integral proteins.

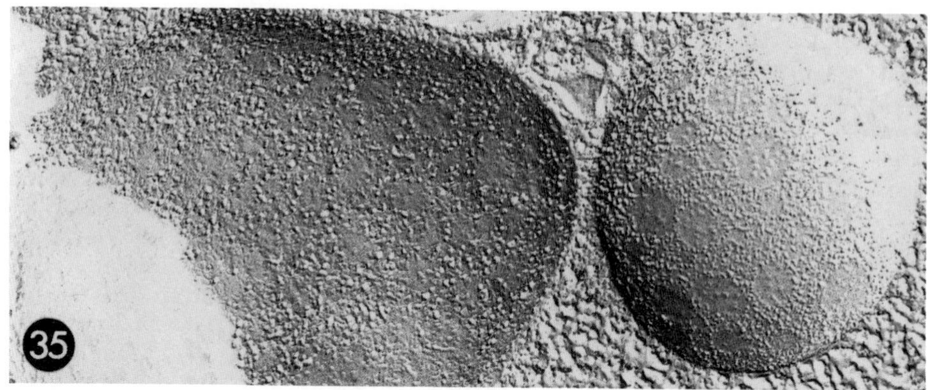

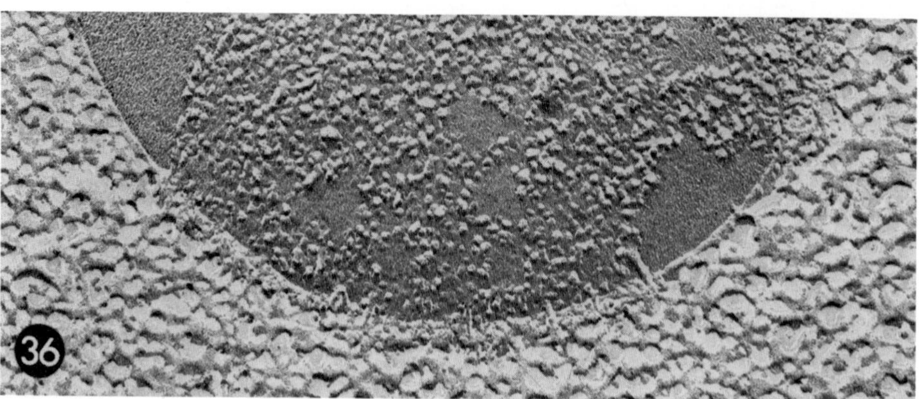

Figs. 35, 36. Lateral mobility of cytochrome *c* oxidase in the energy transducing membrane and its hindrance by immunoglobulin latticing. Rabbit cyt ox Ab was permitted to bind to the membrane surface followed by addition of goat anti-rabbit IgG. After latticing, membranes were cooled to -13°C and then frozen: Fig. 35. Both fracture faces reveal only slight lateral displacement of intramembrane particles, X 67,500 (cf. non-latticed membranes, Fig. 17); Fig. 36. High magnification reveals immunoglobulin lattice in the etched ice surface in register with the integral proteins and no immunoglobulin in regions of smooth gel state bilayer lipid. X 132,000.

Since cytochrome *c* oxidase occupies a completely transmembraneous position and may occur in the membrane as a relatively large polymer, the oxidase likely occurs as a large, rather than small, intramembrane particle and can also be expected to exhibit free lateral translational diffusion.

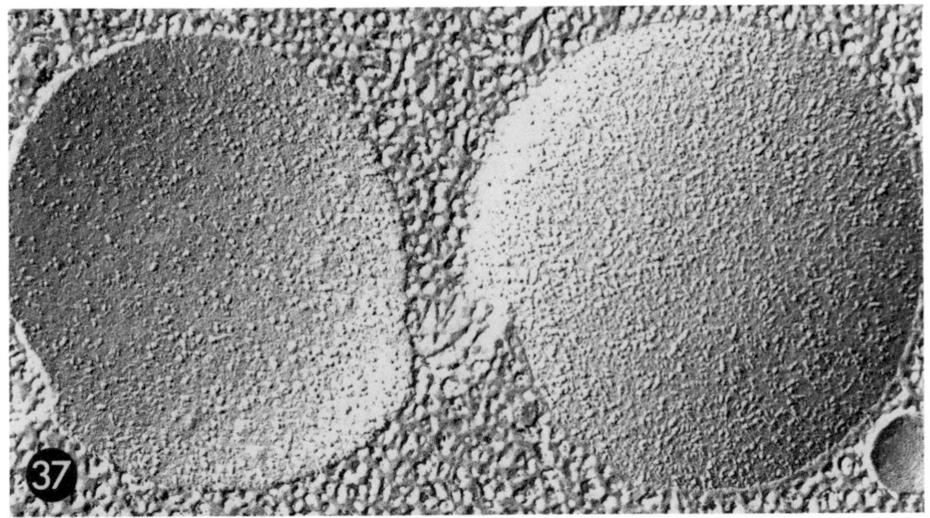

Fig. 37. Free lateral diffusion of integral proteins in the energy transducing membrane. Membranes cooled to -10°C, returned to 25°C and then frozen show completely random intramembrane particles in both fracture faces (cf. membranes at -8 to -13°C, Figs. 16, 17) X 67,000.

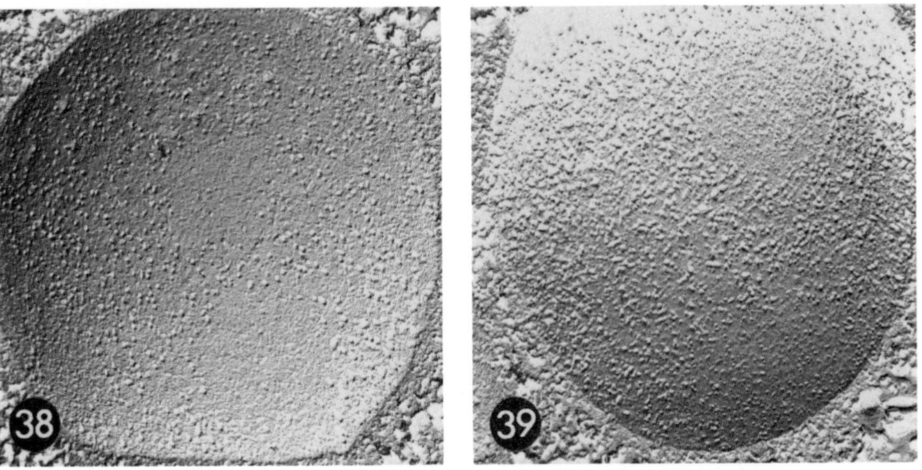

Figs. 38, 39. Free lateral diffusion of integral proteins in the energy transducing membrane independent of cytochrome *c* oxidase mobility. Membranes were cooled to -10°C, then latticed with rabbit cyt ox Ab and goat anti-rabbit IgG, then returned to 25°C and finally frozen: Fig. 38. Concave fracture face shows smooth, particle-free regions at 25°C; Fig. 39. Convex fracture face shows large areas containing small but not large particles at 25°C. X 67,500.

A significant observation is that the large intramembrane particles can be completely immobilized through a continuous lattice composed of affinity purified rabbit cyt ox Ab and goat anti-rabbit IgG, while small intramembrane particles are not inhibited in their free lateral translational diffusion by such latticing. In these experiments, the energy transducing membrane was cooled to -8 or -10°C to induce thermotropic lateral separation between integral proteins and gel state bilayer lipid (Fig. 16). This was followed by the addition of affinity purified rabbit cyt ox Ab and subsequently of goat anti-rabbit IgG at -10°C. Since, at this temperature the integral membrane proteins are excluded from the gel state bilayer lipid and thus become highly aggregated, we can reasonably assume that the intermolecular distances of cytochrome *c* oxidase were decreased, thus permitting complete crossbridging of the oxidases in the plane of the membrane through the immunoglobulin lattice.

After such immunospecific latticing, it was determined that the large intramembrane particles did not disaggregate to randomize through free lateral translational diffusion when the temperature was raised from -10 to 25°C, which is well above the phase transition of the membrane bilayer lipid (Figs. 38,39). Nonspecific IgG did not inhibit diffusion. At 25°C, the concave fracture face of the energy transducing membrane exhibited smooth, particle-free regions of liquid crystalline bilayer lipid (Fig. 38). The convex fracture face revealed similar regions free of large intramembrane particles, but contained numerous small intramembrane particles ranging in size from 4.4 to 7.4 nm in diameter (Fig. 39). Thus, we determined that various small integral proteins in the energy transducing membrane can undergo free lateral translational diffusion independent of cytochrome *c* oxidase.

On the basis of these observations, it may be concluded that cytochrome *c* oxidase is represented by large intramembrane particles and exhibits free lateral translational diffusion in the energy transducing membrane. Such diffusion can be inhibited by crossbridging through an immunoglobulin lattice composed of an IgG monospecific for the oxidase. If all the integral proteins in the energy transducing membrane are associated through a continuous, natural protein-protein lattice, then such a lattice is far less effective in immobilizing cytochrome *c* oxidase than the immunoglobulin lattice that we have used. However, it is important to note that the number of intramembrane particles inhibited in free lateral translational diffusion by the immunoglobulin lattice is greater than can be accounted for by the estimated (17,000) or determined (2,000) number of oxidase molecules or molecule polymers located in the membrane. Thus, we may further conclude that when cytochrome *c* oxidase diffuses laterally in the membrane, other as yet unidentified integral proteins diffuse with it. These may consist of cytochrome $c_1$, indeed most of the cytochrome chain, and/or the hydrophobic portion of the ATPase complex. As we have seen, those detectable integral proteins which can diffuse independently of the oxidase are restricted to the inner half bilayer of the membrane, i.e., that half of the membrane in apposition with the mitochondrial matrix. These proteins may consist of various membrane-bound dehydrogenases including succinate dehydrogenase, the *b* cytochromes, ion and substrate translocating integral proteins, and/or the hydrophobic portion of the ATPase complex. Whether or not ATPase undergoes free lateral translational diffusion independently of cytochrome *c* oxidase and other redox components of the energy transducing membrane is a particularly important consideration, since the physical relationship between ATPase and the redox components is intimately related to the mechanism of energy conservation in the mitochondrion.

## CYTOCHROME *c* OXIDASE MEMBRANE

To observe cytochrome *c* oxidase in the bilayer lipid, essentially free of other integral proteins, we prepared cytochrome *c* oxidase membranes from rat liver mitochondria using the non-ionic detergent, Triton X-114 (Jacobs *et al.*, 1966). We also purified the liver oxidase from such membranes after solubilization in 4% sodium cholate in the presence of 25% ammonium sulfate at pH 7.6-7.8. The purified enzyme which precipitates in the presence of between 25 and 32% ammonium sulfate was finally resuspended in 0.25% Tween 20 at pH 7.4.

The absolute absorption spectra of the liver cytochrome *c* oxidase is demonstrated in Figure 40 for both the purified oxidase and the oxidase membrane. Both preparations show the $\alpha$ band of the oxidized enzyme to shift from 598 to 604 nm with reduction by dithionite. Reaction of the dithionite-reduced oxidase plus carbon monoxide resulted in a decreased absorption at 604 nm and a shift in the Soret region from 444-445 to 430-432 nm, indicative of the complex of ferrocytochrome $a_3$ with carbon monoxide. Hemes other than $aa_3$ were not detected in either the oxidase or the oxidase membrane preparation. Electrophoresis of the purified oxidase and the oxidase membrane preparation in polyacrylamide disc gels containing sodium dodecyl sulfate revealed six and eight major polypeptide subunits respectively (Fig. 41). The apparent molecular weights of the six subunits common to both preparations estimated by comparison to the relative positions of standard proteins were: I, 44,500; II, 24,500; III, 15,500; IV, 12,000; V, 11,000; and VI, 9,500 (Fig. 41). Based on these subunits, a minimum molecular weight for the liver oxidase monomer is 117,000. Using a millimolar extinction coefficient of 12.0, the heme $aa_3$ content of the purified liver oxidase was approximately 10 nmol/mg while the iron content was also determined to be 10 nmol/mg protein. Thus, the purified, functional liver oxidase appears to exist as a dimer closer to 200,000 molecular weight.

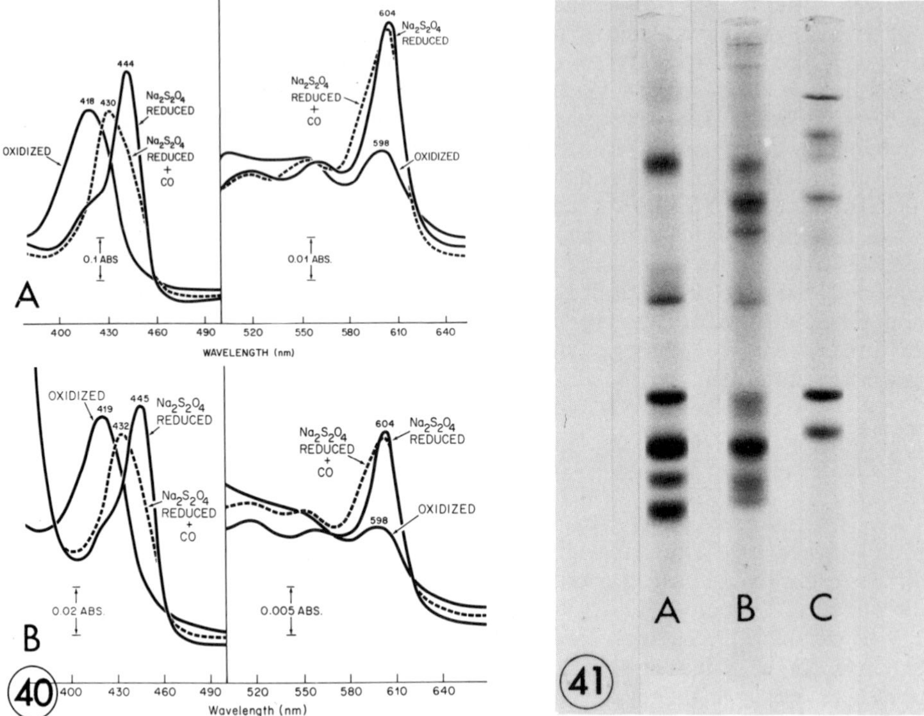


Fig. 40. Absolute absorption spectra of oxidized, dithionite reduced, and dithionite reduced plus CO purified cytochrome *c* oxidase (A) and cytochrome *c* oxidase membranes (B) prepared from rat liver mitochondria.

Fig. 41. Electrophoresis of purified cytochrome *c* oxidase (A) and cytochrome *c* oxidase membranes (B) prepared from rat liver mitochondria in sodium dodecyl sulfate-12% polyacrylamide gels. Gel C contains catalase, bovine serum albumin, glyceraldehyde phosphate dehydrogenase, myoglobin and cytochrome *c* standards.

Both the oxidase and oxidase membrane preparations catalyzed ascorbate-TMPD oxidation and oxygen reduction, while the purified enzyme was totally inhibited by affinity purified cyt ox Ab prepared against bovine cytochrome *c* oxidase. The ferritin conjugate of the cyt ox Ab clearly revealed the lateral distribution of the enzyme in the oxidase membrane (Fig. 42). Both fracture faces of the cytochrome *c* oxidase membrane exhibited large intramembrane particles with a frequency size distribution ranging from 6 to 24 nm and peaking at 15 nm in diameter (Fig. 43). Less than 5% of

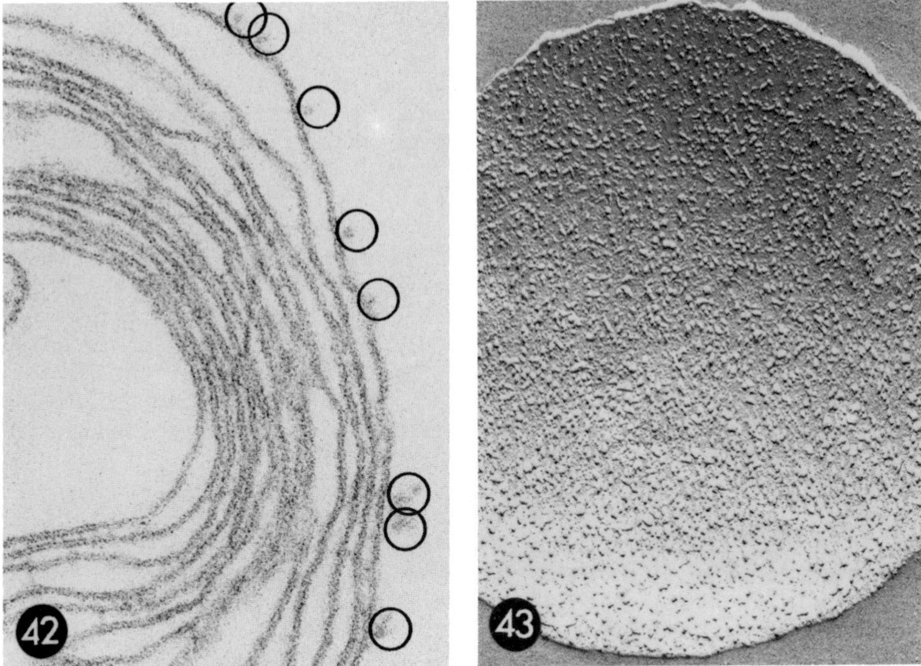


Figs. 42, 43. Cytochrome *c* oxidase membranes prepared from rat liver mitochondria: Fig. 42. Thin section showing surface distribution of the oxidase by the ferritin conjugate of cyt ox Ab, X 192,000; Fig. 43. Concave fracture face showing large intramembrane particles, X 67,500.

the intramembrane particles fell within the size of a dimeric cytochrome *c* oxidase of the mass of 200,000 Daltons (= 7.6 nm in diameter). The large size of the intramembrane particles in the cytochrome *c* oxidase membrane supports the conclusion made for the natural energy transducing membrane, i.e., that the larger intramembrane particles are representative of a large polymeric cytochrome *c* oxidase.

## FLUIDITY AND FUNCTION IN THE ENERGY TRANSDUCING MEMBRANE

Most of the integral proteins in the energy transducing membrane of the mitochondrion are nuclear gene products, i.e., other than several polypeptide subunits of cytochrome *c* oxidase and the ATPase complex and a cytochrome *b* component. We have suggested previously that nuclear encoded polypeptides destined for the energy transducing membrane enter the membrane at the mitochondrial inner membrane-outer membrane contact sites (Hackenbrock, 1968b; Hackenbrock & Miller, 1975). Such sites of contact between the two mitochondrial membranes can be observed in Figure 1 (arrows). This suggestion finds support in the finding that cytoplasmic ribosomes anchor to the outer mitochondrial membrane surface, but only in regions of contact sites (Kellems *et al.*, 1975). Other studies reveal that newly synthesized polypeptides, transcribed by the mitochondrial genome and destined for the energy transducing membrane, appear first in the inner boundary portion rather than the cristal membrane portion of the energy transducing membrane (Werner & Neupert, 1972). Thus, once a particular integral protein is assembled in the outer membrane (nuclear gene product) or is assembled in the inner membrane (mitochondrial gene product) at the membranes' contact sites, the protein may assume its ultimate functional location in a fluid energy transducing membrane through free lateral translational diffusion.

With regard to metabolic activity, as alluded to in my introduction the half time relationship between the oxidations of some redox components in the energy transducing membrane and ATP synthesis is difficult to reconcile in terms of a physical coupling of redox components to ATPase. In the case of the half time oxidation of cytochrome *c* oxidase (0.5 msec), ATP synthesis coupled to this oxidation shows considerable delay in rate and half time (100 msec) (Lemasters & Hackenbrock, 1975; Thayer & Hinkle, 1975). Such delays may be rationalized in terms of the time required for conformational changes to occur in the redox and ATPase components during conformational coupling in the membrane, or the time required to establish a protonmotive gradient across the membrane adequate to generate ATP. Considering the fluid nature of the energy transducing membrane, such delays may also be accounted for by some component of macromolecular diffusion in the overall reaction during which oxidative energy is transduced. Macromolecular diffusion in the energy transducing membrane may also account for the interaction of cytochrome *c* with cytochrome *a* of more than one respiratory chain (Wohlrab, 1970), as well as the extended oxidation half times of cytochrome *b* and of flavoproteins in oxygen pulse experiments (Chance, 1967). Delay times in the transfer of reducing equivalents to the cytochrome chain from succinate and NADH dehydrogenase is consistent with the ratio of the dehydrogenase to cytochrome chain content of the membrane (1:10). In this case, ubiquinone appears to function as a redox component shuttling between the dehydrogenases and *b* cytochromes to deliver reducing equivalents to the cytochrome chain (Kröger *et al.*, 1973a, 1973b).

The fluid nature of the energy transducing membrane provides for the motional freedom required in these metabolic events. A highly fluid lipid environment is consistent with the low viscosity of approximately 0.1 P at 30°C determined for the lipid phase of mitochondria (Keith *et al.*, 1970), the high mobility of phospholipid acyl chains detected by Carbon-13 NMR spectra (Keough *et al.*, 1973), and the lack of cholesterol as mentioned earlier. For integral membrane proteins in natural membranes studied so far, lateral translational mobilities due to Brownian motion occur with diffusion coefficients centering at approximately $5 \times 10^{-9}$ $cm^2 \cdot sec^{-1}$ at 20°C (Edidin & Fambrough, 1973; Poo & Cone, 1974; Liebman & Entine, 1974) which is equal to an approximate molecular displacement (root-mean-square) of 100 nm in one second. Thus, as a first approximation, the potential for free lateral translational diffusion of integral proteins in the energy transducing membrane of the mitochondrion can be greater than 10 nm•100 $msec^{-1}$ at physiological temperature.

## SUMMARY AND CONCLUSIONS

The energy transducing membrane of the mitochondrion is distinct from other membranes of eukaryote cells in that it supports a number of rapid macromolecular interactions effecting the transduction and conservation of metabolic energy. Also in distinction is the membrane's unusually high content of protein, high degree of unsaturated phospholipids, and lack of cholesterol. The picture that emerges is that of a highly effective concentration of integral proteins partitioned in a polar bilayer phospholipid environment of relatively low viscosity and high fluidity. The polar environment provides for a precise vertical orientation of the integral metabolically active membrane proteins. The fluid environment provides for lateral translational and rotational mobility of the integral metabolically active membrane proteins which can diffuse laterally, depending on the specific metabolic role, either independent of or in association with other integral proteins. Cytochrome *c* oxidase, a major integral metabolically active protein which occurs in the membrane as a completely transmembraneous polymer, can undergo lateral translational diffusion independent of some, and in association with other, integral proteins. In biogenesis, lateral motional freedom of integral proteins in the energy transducing membrane may permit diffusion of newly incorporated proteins to sites of functional activity. In metabolic functions, lateral motional freedom can account for a diffusional component in the mechanism of electron transport in various segments of the respiratory chain as well as in the mechanism of oxidative phosphorylation.

I wish to thank my associates who have contributed significantly to the research presented here: Drs. Matthias Höchli, Luci Höchli, and Peter Andrews; also to Katy Hammon, Mary Tobleman and Jeffrey Day. These studies were supported by the U. S. National Science Foundation and National Institutes of Health.

## REFERENCES

P. Andrews & C. R. Hackenbrock, Exp. Cell Res. 90,127 (1975).

P. D. Boyer, *in* Oxidases and Related Redox Systems, eds. T. E. King, H. S. Mason & M. Morrison, Wiley, New York, p. 994 (1965).

P. D. Boyer, B. O. Stokes, R. G. Wolcott & C. Degani, Fed. Proc. 34, 1711 (1975).

R. A. Capaldi & D. E. Green, FEBS Lett. 25, 205 (1972).

R. A. Capaldi & P-F. Tan. Fed. Proc. 33, 1515 (1974).

B. Chance, D. DeVault, V. Legallais, L. Mela & T. Yonetani, *in* Nobel Symposium 5, Fast Reactions and Primary Processes in Chemical Kinetics, ed. S. Claesson, Interscience Pub., New York, p. 437 (1967).

D. Chapman, J. Urbina & K. M. Keough, J. Biol. Chem. 249, 2512 (1974).

Y. S. Chen & W. L. Hubbell, Exp. Eye Res. 17, 517 (1973).

A. Colbeau, J. Nachbaur & P. M. Vignais, Biochim. Biophys. Acta, 249, 462 (1971).

M. Edidin & D. Fambrough, J. Cell Biol. 57, 27 (1973).

S. Fleisher, B. Fleisher & W. Stoeckenius, J. Cell Biol. 32, 193 (1967).

C. W. M. Grant & H. M. McConnell, Proc. Natl. Acad. Sci. USA 71, 4653 (1974).

C. R. Hackenbrock, J. Cell Biol. 30, 269 (1966).

C. R. Hackenbrock, J. Cell Biol. 37, 345 (1968a).

C. R. Hackenbrock, Proc. Natl. Acad. Sci. USA 61, 598 (1968b).

C. R. Hackenbrock, Ann. N.Y. Acad. Sci. 195, 492 (1972a).

C. R. Hackenbrock, J. Cell Biol. 53, 450 (1972b).

C. R. Hackenbrock, *in* Mechanisms in Bioenergetics, eds. G. F. Azzone, L. Ernster, S. Papa, E. Quagliariello & N. Siliprandi, Academic Press, New York, p. 77 (1973).

C. R. Hackenbrock, Arch. Biochem. Biophys. 170, 139 (1975).

C. R. Hackenbrock & K. J. Miller, J. Cell Biol. 65, 615 (1975).

C. R. Hackenbrock & K. Miller Hammon, J. Biol. Chem. 250, 9185 (1975).

C. R. Hackenbrock, T. G. Rehn, E. C. Weinbach & J. J. Lemasters, J. Cell Biol. 51, 123 (1971a).

C. R. Hackenbrock, T. G. Rehn, E. C. Weinbach & J. J. Lemasters, *in* Energy Transduction in Respiration and Photosynthesis, eds. E. Quagliariello, S. Papa & C. S. Rossi, Adriatica Editrice, Bari, p. 285 (1971b).

H. J. Harmon, J. D. Hall & F. L. Crane, Biochim. Biophys. Acta. 334, 119 (1974).

E. E. Jacobs, E. C. Andrews, W. Cunningham & F. L. Crane, Biochem. Biophys. Res. Commun. 25, 87 (1966).

P. C. Jost, O. H. Griffith, R. A. Capaldi & G. Vanderkooi, Proc. Natl. Acad. Sci. USA 70, 480 (1973).

R. E. Kellems, V. Allison & R. A. Butow, J. Cell Biol. 65, 1 (1975).

A. Keith, G. Bulfield, & W. Snipes, Biophysic. J. 10, 618 (1970).

K. M. Keough, E. Oldfield, D. Chapman & P. Beynon, Chem. Phys. Lipids 10, 37 (1973).

W. Kleeman, C. W. M. Grant & H. M. McConnell, J. Supramol. Struct. 2, 609 (1974).

M. Klingenberg, *in* Biological Oxidations, ed. T. P. Singer, Wiley, New York, p. 3 (1968).

A. Kröger, M. Klingenberg & S. Schweidler, Eur. J. Biochem. 34, 358 (1973a).

A. Kröger, M. Klingenberg & S. Schweidler, Eur. J. Biochem. 39, 313 (1973b).

B. D. Ladbrooke & D. Chapman, Chem. Phys. Lipids 3, 304 (1969).

M. P. Lee & A. R. L. Gear, J. Biol. Chem. 249, 7541 (1974).

A. L. Lehninger, Biochemistry, Worth Publishers, Inc., New York (1970).

J. J. Lemasters & C. R. Hackenbrock, Fed. Proc. 34, 596 (1975).

P. A. Liebman & G. Entine, Science 185, 457 (1974).

H. M. McConnell, K. L. Wright & B. G. McFarland, Biochem. Biophys. Res. Commun. 47, 273 (1972).

D. Papahadjopoulos, W. J. Vail & M. Moscarello, J. Membrane Biol. 22, 143 (1975).

D. Parsons, G. R. Williams & B. Chance, Ann. NY Acad. Sci. 137, 643 (1966).

M-M. Poo & R. A. Cone, Nature 247, 438 (1974).

J. K. Raison, J. M. Lyons & W. W. Thomson, Arch. Biochem. Biophys. 142, 83 (1971).

F. S. Sjöstrand & L. Barajas, J. Ultrastruct. Res. 32, 293 (1970).

J. M. Steim, M. E. Tourtellotte, J. C. Reinert, R. N. McElhaney & R. L. Rader, Biochemistry 63, 104 (1969).

W. S. Thayer & P. C. Hinklė, J. Biol. Chem. 250, 5336 (1975).

H. Trauble & P. Overath, Biochim. Biophys. Acta 307, 491 (1973).

G. Vanderkooi, Biochim. Biophys. Acta 344, 307 (1974).

S. Werner & W. Neupert, Eur. J. Biochem. 25, 369 (1972).

H. Wohlrab, Biochemistry 9, 474 (1970).

# PHOSPHOLIPID METABOLISM IN SOME EXCITABLE BIOLOGICAL MEMBRANES

J. N. Hawthorne

Department of Biochemistry, University Hospital and

Medical School, Nottingham NG7 2UH, U.K.

## ABSTRACT

Triphosphoinositide, a constituent of myelin and plasma membranes, may be important in binding calcium ions. The purification to homogeneity of a phosphatase from brain which removes the 4- and 5-phosphate groups of this phospholipid is described. In iris muscle, acetylcholine activates the breakdown of triphosphoinositide, the effect being blocked by atropine but not tubocurarine. The related lipid phosphatidylinositol was most highly labelled *in vivo* in the membrane of transmitter vesicles from brain synaptosomes. Electrical stimulation of synaptosomes caused rapid breakdown of this phosphatidylinositol and of phosphatidic acid in another sub-synaptosomal membrane fraction. Possible functions of these lipids in transmitter release are discussed.

## 1. INTRODUCTION

The first part of this paper deals with recent studies of triphosphoinositide metabolism in nerve and muscle. It was suggested some years ago (Hawthorne and Kemp, 1964) that through its ability to bind calcium this phospholipid might be involved in membrane permeability changes related to nerve conduction.

The second part concentrates upon phosphatidylinositol and phosphatidic acid metabolism in synaptosomes. Transmitter release from the nerve ending is associated with the breakdown of these phospholipids and it is suggested that the diacylglycerol produced may be important in exocytosis. There is also recent evidence that phosphatidylinositol can be associated with tubulin in the brain.

## 2. TRIPHOSPHOINOSITIDE

Triphosphoinositide is formed by the phosphorylation of phosphatidylinositol and has the structure 1-phosphatidyl (myo-inositol 4,5 bisphosphate). Dawson (1954) first showed that the monoesterified 4- and 5-phosphate groups on the inositol ring had an unusually rapid turnover. Increased turnover of these phosphate groups may accompany conduction of impulses in vagus and other nerves (Birnberger *et al.*, 1971; Salway and Hughes, 1972; White *et al.*, 1974). Conduction is known to involve release of calcium ions from the axolemma and this may be accompanied by changes in triphosphoinositide metabolism. In the erythrocyte membrane, this lipid and the closely related diphosphoinositide can regulate calcium binding. Tightly bound $Ca^{2+}$ ($K_{ass.}$ $4 \times 10^4$ litres/mol) increased in 1:1 molar ratio with polyphosphoinositide monoester phosphate (Buckley and Hawthorne, 1972). Garrett *et al.* (1975) have shown that entry of $Ca^{2+}$ into the red cell leads to very rapid breakdown of triphosphoinositide. Analogous events may well take place in nerve tissue when depolarization or transmitter action cause entry of calcium ions.

One of the interesting features of triphosphoinositide metabolism in brain is that the enzymes hydrolysing this lipid are at least two orders of magnitude more active than the kinases forming it from phosphatidylinositol. Regulation of these hydrolytic enzymes must therefore be of considerable importance, and with this in mind Dr. M. S. Nijjar (unpublished work) has recently purified one of them in our laboratory. This is the phosphatase catalysing the following reactions:

triphosphoinositide + $H_2O \rightarrow$ diphosphoinositide + $P_i$

diphosphoinositide + $H_2O \rightarrow$ phosphatidylinositol + $P_i$

Studies of a similar enzyme in kidney suggest that it can remove both 4- and 5-phosphates from triphosphoinositide (Cooper and Hawthorne, 1975). Table 1 outlines the purification of the brain enzyme.

The purified phosphatase had a $K_m$ of 25 μM for triphosphoinositide in the absence of divalent ions, and did not appear to be dependent on the latter. Calcium ions increased activity 4-fold however, the optimum $Ca^{++}$ concentration being 100 μM. Magnesium ions were much less effective activators, 50 mM being required for the maximum effect. It seems likely that a sudden influx of calcium ions would activate the hydrolysis of triphosphoinositide, but the enzyme is probably subject to strong

Table 1. Purification of triphosphoinositide phosphatase from rat brain

| | Protein (mg) | Specific activity ( mol $P_i$/mg protein/h) | Recovery (%) | Purification |
|---|---|---|---|---|
| Brain homogenate | 3498 | 1.03 | 100 | 1.0 |
| Supernatant (105,000 x *g*, 60 min) | 660 | 2.47 | 45.2 | 2.4 |
| DEAE-cellulose | 154 | 6.34 | 27.1 | 6.2 |
| Calcium phosphate gel | 3.4 | 164.7 | 15.7 | 160.1 |
| Sephadex G-100 | 0.38 | 443.6 | 4.7 | 430.7 |

inhibition under normal conditions.

It has long been known that activation of many cells through surface receptors leads rapidly to changes in the metabolism of phosphatidylinositol and phosphatidic acid. Examples range from the action of neurotransmitters to the activation of platelets and lymphocytes (Michell, 1975). No consistent changes in the turnover of triphosphoinositide or its phosphate groups had been described in such systems, however, until Dr. A. A. Abdel-Latif, working as visiting professor in our Nottingham laboratory, showed that acetylcholine causes the hydrolysis of this lipid in the rabbit iris muscle. In this recent study (Abdel-Latif and Akhtar, 1976) 0.05 mM acetylcholine was added to muscle pre-labelled for 30 min *in vitro* with $^{32}P_i$ or $[^3H]$inositol. In the $^{32}P$ experiments there was a 32% loss of label from triphosphoinositide after 15 min incubation with acetylcholine and this was accompanied by a 20% increase in labelling of phosphatidylinositol and a 36% increase in phosphatidate labelling. Changes were readily detectable within 2.5 min of adding the acetylcholine. Atropine blocked the changes in labelling caused by acetylcholine, but tubocurarine had no effect, suggesting that muscarinic receptors are involved. In all these experiments the 10 mM 2-deoxyglucose was added after the pre-incubation period to trap newly formed ATP. Under these conditions there should be little phospholipid synthesis once the high-energy intermediates such as CDP-diacylglycerol formed during pre-incubation have been used up. Abdel-Latif and Akhtar suggest that acetylcholine activates the hydrolysis of triphosphoinositide by the phosphomonoesterase discussed already, thus increasing the pool of labelled phosphatidylinositol. It cannot be excluded that further phosphatidylinositol synthesis *de novo* from labelled intermediates already present was also taking place. If the triphosphoinositide phosphatase of muscle resembles that of brain in being activated by calcium ions, acetylcholine might act by making such ions available to it. Schacht and Agranoff (1972) found that labelling of diphosphoinositide and triphosphoinositide with $^{32}P$ *in vitro*, using subcellular fractions from guinea pig brain, was 20% inhibited by acetylcholine. The effect was not reversed by atropine however, and was not specifically associated with nerve endings.

## 3. PHOSPHATIDIC ACID AND PHOSPHATIDYLINOSITOL

In nervous tissue it seems clear that excitation electrically or by transmitters such as acetylcholine increases the turnover of phosphatidylinositol and phosphatidic acid and that the phospholipid changes accompany both pre-synaptic and post-synaptic events. The work described here makes use of synaptosomes (nerve-ending particles) electrically stimulated *in vitro*, using the technique of De Belleroche and Bradford (1972), who showed that this stimulation caused transmitter release with little membrane damage. Such experiments provide a model of pre-synaptic events in chemical transmission of the nerve impulse and are likely to be more meaningful than the incubation of synaptosomes with transmitters, the technique previously used. Much of the earlier work on these and similar phospholipid changes involved studies of $^{32}P$ incorporation, a process dependent on phospholipid biosynthesis. The initial response to excitation could be hydrolysis of phosphatidylinositol, however. In this connection it is significant that the only phospholipase C acting on glycerophosphatides of brain is specific for inositol lipids. It hydrolyses phosphatidylinositol to diacylglycerol and cyclic inositol phosphate. With this in mind Mr. M. R. Pickard in my laboratory has studied changes in electrically stimulated synaptosomes in which the phosphatidylinositol had been pre-labelled *in vivo*. His work is described in the following three paragraphs.

Conditions for synaptosomal labelling *in vivo*. It was necessary first to establish optimum conditions for the labelling of synaptosomal phosphatidylinositol *in vivo*. Guinea pigs were anaesthetised for intraventricular injection of carrier-free $^{32}P_i$ by the method of Lunt and Pickard (1975). Animals were killed at various times after injection and synaptosomes prepared from brain cortex as described by Yagihara *et al.* (1973). A purer synaptosomal fraction was obtained by the inclusion of an additional sucrose density gradient centrifugation as the final stage. Material from the 0.8 M/1.2 M sucrose interface was layered above a finer gradient of 1.0 M and 1.2 M sucrose and centrifuged for 2 h at 53,000 x *g*. Electron micrographs showed that this additional step greatly reduced contamination by mitochondria and myelin fragments. A marked peak of phosphatidate and phosphatidylinositol labelling was seen in the synaptosomal fraction after 2 h *in vivo*. No such peak was seen in the labelling of these lipids in whole brain or in synaptosomal phosphatidylcholine, phosphatidylserine or phosphatidylethanolamine. At this time the specific radioactivity of synaptosomal phosphatidate was about 12 times higher than that of synaptosomal phosphatidylcholine and roughly twice that of phosphatidylinositol. Synaptosomal phosphatidate was 7 times more radioactive than whole brain phosphatidate.

Localisation of phospholipid changes within the synaptosome. Synaptosomes prepared from brain labelled for 2h in this way were subjected to osmotic shock for the preparation of sub-synaptosomal fractions according to Whittaker *et al.* (1964). Specific radioactivities of individual phospholipids in these fractions were then determined as previously described (Yagihara *et al.*, 1973). Fraction D, as confirmed by catecholamine assay, is rich in synaptic vesicles; fraction E contains microsomal membranes, fraction H damaged synaptosomes. Using the method of Whittaker *et al.* (1964) fraction I contains mostly mitochondria but the purer synaptosomal fraction used in the present study gave a fraction I with large membrane fragments and relatively few mitochondria. Table 2 shows that the most highly labelled phosphatidylinositol is in the

Table 2. Labelling of sub-synaptosomal membranes *in vivo*

| Fraction | Relative specific activity | | |
|---|---|---|---|
| | Phosphatidate | Phosphatidyl-inositol | Phosphatidyl-ethanolamine |
| D | 84 | 249 | 89 |
| E | 210 | 83 | 74 |
| F | 55 | 32 | 87 |
| G | 88 | 62 | 79 |
| H | 69 | 125 | 100 |
| I | 130 | 56 | 74 |

Relative specific activity represents specific activity (c.p.m./µmolP) of the phospholipid in a particular fraction as a percentage of the specific activity of that phospholipid in the original synaptosomal fraction. Results are means of 4 experiments or 2 experiments in the case of Fraction I.

synaptic vesicle fraction. There is also pronounced labelling in fraction H which contained synaptosome 'ghosts' and possibly some intact synaptosomes. Phosphatidate on the other hand was most active in the microsomal fraction E and fraction I. Labelling of phosphatidylethanolamine was much the same in all fractions.

Electrical stimulation of pre-labelled synaptosomes. For these experiments labelled synaptosomes were prepared as described in the previous paragraph. They were then washed in the Bradford medium, incubated at 37° for 15 min in fresh medium of this composition and stimulated electrically by the method of Bleasdale and Hawthorne (1975). Subsynaptosomal fractions were then prepared as already outlined and phospholipid labelling was measured in both stimulated and control fractions. Table 3 summarises the results. The stimulation produced a marked loss of labelled phosphatidylinositol from the vesicle fraction and a similar loss of phosphatidate from the microsomal fraction E. No changes were seen in the phosphatidylethanolamine of any fraction.

Table 3. Changes caused by electrical stimulation of synaptosomes labelled *in vivo* with $^{32}P$

| Fraction | Relative specific activity | | | | | |
|---|---|---|---|---|---|---|
| | Phosphatidate | | Phosphatidyl-inositol | | Phosphatidyl-ethanolamine | |
| | C | S | C | S | C | S |
| Synaptosomes | 100 | 56 | 100 | 17 | 100 | 95 |
| D | 36 | 38 | <u>301</u> | <u>78</u> | 111 | 127 |
| E | <u>124</u> | <u>11</u> | 47 | 44 | 83 | 106 |
| F | 22 | 24 | 69 | 76 | 90 | 115 |
| G | 38 | 50 | 37 | 62 | 96 | 113 |
| H | 19 | 32 | 120 | 97 | 98 | 108 |
| I | 97 | 88 | 56 | 53 | 112 | 103 |

C, control; S, stimulated. Relative specific activity is defined as the specific radioactivity of a phospholipid in a particular fraction expressed as a percentage of the specific radioactivity of that phospholipid in the control synaptosomes.

Significance of the phospholipid changes in relation to transmitter release. Previous work in which unlabelled synaptosomes were stimulated electrically in the presence of $^{32}P_i$ showed that there was increased labelling of phosphatidate in the synaptic vesicle fraction (Bleasdale and Hawthorne, 1975). There was no consistent increase in phosphatidylinositol labelling under these conditions. Phosphatidate changes were detectable within 2 min of the onset of stimulation, were dependent on external calcium ions and were also seen when depolarization was produced by a high concentration of external potassium ions. This labelling *in vitro* can be compared with the labelling *in vivo* of Table 2, where phosphatidylinositol not phosphatidate is the highly labelled vesicle lipid. Since phosphatidate is a precursor of phosphatidylinositol (reactions 1 and 2) one explanation would be that the conversion of vesicle phosphatidate to vesicle

$$\text{phosphatidate} + \text{CTP} \rightarrow \text{CDP-diacylglycerol} + \text{P-P}_i \quad (1)$$

$$\text{CDP-diacylglycerol} + \text{inositol} \rightarrow \text{phosphatidylinositol} + \text{CMP} \quad (2)$$

phosphatidylinositol requires a system present in intact neurones but not in synaptosomes. Whatever the explanation, it seems that turnover of phosphatidylinositol occurs in the membranes of synaptic vesicles. Phosphatidate synthesis *in vivo* is apparently most active in other membranes (Table 2, fraction E). These could be from the endoplasmic reticulum or plasma membrane.

Electrical stimulation of pre-labelled synaptosomes (Table 3) in a non-radioactive medium causes loss of labelled phosphatidylinositol from the vesicles. The following reaction is most likely to be occurring:

$$\text{phosphatidylinositol} + H_2O \rightarrow \text{diacylglycerol} + \text{cyclic inositol phosphate}$$

Conversion of phosphatidylinositol to diacylglycerol will alter the properties of the vesicle membrane and could promote the fusion with the plasma membrane required for exocytosis and transmitter release. Allan and Michell (1975) have shown that phospholipase C treatment of erythrocytes produces diacylglycerol in the membrane. They attributed the resulting vesiculation to the membrane-fusing ability of the diacylglycerol.

Stimulation also causes loss of phosphatidate from fraction E which is harder to explain. The first problem is that the origin of the membranes which make up this fraction is not known, though endoplasmic reticulum and plasma membrane are most probable. Cotman *et al.* (1971) showed that phosphatidate phosphatase is present in the synaptosomal plasma membrane. If depolarization activates this enzyme, diacylglycerol would be produced in the plasma membrane and could contribute to exocytosis as indicated in the previous paragraph. Diacylglycerol kinase is active in brain (Lapetina and Hawthorne, 1971) and would convert this diacylglycerol and that from phosphatidylinositol back to phosphatidate. This enzyme could account for the labelling of vesicle phosphatidate seen *in vitro* (Hawthorne and Bleasdale, 1975).

An alternative explanation for the loss of phosphatidate from fraction E membranes is its conversion to CDP-diacylglycerol and then phosphatidylinositol, to replace loss from the vesicle membranes.

The observations are consistent with a cycle of events similar to that suggested by Hokin and Hokin (1964) in a different context.

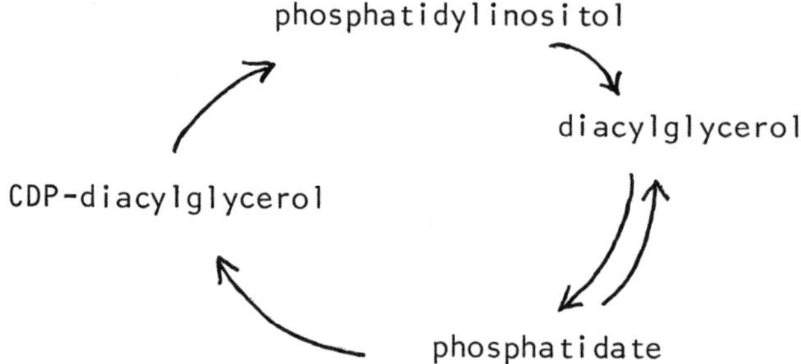


For transmitter release, it is suggested that the key event is diacylglycerol production, promoting membrane fusion. Re-sealing of vesicles would involve the re-synthesis of phosphatidate and phosphatidylinositol. Further work will show whether these speculations are valid.

Finally, another recent observation should be mentioned. Tubulin, the brain protein capable of aggregation into tubular structures, is considered by some workers to be concerned in transmitter release. Dr. J. Lagnado of Bedford College London (personal communication) has shown that small amounts of phosphatidylinositol are specifically bound to tubulin. Daleo *et al.* (1974) have claimed that diglyceride kinase is also associated with microtubules and we have confirmed this for brain tubulin. Work is in progress on the possible relationship between these observations and our own studies with synaptosomes.

## REFERENCES


Abdel-Latif, A.A. and R. A. Akhtar, Acetylcholine causes an increase in the hydrolysis of triphosphoinositide pre-labelled with [$^{32}$P]phosphate or [$^{3}$H]inositol and a corresponding increase in the labelling of phosphatidylinositol and phosphatidic acid in rabbit iris muscle, Biochem. Soc. Trans. *4*, 317, 1976.

Allan, D. and R. H. Michell, Elevation of intracellular calcium ion concentration provokes production of 1,2-diacylglycerol and phosphatidate in human erythrocytes, Biochem. Soc. Trans., *3*, 751, 1975.

Birnberger, A.C., K. L. Birnberger, S. G. Eliasson and P. C. Simpson, Effect of cyanide and electrical stimulation on phosphoinositide metabolism in lobster nerves, J. Neurochem. *18*, 1291, 1971.

Bleasdale, J.E. and J. N. Hawthorne, The effect of electrical stimulation on the turnover of phosphatidic acid in synaptosomes from guinea-pig brain, J. Neurochem. *24*, 373, 1975.

Buckley, J.T. and J. N. Hawthorne, Erythrocyte membrane polyphospho-inositide metabolism and the regulation of calcium binding, J. Biol. Chem. *247*, 7218, 1972.

Cooper, P.H. and J. N. Hawthorne, Phosphomonoesterase hydrolysis of polyphosphoinositides in rat kidney, Biochem. J. *150*, 537, 1975.

Cotman, C.W., R. E. McCaman and S. A. Dewhurst, Subsynaptosomal distribution of enzymes involved in the metabolism of lipids, Biochim. Biophys. Acta, *249*, 395, 1971.

Daleo, G.R., M. M. Piras and R. Piras, The presence of phospho-lipids and diglyceride kinase activity in microtubules from different tissues, Biochem. Biophys. Res. Commun. *61*, 1043, 1974.

Dawson, R.M.C., Labelling of brain phospholipids with radioactive phosphorus, Biochem. J. *57*, 237, 1954.

De Belleroche, J.S. and H. F. Bradford, Metabolism of beds of mammalian cortical synaptosomes: response to depolarizing influences, J. Neurochem. *19*, 585, 1972.

Garrett, N.E., R. J. B. Garrett, R. T. Talwalkar and R. L. Lester, Rapid breakdown of diphosphoinositide and triphosphoinositide in erythrocyte membranes, J. Cell. Physiol. *87*, 63, 1975.

Hawthorne, J.N. and J. E. Bleasdale, Phosphatidic acid metabolism, calcium ions and transmitter release from electrically stimulated synaptosomes, Molec. Cell. Biochem. *8*, 83, 1975.

Hawthorne, J.N. and P. Kemp, The brain phosphoinositides, Advan. Lipid Res. *2*, 127, 1964.

Hokin, M.R. and L. E. Hokin, Interconversions of phosphatidylinositol and phosphatidic acid involved in the response to acetylcholine in the salt gland, in Metabolism and Physiological Significance of Lipids (ed. R. M. C. Dawson and D. N. Rhodes) p.423, John Wiley, London, 1964.

Lapetina, E.G. and J. N. Hawthorne, The diglyceride kinase of rat cerebral cortex, Biochem. J. *122*, 171, 1971.

Lunt, G.C. and M. R. Pickard, The subcellular localization of carbamylcholine-stimulated phosphatidylinositol turnover in rat cerebral cortex *in vivo*, J. Neurochem. *24*, 1203, 1975.

Michell, R.H., Inositol phospholipids and cell surface receptor function, Biochim. Biophys. Acta, *415*, 81, 1975.

Salway, J.G. and I. E. Hughes, The possible role of phosphoinositides as regulators of action potentials: effect of electrical stimulation, tetrodotoxin and cinchocaine on phosphoinositide labelling by $^{32}P$ in rabbit vagus, J. Neurochem. *19*, 1233, 1972.

Schacht, J. and B. W. Agranoff, Effects of acetylcholine on labelling of phosphatidate and phosphoinositides by [$^{32}P$]-orthophosphate in nerve ending fractions of guinea pig cortex, J. Biol. Chem. *247*, 771, 1972.

White, G.L., H. U. Schellhase and J. N. Hawthorne, Phosphoinositide metabolism in rat superior cervical ganglion, vagus and phrenic nerve: effects of electrical stimulation and various blocking agents, J. Neurochem. *22*, 149, 1974.

Whittaker, V.P., I. A. Michaelson and R. J. A. Kirkland, The separation of synaptic vesicles from nerve-ending particles ('synaptosomes'), Biochem. J. *90*, 293, 1964.

Yagihara, Y., J. E. Bleasdale and J. N. Hawthorne, Effects of acetylcholine on the incorporation of [$^{32}P$]orthophosphate *in vitro* into the phospholipids of subsynaptosomal membranes from guinea-pig brain, J. Neurochem. *21*, 173, 1973.

# ASPECTS ON STRUCTURE AND FUNCTION OF SPHINGOLIPIDS IN CELL SURFACE MEMBRANES

Karl-Anders Karlsson

Department of Medical Biochemistry

University of Göteborg, Fack

S-400 33 Göteborg 33, Sweden

## ABSTRACT

Sphingolipids are components of cell surface membranes and are probably exclusively located in the outer half of the lipid bilayer. It is shown by calculation that sphingolipid is a major part of the surface monolayer, in addition to protein, cholesterol and phosphatidylcholine. The lipophilic part, ceramide, has, beside a hydrocarbon chain variation as found for other membrane lipids, characteristic structural features (amide, hydroxyls) with a natural variation (number of hydroxyls), which are suggestive of lateral polar interactions with other surface layer components. This intermediate zone of the bilayer (between the hydrophobic part and the polar head groups) is proposed to be of importance for some of the matrix properties of surface membranes. It has, however, attracted almost no attention so far in model studies. Three surface membrane functions have been found associated with sphingolipid polar head groups composed of carbohydrate: $Na^+$ transport (sulphatide), cholera toxin binding (a ganglioside), and recognition (surface antigens). A molecular model of $Na^+$-$K^+$ translocation is formulated, where sulphatide is postulated to be essential for $K^+$ influx (cofactor site model).

## 1. INTRODUCTION

Sphingolipids are mainly located in surface membranes. When found intracellularly they exist in the endoplasmic reticulum and the Golgi apparatus, from which the surface membranes are biosynthetically derived, possibly through vesicle transport and fusion with the surface membrane.

There is increasing evidence for an asymmetrical arrangement of lipids across the erythrocyte membrane (see Singer, 1974; Renooij et al., 1976). Choline-containing lipids (phosphatidylcholine and sphingomyelin) are preferentially located in the outer half of the bilayer, and amino group-containing lipids (phosphatidylserine and phosphatidylethanolamine) mainly in the inner half. Cholesterol is present in both layers. It is of interest that sphingolipids may exist exclusively in the outer layer, as shown for sphingomyelin (see Renooij et al., 1976), the major phosphosphingolipid, and for globoside (Steck and Dawson, 1974), one of the major glycosphingolipids of human tissues.

What does it mean functionally that these amphipathic lipids with characteristic chemical features are associated with the cell surface? In the following I want to define this question and illustrate some points of relevance for studies in this field. For this purpose it is convenient to consider the two parts of the sphingolipid separately, the ceramide and the polar head group. When seen in the fluid mosaic membrane, as formulated by Singer and Nicolson (see Singer 1974; Nicolson, 1976), the sphingolipid is anchored with ceramide in the fluid lipophilic matrix, and with the polar head group exposed to the cell environment.

## 2. CERAMIDE, THE LIPOPHILIC PART

Sphingolipids are usually a smaller part of <u>total</u> lipids of different tissues, probably due to a rather small contribution of plasma membranes in relation to intracellular membranes. This may explain why these lipids have attracted relatively little attention so far in physicochemical experiments aimed to define membrane properties. However, assuming all sphingolipid to exist in the outer half of the plasma membrane, one may calculate (Table I) that each second to third lipid molecule in the surface layer is a sphingolipid (with a varying ratio of phospho- and glycolipid, see Table I). This is true for representative cells of mammalian tissues and may be a general phenomenon of animal cells. In most microorganisms, however, sphingolipids, and also cholesterol, have not been detected, although they seem to be obligate components of yeast cells and protozoa.

Our postulate, based on the assumptions given above, is that the outer half of the plasma membrane bilayer of animal cells consists of protein and three major lipid components, cholesterol, phosphatidylcholine and sphingolipid. Until more analytical data have gathered, this neglected aspect on sphingolipids may be a fruitful working hypothesis for model experiments.

What has made the cell to select ceramide as an important constituent of the cell boundary to the environment? Some variations

TABLE I. Calculated Sphingolipid Concentration of Outer Layer of Some Plasma Membranes

| Cell type | Per cent sphingolipid of total lipid in outer monolayer[a] | Sphingomyelin: glycolipid | Literature reference |
|---|---|---|---|
| Glia nerve cell (Human)[b] | 63 | 1:5 | O'Brien and Sampson, 1965 |
| Liver cell (Rat) | 54 | 3:1 | Dod and Gray, 1968 |
| Epithelial cell (Rat intestine) | 50 (circa) | 1:10 | Forstner et al., 1968, 1973 |
| Erythrocyte (Human) | 32[c] | 5:1[c] | Sweeley and Dawson, 1969 |
| Erythrocyte (Sheep)[d] | 84 | 3:1 | Rouser et al., 1968 |
| Erythrocyte (Horse) | 60 | 1:3 | Rouser et al., 1968 |

[a] Assuming an equal amount of lipid in the two layers

[b] The figures were calculated from myelin, which is derived from glia cell plasma membrane

[c] Molar Ratio

[d] The very low amount of phosphatidylcholine reported for this cell would mean a dominance of sphingolipid and cholesterol, in an approximate molar ratio of 1.

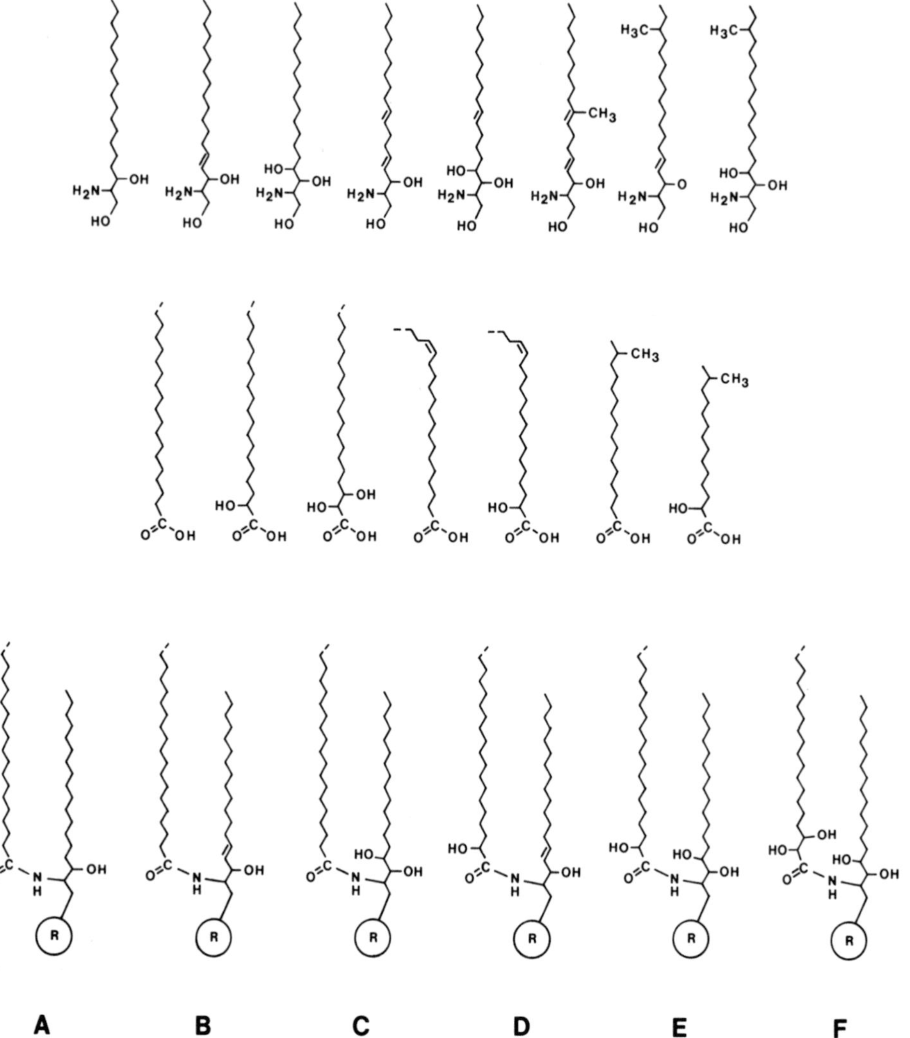


Figure 1. Selected molecular species of sphingolipid long-chain base (above), fatty acid (middle) and some of their combinations in sphingolipid (below).

in ceramide structure apparently related to tissue function have given us some suggestion.

The hydrocarbon chains of ceramide components (fatty acid and long-chain base, see Fig. 1) obviously adjust their fluidity

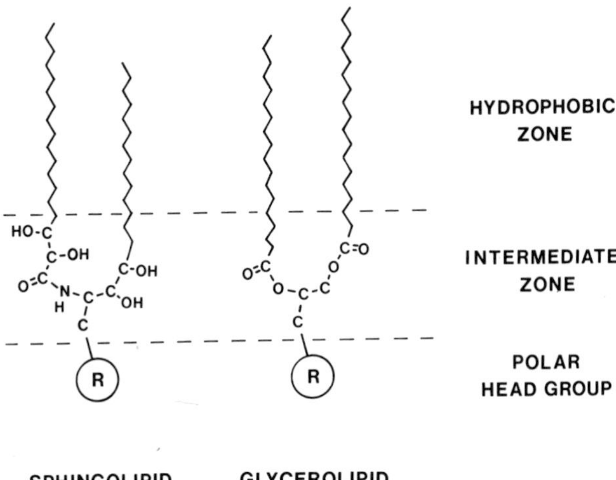


Figure 2. Structural features of sphingolipid and glycerolipid. The intermediate zone is suggested to be of importance for lateral interactions in the surface membrane.

characteristics to tissue requirement as found for membrane lipids in general, except that sphingolipids are consistently more saturated and have longer paraffin chains than glycerolipids. In the region of chemical linkage of the two components, however, there is a variation in polar functional groups of the ceramide (hydroxyls, 'allylic' group), which has no corresponding expression in other lipids, except in the case of plasmalogens (the ester in position one of glycerolipids is exchanged with a vinyl ether group, a change expected to influence hydrogen bonding, see below). This part of the molecules (or membrane) is provisionally named the intermediate zone (placed between the hydrophobic part and the polar head groups, see Fig. 2), to aid the discussion and call attention to potential model experiments.

There is a striking relation in different tissues of the number of ceramide hydroxyl groups and an increased mechanical and chemical stress on the membrane (see Table II), suggesting that the intermediate zone may create stabilizing polar interactions (hydrogen bonding) in a lateral direction of the membrane. This means that the cell may have two separate ways to modify the properties of its surface monolayer matrix through ceramide structure. One is by selection of optimally fluid hydrocarbon chains, a second by a regulation of polar groups in the intermediate zone. In fact, we have found a natural variation in these two molecular parts, apparently independent of each other (some few examples may be extracted from

TABLE II. Ceramide Composition of Some Selected Cells or Tissues. In All Cases a Polar Head Group is Bound at Carbon Atom One of the Long-Chain Base

| Cell type | Major ceramide (Compare Fig. 1) | Number of hydroxyl groups | Literature reference | Comment |
|---|---|---|---|---|
| Lens (Human) | A,B | 1 | Tao and Cotlier, 1975 | This tissue is unique in having more A than B. |
| Erythrocyte (Human) | B | 1 | Karlsson, 1970 | Constant environment (blood plasma). |
| Brain | B,D | 1,2 | Karlsson, 1970 | Local variation in ionic strength within certain limits. |
| Kidney (Human) | B,C,D,E | 1,2,3 | Karlsson, 1970 | Varying composition (pH and ionic strength) of primary urine of kidney tubules. Increasing hydroxyl content and chain length towards papilla. |
| Intestine (Dog) | | | | |
| -Lamina propria | B | 1 | Smith et al., 1975a | Embedded in connective tissue. |
| -Epithelium | E | 3 | Smith et al., 1975a | Exposed to a varying content of intestinal lumen. No evidence for B. Possibly F in small amounts. |
| Yeast cells | E,F | 3,4 | Smith and Lester, 1974 | Unicellular organism in non-regulated environment. |

Fig. 1), supporting the view of a parallel existence of these different functions in the same molecule.

When considering possible interactions of cholesterol, phosphatidylcholine, sphingolipid and protein, the major components in the surface monolayer according to our calculations, it is of interest that the intermediate zone is distinctly different in sphingolipid and glycerolipid (Fig. 2). In the latter, only acceptors of hydrogen bonding exist, while sphingolipid has both acceptors and donors (compare discussion by Sundaralingam, 1972). In a monolayer of these two lipids one would therefore expect hydrogen bonding between their intermediate zones, to lower the free-energy state of the system. Alternatively, but less probable, both lipids may form hydrogen bonding with water, a situation expected to be favoured at equilibrium in a monolayer of pure glycerolipid. Thus one may predict a more dense, and less permeable (for both polar and nonpolar substances), membrane in the presence of sphingolipid. Furthermore, addition of hydroxyls to the ceramide in a specified way (asymmetric groups close to the rigid amide) may create specific interaction points, or at least, increase the probability of an associated before a dissociated state in the fluid membrane.

Unfortunately, very little is known concerning physicochemical properties of sphingolipids compared with the large volume of data for cholesterol and phosphatidylcholine (see Chapman, 1975). Has ceramide a condensing effect on phosphatidylcholine monolayers, similar to the well documented effect of cholesterol? How do liposome characteristics vary with different proportions of cholesterol, phosphatidylcholine and sphingolipid, and which are the effects of a structural variation of the intermediate zone? Is there a lateral phase separation with a unit ratio of these lipids or do they form a network of a three-component unit? Is a lateral diffusion followed by less leakage of the membrane if there is a rapid interchange of hydrogen bonding of the intermediate zone? Does the probability of such interchange vary with the structure of the intermediate zone?

This working hypothesis has evolved (Karlsson and Pascher, in preparation) in a joint project of biological and physical chemists, where we have selected sphingolipids of biological interest for chemical synthesis (Karlsson and Pascher, 1971; Pascher, 1974) and model studies, primarily by X-ray methods (see Abrahamsson et al., 1972). Results of relevance for the present aspects are collecting (see Abrahamsson et al., this volume). Noteworthy is a very low area per molecule in a condensed monolayer of an equimolar mixture of cholesterol, phosphatidylcholine and galactosylceramide. This may in part be due to stereospecific interactions of the intermediate zone, supporting the idea of a relatively dense and impermeable cell surface layer. Much interest is focused on crystal structures with precise information on molecular arrangements, but crystals of good quality are very difficult to obtain for true membrane lipids, although deductions from components are useful (Sundaralingam, 1972; Pascher, 1976). Very recently, however, the structure of a galactosylceramide was solved (see Abrahamsson, et al., this volume). Interesting features of the ceramide conformation of this molecule (see Fig. 4) compared with other data, suggest that a variable structure of the intermediate zone may give precise orientation effects, of importance not only for hydrogen bonding in this part of the membrane, but also for the molecular conformation as a whole.

## 3. THE POLAR HEAD GROUP

Of the known polar head groups of sphingolipids (Fig. 3) only those with sugar residues have been possible to connect with specific cellular functions: galactose-3-sulphate with $Na^+$ transport, a neuraminic acid-containing pentasaccharide with cholera toxin binding, and oligosaccharides in general with recognition phenomena (surface antigens).

### A. Sulphatide and $Na^+$ Transport

We have shown in a series of investigations that sulphatide (Fig. 4) concentration is correlated in a number of tissues (Karlsson et al., 1974) with the activity of the $Na^+$ and $K^+$ dependent ATPase ($Na^+$-$K^+$-ATPase), an enzyme considered to be specifically involved in $Na^+$ translocation at the surface membrane of animal cells (see review by Glynn and Karlish, 1975; and paper by Skou in this volume). This relation (Fig. 5) exists not only for tissues with an elevated $Na^+$ transport, like salt glands, electric organ, kidney and nerve cells, but also for erythrocytes with only a few hundred enzyme molecules per cell (the scale for erythrocytes in Fig. 5 has been multiplied by $10^3$). With an increased demand on $Na^+$ translocation (salt load of domestic duck), there is a parallel

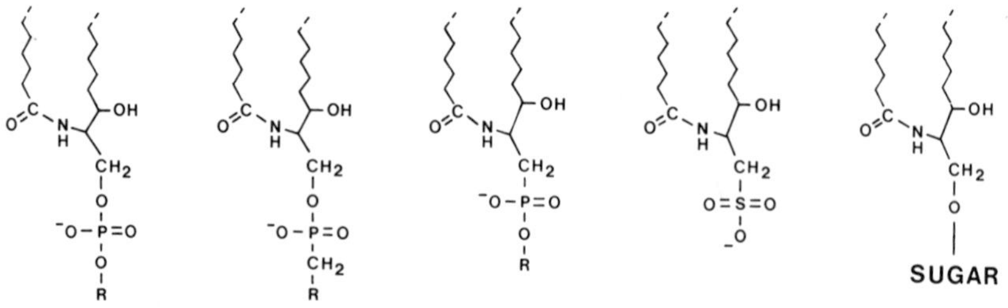

Figure 3. Illustration of partial structures of polar head groups of sphingolipids.

induction for enzyme and sulphatide. This lipid is unique among membrane lipids in showing this relation (Karlsson et al., 1974). There are however two deviations from the symmetric profile of Fig. 5. The too high enzyme activity for kidney cortex is commented upon below. The situation for white matter, where the sulphatide concentration is too high, is obviously explained by the presence of myelin, a multimembrane system specialized for an insulation function around the nerve axons. Myelin is derived from glia cell surface membranes, which have lost their enzyme activities but retained their lipids.

Does this mean that sulphatide with its anionic group is an essential part of the translocation unit? This question cannot be answered adequately until more data from directed reconstitution experiments have been obtained. However, in the following discussion a molecular working hypothesis is presented, where sulphatide is proposed essential for $K^+$ translocation (cofactor site model). The advantage of the model is its accessibility to precise experimental test. A more detailed discussion on this topic will be given elsewhere.

The $Na^+$-$K^+$-ATPase is a membrane-bound lipoprotein that has been isolated in a highly pure state from several tissues in recent time (see review by Jørgensen, 1975). The assay of activity during isolation is ATP hydrolysis, and numerous papers have shown a lipid--dependence of this reaction. A liquid-crystalline state of the

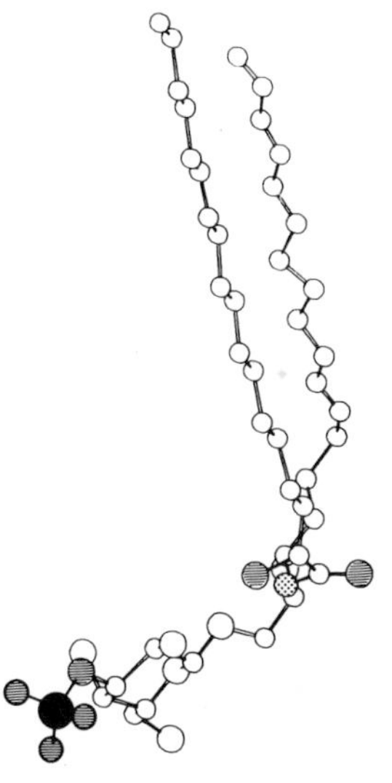

Figure 4. Constructed conformation of sulphatide (β-galactopyranosylceramide with a sulphate in position 3 of the sugar). The detailed conformation shown is the known crystal conformation of a galactosylceramide (Abrahamsson et al., this volume), to which a sulphate group was added (sulphur atom in black, sulphur-bound oxygens and fatty acid oxygens shaded, and nitrogen dotted). A possible spoon--like conformation of sulphatide may be of importance for $K^+$ donation in the translocation process (see text and Figs. 6 and 7). The intermediate zone (compare Fig. 2) may stabilize the membrane trough hydrogen bonding in a lateral direction.

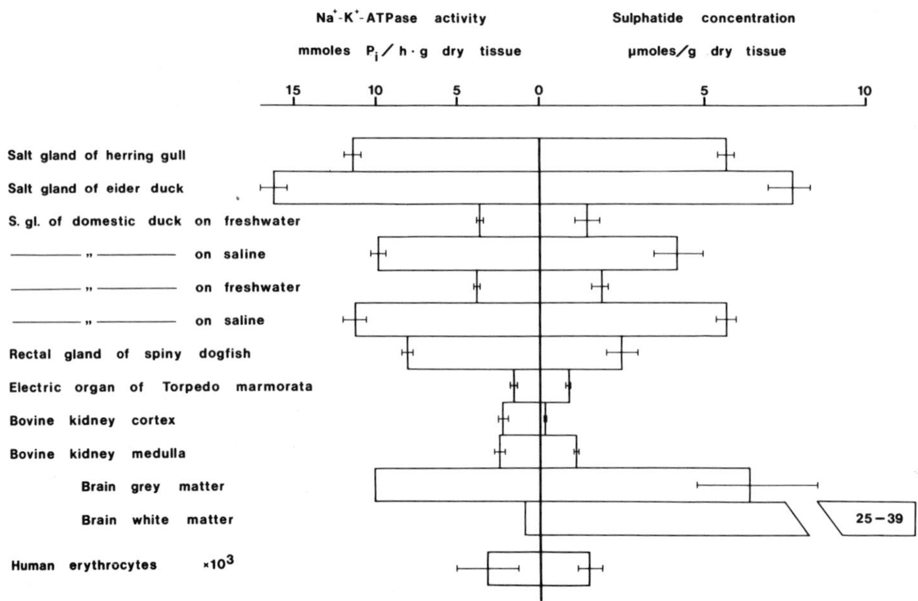


Figure 5. Diagram showing the relation of sulphatide concentration and $Na^+$-$K^+$-ATPase activity in several animal tissues (Karlsson et al., 1974). The data for electric organ and erythrocytes have not yet been published (Hansson et al., in preparation).

lipid is obviously needed for conformational changes in the protein during reaction, but a specific requirement of a certain polar head group has not been convincingly proved. Phosphatidylserine, the most studied in this field, is effective in restoring activity after partial delipidation of the enzyme, but may be substituted with phosphatidylglycerol. Less effective are other anionic lipids like phosphatidylinositol and sulphatide, and these results show an interesting relation to cation binding of the lipids (Kimelberg and Papahadjopoulos, 1972).

Although the ATP hydrolysis is considered as an essential step in $Na^+$ and $K^+$ translocation, the mentioned studies on lipid requirements for ATP hydrolysis do not prove if a certain lipid is essential for the translocation process. More adequate to answer this question are the attempts done during the last year to reconstitute the ion pump by use of purified enzyme and lipid vesicles (liposomes). Some data are collected in Table III.

TABLE III. Some Properties of reconstituted $Na^+$ Pumps

| Paper | Enzyme source | Method of isolation (see Table IV) | Liposome lipid | Ions trans-located | Transport stoichiometry | |
|---|---|---|---|---|---|---|
| Goldin and Tong, 1974 | Canine kidney | Kyte, 1971 | Egg yolk lecithin | $Na^+$ $Cl^-$ | $Na^+$:ATP | 1.0 |
| Sweadner and Goldin, 1975 | Canine brain | Nakao et al., 1973 | Egg yolk lecithin | $Na^+$ $K^+$ | $Na^+$:$K^+$ | 3:1.8 |
| Hilden and Hokin, 1975 | Dogfish rectal gland | Hokin et al., 1973 | Egg yolk lecithin | $Na^+$ $K^+$ | $Na^+$:$K^+$<br>$Na^+$:ATP | 2.8:2<br>1.5 |
| Racker and Fischer, 1975 | Electric organ | Albers et al., 1963 | Egg yolk lecithin<br>Soybean Cephalin | $Na^+$ | | |

In all cases a net transport of $Na^+$ from outside to inside was observed when ATP was added to the dispersion. Ouabain gave a complete inhibition only from the inside, demonstrating a trans orientation of the protein, as known for living cells. The liposomes consisted of natural lecithin, and in one case also of cephalin. Addition of phosphatidylserine had no effect (Racker and Fischer, 1975). A conclusion from these experiments concerning lipid requirement may be that phosphatidylserine and sulphatide are not essential for $Na^+$ translocation.

It is, however, of relevance for the interpretation that the enzyme preparations often contain more lipid than protein and that a completely delipidated protein is impossible to reactivate (see, however, the recent paper by Ottolenghi, 1975). Therefore several hundred, or more, lipid molecules are present per molecule of enzyme, also when the preparation is pure from a protein-chemical point of view. A possible lipid dependence of the translocation process cannot be defined, therefore, until the precise composition of both liposome matrix and enzyme are known.

I have tried to reconstruct, from the few data available, the lipid composition of the enzyme preparations used for reconstitution, see Table IV. The kidney enzyme, used by Goldin and Tong, was documented to be of the same purity as was the original preparation of Kyte, with known lipid composition (Table IV). The canine brain preparation used by Sweadner and Goldin was not analyzed. The preparation method was, however, worked out for pig brain and this enzyme has been analyzed for lipid (Table IV). The rectal gland preparation used by Hilden and Hokin was noted to contain glycolipid. As the major glycolipid of this organ is sulphatide (see Fig. 5), and as sulphatide is present in a similar preparation made by Skou (Table IV), the enzyme used by Hilden and Hokin therefore most probably contained sulphatide. The electric organ membrane reconstituted by Racker and Fischer probably included all plasma membrane lipids of this tissue, which is rich in sulphatide (Fig. 5).

If these assumptions are correct, sulphatide and phosphatidylserine are not needed for $Na^+$ translocation, as the system of Goldin and Tong most probably lacked these lipids. The other systems, however, should all contain these two lipids. This difference in membrane components may explain the absence of $K^+$ translocation in the experiment of Goldin and Tong, where the pump was and "electrogenic" $Na^+$ pump with a cotransport of $Cl^-$. That the absence of $K^+$ transport was not due to the technique of reconstitution (liposome preparation etc.) was shown by Hilden and Hokin, who reproduced the conditions of Goldin and Tong, but now with the rectal gland enzyme, giving both $Na^+$ and $K^+$ transport. Apparently, the ability to transport $K^+$ was lost during the preparation of the kidney enzyme.

TABLE IV. Lipid Composition of $Na^+$-$K^+$-ATPase Preparations

| Method of enzyme preparation | Enzyme source | Activity (μmoles $P_i$ x mg protein$^{-1}$ x $h^{-1}$) | Lipid components[a] and approximate relative amounts | Sulphatide | Reference for lipid analysis |
|---|---|---|---|---|---|
| Kyte, 1971 | Canine kidney | 800 | PE 20, PI 3, PC 1, PS 0, SM 0, Su 0 | Absent | Karlsson, Kyte and Samuelsson, unpublished |
| Nakao, et al., 1973 | Pig brain | 1500 | PE 8, PC 8, PS 2, PI 1, SM 1, Su + | Present | Kawai et al., 1973 |
| Hokin et al., 1973 | Dogfish rectal gland | 1500 | Lipids not analyzed but PAS-positive component in SDS-gel noted to be glycolipid | Probably present | Hokin et al., 1973 |
| Skou, unpublished | Rectal gland | 1200 | PC 18, PE 9, PS 3, SM 3, PI 2, Su 1 | Present | Karlsson, Samuelsson and Skou, unpublished |
| Albers et al., 1963 | Electric organ | | Lipids not analyzed. Particulate fraction not extracted. | Probably present | |

[a] PE= phosphatidylethanolamine, PI= phosphatidylinositol, PC= phosphatidylcholine, PS= phosphatidylserine, SM= sphingomyelin, Su= sulphatide.

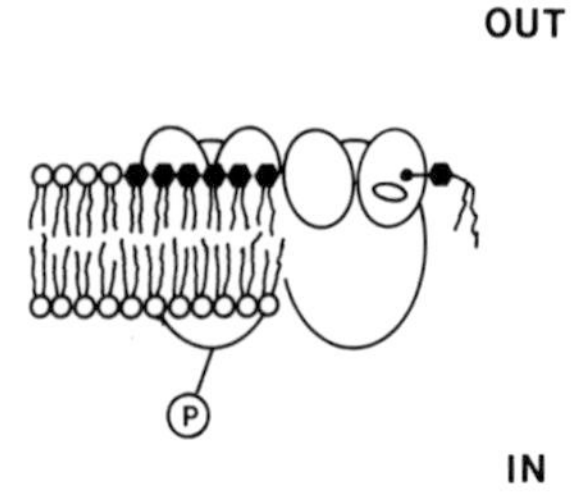


Figure 6. Cofactor site model of $Na^+$-$K^+$-ATPase and sulphatide in the surface membrane. The enzyme is arranged as a dimer of two big subunits, only one reacting at a time with ATP (half-site model). Two small glycopeptide subunits for each large polypeptide carry a cation gate (ellipsoid hole). Sulphatide (black head group) is in the outer half of the bilayer in an amount sufficient to give an annulus around the proteins. The sulphate group (black dot) is in close proximity to the outside of the cation gate (compare Fig. 7).

My suggestion is that sulphatide, which has a unique tissue appearance with this enzyme (Fig. 5), is essential for the $K^+$ translocation and functions as a $K^+$ site with its galactose-3-sulphate group on the outside of the plasma membrane (Fig. 6). This lipid is, however, of no importance for the $Na^+$ translocation step. In the absence of sulphatide the pump becomes electrogenic. A model will be more specifically formulated below.

Why would a sulphate be more suited than other anionic groups as a $K^+$ receptor of the $Na^+$-$K^+$-pump? The known cation binding properties of different anionic groups may provide an answer to this question. As predicted by the field strength theory of Eisenman (Eisenman, 1962), a large anionic radius, as for sulphate and sulphonate, results in a low field strength, and cations are bound with a retained hydration shell, as they prefer water before the anionic site. The selection sequence will be $Rb^+ > K^+ > Na^+$, as the hydrated radius is largest for $Na^+$. With a rather small anionic radius, as for carboxylate, phosphate and phosphonate, with a higher field strength, the cation prefers the anionic site before water, and the cation selection then follows the unhydrated radius, that is $Na^+ > K^+ > Rb^+$. As a consequence, sulphatide has a reversed cation selection compared with all acid phospholipids, with a higher affinity for $K^+$ than for $Na^+$. This has been verified experimentally for sulphatide (Abramson et al., 1967) and phosphatidylserine (Abramson et al., 1964). In liposomes with these lipids, sulphatide created a permeability for $Na^+$ in favour of $K^+$, and phosphatidylserine for $K^+$ in favour of $Na^+$ (Kimelberg and Papahadjopoulos, 1972). Zwitterionic lipids, like phosphatidylcholine and phosphatidylethanolamine, did not discriminate between these two ions.

The following postulates, with some explaining comments, bring together our model for sulphatide (and acid phospholipid) function in the $Na^+$-$K^+$ pump of animal cells (Fig. 6). The two essential features of the model are

a) an asymmetric arrangement in the bilayer of lipids with reversed selection for $Na^+$ and $K^+$,
b) a $K^+$-donating sulphate group in close proximity of the protein translocation site.

1. Sulphatide is located in the outer half of the plasma membrane bilayer.
   -This has not yet been demonstrated, but this localization is probable as sphingomyelin and globoside of erythrocyte (see Introduction) and ganglioside of small intestine (see below) are all in the surface monolayer.

2. Sulphatide has a specific polar interaction through its galactose--3-sulphate group with the glycopeptide subunit of the enzyme.
   -The purified ATPase is composed of a large and a small subunit in a molar ratio of 1:1 or 1:2 (see Jørgensen, 1975). The large unit spans the membrane and is phosphorylated from ATP on the inside and binds ouabain on the outside of the membrane. This polypeptide probably occurs as a noncovalent dimer in the membrane (see Glynn and Karlish, 1975). The smaller polypeptide is a glycoprotein and therefore probably exposed at the cell surface. A polar interaction of sulphatide and protein is indicated by an obligate salt extraction step, in addition to detergent, for those procedures where sulphatide is removed (Kyte, 1971; Ottolenghi, 1975). We have calculated 60-100 moles of sulphatide per mole enzyme (Karlsson et al., 1974). If we use the 'half-site' model (see Glynn and Karlish, 1975), this amount of sulphatide is enough to give an annulus around the surface protein (see Fig. 6). Such large number of specific association sites at the protein for sulphatide is not probable. However, four sites (Fig. 6) with moderate affinity for sulphatide are sufficient to give an enrichment of sulphatide around the protein, even for erythrocyte, where the enzyme and sulphatide levels are very low (compare Fig. 5). It may be noted that sulphatide should create a less permeable environment of the enzyme than glycerolipid, due to less fluid hydrocarbon chains (see Chapman, 1975), thus minimizing cation leakage.

3. The sulphate group is intimately associated with a cation gate of the glycoprotein.
   -Conductance studies on black lipid membranes have revealed an $Na^+$-dependent ionophore as a part of the glycoprotein (Shamoo and Myers, 1974). The order of cation requirement did not simply follow the theory of Eisenman (compare discussion above), indicating steric factors to be important (peptide pore). $NH_4^+$ was almost as effective as $Na^+$, showing the system to be

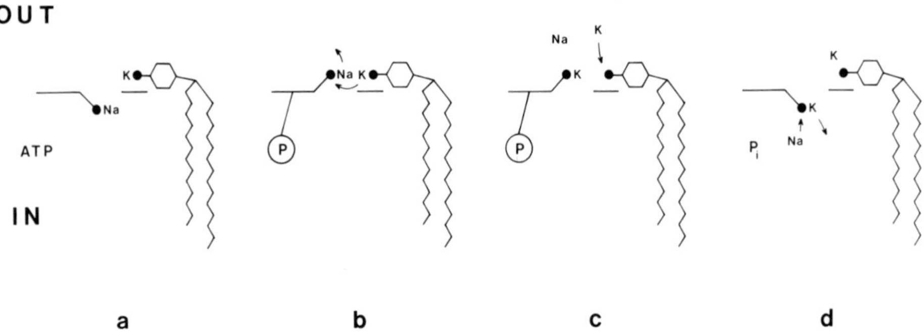


Figure 7. Separate steps in the translocation of $Na^+$ and $K^+$ according to the cofactor site model (see text and Fig. 6). The sulphate group of sulphatide (black dot with $K^+$ in step a ) is in close proximity to the outside of a protein gate with the translocation site (black dot with $Na^+$ in step a ).

similar in properties to the $Na^+$ channel of nerve, defined by Hille (see Hille, 1975), and recently interpreted in molecular models (Smythies et al., 1974). An essential acid group of the channel was suggested to be carboxylate (Hille, 1975).

4. The translocation process is illustrated in Fig. 7. It starts (a) with $Na^+$ (not fully hydrated) at a $Na^+$-selective anionic site (carboxyl) at the gate (inside) and a $K^+$ (hydrated) at the sulphatide (outside). Phosphorylation (b) from ATP of the large subunit (inside) gives a cooperative conformational change (in the order of a few Å) of the gate, moving the binding site to the outside. Concomitantly, through induction from the microenvironment of the site, the field strength of the anion site decreases, favouring hydration and dissociation of $Na^+$. The stability constants now explain a $K^+$ exchange between sulphate and gate site (c). At dephosphorylation (d), the gate site returns with $K^+$ to the original position (inside) with a higher $Na^+$ affinity. Equilibration at both sides (through competition between $Na^+$ and $K^+$) now gives the original situation (a).

5. In the absence of sulphatide (or proximity of gate and sulphate) the pump is electrogenic: there is $Na^+$ efflux but no $K^+$ influx, explained by a rate-limiting step in $K^+$ diffusion from the medium to the $K^+$ selective gate site (Glynn and Karlish, 1975).
   -This assumes that the lifetime of the phosphorylated state (gate site on the outside) is too short (b and c of Fig. 7) for an equilibrium to take place between gate site and the

outside medium. ($Na^+$ is dissociated but no $K^+$ of the high $Na^+$-low $K^+$ medium is able to reach the site by diffusion.) In other words, a cofactor $K^+$ affinity site (sulphate), which by proximity can donate $K^+$ to the shortlived $K^+$ site of the gate, has a higher probability to equilibrate with the medium during steps d and a (longer lifetime) than the gate site alone has during b and c (shorter lifetime). This cofactor site explains the existence of a $K^+$ affinity site on the outside of erythrocytes before phosphorylation (see Glynn and Karlish, 1975). The situation for kidney cortex (Fig. 5), with a too low sulphatide - ATPase ratio compared with all other tissues, is of interest, as it may imply that part of the translocation in the proximal tubules is electrogenic with a direct cotransport of $Cl^-$ (compare discussion of Goldin and Tong, 1974). The uncoupled $Na^+$ efflux long known for erythrocytes after addition of choline or $Mg^{++}$ to the medium, may be explained by a competition with $K^+$ binding at the sulphate group. A rather strong binding of sulphatide to choline phospholipids has been shown by titration changes for phosphate (Abramson and Katzman, 1968). The binding was inhibited by cholesterol, which should forbid a blocking of sulphatide by lecithin or sphingomyelin in the membrane.

6. In the absence of $K^+$ outside, the sulphate binds $Na^+$, giving a $Na^+$ - $Na^+$ exchange, and in the absence of $Na^+$ inside, there will be a $K^+$ - $K^+$ exchange (Glynn and Karlish, 1975). Theoretically, the absence of both $Na^+$ inside and sulphatide will give an electrogenic $K^+$ efflux.

7. Acid phospholipids in the inner half of the membrane bilayer and close to the enzyme protein make the pump more effective by donating $Na^+$ in a high $K^+$-low $Na^+$ cytoplasm, but no particular lipid is essential for the $Na^+$ translocation.
   -Phosphatidylserine and phosphatidylinositol are located in the inner half of the bilayer of erythrocyte membranes (Renooij et al., 1976). In the reconstitution experiment of Goldin and Tong (Table III) only the latter lipid was present. If one postulates a cofactor of the enzyme to be essential for the function, one would expect a regulated biosynthesis of a "translocation unit", with protein <u>and</u> cofactor, including its transport through vesicles and asymmetric fusion with the surface membrane. Of the lipids mentioned, only sulphatide fits this model, with a stoichiometric relation to the enzyme, also after induction (Fig. 5). It would be of interest to know the ATPase activity of sulphatide-rich Golgi fractions isolated from kidney cells (Zambrano et al., 1975). In contrast, phosphatidylserine is one of the major phospholipids of erythrocyte, where only a few hundred enzyme molecules are present per cell. Concerning the effectiveness of acid phospholipids in reactivating delipidated ATPase (ATP hydrolysis assayed),

this may be explained by an optimal $Na^+$ donating mechanism through a proper association of amphipathic protein and lipid. A cation at the gate ($Na^+$ in Fig. 7 a) may be a prerequisite for phosphorylation. This may explain why sulphatide, enriching $K^+$ from a $Na^+$-$K^+$ medium, is less effective in restoring ATP hydrolyzing activity. Similarly, the Eisenman theory predicts that phosphatidylinositol, through induction from inositol oxygens, has less $Na^+$ affinity and should be less effective in restoring activity than phosphatidylglycerol and phosphatidylserine (Kimelberg and Papahajopoulos, 1972), although steric effects may also be important.

8. A $Na^+$-$K^+$ selection inversion is thus created across the membrane through fixed anionic cofactor sites, opposing the physiological concentration gradient, with $Na^+$>$K^+$ at the inside (acid phospholipid), and $K^+$>$Na^+$ at the outside (sulphatide). A regulated inversion of selection is the fundamental characteristic of the protein gate site. The short lifetime of the $K^+$ selective gate site (outside) compared with the $Na^+$ selective gate site (inside) makes sulphatide essential for $K^+$ translocation (compare 5).

9. Translocation of more than one $Na^+$ (maximum 3) or $K^+$ (maximum 2) for each ATP hydrolyzed (see Glynn and Karlish, 1975), is in agreement with this model only if one assumes more than one gate (of smaller subunit) operating for each large subunit phosphorylated. This is possible with a half-site ATP-binding model (see Glynn and Karlish, 1975) and two small subunits for each large subunit (see Fig. 6).

## B. Other Sulpholipids

The dominating anionic groups of all kinds of membranes are carboxylate and phosphate ($Na^+$ selection type), of both protein, lipid and carbohydrate (although 'glycocalyx', surrounding many plasma membranes, contains proteoglycans with sulphate groups). The distinct appearance of the relatively uncommon sulpholipids (see Haines, 1971), may suggest functions related to their anionic properties. One example is the sulphonate-containing glycolipid of plants and Echinodermata, 6-sulpho-α-quinovosyldiglyceride, which has been found correlated with an $Na^+$ and $K^+$ dependent ATPase of sugar beet tissues (see Karlsson et al., 1974). The crystal conformation of the polar part of this lipid (Okaya, 1964) may be speculated to have several traits in common with β-galactose-3-sulphate (compare Fig. 4), suggesting a similar association of these two sulpholipids with protein in a translocation unit.

Most intriguing, however, is the sulphatide of extremely halophilic bacteria (see Haines, 1971; Kates and Deroo, 1973; Karlsson

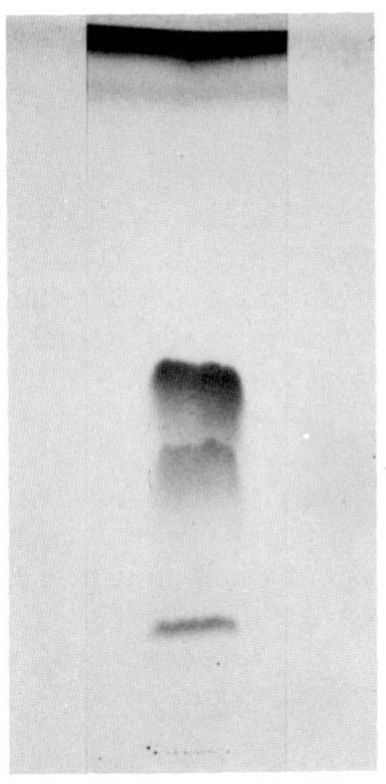

Figure 8. Thin-layer chromatogram of amphipathic lipids of the extremely halophilic bacterium, Halobacterium salinarium (Karlsson, unpublished). The slowest moving band is the sulphatide shown in Fig. 9, and the two major bands are the glyceryldiether analogs of phosphatidylglycerolphosphate (upper) and phosphatidylglycerol (lower), see Kates and Deroo, 1973. The layer was silica gel G, the solvent chloroform:methanol:water 65:25:4 (by vol.), and the reagent of detection cupric acetate.

et al., 1974), making up an important part of total polar lipids of these organisms (Fig. 8 and 9). The purple membranes, which appear to function as a light-driven proton pump of these bacteria, do not contain this sulphatide (Kushwaha et al., 1976). There is a steep concentration gradient for $K^+$ over the surface membrane (500-1000 times) and it is highly suggestive that this sulphatide has a $K^+$ binding role in translocation, possibly at the outside of the membrane, with the two acid phospholipids at the inside (see Fig. 8). The peculiar diether lipophilic part of these lipids (Fig. 9) is probably required for optimal fluidity at the very high ionic strength of the environment (sea water saturated with salt).

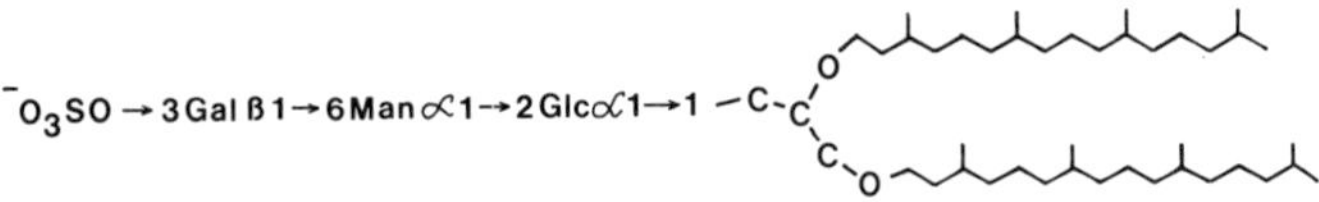


Figure 9. The sulphatide of extremely halophilic bacteria (compare Fig. 8). The β-galactose-3-sulphate is identical with the animal sulphatide (see Fig. 4).

### C. Ganglioside and Cholera Toxin

An already well established function of a sphingolipid is the binding of cholera toxin to ganglioside, discovered some years ago (van Heyningen et al., 1971). Several groups have since demonstrated a very high specificity *in vitro* for a particular ganglioside, see Fig. 10. Recently, evidence was presented that the toxin receptor of small intestine, where the diarrheogenic action of the toxin is found, actually is this ganglioside and not protein-bound carbohydrate (Holmgren et al., 1975). The good correlation in these studies of toxin binding and ganglioside concentration in the mucosa cells of several species may be taken as additional evidence for an exclusive surface localization of glycolipids (see Introduction). A lateral redistribution of toxin receptors (patch and cap formation), possibly moderated by membrane proteins, has also been shown (Craig and Cuatrecasas, 1975; Sedlacek et al., 1976).

These results are of basic importance when considering functions of cell surface antigens of glycolipid nature (see below), because they demonstrate that a ligand may exert a specific cell effect through binding to a membrane component (ceramide-bound carbohydrate), which does not,by all probability, span the membrane. A specific association of glycolipid carbohydrate and protein which penetrates the membrane (as suggested above for sulphatide and ATPase) is a possible part of the mechanism (compare Cuatrecasas, 1974).

### D. Surface Antigens

Although the common mammalian simple sugars are only about 10 compared with 20 amino acids, the number of possible combinations of sugars are amazingly large and far above that of amino acids. A simple calculation for disaccharides of two different hexoses gives about 70 isomers (the glycosidic bond varies in both position and configuration, and the heterocycle may be five- or six-membered), compared with only two peptide possibilities for two amino acids. A second difference of interest, when comparing cell surface carbohydrate and protein, is the volume of genetic information needed for their synthesis. The catalysis of peptidation at the ribosome

Gal ß1→3GalNAc ß1→4Gal ß1→4Glc ß1→1CERAMIDE
3
↑
2
αNeuNAc

Figure 10. The receptor for cholera toxin.

is not specific for a certain type of linkage. In contrast, each glycosidic bond needs its own glycosyltransferase, always with a very high specificity (see Shur and Roth, 1975). Therefore, carbohydrate may be a very potent and specific carrier of an obligate surface information in multicellular systems.

The classical example of specific cell surface antigens are the carbohydrate antigens belonging to the major blood-groups in man and other animals (see review by Hakomori and Kobata, 1974). The general view at present is that membrane-bound activity is based on glycosphingolipid, and that secreted, soluble activity is in the form of glycoprotein.

There is a great number of different oligosaccharides bound to ceramide, although most of these glycolipids usually exist in very small amounts (Hakomori and Kobata, 1974). In fact, in a pioneering chemical study, only a few milligrams of blood group active glycolipid were obtained from 30 liters of human blood (Hakomori and Strycharz, 1968). We have recently devised improved methods for the microscale characterization of this group of substances, allowing a precise comparison of cells of different genetic and functional states. Examples of glycolipid complexity and specificity are illustrated in Fig. 11. Of the non-acid glycolipids of human blood-group $A_1$ erythrocytes (some of them shown in Fig. 11:a) only a few substances have so far been chemically characterized (Hakomori and Kobata, 1974). Of great importance for the microscale structure elucidation of these glycolipids is a novel fingerprinting technique based on mass spectrometry (Karlsson, 1973; 1974), providing information on the type, number and sequence of sugars, as well as ceramide structure. The spot indicated by an arrow in Fig. 11:a contained at least three glycolipids, of the probable structures shown in Fig. 12. The mass spectrum of this mixture (Fig. 13) is of general interest, because it represents the largest organic substances so far analyzed by this sensitive and specific technique.

The nonaglycosylceramide shown in Fig. 12 is, in a strict sense, the largest glycolipid identified conclusively so far, as the structure was supported by a direct method and not only based on reconstruction from degradation data. However, it seems clear from the chromatographic properties that the most slow-moving glycolipids of erythrocyte and other cells may contain up to 20 sugars. This number was suggested to be the upper limit for saccharide chains of blood-group glycoproteins (Rovis et al., 1973). According to our calculations for $A_1$ erythrocyte, the total number of glycolipid species with more than four sugars should be at least 40. The number of lipid-bound carbohydrate species on the cell surface is thus remarkably large.

Why is the cell surface equipped with a large and varying structural information in the form of carbohydrate? There is no

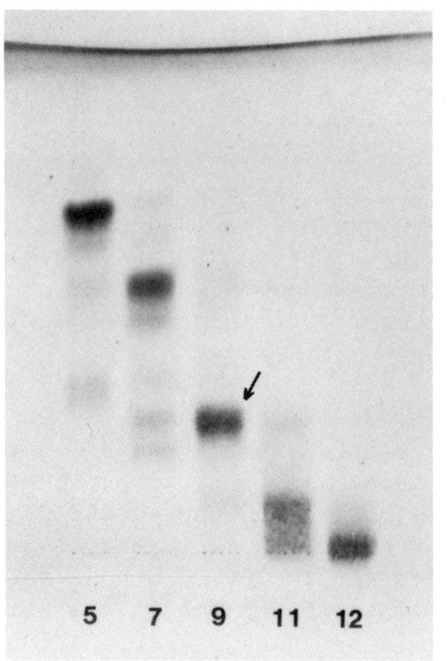


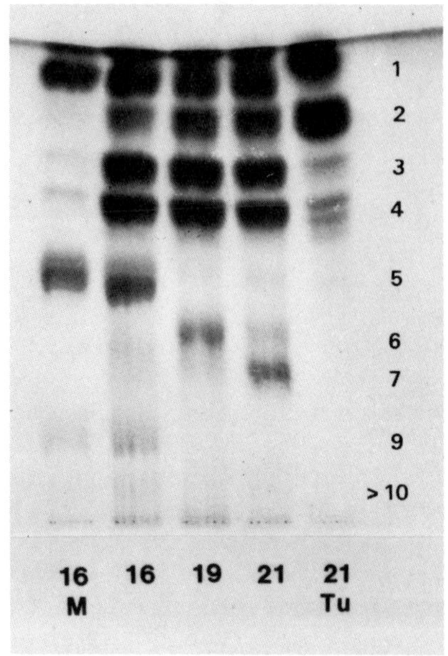


Figure 11:a (left). Thin-layer chromatogram of selected glycosphingolipid fractions from human blood-group $A_1$ erythrocyte membranes. The bands shown contain from 4 (fast-moving) up to possibly 20 sugars (origin) bound to ceramide. The band indicated by an arrow was analyzed by mass spectrometry (Fig. 13) and contained at least three blood-group A active glycolipids (Fig. 12).

Figure 11:b (right). Thin-layer chromatogram of small intestinal glycosphingolipids, demonstrating a molecular variation for three human blood-group A individuals (16, 19 and 21). The figures to the right indicate the number of sugars in the glycolipid. 16 M shows a mucosa scraping from intestine 16, and 21 Tu a malignant melanoma of case 21. The major part of triglycosylceramide shown is galactosylgalactosylglucosylceramide and not of the novel type shown in Fig. 14 (see text).

simple answer to this question. Surface carbohydrates have been considered involved in cellular recognition processes and as regulators of cell behaviour. Specific intercellular association and modification may be mediated by protein-carbohydrate interaction, as formulated by the glycosyltransferase-substrate model of Roseman et al. (see Shur and Roth, 1975), and the lectin-sugar model of Barondes et al. (Reitherman et al., 1975). Normal and abnormal (tumor) differentiation processes may be described in terms of these models, although existing experimental evidence is far from

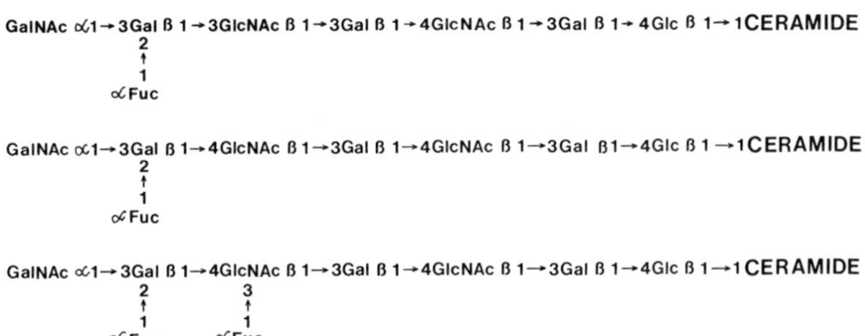


Figure 12. Proposed structures of three blood-group A active glycosphingolipids isolated from the thin-layer chromatographic spot of Fig. 11:a (arrow).

conclusive. Glycosyltransferases and lectins are found on cell surfaces and seem to be developmentally regulated and involved in modification of cell behaviour upon cell-cell contact. Tumor cells are of a primitive type, with less developed transferase patterns and with blocked synthesis of glycolipids in many cases (Hakomori, 1975). This blocked synthesis is often associated with accumulation of precursors. As an example, non-acid glycolipids of a malignant melanoma and small intestine of the same individual (case 21) are shown in Fig. 11:b. (The normal precursor cell of melanoma cells is not available for this analysis.) Glycolipids with more than five sugars are practically absent from the melanoma. On the other hand, there is a substantial rise in lactosylceramide and glucosylceramide, and also of free ceramide and gangliosides with three and four sugars (not shown). Free sphingosine was present, not shown before in animal cells. Of interest was the identification of a novel triglycosylceramide (Fig. 14), not detected in the normal adult tissue (Karlsson, unpublished). The same glycolipid was recently found in erythrocytes of a hereditary disorder of erythropoiesis, HEMPAS (Joseph and Gockerman, 1975), and has also been detected in small amounts in foetal erythrocytes (Karlsson and Larsson, unpublished).

This glycolipid is probably an intermediate in the stepwise elongation of the sugar chain to more complex glycolipids (compare Fig. 12). Its apparent absence in the tissue may be due to a short lifetime, making it difficult to detect, except in rapidly growing cells. Alternatively, a multi-glycosyltransferase complex (Shur and Roth, 1975) may normally not dissociate this intermediate. It is noteworthy that a terminal glucosamine has not been found in soluble blood-group substances (Rovis et al., 1973). It would be of interest

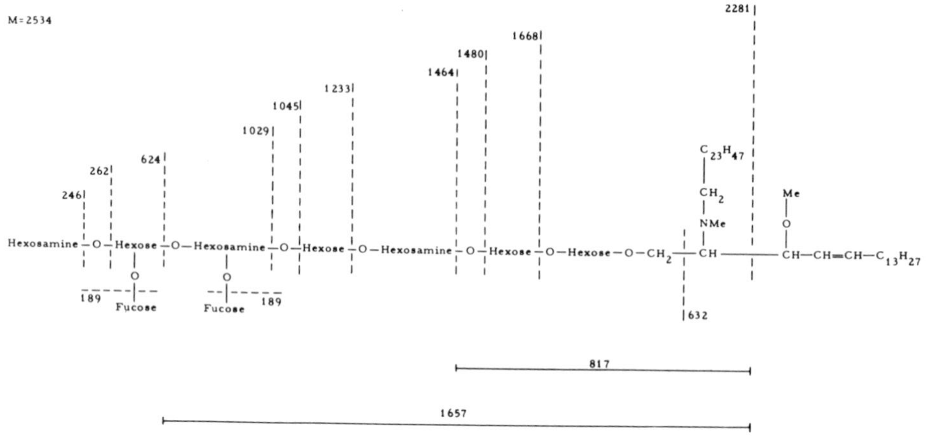


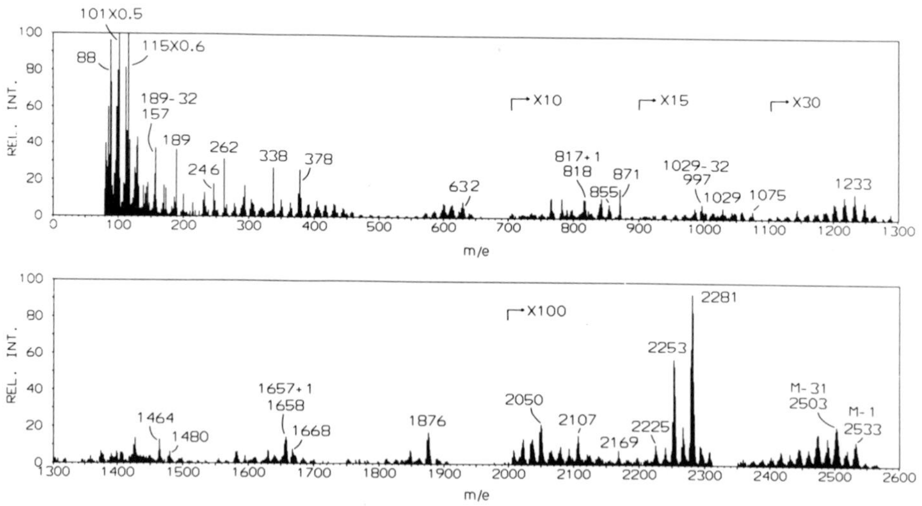


Figure 13. Mass spectrum of the methylated and reduced ($LiAlH_4$) glycolipid fraction indicated by an arrow in Fig. 11:a and composed of at least three glycolipids (Fig. 12). The simplified formula for interpretation corresponds to the major molecular species present. Very intense ions are found for the complete saccharide and the fatty acid (m/e 2281 for fatty acid 24:0, and 2253 for 22:0). For a correct interpretation of sequence, a spectrum of the only methylated derivatives is also needed (not shown). The instrument used was MS 902 (AEI, Ltd.), and the conditions of analysis were: electron energy 58 eV, trap current 500 µA, acceleration voltage 3.1 kV, ion source temperature 350°C, and probe temperature 320°C.

GlcNAc β1→3Gal β1→4Glc β1→1CERAMIDE

Figure 14. "Incomplete" glycolipid present in human malignant melanoma, in erythrocytes of an hereditary human disorder of erythropoiesis, HEMPAS, and in foetal erythrocytes.

to know if the melanoma contains small amounts of more complex glycolipids with a terminal glucosamine (5 or more sugars, compare Fig. 12, and Rovis et al., 1973). Such immature or incomplete structures may be candidates for "tumor-specific" cell surface antigens, which are of potential importance for possible tumor therapy. Specific antisera to these antigens show crossreactions with antigens of foetal cells but not of adult normal cells. No such antigen has so far been chemically identified, but it is of interest that a recent immunological study has indicated a melanoma-specific antigen, behaving on polyacrylamide gel electrophoresis as a glycolipid (Gorodilova and Hollinshead, 1975).

Human erythrocyte is an important reference cell in all these studies because it is by far the best mapped cell concerning immunology. However, due to the very low amount of glycolipid present, cells from several donors have to be pooled for adequate chemical studies (the fractions of Fig. 11:a were from 215 individuals). Of importance was therefore the finding by McKibbin and coworkers (see Smith et al., 1975b, and references therein), that small intestine is a rich source of blood-group glycolipids. This gives a new possibility to study molecular individuality in man. Earlier studies for this purpose have been confined to glycoproteins of ovarial cyst secretions, where often gram quantities may be obtained from one single cyst (see Rovis et al., 1973). However, the complexity of these molecules, where a number of different sugar chains are bound to each polypeptide, makes a comparison of individuals technically much more difficult than for glycolipids, with one defined oligosaccharide per molecule. Three blood-group A positive individuals shown in Fig. 11:b (cases 16, 19 and 21) present a chemical polymorphism in the chromatographic interval of blood group activity (Breimer et al., unpublished). By a comparison of immunological activity and mass spectral data of different fractions, it was shown that the most fast-moving A-active glycolipid of case 19 had 6 sugars, case 21 had 6 and 7 sugars, and case 16 at least 10 sugars. The detailed structural basis of these interesting patterns is presently under investigation.

Also shown in Fig. 11:b is the glycolipid pattern of a mucosa scraping (16 M) compared with whole intestine (Karlsson and Leffler, unpublished). The enrichment of slow-moving glycolipids in epithelial cells of small intestine opens up some new experimental possibilities. These cells present an ideal system for studying differentiative events, because of a segregation of the epithelial cells according to their degree of differentiation (see Lipkin, 1973). The villi are covered with a single layer of cells, which undergo mitosis in the crypt at the base of the villi and then move up the villi until they are extruded from the tips. The total renewal time is 1 - 3 days. Cells of different maturity levels are now possible to prepare (Weiser, 1973) for a detailed study of sphingolipid relations to cell dynamics. The very large total surface area of the cell border to the intestinal lumen is involved not only in a series of specific active transport processes, but also in recognition mechanisms in health and disease, including self-not--self phenomena. Of interest is the important selective uptake of maternal milk antibodies during the first period of life (Rodewald, 1973), binding of bacteria and toxins (see cholera toxin above), and a number of allergic manifestations. The pathogenesis of the gluten-sensitive enteropathy is probably a lectin-like gluten binding to incomplete (immature) oligosaccharide chains of the surface membrane, with a prolonged binding being toxic and leading to cell death (Weiser and Douglas, 1976).

Obviously, some important cell functions may in part depend on sphingolipids. I believe that the intestinal epithelium, with its regeneration characteristics and large surface area with transport and recognition functions, is one of the most potent mammalian systems for extending our knowledge on the structure and function of sphingolipids in surface membranes.

## ACKNOWLEDGMENT

I am grateful to all people at this Institute who have participated in discussions and experiments on sphingolipids and membranes. The work was supported by a grant from The Swedish Medical Research Council (03X-3967).

## REFERENCES

Abrahamsson, S., I. Pascher, K. Larsson and K.-A. Karlsson, Molecular arrangements in glycosphingolipids, Chem. Phys. Lipids 8, 152, 1972.

Abramson, M.B., R. Katzman and H.P. Gregor, Aqueous dispersions of phosphatidylserine. Ionic properties, J. Biol. Chem. 239, 70, 1964.

Abramson, M.B., R. Katzman, R. Curci and C.E. Wilson, The reactions of sulphatide with metallic cations in aqueous systems, Biochemistry 6, 295, 1967.

Abramson, M.B. and R. Katzman, Ionic interaction of sulphatide with choline lipids, Science 161, 576, 1968.

Albers, R.W., S. Fahn and G.J. Koval, The role of $Na^+$ in the activation of Electrophorus electric organ ATPase, Proc. Nat. Acad. Sci USA 50, 474, 1963.

Chapman, D., Phase transitions and fluidity characteristics of lipids and cell membranes, Quart. Rev. Biophys. 8, 185, 1975.

Craig, S.W. and P. Cuatrecasas, Mobility of cholera toxin receptors on rat lymphocyte membranes, Proc. Nat. Acad. Sci. USA 72, 3844, 1975.

Cuatrecasas, P., Membrane receptors, Annu. Rev. Biochem. 43, 169, 1974.

Dod, B.J. and G.M. Gray, The lipid composition of rat-liver plasma membranes, Biochim. Biophys. Acta 150, 397, 1968.

Eisenman, G., Cation selective glass electrodes and their mode of operation, Biophys. J. 2, 259, 1962.

Forstner, G., K. Tanaka and K.J. Isselbacher, Lipid composition of the isolated rat intestinal microvillus membrane, Biochem. J. 109, 51, 1968.

Forstner, G. and J.R. Wherrett, Plasma membrane and mucosal glycosphingolipids in the rat intestine, Biochim. Biophys. Acta 306, 446, 1973.

Glynn, I.M. and S.J.D. Karlish, The sodium pump, Annu. Rev. Physiol. 37, 13, 1975.

Goldin, S.M. and S.W. Tong, Reconstitution of active transport catalyzed by the purified $Na^+$-$K^+$-ATPase from canine renal medulla, J. Biol. Chem. 249, 5907, 1974.

Gorodilova, V.V. and A. Hollinshead, Melanoma antigens that produce cell-mediated immune responses in melanoma patients: joint US-USSR study, Science 190, 391, 1975.

Haines, T.H., The chemistry of sulpholipids, Progr. Chem. Fats Lipids (Holman, R.T. ed.) 11, 297, 1971.

Hakomori, S.-i., Structures and organization of cell surface glycolipids. Dependency on cell growth and malignant transformation, Biochim. Biophys. Acta 417, 55, 1975.

Hakomori, S.-i. and A. Kobata, Blood group antigens, *in* The Antigens. M. Sela, editor. Acad. Press, New York, 1974, vol. 2, p. 79.

Hakomori, S.-i. and G.D. Strycharz, Investigations on cellular blood-group substances. I. Isolation and chemical composition of blood-group ABH and $Le^b$ isoantigens of sphingoglycolipid nature, Biochemistry 7, 1279, 1968.

Hilden, S. and L.E. Hokin, Active $K^+$ transport coupled to active $Na^+$ transport in vescles reconstituted from purified $Na^+$-$K^+$--ATPase from the rectal gland of *Squalus acanthias*, J. Biol. Chem. 250, 6296, 1975.

Hille, B., An essential ionized group in $Na^+$ channels, Fed. Proc. 34, 1318, 1975.

Hokin, L.E., J.L. Dahl, J.D. Deupree, J.F. Dixon, J.F. Hackney and J.F. Perdue, Studies on the characterization of the $Na^+$-$K^+$-ATPase. X. Purification of the enzyme from the rectal gland of *Squalus acanthias*, J. Biol. Chem. 248, 2593, 1973.
Holmgren, J., I. Lönnroth, J.-E. Månsson and L. Svennerholm, Interaction of cholera toxin and membrane $G_{M1}$ ganglioside of small intestine, Proc. Nat. Acad. Sci. USA 72, 2520, 1975.
Jørgensen, P.L. Isolation and characterization of the components of the sodium pump, Quart. Rev. Biophys. 7, 239, 1975.
Joseph, K.C. and J.P. Gockerman, Accumulation of glycolipids containing N-acetylglucosamine in erythrocyte stroma of patients with congenital dyserythropoietic anemia type II (HEMPAS), Biochem. Biophys. Res. Commun. 65, 146, 1975.
Karlsson, K.-A., Sphingolipid long-chain bases, Lipids 5, 878, 1970.
Karlsson, K.-A., Carbohydrate composition and sequence analysis of cell surface components by mass spectrometry. Characterization of the major monosialoganglioside of brain, FEBS Letters 32, 317, 1973.
Karlsson, K.-A., Carbohydrate composition and sequence analysis of a derivative of brain disialoganglioside by mass spectrometry, with molecular weight ions at *m/e* 2245. Potential use in the specific microanalysis of cell surface components, Biochemistry 13, 3643, 1974.
Karlsson, K.-A. and I. Pascher, Thin-layer chromatography of ceramides, J. Lipid Res. 12, 466, 1971.
Karlsson, K.-A., B.E. Samuelsson and G.O. Steen, The lipid composition and $Na^+$-$K^+$-dependent adenosine-tri-phosphatase activity of the salt (nasal) gland of eider duck and herring gull. A role for sulphatides in sodium-ion transport, Eur. J. Biochem. 46, 243, 1974.
Kates, M. and P.W. Deroo, Structure determination of the glycolipid sulphate from the extreme halophile *Halobacterium cutirubrum*, J. Lipid Res. 14, 438, 1973.
Kawai, K., M. Nakao, T. Nakao and M. Fujita, Purfication and some properties of $Na^+$-$K^+$-ATPase. III. Comparison of lipid and protein components of $Na^+$-$K^+$-ATPase preparations at various purification steps from pig brain, Biochem. J. 73, 979, 1973.
Kimelberg, H.K. and D. Papahajopoulos, Phospholipid requirements for ($Na^+$-$K^+$)-ATPase activity: head-group specificity and fatty acid fluidity, Biochim. Biophys. Acta 282, 277, 1972.
Kushwaha, S.C., M. Kates and W. Stoeckenius, Comparison of purple membrane from *Halobacterium cutirubrum* and *Halobacterium halobium*, Biochim. Biophys. Acta 426, 703, 1976.
Kyte, J., Purification of $Na^+$-$K^+$-ATPase from canine renal medulla, J. Biol. Chem. 246, 4157, 1971.
Lipkin, M., Proliferation and differentiation of gastrointestinal cells, Physiol. Rev. 53, 891, 1973.

Nakao, T., M. Nakao, N. Mizuno, Y. Komatsu and M. Fujita, Purification and properties of $Na^+$-$K^+$-ATPase. I. Solubilization and stability of lubrol extracts, J. Biochem. 73, 609, 1973.

Nicolson, G.L., Transmembrane control of the receptors on normal and tumor cells, Biochim. Biophys. Acta 457, 57, 1976.

O'Brien, J.S. and E.L. Sampson, Lipid composition of the normal human brain: gray matter, white matter, and myelin, J. Lipid Res. 6, 537, 1965.

Okaya, Y., The plant sulpholipid: a crystallographic study, Acta Cryst. 17, 1276, 1964.

Ottolenghi, P., The reversible delipidation of a solubilized $Na^+$-$K^+$-ATPase from the salt gland of the spiny dogfish, Biochem. J. 151, 61, 1975.

Pascher, I., Synthesis of galactosylphytosphingosine and galactosylceramides containing phytosphingosine, Chem. Phys. Lipids 12, 303, 1974.

Pascher, I., Molecular arrangements in sphingolipids. Conformation and hydrogen bonding of ceramide and their implication on membrane stability and permeability, submitted, 1976.

Racker, E. and L.W. Fischer, Reconstitution of an ATP-dependent $Na^+$ pump with an ATPase from electric eel and pure phospholipids, Biochem. Biophys. Res. Commun. 67, 1144, 1975.

Renooij, W., L.M.G. van Golde, R.F.A. Zwaal and L.L.M. van Deenen, Topological asymmetry of phospholipid metabolism in rat erythrocyte membranes, Eur. J. Biochem. 61, 53, 1976.

Reitherman, R.W., S.D. Rosen, W.A. Frazier and S.H. Barondes, Cell surface species-specific high affinity receptors for discoidin: developmental regulation in *Dictyostelium discoideum*, Proc. Nat. Acad. Sci. USA 72, 3541, 1975.

Rodewald, R., Intestinal transport of antibodies in the newborn rat, J. Cell Biol. 58, 189, 1973.

Rouser, G., G.J. Nelson, S. Fleischer and G. Simon, Lipid composition of animal cell membranes, organelles and organs, *In* Biological Membranes. D. Chapman, editor. Academic Press, London, 1968, p. 5.

Rovis, L., B. Anderson, E.A. Kabat, F. Gruezo and J. Liao, Structures of oligosaccharides produced by base-borohydride degradation of human ovarian cyst blood group H, $Le^b$ and $Le^a$ active glycoproteins, Biochemistry 12, 5340, 1973.

Sedlacek, H.H., J. Stärk, F.R. Seiler, W. Ziegler and H. Wiegandt, Cholera toxin induced redistribution of sialoglycolipid receptor at the lymphocyte membrane, FEBS Letters 61, 272, 1976.

Shamoo, A.E. and M. Myers, $Na^+$-Dependent ionophore as part of the small polypeptide of the $Na^+$-$K^+$-ATPase from eel electroplax membrane, J. Membrane Biol. 19, 163, 1974.

Shur, B.D. and S. Roth, Cell surface glycosyltransferases, Biochim. Biophys. Acta 415, 473, 1975.

Singer, S.J., The molecular organization of membranes, Annu. Rev. Biochem. 43, 805, 1974.

Smith, E.L., J.M. McKibbin, K.-A. Karlsson, I. Pascher and B.E. Samuelsson, Main structures of the Forssman glycolipid hapten and a $Le^b$-like glycolipid of dog small intestine, as revealed by mass spectrometry. Difference in ceramide structure related to tissue localization, Biochim. Biophys. Acta 388, 171, 1975a.

Smith, E.L., J.M. McKibbin, K.-A. Karlsson, I. Pascher, B.E. Samuelsson, Y.-T. Li and S.-C. Li, Characterization of a human intestinal fucolipid with blood group $Le^a$ activity, J. Biol. Chem. 250, 6059, 1975b.

Smith, S.W. and R.L. Lester, Inositol phosphorylceramide, a novel substance and the chief member of a major group of yeast sphingolipids containing a single inositol phosphate, J. Biol. Chem. 249, 3395, 1974.

Smythies, J.R., F. Bennington, R.J. Bradley, W.F. Bridgers and R.D. Morin, The molecular structure of the $Na^+$ channel, J. Theoret. Biol. 43, 29, 1974.

Steck, T.L. and G. Dawson, Topographical distribution of complex carbohydrates in the erythrocyte membrane, J. Biol. Chem. 249, 2135, 1974.

Sundaralingam, M., Molecular structure and conformations of the phospholipids and sphingomyelins, Ann. NY Acad. Sci 195, 324, 1972.

Sweadner, K.J. and S.M. Goldin, Reconstitution of active ion transport by $Na^+$-$K^+$-ATPase from canine brain, J. Biol. Chem. 250, 4022, 1975.

Sweeley, C.C. and G. Dawson, Lipids of the erythrocyte, In Red Cell Membrane Structure and Function. G.A. Jamieson and T.J. Greenwalt, editors. J.B. Lippincott Co., Philadelphia, 1969, p. 172.

Tao, R.V.P. and E. Cotlier, Ceramides of human normal and cataractous lens, Biochim. Biophys. Acta 409, 329, 1975.

van Heyningen, W.E., C.C.J. Carpenter, N.F. Pierce and W.B. Greenough III, Deactivation of cholera toxin by ganglioside, J. Infect. Dis. 124, 415, 1971.

Weiser, M.M., Intestinal epithelial cell surface membrane glycoprotein synthesis. I. An indicator of cellular differentiation, J. Biol. Chem. 248, 2536, 1973.

Weiser, M.M. and A.P. Douglas, An alternative mechanism for gluten toxicity in coeliac disease, The Lancet 1, 567, 1976.

Zambrano, F., S. Fleischer and B. Fleischer, Lipid composition of the Golgi apparatus of rat kidney and liver in comparison with other subcellular organelles, Biochim. Biophys. Acta 380, 357, 1975.

# THE MEMBRANE OF THE HEN ERYTHROCYTE AS A MODEL FOR STUDIES ON MEMBRANE FUSION

J. A. Lucy

Department of Biochemistry and Chemistry, Royal Free Hospital School of Medicine, University of London, 8 Hunter Street, London WC1N 1BP, U.K.

## INTRODUCTION

This paper is primarily concerned with the molecular mechanisms and ultrastructural changes involved in the fusion of hen erythrocytes with one another to produce bi-, tri- and multinucleated cells. Related studies on liposomes and on monomolecular films of lipids and phospholipids will also be discussed.

As a model system for studying membrane fusion, the hen erythrocyte has much to recommend it. Firstly, the cell does not divide and hence there is no possibility of mistaking a cell that is dividing for two cells that are undergoing fusion. Secondly, under normal circumstances, the cells do not fuse. As a result, there is no background of spontaneous cell fusion that may complicate the interpretation of experimentally-induced fusion occurring under the influence of viral or chemical agents. Also, about 54% of the total membranous material of the hen erythrocyte is plasma membrane, with nearly 46% being nuclear membrane (Zentgraf _et al._, 1971). Mitochondrial, lysosomal and endoplasmic reticulum membranes represent less than 1% of the total membrane. It can therefore be concluded with reasonable confidence that the fusion of erythrocyte membranes observed in our experiments is due to a direct modification of the plasma membrane, and that it does not result from the release of lysosomal enzymes. Finally, not only is it much easier to see if cell fusion has occurred when nucleated erythrocytes are being investigated, as compared with mammalian erythrocytes, it also appears that the ultrastructure of the hen erythrocyte enables this cell to fuse more readily than the human erythrocyte.

Why should the fusion of hen erythrocytes be of interest? In the first instance, we aim to elucidate relationships between membrane structure and membrane function in the process of membrane fusion. Also, in the short term, our work has relevance to erythrocyte pathology, e.g. in membrane budding occurring in spherocytosis (Jacob, 1975), in the entry of $Ca^{2+}$ into cells in sickle cell disease (Eaton *et al.*, 1973), and in the penetration of the erythrocyte by the malaria parasite. The work is also related to the development of procedures for introducing enzymes, metabolites, hormones, etc., into cells, *in vitro* or *in vivo*, by the technique of "micro-injection", - i.e. by fusion with liposomes (Papahadjopoulos *et al.*, 1974a, 1974b; Pagano *et al.*, 1974; Martin & MacDonald, 1974; Batzri & Korn, 1975) or with erythrocyte ghosts (Furusawa *et al.*, 1974; Loyter *et al.*, 1975), containing the molecules in question. In the longer term, our studies are relevant to membrane fusion occurring in both physiology and in disease, at cellular and intracellular levels, as in myoblast fusion, the entry of viruses into cells, malignant diseases (particularly perhaps regarding tumours that secrete excessive quantities of amines, peptides, or proteins), fertilisation (Lucy, 1975), liver function (Lucy, 1976), immunology, adjuvants and rheumatoid arthritis (Ahkong *et al.*, 1974; Whitehouse *et al.*, 1974).

Without prejudice to the differing mechanisms by which membrane fusion may occur, we have suggested that the adjective "fusogenic" and the noun "fusogen" may be appropriately used to describe the behaviour of agents or conditions that give rise to membrane fusion in general, including cell fusion, (Ahkong *et al.*, 1973a). Although it has been known for many years that a number of enveloped viruses cause cells (including erythrocytes) to fuse, the biochemical mechanisms involved remain uncertain. We have concentrated our own attention primarily but not exclusively on the chemically-induced fusion of hen erythrocytes and we have found that the fusion process may be triggered in several different ways. I now propose to discuss these, and to consider the underlying mechanisms that appear to be involved. In the longer term, some of these mechanisms may prove to be important in processes of cell fusion and membrane fusion that occur normally and pathologically *in vivo*.

## LYSOPHOSPHATIDYLCHOLINE

The first simple chemical that we investigated was lysolecithin (Howell & Lucy, 1969). This was studied because it had been known for many years that lysolecithin has a direct effect on the integrity of lipoprotein membranes. In essence it was argued that, for two membranes to fuse both of them must undergo a decrease in structural stability, otherwise membrane fusion would be an energetically unfavourable process. Our initial experiments on the treatment of hen erythrocytes with lysolecithin in aqueous solution were disappointing because cell lysis occurred without evidence of accompanying cell fusion. It was then found that fixation of the treated cells, within 30 seconds of adding lysolecithin, arrested the damage to membranes at a point where fusion between hen erythrocytes could be clearly seen. Subsequently, lysolecithin was shown to fuse a number of different types of cell (Poole *et al*., 1970; Croce *et al*., 1971; Keay *et al*., 1972), including sperm cells (Gledhill *et al*., 1972). Later it was found that the destructive effects of lysolecithin could be minimised by incorporating it into microdroplets of lipid, which localised the action of the lysophospholipid, and allowed mouse fibroblast - hen erythrocyte heterokaryons to be obtained (Ahkong *et al*., 1972). These, and other studies on the use of lysolecithin in the presence of serum albumin (Croce *et al*., 1971) have, however, not been capable of any extensive development as a means of obtaining hybrid cells because of the difficulty of controlling the membrane-damaging effects of lysolecithin.

Initially it was suggested that the formation of globular micelles might be involved in membrane fusion induced by lysolecithin (Lucy, 1970), and this view is consistent with the observed actions of lysolecithin on aqueous dispersions of lecithin (Bangham & Horne, 1964; Howell *et al*., 1973). More recently, Breisblatt and Ohki (1975) have proposed that a semi-micelle membrane configuration, for the hydrocarbon chains of phospholipids, may be involved in the fusion of spherical bilayers with and without lysolecithin; in their experiments, the temperature required for fusion was lowered by 10°C in the presence of lysolecithin. It is relevant that, for alamethicin-mediated fusion of lecithin vesicles, Lau and Chan (1975) have suggested that transient rearrangements in the local structure of the lipids occur, with the formation of an inverted micelle of alamethicin.

Two important points should be noted in relation to the fusogenic properties of lysolecithin. We have used the negative-staining technique of electron microscopy to investigate interactions between phospholipids and a number of lipids that induce hen erythrocytes to fuse into multinucleated cells; of the fusogenic compounds studied, only lysolecithin was observed to cause bilayers of lecithin to become micellar (Howell *et al*., 1973). It therefore seems that, while globular micelles may be involved in the fusogenic actions of lysolecithin, membrane fusion induced by the other lipids probably occurs by different mechanisms. Secondly, although it remains theoretically possible that membrane fusion occurring *in vivo* may involve the transient presence in membranes of lysolecithin (cf., Lucy, 1970; Suzuki & Matsumoto, 1974), produced and subsequently destroyed by membrane-bound enzymes, there is no direct evidence in support of this concept. Indeed, searches for the involvement of lysophosphatides in virus-induced cell fusion have yielded negative results (Falke et al., 1967; Elsbach et al., 1969; Blough et al., 1973; Pasternak & Micklem, 1974; Parkes & Fox, 1975). Further, lysolecithin inhibits the fusion of myoblasts in vitro: the subsequent removal of lysolecithin allows cell fusion to proceed (Reporter & Norris, 1973). Although phospholipase A activity has been found in the plasma membranes of fusing muscle cells, $Ca^{2+}$ was not required for enzyme activity even though $Ca^{2+}$ is needed for muscle cell fusion (Kent & Vagelos, 1975).

## FUSOGENIC LIPIDS AND MEMBRANE FLUIDITY

Many fatty acids and certain of their esters were highly effective in fusing hen erythrocytes into multinucleated cells (Ahkong et al., 1973a). It is noteworthy that, in general, fusion of hen erythrocytes by saturated fatty acids occurred only when they contained between 10 and 14 carbon atoms. Saturated acids with fewer, or more, carbon atoms were inactive. Unsaturated $C_{16}$ and $C_{18}$ fatty acids were, by contrast, active. Among the most effective of lipids inducing cell fusion were the unsaturated compounds, oleic acid and glycerol mono-oleate. Because oleic acid caused cell fusion while stearic acid was inactive, and because the fusion of hen erythrocytes was found to be strongly dependent on temperature, we suggested that unsaturated fatty acids and their derivatives induced erythrocytes to fuse by increasing membrane fluidity. Alternatively benzyl alcohol (40 mM), which increases membrane fluidity (Metcalfe et al., 1968) but is not fusogenic at this concentration, may be used instead of heat to facilitate fusion by fusogenic lipids (Fisher, 1975). Our interpretation of the effects of unsaturated lipids on hen erythrocyte membranes was supported by the observation that these cells can also be fused by heat alone, in the absence of exogenously added lipids, when they are incubated at about 48$^{o}$C in close proximity to one another on a microscope slide (Ahkong et al., 1973b). Papahadjopoulos et al., (1973) have reported that vesicles of phospholipid that are below their transition temperature give less cell fusion than similar numbers of vesicles containing lipids at or above their transition temperature. They also found that the incorporation of equimolar quantities of cholesterol, into vesicles composed of lipids in a liquid-crystalline state, reduced their ability to fuse with one another (Papahadjopoulos et al., 1974c). Kosower et al., (1975) have observed that the membrane mobility agent, $A_2C$, promotes the fusion of hen erythrocytes, and they have concluded that it seems reasonable to suppose that some increase in local fluidity in membranes favours cell fusion.

The rate of fusion of vesicles prepared from dimyristoyllecithin has been found to increase markedly near the hydrocarbon phase transition temperature of the vesicles (Prestegard & Fellmeth, 1974). Kantor and Prestegard (1975) have observed further that dimyristoyllecithin vesicles, containing 2% myristic acid, fuse rapidly at temperatures between 17°C and 20°C. Similar vesicles containing 4% lauric acid fused rapidly between 11°C and 15°C, but those containing 4% palmitic acid did not fuse at an appreciable rate between 17°C and 37°C. These investigators have remarked that their results appear to follow the trends that we have observed with chicken erythrocytes, since the cells were fused at 37°C by lauric and myristic acids but not by palmitic acid (Ahkong *et al.*, 1973a). van der Bosch and McConnell (1975) further found that the fusion of liposomes of dipalmitoylphosphatidylcholine induced by concanavalin A exhibits a maximum at 36°C - the midpoint of the phase transition range of this phospholipid. Lawaczeck *et al.*, (1975) have recently drawn attention to the importance of the history of liposomes, with respect to annealing, in relation to the interpretation of experiments on liposome fusion. It is interesting that the anti-depressant molecule, desipramine, which has been shown to reduce the transition temperature of lipid vesicles (Cater *et al.*, 1974), has also been reported to induce the fusion of phospholipid vesicles (Bermejo *et al.*, 1975).

Glycerol mono-oleate, like lysolecithin, can be used not only to fuse erythrocytes but also to obtain viable hybrid cells from fibroblast cell lines (Cramp & Lucy, 1974). The ester is less lytic to cells than lysolecithin. However, its efficiency is not as great as that of Sendai virus in yielding viable hybrid cells. In passing, it is interesting to note that a number of non-ionic surface active molecules, long chain amines, and straight chain hydrocarbons, all of which have immunological adjuvant properties have been observed to be fusogenic towards hen erythrocytes (Ahkong *et al.*, 1974). This finding indicates that some kind of relationship exists between adjuvant and fusogenic properties, and it seems possible that adjuvant and fusogenic behaviour may both involve the introduction of instability into the structure of cell membranes. Cell fusion might itself play a role in adjuvant function, or alternatively adjuvants may facilitate the secretion of immunological mediators by affecting the fusion of membranes that occurs in exocytosis.

Electron microscopy of negatively-stained preparations of phospholipids and fusogenic lipids showed that treatment of lecithin, sphingomyelin, and phosphatidylserine with either glycerol mono-oleate or oleic acid gave rise to various vesicular and hexagonal macromolecular assemblies that were not given by the non-fusogenic compounds, glycerol monostearate and stearic acid (Howell *et al.*, 1973). Subsequently fusogenic lipids were found to exhibit interactions, which were not shown by non-fusogenic lipids, in mixed monolayers with several species of phospholipid, particularly those containing a choline head group. Fusogenic lipids exhibited negative deviations from the ideality rule for mean molecular areas in mixed monolayers with erythrocyte lipids, natural and synthetic preparations of phosphatidylcholine, sphingomyelin, dipalmitoyl-NN-dimethylphosphatidylethanolamine and phosphatidylserine (Maggio & Lucy, 1975). The fusogenic lipids showed no specific effects in mixed monolayers with phosphatidylethanolamine or galactosylceramide (from which the choline-containing head group is absent). These findings may be related to the fact that diacyl-phosphatidylcholine has a higher affinity for water than has diacyl-phosphatidylethanolamine (Jendrasiak & Hasty, 1974), as well as a different conformation and orientation of the polar head group (Phillips et al., 1972). In the light of our observations on the importance of the choline group in the molecular interactions of low-melting fusogenic lipids with phospholipids, it was suggested that membrane fusion induced by fusogenic lipids may be mediated by the asymmetrical distribution of choline-containing lipids (Maggio & Lucy, 1975), since these are mainly in the outer half of the lipid bilayer of erythrocyte membranes (Zwaal et al., 1973). The outer half of the erythrocyte bilayer might not only be preferentially expanded by fusogenic lipids, but may also have its molecular structure modified in favour of a more fluid, and possibly more permeable, organisation.

Very recently investigations have been undertaken on the surface potentials of mixed monolayers of synthetic phospholipids with lipids that are fusogenic for hen erythrocytes (Maggio & Lucy, 1976). At pH 5.6 and 10, but not at pH 2, mixed monolayers of the fusogenic lipid, glycerol mono-oleate, with phosphatidylcholine showed negative deviations from the ideality rule in surface potential per molecule which accompanied negative deviations in mean molecular area. These changes in surface potential were similarly not seen with chemically related but non-fusogenic lipids, nor were they observed in mixed monolayers of any of the lipids with phosphatidylethanolamine. The fact that negative deviations from ideality in both surface potential and surface area, in mixed monolayers of ionizable (e.g. oleic acid) and non-ionizable (e.g. glycerol mono-oleate) fusogenic lipids with dipalmitoylglycerylphosphorylcholine, were decreased on changing the pH of the subphase from pH 5.6 to 2 indicates that the polar head group of the phospholipids is involved in the observed interactions. It is also significant in this respect that the differing behaviour shown in mixed films by choline-containing phospholipids (negative deviations in mean molecular area) and ethanolamine-containing phospholipids (positive deviations in mean molecular area) was eliminated at pH 2, and that the two species of phospholipid showed almost ideal mixing behaviour at this pH with fusogenic and with non-fusogenic lipids.

We found that the presence in the subphase of bivalent cations, including $Ca^{2+}$, significantly modified the properties of mixed monolayers containing fusogenic lipids and phosphatidylcholine or phosphatidylethanolamine. It was thus apparent that whether or not phosphatidylcholine interacts differentially with fusogenic and non-fusogenic lipids, and whether phosphatidylethanolamine behaves in the same way or differently, depends on the surface pressure of the system and on the degree to which the ion-dipole behaviour of the phospholipid is constrained by the influence of an external bivalent metal ion. For a fusogenic lipid to modify the properties of phosphatidylethanolamine molecules, in which intermolecular ionic interactions are probably greater than with phosphatidylcholine (Phillips et al., 1972), it appears to be necessary to disturb the orientation of the polar moieties of the phospholipid monolayer by adding bivalent cations to the subphase. Some of these possibilities are shown in Plate 1. Alternative structural arrangements of the interacting molecules are possible, however, and those shown should be regarded as illustrative rather than definitive. On the basis of our findings, we have speculated that in the presence of a fusogenic lipid the properties of the inner half of the lipid bilayer of the erythrocyte membrane (relatively rich in phosphatidylethanolamine) could well be altered, and behave like the outer half of the bilayer (relatively rich in phosphatidylcholine), if the concentration of cytoplasmic $Ca^{2+}$ is raised to about 1 mM. This increase in symmetry in the membrane might lower energy barriers for the intermixing of the constituents of membranes in closely adjacent erythrocytes and thus facilitate cell fusion.

## WATER-SOLUBLE FUSOGENS

Fusogenic properties are not restricted to lipid-soluble substances. High concentrations of dimethyl sulphoxide (5 M), or glycerol (1.5 - 2.5 M) also induce hen erythrocytes to fuse (Ahkong et al., 1975a). Other polyols have been investigated for fusogenic properties in our laboratory, and it has been found that sorbitol fuses these cells, numerous multinucleated erythrocytes being present after two hours of incubation in 2.5 or 3 M sorbitol. Higher and lower concentrations of sorbitol were less effective. Multinucleated cells have also been observed with mannitol, ethylene glycol, and sucrose. Long incubation (30-48 hours) with a high molecular weight dextran ($2 \times 10^6$) induced the cells to fuse. Polyethylene glycol (M.W. 6000) was also capable of causing extensive cell fusion with hen erythrocytes. Furthermore, hen erythrocytes can be fused with yeast protoplasts with the aid of polyethylene glycol (Ahkong et al., 1975b).

At first sight, it would seem unlikely that these water-soluble fusogens interact with the hydrophobic lipid bilayer of the erythrocyte membrane. How then are they likely to function?

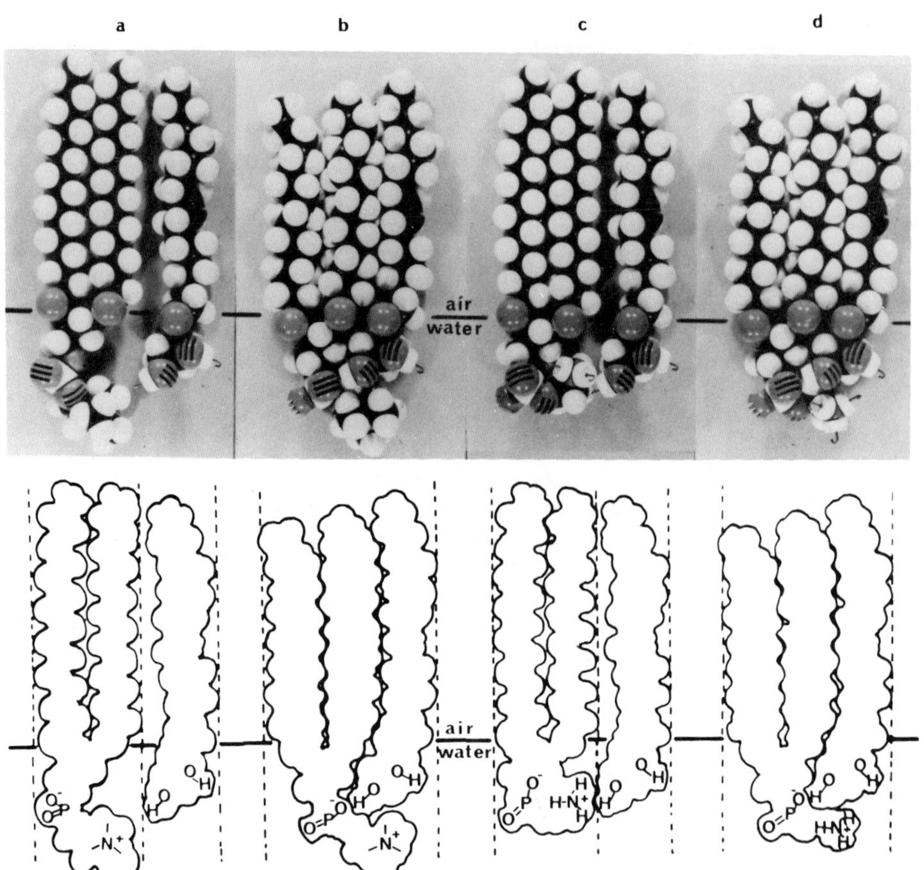


Plate 1. Molecular models of glycerol mono-oleate and phospholipids. Space filling Corey-Pauling-Koltun molecular models illustrating some possible molecular interactions between glycerol mono-oleate and phospholipids. Outline projections of the models are also shown to facilitate the illustration of molecular and polar moiety volumes (vertical dotted lines). The molecules are considered to be in a closely-packed state (i.e. above a surface pressure of 30 $mN.m^{-1}$). Equimolar mixture of glycerol mono-oleate and dipalmitoylglycerylphosphorylcholine: exhibiting theoretical ideal behaviour according to the additivity rule (a), illustrating a possible molecular arrangement giving non-ideal mixing (b). Equimolar mixture of glycerol mono-oleate and dipalmitoylglycerylphosphorylethanolamine: exhibiting ideal behaviour according to the additivity rule (c), illustrating a possible molecular arrangement giving non-ideal mixing (d). In (a) and (c) the phospholipids are represented with the polar head groups orientated according to

Phillips *et al*., (1972) and with their acyl chains in an all-trans configuration. In (b) and (d) three 2 gl kinks (Traüble & Haynes, 1971; Seelig & Seelig, 1974) are illustrated in each of the phospholipid acyl chains that allow cooperative close packing with the fusogenic lipid (shown with three 2 gl kinks) consequent upon ion-dipole interactions between a hydroxyl group of glycerol monooleate and the anionic oxygen in the polar head group of the phospholipid. Interactions of this type do not occur with phosphatidylethanolamine (c), until after the normal orientation of the polar moiety of this phospholipid has been disturbed (d) by bivalent cations in the subphase (see text). The position of the air-water interface shown is intended to be illustrative and not definitive. This plate is reproduced with permission from Maggio & Lucy (1976).

---

One possibility is that their effects are mediated by the aggregation of the intramembranous particles of the erythrocyte membrane since it is known that dimethyl sulphoxide and glycerol aggregate the intramembranous particles of unfixed T and B mouse lymphocytes (McIntyre *et al*., 1974); glycerol-induced aggregation of membrane-intercalated particles has also been noted in *Entamoeba histolytica* (Pinto da Silva & Martinez-Palomo, 1974). Additionally, Maroudas (1975) has drawn attention to the flocculation of the glycocalyx polymers of membranes by cryoprotectant molecules in relation to cell fusion induced by these substances.

One of the water-soluble fusogens should be singled out for special comment, namely polyethylene glycol. Pontecorvo (1975) has investigated the production of indefinitely multiplying mammalian somatic cell hybrids using polyethylene glycol, and he has used this water-soluble fusogen to yield hybrid cells from the fusion of Chinese hamster HGPRT$^-$ cells (strain WG, clone 1) with mouse TK$^-$ cells (strain 3T3). In addition, human skin fibroblast cells have been fused with one another, and with human lymphocytes. The relative lack of cytotoxic effects observed with polyethylene glycol, as compared with fusogenic lipids, and the high incidence of cell fusion and hybrid cell production that have been obtained, indicate that polyethylene glycol may perhaps be routinely used in place of Sendai virus in the future as a tool for the production of hybrid cells.

## CALCIUM IONS

The involvement of $Ca^{2+}$ in the fusion of biological membranes seems to be an almost universal phenomenon. For example $Ca^{2+}$ is necessary in myoblast fusion (Shainberg *et al*., 1969), in cell fusion induced by liposomes (Papahadjopoulos *et al*., 1973) and in the fusion of nascent membranes of the protozoon Echinosphaerium nucleofilum (Vollet & Roth, 1974). In addition, $Ca^{2+}$ is of major importance in many processes of secretion occurring by exocytosis (Carafoli *et al.*, 1975).

Once again, however, it appears that more than one pathway for membrane fusion is possible, and that membrane fusion can also occur in the absence of $Ca^{2+}$. Peretz *et al.*, (1974) have reported that human erythrocytes treated with Sendai virus will fuse in the absence of bivalent metals, or in the presence of EDTA, provided that low concentrations of virus are used to minimise haemolysis. In experiments with chicken erythrocytes, in which a small number of Sendai virions were used, it has been found in my laboratory that maximum cell fusion occurs in the presence of EGTA, and that fusion is significantly decreased by the presence of $Ca^{2+}$, even at a concentration of 0.2 mM (C. A. Hart, D. Fisher, T. Hallinan & J. A. Lucy, unpublished work). This is in marked contrast to the requirement for $Ca^{2+}$ in the fusion of hen erythrocytes by lipid molecules (Ahkong *et al.*, 1973a), and by water-soluble fusogens (Ahkong *et al.*, 1975a). It is also interesting that the general hypothesis for the mechanism of membrane fusion proposed by Poste and Allison (1973), which involves the activity of a $Ca^{2+}$-dependent ATPase, would seem not to apply to the fusion of human and chicken erythrocytes by Sendai virus.

The importance of the concentration of cytoplasmic $Ca^{2+}$ has been emphasized within recent years by numerous demonstrations that the bivalent cation ionophores X537A and A23187, which are known to facilitate the movement of $Ca^{2+}$ through membranes (Reed & Lardy, 1972; Pressman, 1973), trigger secretion in the presence of extracellular $Ca^{2+}$ (Cochrane & Douglas, 1974; Foreman *et al.*, 1973). Significantly, hen erythrocytes will fuse into multinucleated cells on treatment with these bivalent cation ionophores under appropriate conditions (Ahkong *et al.*, 1975c). Light microscope experiments provided no evidence that either valinomycin, with or without 2,4-dinitrophenol, or 2,4-dinitrophenol alone caused fusion. Increasing the concentration of $Ca^{2+}$ from 1 to 3 mM in the presence of X537A and A23187 allowed cell fusion to occur more rapidly, but this was without effect on cells treated with valinomycin or valinomycin and DNP. The actions of both X537A and A23187 with 1 mM $Ca^{2+}$ were inhibited by EGTA and EDTA (2 mM); inhibition by the chelating agents was, however, overcome by excess of $Ca^{2+}$ (final concentration 3 mM). It was therefore concluded - as with secretory systems - that the observed fusogenic action of the two bivalent cation ionophores is mediated by entry of $Ca^{2+}$ into the cells. It seems possible, although it has yet to be established, that the actions of the fat-soluble and water-soluble fusogens discussed above may be mediated by allowing the entry of $Ca^{2+}$ into the interior of the treated erythrocytes. This would also seem to be contrary to the views of Poste and Allison (1973), who have postulated that membranes can exist in two states, (1) the normal $Ca^{2+}$-associated state; (2) the "fusion susceptible" state in which $Ca^{2+}$ has been displaced from the membrane.

N-ethylmaleimide inhibits the fusion of hen erythrocytes induced by the bivalent cation ionophore A23187, and also fusion caused by low-melting lipid molecules (Hart *et al.*, 1975). The inhibitory action of the thiol reagent may be on membrane proteins that participate in a sequence of events initiated by $Ca^{2+}$ that lead to cell fusion. N-ethylmaleimide added to erythrocytes prior to $Ca^{2+}$ inhibits cell fusion, whereas addition of the thiol reagent subsequently to incubation with $Ca^{2+}$ has little effect (Q. F. Ahkong, W. Tampion & J. A. Lucy, unpublished work).

## INTRAMEMBRANOUS PROTEIN PARTICLES

Extracted mixtures of spectrin and actin, from human erythrocyte ghosts, precipitate in the presence of low concentrations of $Ca^{2+}$, and the conditions that lead to the precipitation of spectrin and actin also induce aggregation of the intramembranous particles in spectrin-depleted erythrocyte ghosts (Elgsaeter *et al.*, 1976). $Ca^{2+}$ has important effects on membrane phospholipids also: the binding of $Ca^{2+}$ to phosphatidylserine molecules in a membrane yields solid aggregates and allows other phospholipids to form fluid clusters (Ohnishi & Ito, 1974). If this should occur at the cytoplasmic surface of erythrocytes, where most of the phosphatidylserine is located (Zwaal *et al.*, 1973), it would probably again lead to an aggregation of intramembranous proteins (Grant & McConnell, 1974).

It has been suggested that the aggregation of membrane proteins (Poste & Allison, 1973), and the clustering of lipid molecules (Ahkong *et al.*, 1975) are important in membrane fusion. The importance of the aggregation of intramembranous protein particles in cell fusion is supported by our finding that dimethyl sulphoxide and glycerol induce hen erythrocytes to fuse (Ahkong et al., 1975a), since these chemicals are known to aggregate the intramembranous particles in the membranes of other cells.

Recently, we have undertaken freeze-fracture studies on suspensions of hen erythrocytes during cell fusion induced by the ionophore A23187, in the hope of obtaining direct information on the movement of the intramembranous protein particles in membrane fusion (J. Vos and colleagues, unpublished work). As already discussed membrane fluidity is important in cell fusion, and fusion induced by A23187 and X537A is markedly accelerated by raising the temperature of incubation from 37 to 47°C. (Thermally-induced cell fusion at 48-50°C and pH 5.6 occurring in the absence of any exogenous agent, which is mentioned above, occurs only with cells that are in close contact on a heated microscope stage and not with cells in free suspension.) Incubation at 37°C of neuraminidase-treated hen erythrocytes for 3 hours in the presence of A23187, 1 mM $Ca^{2+}$, and dextran, gave rise to extensive clumping of the cells but to only limited cell fusion (Ahkong *et al.*, 1975c). Freeze-fracture studies on hen erythrocytes incubated for 1 hour at 37°C under these conditions, without any significant cell fusion occurring, showed that the intramembranous particles of the plasma membranes were more closely packed than those in the membranes of control, untreated cells. Numerous, well-defined circular areas were also observed on the A fracture face of the treated cells, ranging in diameter from about 100 to 300 nm, that were almost completely free from intramembranous particles. Raising the temperature to 47°C, and thereby inducing extensive cell fusion, caused a marked change in the distribution of the intramembranous particles of the A fracture face. Characteristically, it was found that the particles were then clustered in random aggregates, with disruption of the previously-bare circular areas. Particle aggregation and the random disorganisation of the arrangement of intramembranous particles in membranes both therefore seem to be important in cell fusion.

The human erythrocyte is not readily susceptible to fusion by treatment with the ionophore A23187, apparently because the intramembranous proteins of intact human erythrocytes are much less free to move than their counterparts in chicken cells (cf., the membranes of fresh ghosts from human erythrocytes, Elgsaeter *et al.*, 1976). However, studies on the membrane proteins of human erythrocytes during fusion induced by glycerol mono-oleate, to which the cells are susceptible, have revealed interesting phenomena (S. Quirk and colleagues, unpublished work). In particular, gel electrophoresis of the membrane proteins showed a decrease in band 3 protein (cf., Steck, 1974), simultaneously with an increase in proteins of lower molecular weight, indicating that a limited degree of proteolysis may occur during cell fusion under these conditions. The observed changes, like cell fusion itself, were inhibited by EDTA. Parallel freeze-fracture studies on the cells indicated that, with glycerol mono-oleate as the fusogen, fusion was not mediated by the intermediate formation of circular areas of protein-free lipid bilayer. Instead, the small extent of microaggregation of intramembranous particles that is normally observed in preparations of membranes from untreated human erythrocytes was lost, and the particles were distributed in a random array. This observation again appears to underline the importance of the disorganisation of the arrangement of membrane proteins in the process of cell fusion.

Whether or not membrane fusion is mediated by the interdigitation of membrane proteins as proposed by Poste and Allison (1973), or by the intermingling of membrane lipids after the emergence of protein-free areas of lipid bilayer following protein aggregation as we have proposed (Ahkong *et al.*, 1975a), is a matter of debate at the present time. However, observations in support of the latter view, albeit only indirectly so, are accumulating. We have found that membranous vesicles, which bud from chicken erythrocytes on treatment with A23187 at 37°C, appear to be essentially particle-free in freeze-fracture preparations (J. Vos and colleagues, unpublished work), while chemical analyses of similar vesicles from human erythrocytes have shown them to be markedly deficient in protein relative to the original erythrocyte membrane (Allan *et al.*, 1976). Protein is also absent from the vesicles that bud from fresh ghosts of human erythrocytes on exposure to conditions that precipitate extracts of spectrin and actin (Elgsaeter *et al.*, 1976). During mucocyst secretion by exocytosis in *Tetrahymena*, fusion occurs between a region of plasma membrane, sequestered within a rosette of particles, and a smooth area of mucocyst membrane free from particles (Satir *et al.*, 1973). Freeze-fracturing of rat posterior pituitary lobes has shown that, prior to secretion, neurosecretory granules cause bulging of the nerve cell membrane: membrane associated particles are absent from these bulges (Dempsey *et al.*, 1973). Similarly, fusion between the membranes of secretory granules and the plasma membrane in mast cells, during the release of histamine by exocytosis, appears to occur in regions of membrane that are free from intramembranous particles (Lawson *et al.*, 1976). Finally, the experiments of Haywood (1974) on the fusion of Sendai virus with liposomes have shown that there is no absolute requirement for cellular membrane proteins in membrane fusion.

## CONCLUSIONS

On the basis of our various findings, the following tentative working hypothesis has been formulated for the fusion of erythrocytes, *viz.*, - "Fusogenic lipids act initially on erythrocyte membranes to modify the organisation of the polar moieties of their phospholipid molecules, and to increase the permeability of the membranes to $Ca^{2+}$. When $Ca^{2+}$ enters the erythrocyte it further modifies the properties of membrane phospholipids, possibly eliminating the asymmetrical character of the lipid bilayer. $Ca^{2+}$ entry also initiates alterations in membrane proteins that may involve spectrin and band 3 protein. These changes lead to an altered distribution of intramembranous particles that facilitates membrane fusion. It seems that once $Ca^{2+}$ has entered the cell, membrane fusion occurs provided that the membrane lipids are sufficiently fluid - due for example to the presence of an unsaturated fusogenic lipid, or to an increase in temperature, or to the presence of benzyl alcohol."

I wish to emphasize that much of this working hypothesis is

only tenously based on experimental observations. It is hoped that our current and future experiments will indicate more clearly whether or not the substance of this hypothesis is correct.

Work undertaken in this laboratory that is referred to in this article was supported by the Medical Research Council and the British Council: the electron microscope was provided by the Science Research Council.

## REFERENCES

Ahkong, Q.F., Cramp, F.C., Fisher, D., Howell, J.I. & Lucy, J.A. (1972) J. Cell Sci. 10, 769-787

Ahkong, Q.F., Fisher, D., Tampion, W. & Lucy, J.A. (1973a) Biochem. J. 136, 147-155

Ahkong, Q.F., Cramp, F.C., Fisher, D., Howell, J.I., Tampion, W., Verrinder, M. & Lucy, J.A. (1973b) Nature New Biol. 242, 215-217

Ahkong, Q.F., Howell, J.I., Tampion, W. & Lucy, J.A. (1974) FEBS Lett. 41, 206-210

Ahkong, Q.F., Fisher, D., Tampion, W. & Lucy, J.A. (1975a) Nature 253, 194-195

Ahkong, Q.F., Howell, J.I., Lucy, J.A., Safwat, F., Davey, M.R. & Cocking, E.C. (1975b) Nature 255, 66-67

Ahkong, Q.F., Tampion, W. & Lucy, J.A. (1975c) Nature 256, 208-209

Allan, D., Billah, M.M., Finean, J.B. & Michell, R.H. (1976) Nature, in press

Bangham, A.D. & Horne, R.W. (1964) J. Mol. Biol. 8, 660-668

Batzri, S. & Korn, E.D. (1975) J. Cell Biol. 66, 621-634

Bermejo, J., Eirin, T. & Barbadillo, A. (1975) FEBS Lett. 58, 289-292

Blough, H.A., Gallaher, W.R. & Weinstein, D.B. (1973) in Membrane Mediated Information (Kent, P.W., ed), Vol. 1, pp. 183-199, Medical and Technical Publishing Co. Ltd., Lancaster

Breisblatt, W. & Ohki, S. (1975) J. Membrane Biol. 23, 385-401

Carafoli, E., Clementi, F., Drabikowski, W. & Margreth, A., eds., (1975) Calcium Transport in Contraction and Secretion, pp. 588, North-Holland Publishing Co., Amsterdam

Cater, B.R., Chapman, D., Hawes, S.M. & Saville, J. (1974) Biochim. Biophys. Acta 363, 54-69

Cochrane, D.E. & Douglas, W.W. (1974) Proc. Nat. Acad. Sci. U.S. 71, 408-412

Cramp, F.C. & Lucy, J.A. (1974) Exp. Cell Res. 87, 107-110

Croce, C.M., Sawicki, W., Kritchevski, D. & Koprowski, H. (1971) Exp. Cell Res. 67, 427-435

Dempsey, G.P., Bullivant, S. & Watkins, W.B. (1973) Z. Zellforsch 143, 465-484

Eaton, J.W., Skelton, T.D., Swofford, H.S., Kolpin, C.E. &

Jacob, H.S. (1973) Nature 246, 105-106
Elgsaeter, A., Shotton, D.M. & Branton, D. (1976) Biochim. Biophys. Acta 426, 101-122
Elsbach, P., Holmes, K.V. & Choppin, P.W. (1969) Proc. Soc. Exp. Biol. Med. 130, 903-908
Falke, D., Schiefer, H-G. & Stoffel, W. (1967) Z. Naturforsch 22b, 1360-1362
Fisher, D. (1975) in NATO Symposium on Lipids, Proteins and Receptors (Burton, R.M. & Packer, L., eds.), pp. 75-93, Bi-Science Publications Division, Webster Groves, Missouri
Foreman, J.C., Mongar, J.L. & Gomperts, B.D. (1973) Nature 245, 249-251
Furusawa, M., Nishimura, T., Yamaizumi, M. & Okada, Y. (1974) Nature 249, 449-450
Gledhill, B.L., Sawicki, W., Croce, C.M. & Koprowski, H. (1972) Exp. Cell Res. 73, 33-40
Grant, C.W.M. & McConnell, H.M. (1974) Proc. Nat. Acad. Sci. U.S. 71, 4653-4657
Hart, C.A., Ahkong, Q.F., Fisher, D., Hallinan, T., Quirk, S.J. & Lucy, J.A. (1975) Biochem. Soc. Trans. 3, 734-736
Haywood, A.M. (1974) J. Mol. Biol. 87, 625-628
Howell, J.I., Fisher, D., Goodall, A.H., Verrinder, M. & Lucy, J.A. (1973) Biochim. Biophys. Acta 332, 1-10
Howell, J.I. & Lucy, J.A. (1969) FEBS Lett. 4, 147-150
Jacob, H.S. (1975) in Cell Membranes: Biochemistry, Cell biology and Pathology (Weissmann, G. & Claiborne, R., eds.), pp. 249-255, H.P. Publishing Co. Inc. New York
Jendrasiak, G.L. & Hasty, J.H. (1974) Biochim. Biophys. Acta 337, 79-91
Kantor, H.L. & Prestegard, J.H. (1975) Biochemistry 14, 1790-1795
Keay, L., Weiss, S.A., Circulis, N. & Wildi, B.S. (1972) In vitro 8, 19-25
Kent, C. & Vagelos, P.R. (1975) Fed. Proc. 34, 525
Kosower, N.S., Kosower, E.M. & Wegman, P. (1975) Biochim. Biophys. Acta 401, 530-534
Lau, A.L.Y. & Chan, S.I. (1975) Proc. Nat. Acad. Sci. U.S. 72, 2170-2174
Lawaczek, R., Kainosho, M., Girardet, J-L. & Chan, S.I. (1975) Nature 256, 584-586
Lawson, D., Gilula, N.B., Fewtrell, C., Gomperts, B.D. & Raff, M.C. (1976) in 33rd Nobel Symposium: Molecular and biological aspects of the acute allergic reaction, in press, Plenum Publishing Co. Ltd., London
Loyter, A., Zakai, N. & Kulka, R.G. (1975) J. Cell Biol. 66, 292-304
Lucy, J.A. (1970) Nature 227, 814-817
Lucy, J.A. (1975) J. Reprod. Fert. 44, 193-205
Lucy, J.A. (1976) in 2nd NATO Advanced Study Institute on the Biliary System (Taylor, W., ed), in press, Plenum Publishing Co. Ltd., London

Maggio, B. & Lucy, J.A. (1975) Biochem. J. 149, 597-608
Maggio, B. & Lucy, J.A. (1976) Biochem. J. 155, 353-364
Maroudas, N.G. (1975) Nature 254, 695-696
Martin, F. & MacDonald, R. (1974) Nature 252, 161-163
McIntyre, J.A., Gilula, N.B. & Karnovsky, M.J. (1974) J. Cell Biol. 60, 192-203
Metcalfe, J.C., Seeman, P. & Burgen, A.S.V. (1968) Mol. Pharmacol. 4, 87-95
Ohnishi, S-I. & Ito, T. (1974) Biochemistry 13, 881-887
Pagano, R.E., Huang, L. & Wey, C. (1974) Nature 252, 166-167
Papahadjopoulos, D., Poste, G. & Schaeffer, B.E. (1973) Biochim. Biophys. Acta 323, 23-42
Papahadjopoulos, D., Poste, G. & Mayhew, E. (1974a) Biochim. Biophys. Acta 363, 404-418
Papahadjopoulos, D., Mayhew, E., Poste, G., Smith, S. & Vail, W.J. (1974b) Nature 252, 163-165
Papahadjopoulos, D., Poste, G., Schaeffer, B.E. & Vail, W.J. (1974c) Biochim. Biophys. Acta 352, 10-28
Parkes, J.G. & Fox, C.F. (1975) Biochemistry 14, 3725-3729
Pasternak, C.A. & Micklem, K.J. (1974) Biochem. J. 140, 405-411
Peretz, H., Toister, Z., Laster, Y. & Loyter, A. (1974) J. Cell Biol. 63, 1-11
Phillips, M.C., Finer, E.G. & Hauser, H. (1972) Biochim. Biophys. Acta 290, 397-402
Pinto da Silva, P. & Martinez-Palomo, A. (1974) Nature 249, 170-171
Pontecorvo, G. (1975) Somatic Cell Genetics 1, 397-400
Poole, A.R., Howell, J.I. & Lucy, J.A. (1970) Nature 227, 810-813
Poste, G. & Allison, A.C. (1973) Biochim. Biophys. Acta 300, 421-465
Pressman, B.C. Fed. Proc. (1973) 32, 1698-1701
Prestegard, J.H. & Fellmeth, B. (1974) Biochemistry 13, 1122-1126
Reed, P.W. & Lardy, H.A. (1972) in The Role of Membranes in Metabolic Regulation (Mehlman, M.A. & Hanson, R.W., eds.), pp. 111-131, Academic Press, New York
Reporter, M. & Norris, G. (1973) Differentiation 1, 83-95
Satir, B., Schooley, C. & Satir, P. (1973) J. Cell Biol. 56, 153-176
Seelig, A. & Seelig, J. (1974) Biochemistry 13, 4839-4845
Shainberg, A., Yagil, G. & Yaffe, D. (1969) Exp. Cell Res. 58, 163-167
Steck, T.L. (1974) J. Cell Biol. 62, 1-19
Suzuki, Y. & Matsumoto, M. (1974) Biochem. Biophys. Res. Commun. 41, 505-512
Traüble, H. & Haynes, D.H. (1971) Chem. Phys. Lipids 7, 324-335
van der Bosch, J. & McConnell, H.M. (1975) Proc. Nat. Acad. Sci. U.S. 72, 4409-4413
Vollet, J.J. & Roth, L.E. (1974) Cytobiologie 9, 249-262
Whitehouse, M.W., Orr, K.J., Beck, F.W.J. & Pearson, C.M. (1974) Immunology 27, 311-330
Zentgraf, H., Deumling, B., Jarasch, E-D. & Franke, W.W. (1971)

J. Biol. Chem. 246, 2986-2995
Zwaal, R.F.S., Roelofsen, B. & Colley, C.M. (1973) Biochim. Biophys. Acta 300, 159-182

# SOME ELECTRICAL AND STRUCTURAL PROPERTIES OF LIPID-WATER SYSTEMS

Ingemar Lundström

Research Laboratory of Electronics

Chalmers University of Technology, Gothenburg, Sweden

## SUMMARY

The rather general title of this paper hides the results of some experiments which have been performed at our laboratory during the last two years. I will describe measurements of the lateral conductivity of a lamellar lipid-water system and relate these results to Raman spectroscopic studies on the same system. Furthermore I will discuss inelastic light scattering experiments on monomolecular lipid films on water. All these experiments appear to yield new information on biologically interesting properties of lipid-water systems.

## LATERAL CONDUCTIVITY OF LAMELLAR LIPID-WATER SYSTEMS

Lamellar lipid water systems consist of bimolecular lipid lamellae separated by thin water layers. The thin water layers are in some respects rather similar to the interior of the ionic channels in nerve membranes. The exact nature of the ionic channels is not known, but they are believed to have a hydrophilic interior consisting of carbonyl oxygen or carboxyl groups (1-3). Furthermore the lipid-water interfaces in the lamellar systems are similar to a number of interfaces found in biological systems. We therefore thought it worthwhile to investigate the ionic conductivity along the water channels in a well characterized lamellar lipid water system, namely the di-2,2'-ethylhexyl sodium sulphosuccinate (Aerosol OT)-water system (4,5). The Aerosol OT molecule has a molecular structure that is rather similar to a short chain lecithin:

$$
\begin{array}{l}
\qquad\qquad\qquad\quad C_2H_5 \qquad\qquad\quad O \\
\qquad\qquad\qquad\quad | \qquad\qquad\qquad\quad \| \\
CH_3CH_2CH_2CH_2CH - CH_2 \cdot O{-}C{-}CH_2 \\
\qquad\qquad\qquad\qquad\qquad\qquad\qquad\quad | \\
CH_3CH_2CH_2CH_2CH - CH_2 \cdot O{-}C{-}CH\ SO_3Na \\
\qquad\qquad\qquad\quad | \qquad\qquad\qquad\quad \| \\
\qquad\qquad\qquad\quad C_2H_5 \qquad\qquad\quad O
\end{array}
$$

The phase diagram of the Aerosol OT-water system has been investigated by Rogers and Winsor (6,7) and Fontell (8,9). The lamellar repeat distances and the molecular packing within the lamellar mesophase region (10 - 70% Aerosol OT) were determined by Fontell (8).

The lateral electrical conductivity, i.e. the conductivity along the water layers, was measured on samples obtained by pressing a small amount of an Aerosol OT-water preparation between two microscope slides using thin gold wires as spacers and electrodes (see fig. 1). The samples were checked in a polarizing microscope and those not appearing dark were disregarded. A square wave voltage was applied across the sample and a series resistor.

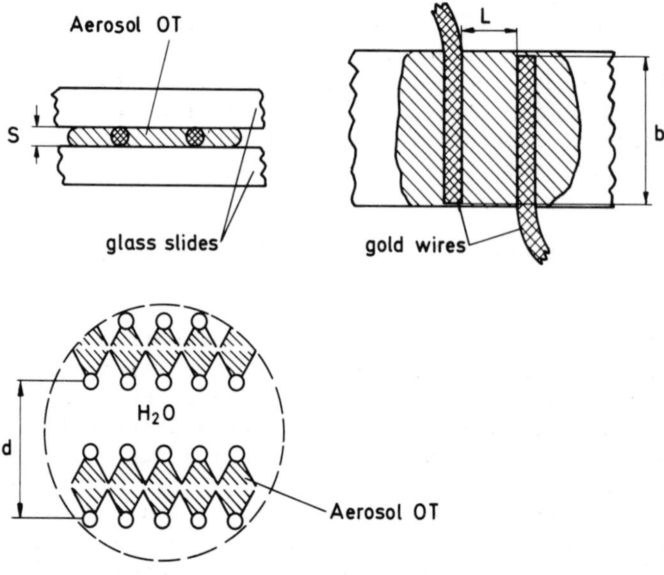


Fig. 1. Schematic illustration of the prepared thin Aerosol OT-water samples. The gold wires had a diameter of 50 µm. The distance L was between 1 and 6 mm. The insert shows the arrangement of the lipid lamellae within the sample.

The current change each time the voltage changes sign is a measure of the true conductance of the sample (4,5). Polarization effects at the electrodes and at "defects" within the samples were avoided in this way. It was, however, observed that the extra series resistance due to the disturbed regions close to the gold wires could be relatively large especially at low water contents. Samples of a given preparation with different distances between the gold wires were therefore studied to enable compensation for the extra series resistance (4,5).

The room temperature conductivity determined from these measurements is shown in fig. 2 versus water layer thickness, $d_w$ (or Aerosol OT concentration). The most interesting feature is the sharp maximum in conductivity around $d_w$ = 16 Å. Since Aerosol OT is a sodium salt it is most natural to attribute the conductivity to sodium ions moving along the water channels. The dashed line in fig. 2 is the conductivity expected if all the sodium ions were free and had the same mobility as in bulk water, $5\times10^{-8}$ $m^2/Vs$. The peak value 3.3 $1/\Omega m$, is about half of that expected for sodium with the same mobility as in bulk water. The influence of the lipid-water interfaces and possibly of the structure of the water on the sodium ions is evident from fig. 2.

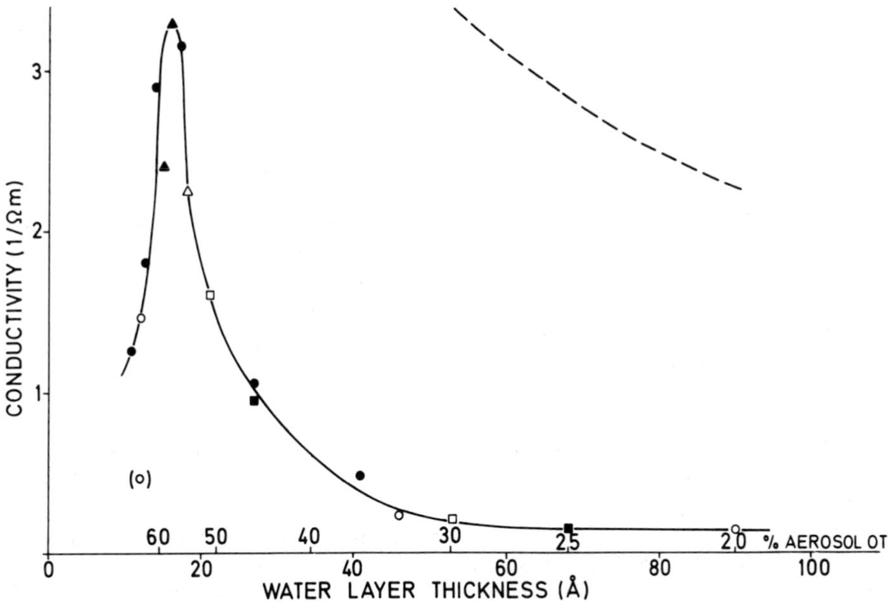


Fig. 2. Lateral conductivity of lamellar Aerosol OT-water systems versus water layer thickness. The different symbols represent Aerosol OT preparations made at different occasions. One preparation (65.1% Aerosol OT) was bistable showing either high or low conductivity. The dashed line is the conductivity expected if all sodium ions had the same mobility as in ordinary water.

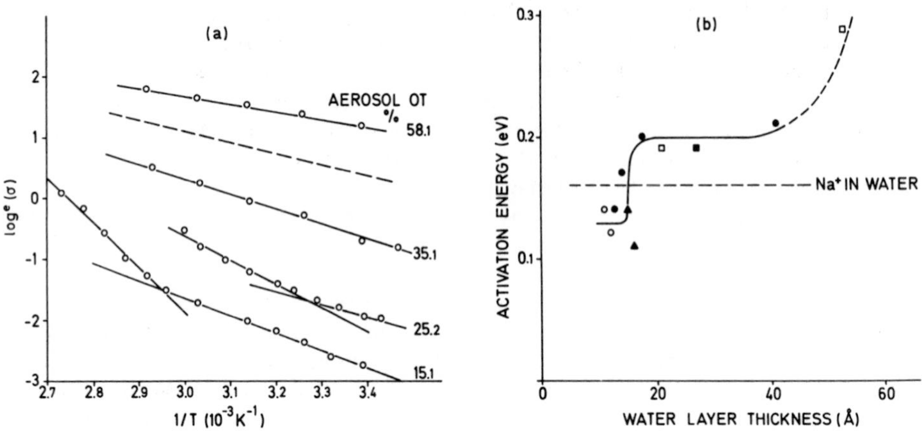


Fig. 3(a). Temperature dependence of the conductivity of lamellar Aerosol OT water systems at different Aerosol OT concentrations (Arrhenius plots). The dashed line indicates the temperature dependence of the conductivity due to sodium ions in ordinary water.

(b) Activation energy (slope of lines like these in (a)) versus water layer thickness.

The temperature dependence of the conductivity was also investigated. Examples of Arrhenius plots are shown in fig. 3(a). Breaking points in the Arrhenius plots are seen at high water contents (thick water layers). Furthermore the activation energy of the conductivity appears to change from a low to a high value at the peak in fig. 2 as indicated in fig. 3(b).

A tentative explanation to the results above is as follows (5). When the water layers are thick the different lipid-water interfaces do not interact. The sodium ions belong to the so called outer Helmholtz plane (OHP) of each lipid-water interface where they are rather immobile (10). This situation is indicated in fig. 4(a). The conductivity in this region is due to a few freely moving ions. The break in the Arrhenius plots may be due to the "bound" ions leaving the OHP, the higher activation energy being the binding energy. When the water layer thickness decreases the two lipid-water interfaces on each side of a water layer start to interact. The ions become less bound to the OHP's and the conductivity increases. We expect a maximum number of mobile sodium ions and/or a maximum mobility when the OHP's of the two interfaces coincide as indicated in fig. 4(b). In this situation there is a minimum electrostatic force acting on the ions towards any of the interfaces. When the water layer thickness decreases even further "full" hydration of the sulphate groups and the sodium ions is no longer possible. A situation like that in fig. 4(c) may arise where some of the ions are firmly bound to the polar part of the lipid.

The explanation above is rather schematic. Changes in water structure as well as in hydration with water layer thickness and not only the electrostatic interaction may change the mobility of the ions.

Some simple numerical estimates can be made. The point of maximum conductivity corresponds to about 58% Aerosol OT or about 18 water molecules per Aerosol OT molecule. Assuming that a water molecule has a diameter of about 2.5 Å the situation in fig. 4(b) could correspond to something like two rows of bound water molecules at each interface and a hydrated sodium ion with four water molecules and a diameter of 6 Å.

Fig. 4 suggests an interpretation of the difference in permeability of a given ionic channel for different ions. If the hydrated ion is too large (needs too many water molecules) a situation according to fig. 4(c) would be appropriate. If on the other hand the hydrated ion is to small it will get "stuck" in the OHP

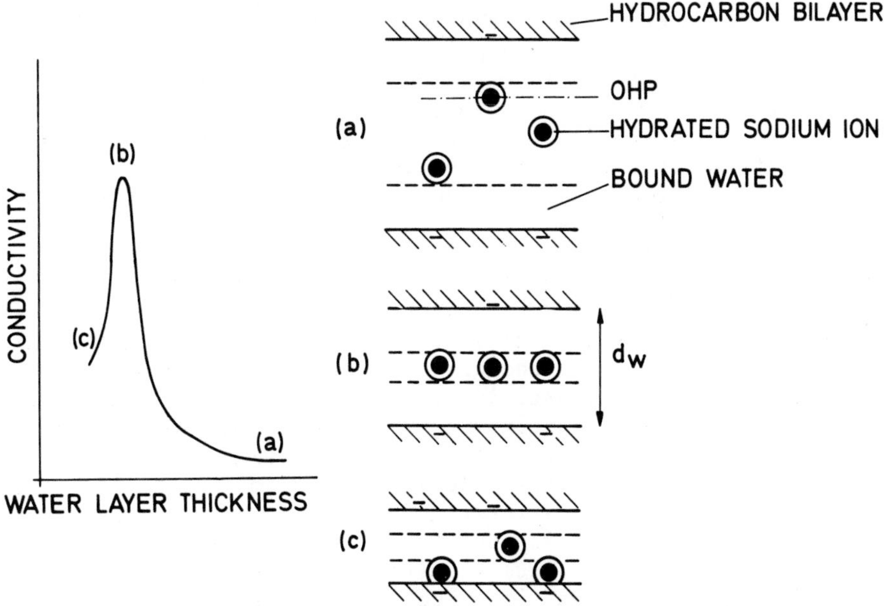


Fig. 4. Tentative explanation to the behaviour of the conductivity in lamellar Aerosol OT systems. See text for full explanation.

according to fig. 4(a). The ion with the right hydration placing it in a situation like fig. 4(b) would have an average mobility much larger than the other two. The ratio between maximum and minimum permeability is about 24 for an ionic channel of "Aerosol OT type" according to fig. 2. It can be mentioned that the ratio of the permeability of sodium and potassium through a sodium channel in a nerve axon is about 12-20 (1,11). It should remembered that the model above is a model for permeability differences and not for ionic selectivity of an ionic channel.

A preliminary note on the lateral conductivity of Aerosol OT-water systems has been published earlier (4). A detailed description of the experiments is given elsewhere (5). Most of the material in this section is taken from ref. (5) with slight changes in presentation.

Lange and Gary-Bobo made some studies on ion diffusion in lecithin-water lamellar phases using radioactive tracers (12,13). They observed an increase in diffusion constant for sodium with water content up to about 24% water where a sudden drop occurred. This drop was attributed to a macroscopic change of the phase at this water content. This structural change could also be observed in a polarizing microscope. They furthermore observed that the relative size of the diffusion constants for $Na^+$ and $C\ell^-$ changed at the phase transition, which was attributed to a change in configuration of the polar head of the lecithin molecule. There are certain differences between the lecithin and the Aerosol OT molecule. The lecithin polar group is zwitterionic, with the possibility of having different configurations, whereas the Aerosol OT molecule becomes negatively charged when it loses its sodium ion. We did not observe any macroscopic structural changes (in a polarizing microscope) with water content in the lamellar region of the Aerosol OT-water system. The change in conductivity was also continuous with water layer thickness as seen in fig. 2 and not abrupt as expected for a phase transition. Furthermore our method of measurement (5) is less sensitive to "defects" or "grain boundaries" inside the sample than diffusion studies. We believe therefore that in case of Aerosol OT the sharp maximum in conductivity depends on the interaction between the polar end groups and the sodium ions as indicated in fig. 4. Electrostatic interactions can also explain the sudden change in the relative magnitude of the diffusion constant for $Na^+$ and $C\ell^-$ in the case of the lecithin-water phases. If the polar group of the lecithin molecule becomes more extended at the phase transition (13) the interaction between $C\ell^-$ and the choline group becomes stronger and that between $Na^+$ and the phosphate group weaker than before the phase transition. Therefore the effective diffusion constant of $Na^+$ is expected to increase and that of $C\ell^-$ to decrease at the phase transition, as observed (13).

## RAMAN SPECTROSCOPIC STUDIES ON THE AEROSOL OT-WATER SYSTEM

R. Faiman, K. Fontell and I recently performed some Raman spectroscopic studies on the Aerosol OT-water system. A full account of this work will be published elsewhere (14). Interestingly enough these studies indicate a "structural change" within the hydrocarbon region of the lipid lamellae in the concentration region where the conductivity has a maximum. The change in molecular organisation within the lipid lamellae was observed as a change in the magnitude of the Raman peak at 1460 $cm^{-1}$. This peak was attributed to antisymmetric methylene rocking modes. The assignment was mainly made from polarization measurements on a viscous isotropic cubic phase (14). A peak at 1734 $cm^{-1}$, assigned to symmetrical C=O stretching modes, was used as a reference, since this peak appeared to be uneffected by a change in Aerosol OT concentration. Fig. 5 shows $I_{1460}/I_{1734}$ versus water layer thickness, where the I's are integrated peak areas (14).

According to the Raman results in fig.5 some rearrangement of the hydrocarbon chains occurs when the water layer thickness decreases. The correlation between the conductivity and the Raman data is not completely understood at present. It is not inconceivable, however, that electrostatic and structural interactions within the water layers influence the molecular arrangement within the lipid bilayers.

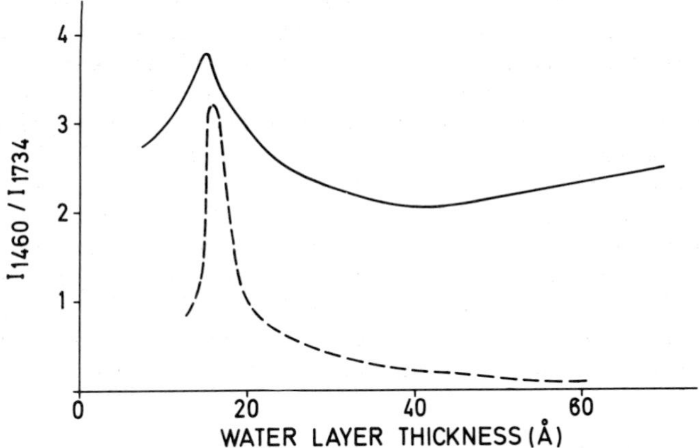


Fig. 5. Relative magnitude of the Raman peak at 1460 $cm^{-1}$ versus water layer thickness for lamellar Aerosol OT water systems (solid line, adapted from Faiman, Lundström and Fontell (14)). The dashed line is a schematic of the conductivity results in fig. 2. The band at 1734 $cm^{-1}$ due to C=O stretching modes (14) was insensitive to Aerosol OT concentration and therefore chosen as a reference. The band at 1460 $cm^{-1}$ is attributed to antisymmetric methylene rocking modes (14).

## ELASTICITY AND VISCOSITY OF MONOMOLECULAR LIPID FILMS

Some years ago we introduced inelastic light scattering techniques to the study of lipid films on water surfaces. In such experiments inelastically scattered light from thermal fluctuations of the liquid surface is frequency analysed (15-18). The spectrum contains information about the surface tension and viscosity of the liquid and, in the case of a film covered surface, also information about the rheological properties of the film. The technique is interesting since no external perturbation is placed on the film. Furthermore motions in the frequency range 5-25 kHz can easily be studied.

It was soon realized, however, by us and others, that using conventional optical heterodyning the instrumental bandwidth was too large to allow accurate quantitative determinations of film and liquid parameters (19). Therefore a new kind of heterodyne apparatus was developed at our laboratory (17). The important difference between the new experimental setup and the earlier ones is that a diffraction grating is used to give a well defined "local oscillator". The setup which is schematically shown in fig. 6 lends itself to quantitative analysis which allows an optimal adjustment in each experimental situation (17). The instrumental bandwidth has also been reduced considerably compared to earlier setups allowing quantitative determinations of surface tension, viscosity and film parameters. The apparatus is thoroughly described in ref. (17). A short schematic description will be given with reference to fig. 6. A laser light beam is reflected at the liquid surface and the diffraction grating gives rise to diffracted light beams (the first order diffraction is indicated in fig. 6). The capillary waves on the liquid surface may be regarded as very weak, oscillating diffraction gratings. They give rise to very weak light beams deflected (or inelastically scattered) a small angle from the reflected main beam. The beam, which coincides with the chosen diffraction spot from the grating (see fig. 6), causes a modulation of the amplitude of the photomultiplier current. The power spectrum of this modulation (fig. 6(b)) reflects the amplitude modulation of the capillary waves. For a free surface of a normal liquid $f_o$ and $\Delta f$ contain information about the surface tension, $\gamma$, and the kinematic viscosity, $\nu$,

$$f_o \approx \frac{1}{2\pi}\sqrt{\frac{\gamma}{\rho}\,k_o^{\,3}} \qquad [1]$$

$$\Delta f_a = \frac{2\nu k_o^{\,2}}{\pi} \qquad [2]$$

where $\rho$ is the liquid density, $\Delta f_a$ is $\Delta f$ corrected for the instrumental bandwidth, $\Delta f_i$, and $2\pi/k_o$ is the wavelength of the chosen capillary waves ($\approx a/\cos\Theta$ for the first order diffraction spot).

For a film covered liquid surface the capillary waves are influenced also by the mechanical properties of the film. In this case accurate expressions of $f_o$ and $\Delta f_a$ can not be given explicitly. An expression for the power spectrum can, however, be obtained, which, neglecting diffusion, reads (18,20)

$$\frac{k_B T}{\pi\omega}\mathrm{Im}\left\{-\frac{k_o}{\rho}\,\frac{\omega^2-\frac{\varepsilon^*}{\gamma}\omega_o^2\left(\frac{\ell}{k_o}-1\right)}{[\omega^2+2i\nu k_o^2\omega]^2-\omega^2\omega_o^2+4\nu^2 k_o^3\ell\omega^2+\frac{\varepsilon^*}{\gamma}\omega_o^2\left[\frac{\ell}{k_o}(\omega_o^2-\omega^2)-\omega_o^2\right]}\right\} \quad [3]$$

where $\omega = 2\pi f$, $k_B$ is Boltzmann's constant, T is the absolute temperature, $\omega_o = \sqrt{\gamma k_o^3/\rho}$, and $\ell = {}^+\sqrt{k_o^2-i\omega/\nu}$. $\varepsilon^*=\varepsilon-i\omega\kappa$, where $\varepsilon$ is the film elasticity and $\kappa$ the film (horizontal) viscosity. The power spectrum [3] is approximately Lorentzian in shape. The static value of $\varepsilon$, $\varepsilon_o$, is obtained from the static surface tension, $\gamma$, versus area per molecule, A, as

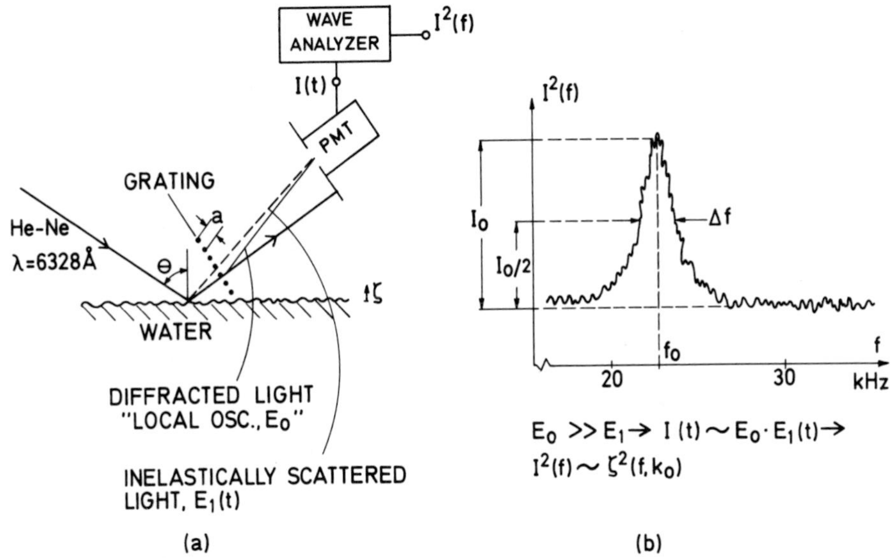


Fig. 6(a). Schematic picture of the new light scattering apparatus. $\zeta$ is the fluctuation of the liquid surface. The (weak) grating was fabricated with photographic techniques (17).

(b) Schematic power spectrum of the current in the photomultiplier tube. The full width at half peak height $\Delta f$ corrected for a small instrumental bandwidth $\Delta f_i$ gives the measured width $\Delta f_a$ due to the capillary waves. $f_o$ and $\Delta f_a$ depend on film and liquid parameters and on the wavelength of the capillary waves studied. The wavelength is determined by $\Theta$, a and the order of the diffraction spot chosen. The quadratic characteristic of the photomultiplier mixes the signals due to the diffraction grating and the capillary waves, respectively. The photomultiplier current contains therefore a component with the frequency $\omega$ determined by the capillary waves.

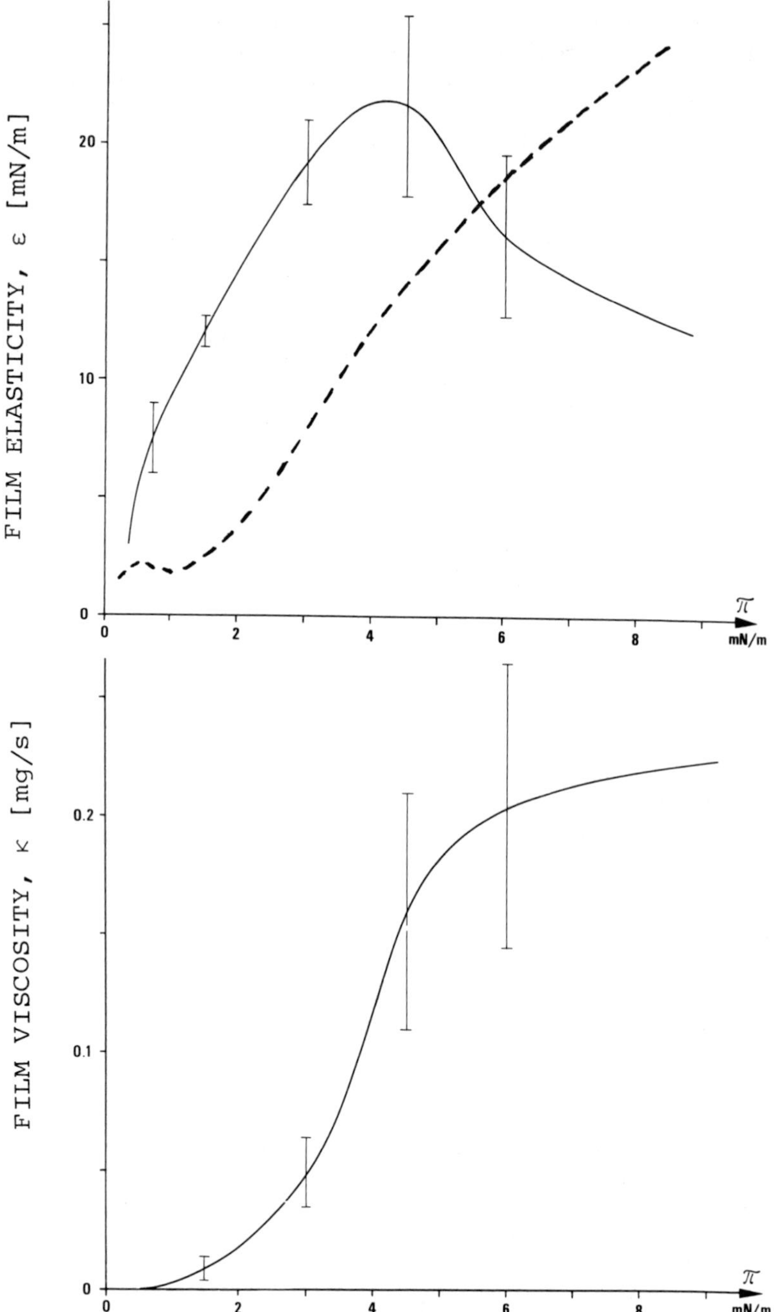


Fig. 7. ε and κ versus surface pressure, π, for propyl stearate monolayers on 0.01 N HCℓ, determined from experiments with $k_o$ = 50600 $m^{-1}$. The error bars are estimated uncertainties in ε and κ (18). The dashed line is $\varepsilon_o$ calculated from eqn. [4].

$$\varepsilon_o = A \frac{d\gamma}{dA} \qquad [4]$$

During rapid compression or expansion of a surface, the surface tension may be measurably different from that of a surface at rest with the same surface coverage. This is due to time dependent phenomena such as diffusion of molecules between bulk solution and film, diffusion within the film, changes in molecular orientation etc. (21). In short the film behaves like a viscoelastic medium and $\varepsilon$ is expected to be different from $\varepsilon_o$. Furthermore $\kappa$, which is due to both film viscosity and diffusional effects, may be different from $\kappa$ determined by other (low frequency) methods.

We have recently used the new light scattering apparatus (17) to study monolayers of propyl stearate on 0.01 N HCℓ aqueous substrates (18). Capillary waves in the frequency range 7-16 kHz were studied. The experiments are fully described in ref. (18). A discussion of the results is given below. $\Delta f_a$ and $f_o$ were determined from experimental power spectra (like that in fig. 6(b)) at different surface pressures and for two different $k_o$. The surface pressure was independently measured with the Wilhelmy method. $\Delta f_a$ and $f_o$ for the larger $k_o$ (50600 $m^{-1}$) were compared with those obtained from plots of eqn. [3] with different values of $\varepsilon$ and $\kappa$. In this way these parameters were determined as a function of surface pressure. They are shown in fig. 7. The so obtained $\varepsilon$ and $\kappa$ were used to calculate "theoretical" $\Delta f_a$ and $f_o$ for the smaller $k_o$ (33530 $m^{-1}$). They are compared with experimental results in fig. 8. From the agreement between these theoretical values and the experimental results we conclude that eqn. [3] describes the wavenumber ($k_o$) dependence of the capillary waves quite well. We see, however, in fig. 7 that $\varepsilon$ is not identical to the static elasticity $\varepsilon_o$ determined from the static surface tension and eqn. [4].

This difference is probably not related to diffusion for the following reasons. Bulk diffusion is generally negligible at high frequencies and furthermore propyl stearate is "insoluble" in the substrate used. Surface diffusion (i.e. diffusion of molecules within the monolayer) gives rise to an effective $\varepsilon^*$ of the form (21)

$$\varepsilon^* = \frac{\varepsilon - i\omega\kappa}{1 + i\frac{k_o^2}{\omega} D_s} \qquad [5]$$

and diffusion generally makes $\varepsilon < \varepsilon_o$ and not $\varepsilon > \varepsilon_o$ as observed for small surface pressures (fig. 7). The agreement between the theoretical and experimental results indicates also that there is no major $k_o$ dependence of $\varepsilon^*$. It was thus concluded that $D_s < 10^{-5}$ $m^2/s$. The dependence of $\varepsilon$ (and $\kappa$) on surface pressure is therefore to a

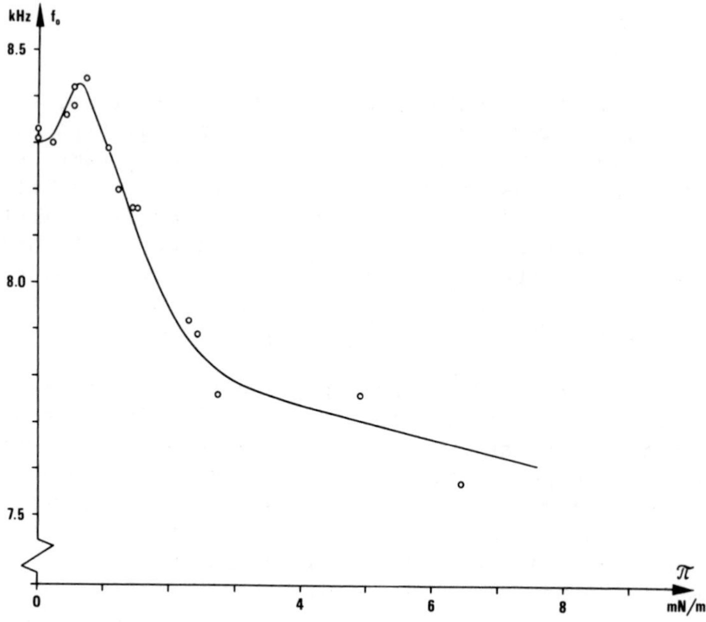


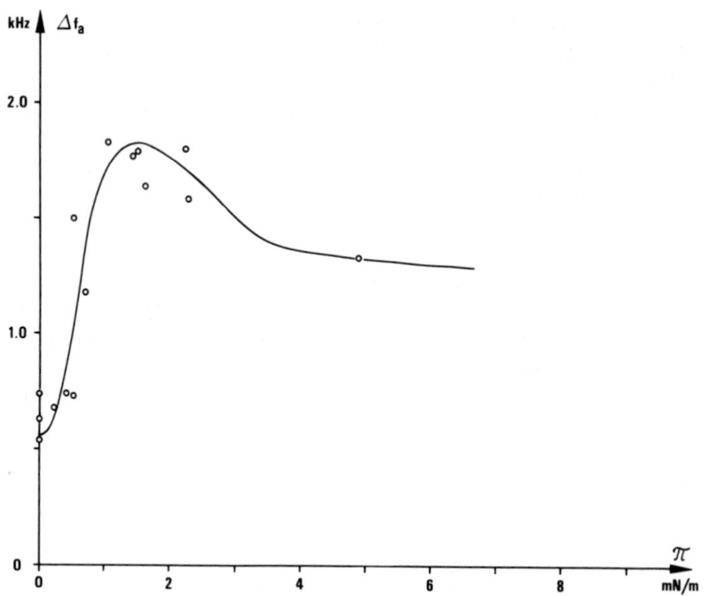


Fig. 8. Experimental results ($f_o$ and $\Delta f_a$) for $k_o$ = 33530 $m^{-1}$. The solid lines were obtained from eqn. [3] using $\varepsilon$ and $\kappa$ from fig. 7.

large extent caused by changes in the "viscoelastic" properties of the film itself.

An observation which can be made from eqn. [3] is that the capillary waves are sensitive to changes in film properties in a certain range of $\varepsilon^*$ only, about

$$0.01 < |\varepsilon^*/\gamma| < 1 \quad [6]$$

This puts certain constraints on for example the elasticity $\varepsilon$ which can be measured.

Preliminary measurements on dimyristoyl-phosphatidyl-choline (DML) showed that $\Delta f_a$ remained at its value for a free surface until the film density passed a critical value of $\approx$ 100 Å$^2$/molecule, where $\Delta f_a$ suddenly jumped to its maximum value (18). This could reflect a transition between two different film states.

## DISCUSSION

The main purpose of this paper was to describe two of the most recent studies on lipid-water systems done at our laboratory. Although the experiments were not made on biological membranes, they appear to have biological significance. The water channels in the Aerosol OT-water system are for example to a certain extent similar to the ionic channels in nerve membranes. The Aerosol OT-water interfaces are similar to the interface between charged surfaces and water in biological systems. The lateral conductivity of the Aerosol OT-water system yields therefore information about charge movements in biologically interesting surroundings.

We have also performed experiments on the electrical properties of fatty acid multilayers (22). These experiments which give information on the charge storage properties of lipids are reviewed elsewhere (23) and were therefore not touched upon here.

The study of capillary waves on film covered water surfaces with the new laser light scattering apparatus is only in its beginning. It should be clear, however, that interesting information is obtianed, complementary to that obtained by other methods. Mechanical impulses may for example, accompany the propagation of electrical nerve impulses (24). Typically a nerve impulse contains frequencies of order 1-100 kHz. Therefore a knowledge of the rheological properties of nerve membranes in this frequency range is of large interest. Inelastic light scattering experiments on capillary waves on film covered water surfaces might then be good model experiments.

## ACKNOWLEDGMENTS

The work described in this paper has been done in collaboration with a number of people both within and outside the Research Laboratory of Electronics: R Faiman, K Fontell, Y Hamnerius, S Hård, K Larsson, H Löfgren, O Nilsson, D McQueen, L Rosén and M Stenberg. I acknowledge their contributions and also like to thank them and T Wallmark for many stimulating discussions on the present topics. The work was partly sponsored by the Swedish Natural Science Research Council.

## REFERENCES

(1) W Ulbricht, Biophys, Struct. Mechanism, 1, 1 (1974)
(2) J R Smythies, F Benington, R J Bradley, W F Bridgers and R D Morin, J.theor.Biol. 43, 29 (1974)
(3) G Baumann and P Mueller, J.Supramol.Struct. 2, 538 (1974)
(4) I Lundström and K Fontell, Chem.Phys.Lipids 15, 1 (1975)
(5) I Lundström and K Fontell, submitted for publication
(6) J Rogers and P A Winsor, Nature 216, 477 (1967)
(7) P A Winsor, Mol.Cryst.Liq.Cryst. 12, 141 (1971)
(8) K Fontell, J.Colloid Interface Sci. 44, 318 (1974)
(9) K Fontell in ACS Symposium Series 9, 270 (1975)
(10) J O'M Bockris and A K N Reddy, "Modern Electrochemistry", vol. 2 (Plenum Press, New York 1970), chapter 7.
(11) B Frankenhaueser and L E Moore, J.Physiol. (London) 169, 438 (1963)
(12) Y Lange and C M Gary-Bobo, Nature New Biol. 246, 150 (1973)
(13) Y Lange and C M Gary-Bobo, J.Gen.Physiol. 63, 690 (1974)
(14) R Faiman, I Lundström and K Fontell, submitted for publication
(15) R H Katyl and U Ingard, Phys.Rev.Lett. 19, 64 (1967)
R H Katyl and U Ingard, Phys.Rev.Lett. 20, 248 (1968)
(16) R H Katyl and U Ingard, "Light Scattering from Thermal Fluctuations of a Liquid Surface", in honor of Philip M Morse, ed. H Feshbach and K U Ingard (M I T Press, Cambridge, Mass. 1969) 70.
(17) S Hård, Y Hamnerius and O Nilsson, J.Appl.Phys., May 1976
(18) S Hård and H Löfgren, submitted for publication.
(19) I Lundström and D McQueen, J.Chem.Soc.Far.Trans.I 70, 2351 (1974)
(20) D Langevin and M-A Bouchiat, C.R.Acad.Sc.Paris, Series B 272, 1422 (1971)
(21) J A Mann and G Du, J.Colloid Interface Sci. 37, 2 (1971)
(22) I Lundström and M Stenberg, Chem.Phys.Lipids 11, 287 (1974)
(23) K Larsson and I Lundström in "Liquid Crystals in Biological Systems", ed. S Friberg (Springer Verlag, 1976)
(24) I Lundström, J.theor. Biol. 11, 487 (1974)

# X-RAY SCATTERING STUDIES OF PROTEIN-LIPID SYSTEMS IN SOLUTION : THE EXAMPLE OF BOVINE RHODOPSIN

Annette TARDIEU, Christian SARDET and Vittorio LUZZATI

Centre de Génétique Moléculaire du C.N.R.S.

91190 GIF-sur-YVETTE, France

## INTRODUCTION

Small-angle X-ray scattering, one of the most powerful techniques for the study of macromolecules in solution, has been applied to a variety of systems of biological interest. A shortcoming of this technique, especially if compared to a conventional crystallographic approach, is its poor resolution, orders of magnitude short of atomic resolution ; its main advantage is the possibility of easily varying physical and chemical conditions and exploring structural transitions. Furthermore, biological systems are not always amenable to high resolution crystallographic analysis.

Scattering studies of biological systems in solution - both X-ray and neutron scattering studies - are undergoing a remarkable renewal, under the impact of two technological developments. One is the recent introduction of position sensitive proportional counters into X-ray diffraction technology (Gabriel & Dupont, 1972), which has shortened exposure times by a factor of 100 to 1000 and has greatly simplified the design of the cameras. The other is the construction of reactors with high neutron fluxes and the use of multiple detectors (Schmatz et al, 1974).

A particularly useful procedure for exploring the structure of particles in solution, especially when the electron density distribution inside the particles displays conspicuous fluctuations, is to vary the electron density of the solvent (usually by adding sucrose or salt). We have applied this technique to several systems containing proteins and lipids, and we have shown how in these systems the contribution of the polar groups can be disentangled from that of the hydrocarbon chains (Stuhrmann et al, 1975 ; Luzzati et al, 1976 ; Tardieu et al, 1976).

Membrane proteins are particularly interesting candidates for this technique, since in spite of extensive physiological and biochemical studies little is known about their structure, with the remarkable exception of bacteriorhodopsin whose structure was recently determined at 7 Å resolution (Henderson & Unwin, 1975). One reason for this general lack of information is the difficulty of dispersing membrane proteins in solvents or of ordering them in crystalline lattices. A number of membrane proteins, however, have been purified and solubilized in the presence of detergents, and particles have been obtained which contain one or more protein molecules associated with stoichiometric amounts of detergent. These protein-detergent complexes are ideal objects for a small-angle X-ray scattering study using solvents of variable density.

We describe here the small-angle X-ray scattering study of a rhodopsin-detergent complex. Rhodopsin is the major protein component of the membranes of rod outer segments. It contains a chromophore, 11-cis retinal, and an apoprotein, opsin : its molecular weight is 39,100 (Daemen et al, 1972). Rhodopsin can be extracted from the rod outer segments by several detergents and can be obtained free of lipids in the form of a complex containing one molecule of rhodopsin and a large number of molecules of detergent (Osborne et al, 1974). In this work we use dodecyl-dimethylamine oxide (DDAO) in its non-ionic form (Applebury et al, 1974 ; Ebrey, 1971). The biochemical part of our study deals with the characterization of the rhodopsin-DDAO complex, namely the determination of its homogeneity, chemical composition, weight and partial specific volume. The X-ray scattering study is performed on the detergent micelles and on the protein detergent complex, sucrose being used to raise the electron density of the solvent. Only the very small angle region of the X-ray scattering curves is explored in this work, the experiments being carried on an absolute scale. We can anticipate that the rhodopsin molecules will turn out to be thin elongated objects, whose maximum length is at least 80 Å, which span a thin and flat detergent micelle.

We have presented elsewhere a more detailed account of this work (Sardet et al, 1976).

## BIOCHEMICAL AND PHYSICO-CHEMICAL STUDY

The detergent used in this work is dodecyl-dimethylamine oxide (DDAO). $^{14}C$ labeled DDAO was also available (CEA, Saclay, France). This detergent is non-ionic, in the conditions of our work (pH 7.0). A commercial version of DDAO, Ammonyx LO (Onyx Chemical), which contains ca 30% of tetradecylamine oxide, was used in some experiments.

Rod outer segments (ROS) membranes were prepared from retinas of the eyes of freshly slaughtered cows (Osborne et al, 1974). Rhodopsin was purified in Ammonyx LO by a procedure inspired by Applebury et al (1974) (Sardet et al, 1976). The rhodopsin-Ammonyx LO solutions were deposited on hydroxylapatite columns, Ammonyx LO was exchanged for $^{14}C$.DDAO and rhodopsin was eluted with 150 mM Pi. Absorbance (650 to 260 nm) and scintillation counting determined the amounts of rhodopsin and DDAO in each fraction.

This procedure gives good yields of purified rhodopsin (spectral ratios $A_{280}/A_{500} < 1.75$ and $A_{400}/A_{500} < 0.20$), free of phospholipids. The concentration of DDAO is 2 mg/ml (more than 4 times the critical micellar concentration of DDAO). Under these conditions 156 molecules of detergent are bound to one molecule of rhodopsin. The binding experiment was also carried out in the presence of 50% sucrose (and 2 mg/ml DDAO) ; the binding ratio was found to be 143 molecules of DDAO per molecule of rhodopsin.

Agarose gel chromatography allowed the determination of the Stokes'radius of the detergent micelles ($R_s$ = 23 Å) and of the rhodopsin-DDAO complex ($R_s$ = 42 Å). The sedimentation coefficient of the complex was observed to be 1.56 Svedbergs. The partial specific volume of the DDAO micelles was measured ($\bar{v}$ = 1.122 cm$^3$/g). The partial specific volume of the complex could not be measured with great accuracy ($0.905 < \bar{v} < 0.910$ cm$^3$/g); it was also estimated from the amino acid composition of rhodopsin, binding ratio and partial specific volume of DDAO ($0.908 < \bar{v} < 0.924$ cm$^3$/g). Using these parameters the molecular weight of the complex can be established to lie within the limits $76.900 < M < 81.100$, in agreement with one rhodopsin and 156 DDAO molecules per complex particle.

## X-RAY SCATTERING STUDY

The X-ray camera and the experimental procedure are described elsewhere (Tardieu et al, 1976). Calibrated nickel filters were used to attenuate the incident beam and make it commensurable with the scattered intensity (Luzzati et al, 1963). The sample is contained in quartz capillary tubes ($\phi$ 1 mm) and the volume of the sample used in each experiment is approximately 30 $\mu$l. The diameter of the capillary and thus the thickness of the sample is determined from the X-ray absorption measured with the capillary filled with water. The diameter is also checked by direct measurement with a microscope. All the experiments were performed under dim red light, at 20°C ; no bleaching occured during the experiments.

The X-ray scattering study was performed on the DDAO micelles and on the rhodopsin-DDAO complex, as a function of sucrose concentration. In each experiment the concentrations of rhodopsin, DDAO and sucrose were determined. The background scattering, subtracted from the scattering curves, consists of the intensity scattered by free DDAO micelles (in the case of the complex), solvent, capillary tube, etc... The background subtracted curves were corrected for collimation distortions. The final, normalized curves, have the expression :

$$i_n(\rho_o, s)/c_e = I(s)/\eta \nu E_o c_e \qquad (1)$$

where $\rho_o$ is the electron density of the solvent, s is the scattering angle ($s = 2 \sin \theta / \lambda$), $c_e$ is the electron concentration (solute/solution), I(s) is the corrected experimental curve, $\eta$ is the thickness of the sample (electrons/$cm^2$), $\nu$ is a physical constant ($\nu = \lambda^2 \times 7.9 \times 10^{-26}$), $E_o$ is the energy of the incident beam.

The rhodopsin-detergent complex and the detergent micelles are poor scatterers ; moreover under our experimental conditions we were unable to raise the concentration beyond 10 mg rhodopsin per ml. Therefore only the intensity at very small angles is usable and only two parameters can be extracted from each experimental curve : the value and curvature at s = 0. These parameters were determined using Guinier's plots.

The whole of the theoretical treatment is based upon two assumptions : 1) - the sample is a perfect solution of identical particles ; 2) - the electron density inside the particles - namely over the volume impenetrable to the sucrose used to raise the density of the solvent - is independent of the electron density of the solvent. The validity of these assumptions is discussed below.

### a) Zero-Angle Intensity

In the absence of sucrose, the zero angle intensity is proportional to molecular weight (Luzzati, 1960).

$$i_n(0, \rho_{H_2O}) = m\, c_e\, (1 - \rho_{H_2O} \psi)^2 \qquad (2)$$

where m is the number of electrons per particle, $\psi$ is the partial electronic volume. Eq. 2, applied to the DDAO micelles, leads to M = 16,650 daltons, namely 73 DDAO molecules per micelles. In the case of the complex the partial specific volume is known less accurately, and consequently M turns out to be in the range 66.000-75.300 daltons, for $0.905 < \bar{v} < 0.910$ $cm^3/g$, in excellent agreement with the chemical dermination (M = 74,824 for one molecule of rhodopsin associated to 156 molecules of DDAO).

The solvent density dependence of the zero-angle intensity takes the form (Sardet et al, 1976) :

$$\left[i_n(0,\rho_o)/c_e\right]^{1/2} = v_1\,(\bar{\rho}_1 - \rho_o)\; m^{-1/2} \qquad (3)$$

where $v_1$ and $\bar{\rho}_1$ are the volume of one particle (namely the volume from which sucrose is excluded) and the average electron density in that volume. Therefore the plot of the experimental points $[i_n(0,\rho_o)/c_e]^{1/2}$ vs $\rho_o$ should yield a straight line, whose intercept and slope define $\bar{\rho}_1$ and $v_1 m^{-1/2}$. These plots are shown in Fig. 1. The experimental points barely depart from straight lines ; therefore the values of $\bar{\rho}_1$ and $v_1 m^{-1/2}$ are determined. Since m is known $v_1$ is defined.

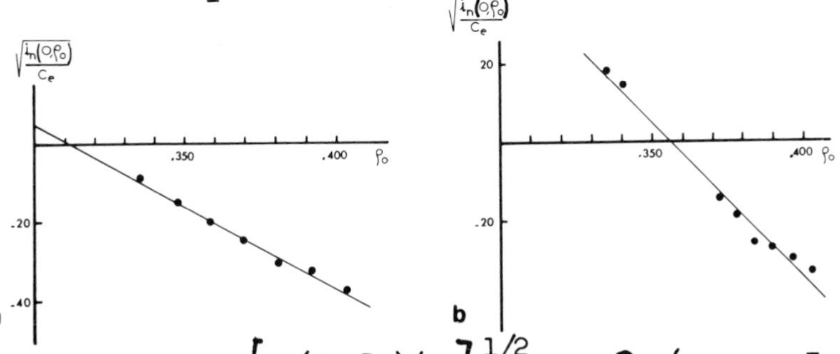


Fig. 1 : Plot $[i_n(0,\rho_o)/c_e]^{1/2}$ vs $\rho_o$ (see eq. 3 and table). The points barely depart from straight lines. a) DDAO micelles. b) Rhodopsin-DDAO complex.

When $v_1$, $\bar{\rho}_1$ and m are known the hydration α, defined by the number of electrons of water associated with one particle, divided by the number of electrons of one particle, becomes :

$$\alpha = v_1\,\bar{\rho}_1 / m - 1 \qquad (4)$$

The hydration $\alpha_p$ of rhodopsin in the complex can also be determined on the assumption that the hydration $\alpha_d$ of the DDAO molecules is the same in the micelles as in the complex :

$$\alpha_p = (\alpha m - \alpha_d m_d) / m_p \qquad (5)$$

where $m_p$ and $m_d$ are the number of electrons of one rhodopsin molecule and of the detergent moiety of the complex. Similarly the volume $v_d$ of the detergent moiety can be determined ($v_d = 41.170 \times 156/73$, see table) as well as the volume $v_p$ and the average electron density $\bar{\rho}_p$ of the protein part of the complex :

$$v_p = v_1 - v_d \qquad (6)$$

$$\bar{\rho}_p = m_p(1 + \alpha_p) / v_p \qquad (7)$$

The values of all these parameters are reported in the table.

b) Radii of Gyration

According to Guinier's law, the curvature of the intensity curves at zero angle defines the value of the radius of gyration of the electron density contrast between solute and solution :

$$i(s, \rho_o) = i(0, \rho_o) \left[1 - \frac{4}{3}\pi^2 R^2 (\rho_o) s^2 + \ldots\right] \qquad (8)$$

The expression of the radius of gyration as a function of the density of the solvent is (Stuhrmann and Kirste, 1967 ; Luzzati et al, 1976) :

$$R^2 (\rho_o) = R_v^2 - a/\Delta\rho_o - b/(\Delta\rho_o)^2 \qquad (9)$$

where $\Delta\rho_o$ is the electron density contrast $\rho_o - \bar{\rho}_1$, $R_v$ is the radius of gyration of the volume excluded to sucrose, a is the second moment $m_{2\rho}$ of the electron density contrast associated with one particle $[\rho_1(\underline{r}) - \bar{\rho}_1]$ with respect to the center of gravity of the volume $v_1$ divided by $v_1$ :

$$v_1 a = m_{2\rho} = \int r^2 \left[\rho_1(\underline{r}) - \bar{\rho}_1\right] dv_{\underline{r}} \qquad (10)$$

b is the square of the first moment of $\left[\rho_1(\underline{r}) - \bar{\rho}_1\right]$ divided by $v_1^2$ :

$$v_1^2 b = \left| \int \underline{r} \left[\rho_1(\underline{r}) - \bar{\rho}_1\right] dv_{\underline{r}} \right|^2 \qquad (11)$$

The terms a and b depend on the internal structure of the particle. a takes into account the relative distribution of the high and the low density regions ; for example a positive (see below) indicates that the high density regions are located preferentially in the outer region ; b is a measurement of the separation $r_{\alpha\beta}$ of the centres of gravity of specific regions $\rho_\alpha(\underline{r})$ and $\rho_\beta(\underline{r})$ of the particles :

$$v_1^2 b = |\underline{r}_{\alpha\beta}|^2 \left( \int \rho_\alpha(\underline{r})\, dv_{\underline{r}} \right)^2 \qquad (12)$$

The plots $R^2(\rho_o)$ vs $(\Delta\rho_o)^{-1}$ are shown in Fig. 2. The experimental points barely depart from straight lines. In the case of the complex this observation shows that the centres of gravity of the rhodopsin and of the detergent moieties are very near to each other (the separation is smaller than 8 Å, see Sardet et al, 1976). The intercepts and slopes of the straight lines define the values of $R_v^2$ and of $m_{2\rho}/v_1$.

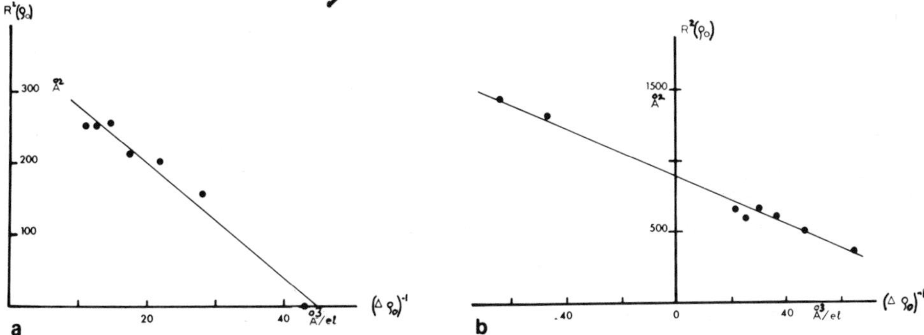


Fig. 2 : Plot $R^2(\rho_o)$ vs $(\Delta\rho_o)^{-1}$ (see eq. 9 and table). The points barely depart from straight lines, showing that the term b is very small. a) DDAO micelles ; b) rhodopsin-DDAO complex.

The analysis of the second moment $m_{2\rho}$ allows us to estimate the radius of gyration of the rhodopsin molecule in the complex. $m_{2\rho}$ can be decomposed into the contributions of the average electron densities of the protein and detergent moieties, $\bar{\rho}_p$ and $\bar{\rho}_d$, and the fluctuations around these average values (Sardet et al, 1976).

$$m_{2\rho} = v_p (R_v)_p^2 \left( \bar{\rho}_p - \bar{\rho}_1 + \varphi_p \right) + v_d (R_v)_d^2 \left( \bar{\rho}_d - \bar{\rho}_1 + \varphi_d \right) \qquad (13)$$

where $(R_v)_p$ and $(R_v)_d$ are the radii of gyration of the volumes of one complex particle occupied by protein and detergent respectively. $\varphi_d$ (and $\varphi_p$) have the expression :

$$\varphi_d = \left\{ \int r^2 \left[ \rho_d(\underline{r}) - \bar{\rho}_d \right] dv_{\underline{r}} \right\} / \; v_d (R_v)_d^2 \qquad (14)$$

Taking into account the equation :

$$v_1 R_v^2 = v_p (R_v)_p^2 + v_d (R_v)_d^2 \quad (15)$$

and the fact that $m_2\rho$, $v_p$, $v_d$, $\bar{\rho}_1$, $\bar{\rho}_p$ and $\bar{\rho}_d$ are known (see table), an expression of $(R_v)_p$ is obtained which depends on $\varphi_p$ and $\varphi_d$. We have discussed elsewhere (Sardet et al, 1976) how to set limits to these parameters : the final result is $(R_v)_p = 30\ (\pm\ 2.5)$ Å, $(R_v)_d = 29\ (\pm\ 2.5)$ Å.

## STRUCTURE OF THE PARTICLES

### a) The Detergent Micelles

The analysis of the X-ray scattering experiments leads to five parameters : m, $R_v$, $m_2\rho$, $\bar{\rho}_1$ (see table). The number of electrons m defines the number of detergent molecules in the micelles. The sign of $m_2\rho$ indicates that the polar groups are concentrated on the outer part of the micelles.

The pair of geometrical parameters $v_1$ and $R_v$ can be used to define the size of the micelles if the shape is assumed to be an ellipsoid of revolution (Luzzati et al, 1961). Two ellipsoids are possible, whose axes are 25.8 x 55.2 x 55.2 Å or 34.4 x 34.4 x 66.4 Å.

### b) The Rhodopsin-DDAO Complex

The chemical and physico-chemical studies show that each particle of the complex consists of one molecule of rhodopsin and 156 molecules of DDAO. The X-ray scattering study defines five parameters of the complex : m, $R_v$, $m_2\rho$, $\bar{\rho}_1$ and $v_1$. Moreover the hydration, density and volume of the rhodopsin and detergent moieties can be determined (see above and table).

The coincidence of the centres of gravity of the protein and detergent regions indicates that the two regions are highly symmetrical. Some hypotheses on the magnitude of the electron density fluctuations in the protein and detergent regions allows us to set limits to the radius of gyration of the protein and detergent volumes (see above).

The very large value of the ratio $(R_v)_p^3 / v_p$ shows that rhodopsin is highly anisometric : $(R_v)_p^3 / v_p \cong 0.36 (\pm\ 0.09)$ for rhodopsin, 0.111 for a sphere. We can envisage a few simple shapes, and use the values of $(R_v)_d$ and $v_p$ to determine their size : for example prolate and oblate ellipsoids, prolate and oblate circular cylinders, dumb-bell. The length of the axes of the ellipsoids would be 33.7, 33.7 ($\mp$ 3.5), 125 ($\pm$ 25) Å and 16.0 ($\mp$ 3), 94.3, 94.3 ($\pm$ 9) Å ; the diameter and height of the cylinders would be 31.4 ($\mp$ 3.5), 96.6 ($\pm$ 20) Å and 84.2 ($\pm$ 8), 13.4 ($\mp$ 2.5) Å ; the maximum length and the diameter of the spheres of the dumb-bell would be 92 ($\pm$ 6), 41.4 Å. It is clear that at least one dimension of the rhodopsin molecule is bound to exceed 80 Å.

We can now try to make a choice between the prolate and the oblate classes of model. The possibility of rhodopsin being totally embedded in the detergent with the whole of its surface covered by the DDAO molecules is easily excluded, since in this case $(R_v)_d$ would be larger than $(R_v)_p$. We are thus led to assume that the rhodopsin molecules span a flat detergent micelle with one long axis normal, or almost normal to the plane of the micelle, and that the DDAO molecules cover the outer surface of the cross-section of the rhodopsin molecule. Under these conditions, if the rhodopsin molecules were oblate in shape the volume occupied by the detergent would extend radially beyond the volume occupied by rhodopsin : the relative values of $(R_v)_p$ and $(R_v)_d$ show that this is not the case. Moreover the small diameters of the oblate models seem too small for a protein. Therefore we conclude that rhodopsin is prolate in shape, with one dimension conspicuously larger than the others.

For the sake of the argument we adopt a cylindrical model of 31.4 Å diameter 96.6 Å height. In this case the dimensions of the detergent moiety spanned by the rhodopsin molecule agree with an ellipsoid of axes 34.0 x 80.2 x 80.2 Å.

## DISCUSSION

This work provides an illustration of the information which can be gained from a small-angle X-ray scattering study of a protein-detergent complex. The presence of free detergent micelles hinders the analysis of the experimental curves beyond the central maximum ; in the case of serum lipoproteins (Tardieu <u>et al</u>, 1976) we have shown the wealth of information which can be obtained from an analysis of the whole of the intensity curves. Such a low resolution study is rewarding when applied to a system whose structure is poorly known ; with the new technical developments the small-angle X-ray scattering experiments are fast (typically 10 hours per experiment with the rhodopsin-DDAO complex) and require small samples (typically 0.04 mg of rhodopsin per experiment).

In the work described here the experimental information was limited to the value and curvature of the intensity curves at zero angle ; it is essential in this case that the experimental curves be measured on an absolute scale. The experimental results are condensed into two straight lines (Figs 1 and 2) and therefore lead to the determination of four parameters : $\bar{\rho}_1$, $v_1(m)^{-1/2}$, $R_v$, $m_2\rho/v_1$. Since the partial specific volume is known m, and thus $v_1$ and $m_2\rho$ are also determined.

TABLE

Parameters Obtained from the X-ray Study

$\psi$-electronic partial specific volume ; m - number of electrons per particle ; $m_2\rho/v_1$ - see text ; $v_1$, $R_v$ - volume and radius of gyration of the particle volume ; $\bar{\rho}_1$ - average electron density in $v_1$ ; $\alpha$ - hydration, defined by the number of electrons of water per electron of one particle. Suffices p and d refer to protein and detergent. The molecular weight and number of electrons adopted for rhodopsin and for DDAO are $M_p = 39{,}100$, $m_p = 20{,}909$ and $M_d = 229$, $m_d = 130$.

| Parameter | | Detergent micelles | Rhodopsin-detergent complex |
|---|---|---|---|
| $\psi$ | (Å$^3$ per electron) | 3.283 | (2.745) |
| m | (electrons) | 9,453 | 41,189 |
| $\nu$ | DDAO molecules per particle | 73 | 156 |
| $m_2\rho/v_1$ | (electrons Å$^{-1}$) | 6.87 | 8.74 |
| $R_v$ | (Å) | 18.4 | 29.5 |
| $\bar{\rho}_1$ | (electrons Å$^{-3}$) | 0.3118 | 0.3562 |
| $v_1$ | (Å$^3$) | 41,170 | 162,900 |
| $\alpha$ | | 0.358 | 0.409 |
| $v_p$ | (Å$^3$) | | 74,600 |
| $\bar{\rho}_p$ | (electrons Å$^{-3}$) | | 0.4087 |
| $\alpha_p$ | | | 0.458 |
| $v_d$ | (Å$^3$) | | 88,300 |
| $\bar{\rho}_d$ | (electrons Å$^{-3}$) | | 0.3118 |
| $\alpha_d$ | | | 0.358 |

The conclusions of this work are heavily dependent upon the assumption that the DDAO micelles and the rhodopsin-DDAO particles are monodisperse and that their structure is independent of sucrose concentration. To some extent the two assumptions lend themselves to experimental verification : 1) the sedimentation and gel filtration behaviour of the complex do not show any indication of polydispersity ; 2) the binding ratio is independent of sucrose ; 3) the molecular weight determined by the X-ray and the biochemical studies are so similar that a large polydispersity in weight can be excluded ; 4) the linear relationships of eqs 3 and 9 (see Figs. 1 and 2) are consistent with the internal structure being independent of sucrose ; 5) more generally the internal consistency of the results of the X-ray study and the agreement of the various parameters with the chemical data support the hypotheses made in their derivation. Moreover the presence of morphological heterogeneities do not have serious consequences on the formal analysis of the experimental results ; density heterogeneities may have more serious effects, although a statistical analysis allows us to set limits to this type of heterogeneity (Sardet _et al_, 1976).

It must be noted that the nature of the detergent may be critical for this type of X-ray scattering study. For example Triton X-100 (Osborne _et al_, 1974) cannot be used, since the micelles of this detergent turn out to be sucrose dependent and thus one suspects that the structure of the rhodopsin-Triton X-100 particles is also sensitive to sucrose.

An important parameter in the determination of the size and shape of the rhodopsin molecules in the complex is the amplitude of the electron density fluctuations in the protein and the detergent regions. We have been able to set limits to the effects of these fluctuations ; it may be noted that these effects are likely to be more serious for neutron than for X-ray scattering.

The major conclusion of our work is that the rhodopsin molecules are thin elongated objects, at least 80 Å long, which span a flat detergent micelle.

An important question related to the possible biological implications of this work is the extent to which rhodopsin in the DDAO complex retains its native conformation. One piece of information is that the spectral properties and transitions of rhodopsin are unaltered in the presence of DDAO (Applebury _et al_, 1974). Moreover proteolytic enzymes cleave rhodopsin into similar fragments when acting on the disc membrane and on the rhodopsin-Triton X-100 complex (Pober & Stryer, 1975 ; Sardet, unpublished). On the other hand, DDAO-solubilized rhodopsin looses the ability to regenerate after bleaching (Stubbs _et al_, 1976), although Hong and Hubbell (1973) have shown that detergent solubilized rhodopsin regains the ability to regenerate upon bleaching when put back into a phospholipid environment. In the absence of more sensitive and specific tests the question of the possible denaturation of rhodopsin in the presence of DDAO is thus difficult to answer.

It is currently accepted that detergents upon interacting with membrane proteins cover the hydrophobic areas which _in situ_ are in contact with the hydrocarbon chains of the lipids (Helenius & Simons, 1975 ; Robinson & Tanford, 1975). In accordance with these ideas the results of this work suggest that in the disc membranes the rhodopsin molecule spans the lipid bilayer and penetrates deeply into the intra-and inter-disc space, with a large area exposed to the aqueous environment.

## Acknowledgments

We thank Drs. M. Chabre, A. Helenius, B. Osborne and K. Simons for stimulating discussions, Dr. P. Letellier for help in synthesizing DDAO, N. Pasdeloup and M.O. Mossé for technical collaboration. One of us (C.S.) was supported by an EMBO fellowship and lately by the Commissariat à l'Energie Atomique. This work was supported in part by a grant of the Délégation Générale à la Recherche Scientifique et Technique.

## REFERENCES

APPLEBURY, M.L., ZUCKERMAN, D.M., LAMOLA, A.A. and JOVIN, T.M. (1974), Biochem., 13, 3448-3458.

DAEMEN, F.J.M., DE GRIP, W.J. and JANSEN, P.A.A. (1972), Biochim. Biophys. Acta, 271, 419-428.

EBREY, T.G. (1971), Vision Res., 11, 1007-1009.

GABRIEL, A. and DUPONT, Y. (1972), Rev. Sci. Instrum., 43, 1600-1603.

HELENIUS, A. and SIMONS, K. (1975), Biochim. Biophys. Acta, 415, 29-79.

HENDERSON, R. et UNWIN, P.N.T. (1975), Nature, 257, 28-32.

HONG, K., HUBBELL, W.L. (1973), Biochemistry, 12, 4517-4523.

LUZZATI, V. (1960), Acta Cryst., 13, 939-945.

LUZZATI, V., TARDIEU, A., MATEU, L. and STUHRMANN, H.B. (1976), J. Mol. Biol., 101, 115-127.

LUZZATI, V., WITZ, J. and BARO, R. (1963), Journal de Physique, 24, 141-146A.

LUZZATI, V., WITZ, J. and NICOLAIEFF, A. (1961), J. Mol. Biol., 3, 367-378.

OSBORNE, B.H., SARDET, C. and HELENIUS, A. (1974), Eur. J. Biochem., 44, 383-390.

POBER, J. and STRYER, L. (1975), J. Mol. Biol., 95, 477-481.

ROBINSON, N.C. and TANFORD, C. (1975), Biochem., 14, 369-378.

SARDET, C., TARDIEU, A. and LUZZATI, V. (1976), J. Mol. Biol., in press.

SCHMATZ, W., SPRINGER, T., SCHELTEN, J. and IBEL, K. (1974), J. Appl. Cryst., 7, 96-116.

STUBBS, G.W., SMITH, H.G. and LITMAN, B.J. (1976), Biochim. Biophys. Acta, 425, 46-56.

STUHRMANN, H.B. and KIRSTE, R.G. (1967), Z. Physik. Chemie, Nune Folge, 56, 334-337.

STUHRMANN, H.B., TARDIEU, A., MATEU, L., SARDET, C., LUZZATI, V., AGGERBECK, L. and SCANU, A.M. (1975), Proc. Nat. Acad. Sci. USA, 72, 2270-2273.

TARDIEU, A., MATEU, L., SARDET, C., WEISS, B., LUZZATI, V.,

AGGERBECK, L. and SCANU, A.M. (1976), J. Mol. Biol., 101, 129-153.

# IMMUNOCHEMISTRY OF MODEL MEMBRANES CONTAINING SPIN-LABELED HAPTENS

P. Brûlet, G.M.K. Humphries* and H.M. McConnell

Stauffer Laboratory for Physical Chemistry

Stanford University, Stanford, California 94305

## INTRODUCTION

A previous study reported that liposomes composed of L-α-dipalmitoylphosphatidylcholine (DPPC), cholesterol and 3 mole % cardiolipin fix complement in the presence of specific anticardiolipin antibodies (1). It was discovered that at both the temperatures employed (6°C and 37°C) complement fixation increased strongly with increasing cholesterol concentration for cholesterol concentrations above 33 mole %. Similar results were obtained when the DPPC was replaced by L-α-dimyristoylphosphatidylcholine (DMPC). Since several physical properties of binary mixtures of phosphatidylcholines and cholesterol changes at cholesterol concentrations of the order of 33 mole % (2) the above study (1) indicates that one or more of these physical properties of the liposomal membrane may play an important role in complement fixation. For example, the physical properties of phosphatidylcholine-cholesterol lipid mixtures might strongly affect the lateral motion and distribution of this lipid hapten (cardiolipin) and thereby modify the immunochemical properties of the membrane.

In order to pursue the possibility that hapten or antigen containing lipid bilayer membranes having known physical properties might be useful in probing various functions of the immune system, we have turned to the use of nitroxide spin labels as haptenic

------------------------------

Present address: Tumour Immunology Unit, Department of Zoology, University College London, Gower Street, LONDON WCIE 6BT.

groups. We believe that such spin label haptens may prove to be particularly useful in the study of the immunochemistry of synthetic membranes since the paramagnetic resonance spectra of such groups can provide information on the rates of hapten lateral diffusion, local molecular motion, hapten clustering, inside-outside hapten distribution, and specific antibody binding. The purpose of the present paper is to summarize briefly some of our preliminary studies; a more detailed report will be published elsewhere. (Brûlet and McConnell, to be published.) The present work should be distinguished from earlier immunochemical studies (3,4) using nitroxide spin labels with respect to both the objectives and the nature of the haptenic group. In earlier studies the antibodies were directed against a non-paramagnetic haptenic group (such as dinitrophenyl) to which a nitroxide spin label was bound. In the present work the paramagnetic nitroxide is the hapten. Our long-range objective is to relate structural properties of the lipid membranes to the distribution and motion of the haptens, and to relate the latter to membrane immunochemistry.

## MATERIALS AND METHODS

### Anti-nitroxide Antibodies

A brief report of the preparation of rabbit anti-nitroxide antibodies has been given earlier (5). Rabbits were immunized by subcutaneous injection of 500μg of Keyhole Limpet hemocayanin, alkylated with N-(1-oxyl-2,2,6,6-tetramethyl)-4-iodoacetamide, together with Freund's complete adjuvant, and then boosted by several intravenous injections of 100μg protein in saline. Immunoglobulins were separated by fractionation with $(NH_4)_2SO_4$ and Sepharose 6B. A symmetric peak of protein eluting at a volume corresponding to MW 150,000 was identified as IgG by gel diffusion against goat anti-rabbit IgGFc. (In other experiments carried out in collaboration with Dr. P. Rey, specific IgG anti-nitroxide antibodies have been prepared by affinity chromatography. The affinity column matrix was prepared from the reaction product of N-(1-oxyl-2,2,6,6-tetramethylpiperidinyl)-4-amine and activated Sepharose 4B. Bound antibodies are eluted with TEMPO (2,2,6,6-tetramethylpiperidine-1-oxyl), followed by dialysis.)

The specific binding of anti-nitroxide antibodies to various spin labels can be detected using paramagnetic resonance (5), and using complement fixation.

The micro C' fixation procedure of Wasserman and Levine (6) was employed. Using this procedure, indicator sheep red blood cell lysis is essentially linear with C' dose for values of

10-90% lysis. An acceptable parameter of C' fixation is, therefore, the difference in absorbance between the supernatant of interest and a suitable control value, usually obtained from antibody, antigen and hemolytic controls showing an equal degree of lysis at a level corresponding to 90% or less. Complement fixation was generally measured for a series of dilutions of hapten-containing liposomes. In many experiments the liposomal membranes were not separated from the depleted complement solution, before addition of indicator red blood cells. Since liposomes, particularly those rich in cholesterol, can be lytic towards red blood cells, complement fixation test involving the simultaneous presence of indicator red blood cells and large concentrations of liposomal lipids are quantitatively unreliable. The source of complement and other reagents were the same as in reference (1).

Spin Label Lipid Haptens

$$\begin{array}{l} \qquad\qquad\qquad\qquad\qquad\quad H_2C - O - CO - (CH_2)_{14}\,CH_3 \\ \qquad\qquad\qquad\qquad\qquad\qquad | \\ CH_3(CH_2)_{14} - CO - O - CH \\ \qquad\qquad\qquad\qquad\qquad\qquad | \\ \qquad\qquad\qquad\qquad\qquad\quad H_2C - O - \overset{O}{\overset{\|}{P}}(O^-) - O - R \end{array}$$

where R is I, II and III below.

$$- CH_2 - CH_2 - \overset{+}{N}(CH_3)_2 - \text{(2,2,6,6-tetramethylpiperidine ring)}\,N - O \qquad \text{I}$$

$$- CH_2 - CH_2 - \overset{H}{N} - CH_2 - \overset{O}{\overset{\|}{C}} - \overset{H}{N} - \text{(2,2,6,6-tetramethylpiperidine ring)}\,N - O \qquad \text{II}$$

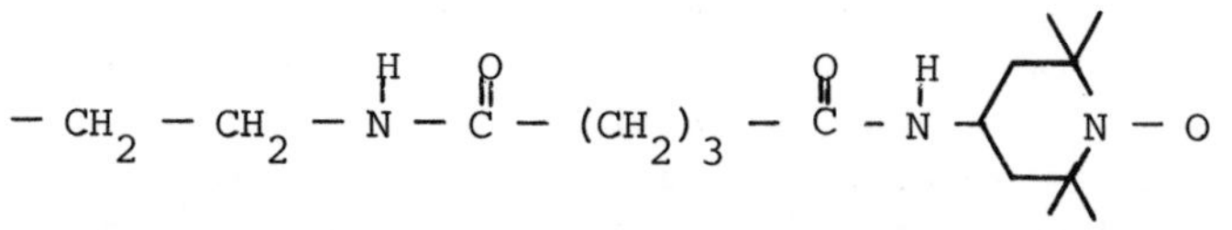

$$-CH_2-CH_2-\overset{H}{N}-\overset{O}{\overset{\|}{C}}-(CH_2)_3-\overset{O}{\overset{\|}{C}}-\overset{H}{N}-\langle\ \ N-O \qquad III$$

Spin label I was prepared as described previously (7). Label II was prepared by the alkylation of L-α-dipalmitoylphosphatidylethanolamine (DPPE) with N-(1-oxyl-2,2,6,6-tetramethylpiperidinyl)-4-iodoacetamide. Label III was prepared by the dicyclohexylcarbodiimide-mediated condensation of DPPE with the appropriate glutaric acid monoamide, the latter being prepared from the appropriate methylated acylmonochloride and spin label amine. All labeled lipids were purified by column chromatography.

## RESULTS

The present section describes some of our preliminary results. A more detailed report will be given elsewhere. (Brûlet and McConnell, to be published.)

### Spin Label Hapten Resonance Spectra

Figure 1a gives a schematic representation of the disposition of labels such as I-III in a phospholipid bilayer membrane. Figure 1b shows the paramagnetic resonance spectra of DMPC liposomes containing 2.5 mole % of label III. The solid ↔ fluid phase transition temperature of this lipid host is ~23°C. The spectrum at 22°C is typical of those taken below 22°C and the spectrum taken at 24°C is typical of those taken above 24°C. Note particularly the width and shape of the high field resonance signal; the width and Lorentzian shape is consistent with the previously reported lateral diffusion constant $D \sim 10^{-8}$ $cm^2$/sec, given that the concentration of III is 2.5 mole % (8). At temperatures of 22°C and below, this signal has a Gaussian shape, consistent with a much lower rate of lateral diffusion when III is included in a "solid solution" of DMPC. Unfortunately, the relation between diffusion rate and lineshape is not generally as simple as implied above. As pointed out earlier (8), increasing temperature decreases line broadening due to dipole-dipole interactions, and increases line broadening due to the increasing frequency of spin exchange collisions. Experiments on the resonance line-widths of spin label hapten I as a function of the concentration of I in binary mixtures of DPPC and cholesterol are consistent with the possibility that there is a transition between collisional spin exchange broadening (low cholesterol) to dipolar broadening (high cholesterol). (P. Brûlet and H.M. McConnell, unpublished.)

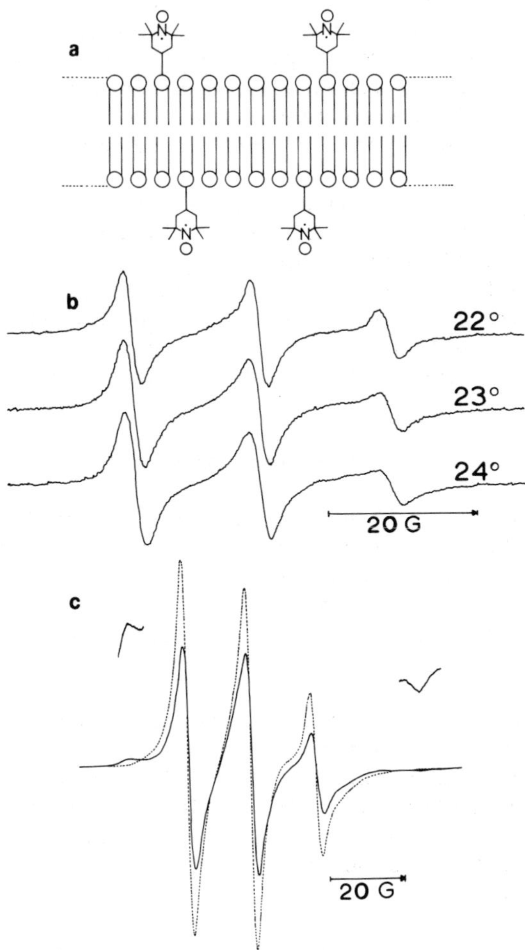


Figure 1. a) Diagramatic representation of phosphatidylcholine bilayer containing spin label lipid hapten. b) Paramagnetic resonance spectra of DMPC bilayers containing 2.5% III. The spectrum taken at 22°C is also typical of those taken below 22°C; the spectrum taken at 24°C is also typical of those taken above 24°C. c) Paramagnetic resonance spectra of large single compartmented DMPC vesicles containing 2.5% III together with IgG prepared from rabbit anti-TEMPO hemocyanin serum, ( ——— ), or IgG prepared from anti-human erythrocyte serum, ( ---- ).

Figure 1c shows the paramagnetic resonance spectra of large single (or nearly single) compartment DMPC vesicles containing 2.5 mole % III together with IgG prepared from rabbit anti-nitroxide serum; for a control a spectrum is given for III in the presence of IgG from rabbit anti-human erythrocyte serum. The specific binding of the anti-nitroxide antibodies is manifest, firstly by the decrease in the weakly immobilized signal from III, and secondly, by the appearance of the strongly immobilized outer wings.

In liposomes containing spin label lipid haptens such as I-III, the fraction of hapten on the outer surface of the liposome is expected to be small, and is found to be small judging by the small or negligible effect that anti-nitroxide antibodies have on the resonance spectra of these liposomes. Even so, the specific binding of anti-nitroxide antibodies to the outer surface of these liposomes can be assayed by depletion of TEMPO binding activity in the solution.

## Complement Fixation

Figure 2 gives an illustrative study of complement fixation, for DMPC and DPPC liposomes containing 0.5 mole % III. The temperature used for the fixation assay, 32°C, is midway between the phase transition temperatures of DMPC (23°C) and DPPC (42°C), so that DMPC is "fluid" at the assay temperature, and DPPC is "solid" at the assay temperature.

The data in Figure 2 indicate the complement fixation is enhanced in the region of antibody excess, for the fluid membrane relative to the rigid membrane. These data might be taken to signify that complement fixation (at least with IgG antibodies) is enhanced when the membrane haptens are mobile. At the present time we have at least two reasons to be cautious about the validity of this conclusion.

In the first place, antigens in a fluid membrane may not have lateral motional freedom if one membrane is cross-linked to another by antibodies. This cross-linking could easily lead to a relatively immobilized "patching" of the haptens. In this connection we have found that single compartment DMPC or DPPC vesicles containing 2.5 mole % III are visibly aggregated by IgG from rabbits immunized with spin-labeled protein, and the aggregation is enhanced by raising the temperature of the reaction mixture to above the phase transition temperature of the carrier lipid. Control IgG from rabbits immunized with human erythrocyte stroma does not aggregate vesicles containing III.

A second reason for being cautious concerning the data in Figure 2 is that we have carried out similar experiments with II at

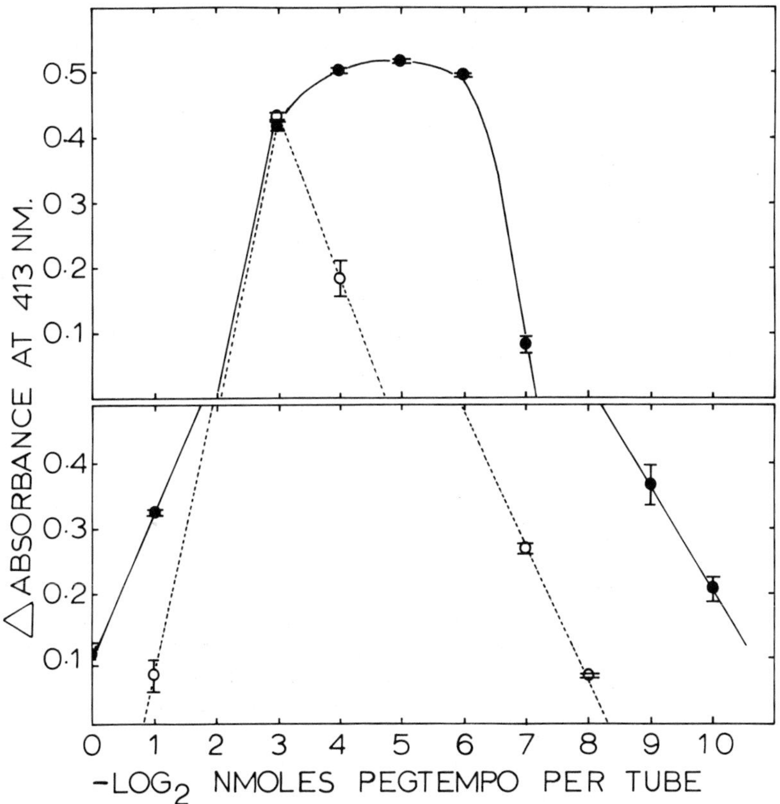


Figure 2. C' fixation at 32°C for 2 h, measured by subsequent inhibition of lysis of indicator sheep cells (Δ Absorbance 413 nm, see ref 1), as a function of amount of III supplied. A constant quantity of IgG, approximately equivalent to that in 1μl serum, was supplied to each tube. DMPC ( ——o—— ) or DPPC ( ---o--- ) liposomes containing 0.5% III were used an antigen. The lower part of this graph was obtained using a limiting quantity of C' such that the effect of omitting antigen, antibody, or both, could be judged; the upper part required the use of about four times as much C'. At the concentrations used, neither the antibodies nor the liposomes were anticomplementary, nor were there any indications that they were lytic towards sheep cells.

32°C, and a much smaller difference between fluid and solid membranes was found in the region of antibody excess. It is possible that the differences between II and III in this regard are due to differences in the strength of antibody binding.

We take this opportunity to summarize certain other preliminary experiments we have carried out that will be described in detail elsewhere (P. Brûlet and H.M. McConnell, to be published).

(1) In general, it appears that antibody binding, and complement fixation, increase in the series I, II, III.
(2) The inclusion of cholesterol in a phosphatidylcholine membrane enhances complement fixation for all the lipid haptens, the enhancement being most dramatic for I, where the nitroxide group is close to the membrane surface.
(3) In the region of antibody excess the cholesterol enhancement of complement fixation by I is similar to that reported earlier for cardiolipin (1).

## Unsolved Problems

The preliminary study described above leaves a number of important, unsolved problems. Some of these are listed below.

(i) It is likely that magnetic resonance and electron microscopic techniques can provide quantitative information on hapten motion, <u>and</u> distribution, but such studies are far from complete at present. Lipid mixtures can yield remarkable ordering effects on membrane antigens (2), and each system must be examined carefully in this respect.
(ii) It is desirable to know the antibody binding constants for haptens on the outer membrane surface, and this in turn requires an accurate knowledge of the number of haptens on this surface.
(iii) Some method must be devised to avoid or control agglutination of membranes by antibodies.
(iv) Our central purpose in using model membranes for studies of immunochemistry is not to study membrane damage, or complement depletion, but to delineate those factors that are critical in recognition and triggering immunochemical events at a membrane surface. For the special case of the complement system, this requires a determination of the relation of $C_1$ activation to antibody binding and hapten structure, distribution, and mobility.

## ACKNOWLEDGMENTS

The present work was supported by the National Science Foundation by a Grant-in-Aid from Sigma Xi and also under Grant NSF BMS 75-02381 We are indebted to Dr. Paul Rey for helpful discussions concerning his related experimental work, and to Dr. Michael Iverson, Stanford Medical Center, for serological characterizations of IgG fractions. Gillian M. Kitch Humphries is the recipient of an Institutional

Research Fellowship from the National Institute of General Medical Science Grant No. GM-07026.

## REFERENCES

1) G.M.K. Humphries and H.M. McConnell, Proc. Nat. Acad. Sci. USA 72, 2483 (1975).
2) See W. Kleemann and H.M. McConnell, Biochim. Biophys. Acta 419, 206 (1976) and references contained therein.
3) See A.I. Käiväräinen and R.S. Nezlin, Biochem. Biophys. Res. Comm. 68, 270 (1976) and references contained therein.
4) L. Stryer and O.H. Griffith, Proc. Nat. Acad. Sci. USA 54, 1785 (1965).
5) G.M.K. Humphries and H.M. McConnell, Biophys. J. 16, 275 (1976).
6) E. Wasserman and L. Levine, J. Immunol. 87, 290 (1961).
7) R.D. Kornberg and H.M. McConnell, Biochem. 10, 1111 (1971).
8) P. Devaux, C.J. Scandella and H.M. McConnell, J. Mag. Res. 9, 474 (1973).

# ON THE COUPLING OF THE GLUCAGON RECEPTOR TO ADENYLATE CYCLASE

M.D. Houslay, A. Johannsson, G.A. Smith, T.R. Hesketh
G.B. Warren and J.C. Metcalfe

Department of Biochemistry, Tennis Court Road
Cambridge, England

## Introduction

Two main classes of model have been proposed for the coupling of hormone receptors to adenylate cyclase across the membrane, which both assume that the catalytic unit and the receptor are distinct entities. The original model, proposed by Sutherland and coworkers,[1] postulated a permanent association between the two components, in which the catalytic unit is activated by conformational changes originating in the receptor when the hormone is bound. More recently, the realisation that membrane proteins may be able to undergo fast lateral diffusion has prompted a second type of model in which the receptor and adenylate cyclase moieties can migrate independently in the plane of the membrane, until binding of the hormone to the receptor causes a locking interaction with the catalytic unit, which is then activated.[2,3,4] The converse model, in which the catalytic unit is released from the receptor on binding hormone has also been proposed.[5] At present it is not possible to distinguish these models with any confidence from the available biochemical data.

In this paper we summarise the results from two approaches to the mechanism of trans-membrane coupling. We have used the asymmetry of the membrane bilayer to distinguish the response of the catalytic unit in the coupled and uncoupled states to the lipid environment. We then extend this comparison to modified membranes in which defined changes in the phospholipid composition have been made. A second approach has been to estimate the apparent molecular size of the two components of the complex in the coupled and uncoupled states *in situ*.

## Coupling across an asymmetric bilayer

It has recently been suggested by Fox and coworkers[6] that the bilayer of plasma membranes from mammalian cells is asymmetric to the extent that distinct lipid phase separations occur in the two halves of the bilayer independently. The phase separations are identified by characteristic upper and lower temperatures which are at about 32°C and 16°C for the outer half of the bilayer and at about 37°C and 21°C for the inner half. It is supposed that penetrant membrane proteins which span the bilayer will sense both of these phase separations, so that the Arrhenius plots of activity are expected to contain at least four breaks, at the characteristic temperatures. It is implied that a range of functional proteins in different membranes will show these breaks at about the same temperatures, because it is suggested that similar phase separations are a common feature of the different membrane structures. A comparison of the response to lipid phase separations of uncoupled adenylate cyclase with adenylate cyclase coupled to the receptor, might be expected to indicate their dispositions within the bilayer.

When adenylate cyclase activity in rat liver plasma membrane is stimulated directly through the catalytic unit by fluoride, or by 5'-guanylyl-imidodiphosphate (GMP - P(NH)P), a linear Arrhenius plot is obtained[7] (Fig. 1). In contrast, stimulation of the enzyme by glucagon through the receptor gives a biphasic Arrhenius plot with a well-defined break at about 29°C. Clearly these profiles do not fit into the pattern described above for the two halves of the bilayer, although we have not attempted to define any smaller inflections which may be beyond the resolution of our data. Qualitatively similar results have been obtained by Kreiner et al[8] who obtained a linear Arrhenius plot in the presence of fluoride and a biphasic plot for glucagon-stimulated activity. However in their data, the break occured at 32°C and the activation energy increased rather than decreased at higher temperatures.

To determine whether the break could reasonably be attributed to lipid phase separation in the membranes, we have measured the temprature-activity profiles for several other enzymes in the rat liver plasma membrane. The results in Fig. 1 show that uncoupled adenylate cyclase and phosphodiesterase which both have their active sites exposed on the inner surface of the membrane, both yield linear Arrhenius plots. On the other hand, $Na^+$,$K^+$ ATPase which spans the membrane, $Mg^{2+}$ ATPase, and 5' nucleotidase which has its active site on the outside surface of the membrane all show breaks in their Arrhenius plots at 25 - 30°C, similar to the coupled adenylate cyclase-receptor complex. Although limited, these results suggest that enzymes spanning the bilayer and/or facing outwards, share a common response to the lipid environment

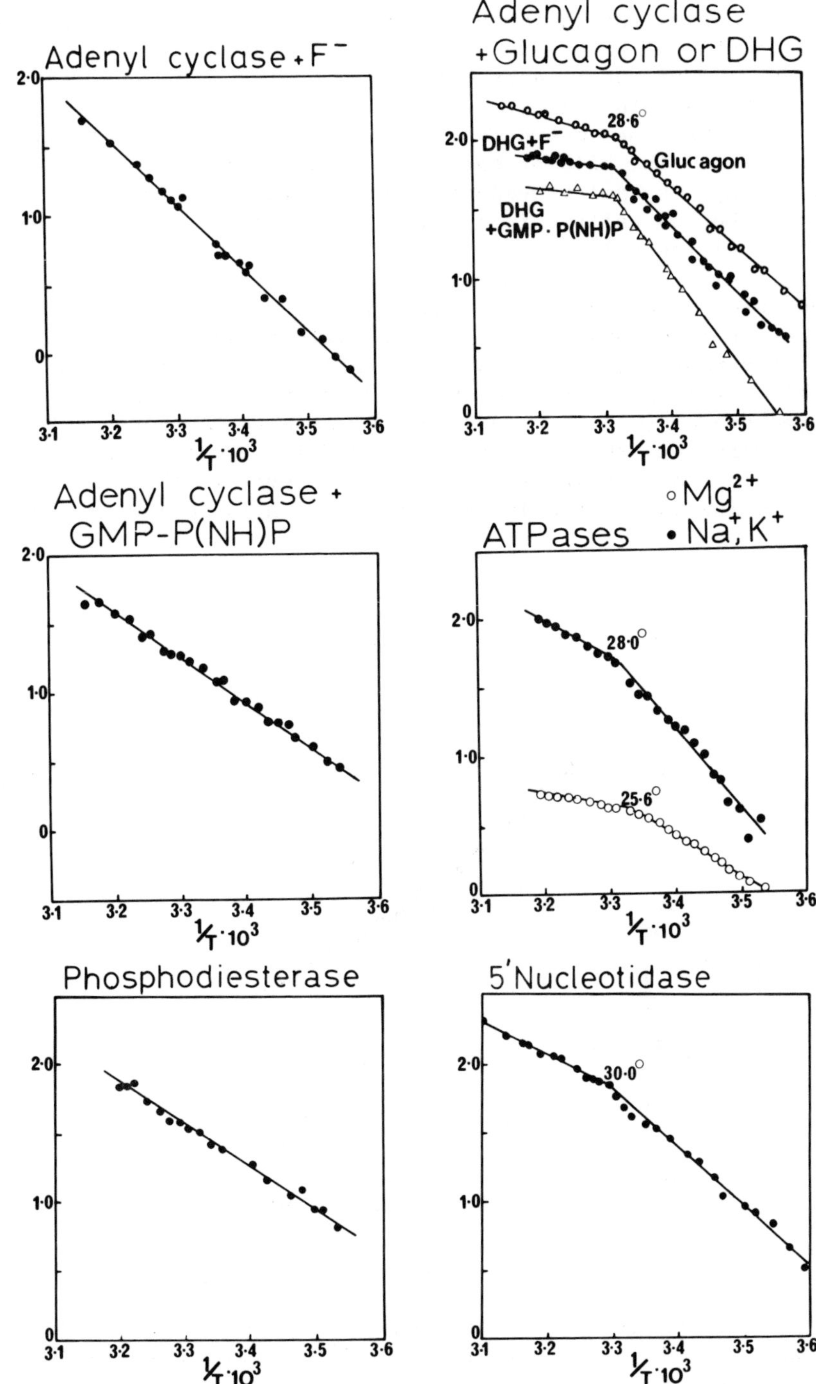


Fig. 1 Arrhenius plots of enzyme activities in rat liver plasma membranes

which is clearly distinguishable from the enzymes facing inwards. This is consistent with a phase separation that occurs only in the outer half of the bilayer and is responsible for the breaks in the Arrhenius plots at 25 - 30°C. However we cannot exclude the possibility that the inward-facing enzymes do not interact with the lipid chains of the bilayer because they are bound to the membrane mainly through the polar headgroups of the inner lipids. They might then be insensitive to a phase separation occuring in the inner half of the bilayer.

Some preliminary spin label data also indicate that the break in the Arrhenius plot is attributable to the lipid bilayer. The order parameter(S) for 5-doxyl stearic acid was measured as a function of temperature and showed a break at about 28°C in the native membranes (Fig. 2). However no break in the corresponding plot for the extracted lipids was observed. This result is consistent with a lipid phase separation at 28°C but suggests that it is due to an organisation of the lipid bilayer which is lost when the lipid

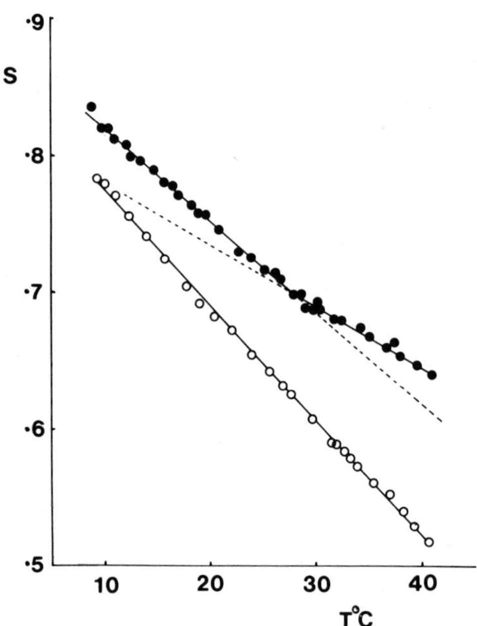


Fig. 2 Order parameter(S) for 5-doxyl stearic acid as a function of temperature in rat liver plasma membrane (o) and extracted lipid bilayers (o).

is extracted and reformed as vesicles of bilayer without protein present.

As a working hypothesis, we assume that the biphasic plot of glucagon stimulated activity reflects the coupling of adenylate cyclase to the receptor, by confering sensitivity to the lipid environment of the receptor onto the catalytic unit. The relevant phase separation may occur only in the outer half of the bilayer A corollary of this analysis is that the competitive antagonist des His glucagon (DHG) must also cause a structural interaction of the receptor with adenylate cyclase, because it produces biphasic Arrhenius plots of activity in combination with either fluoride or GMP-P(NH)P which are very similar in form to those obtained with glucagon alone or in combination with these ligands[7] (Fig. 1, Table I). Des His glucagon does not cause any activation of adenylate cyclase by itself, yet it renders the catalytic unit sensitive to the lipid environment of the receptor. Independent confirmation of the structural interaction caused by des His glucagon

Table 1 Inflection temperatures and activation energies from Arrhenius plots of adenylate cyclase activity stimulated by various ligands.

<table>
<tr><th rowspan="2">Ligands</th><th rowspan="2">Inflection temperature ($^{o}C$)</th><th colspan="2">Activation Energy (Kcal $mole^{-1}$)</th></tr>
<tr><th>Above break</th><th>Below break</th></tr>
<tr><td>Glucagon</td><td>28.6±1</td><td>6±2</td><td>20±3</td></tr>
<tr><td>Fluoride</td><td>-</td><td colspan="2">21±2</td></tr>
<tr><td>Fluoride + glucagon</td><td>28.5±0.5</td><td>6±2</td><td>22±3</td></tr>
<tr><td>Fluoride + des His glucagon</td><td>29.0±0.4</td><td>5±2</td><td>22±3</td></tr>
<tr><td>GMP-P(NH)P</td><td>-</td><td colspan="2">15±4</td></tr>
<tr><td>GMP-P(NH)P + glucagon</td><td>29.0±0.5</td><td>5±2</td><td>21±2</td></tr>
<tr><td>GMP-P(NH)P + des His glucagon</td><td>28.8±0.6</td><td>3±1</td><td>29±4</td></tr>
</table>

is provided from the apparent molecular weight of the coupled complex, described later.

Coupling is sensitive to changes in the lipid environment of the receptor

A more stringent test of the hypothesis outlined above is to make substantial changes in the lipid environment of the membrane and to compare the response of the uncoupled catalytic unit with the receptor-enzyme complex. We have used the technique of lipid-substitution or fusion to introduce up to 60% of synthetic lecithins in the lipid bilayer associated with the membrane.[9] In lipid-substitution experiments, the endogenous lipids of rat liver plasma membranes were replaced by added exogenous lecithins using cholate to equilibrate the lipid pools followed by centrifugation and dialysis to remove the cholate. The details of the technique are described elsewhere for sarcoplasmic reticulum and mitochondrial membranes.[10, 11] To prepare complexes of membranes fused with exogenous lecithins, highly sonicated suspensions of lecithin were added to rat liver plasma membranes, and the fused complexes were separated from unfused lecithin vesicles by centrifugation (see ref. 9 for details). The compositions and activities of the various partially substituted or fused preparations are given in Table 2. The lipids used were dioleoyl-; dimyristoyl; and dipalmitoyl lecithin (DOL; DML; DPL).

In all of the complexes prepared by substitution the residual fluoride or glucagon-stimulated activities are between 5% and 20% of the corresponding activities in the native membrane (Table 2). The activity could not be restored by back-titration with endogenous lipid from rat liver plasma membranes and we conclude that this irreversible loss of activity is due to the use of cholate to equilibrate the lipid pools.[12] The fusion technique caused relatively little loss of fluoride stimulated activity and substantial glucagon-stimulated activity was also retained.

The effects of lipid substitution or fusion on the Arrhenius plots of adenylate cyclase activity are clear-cut. The linear form of the Arrhenius plots for activity stimulated by fluoride or GMP-P(NH)P was unaltered in all of the membrane preparations, with similar activation energies to those observed for the corresponding activity in the native membrane (Fig. 3, Table 3). The activity profile for the enzyme therefore appears to be very insensitive to its lipid environment when stimulated directly by fluoride or GMP-P(NH)P.

In contrast, the break at 28.5°C in the Arrhenius plot of glucagon-stimulated activity was shifted upwards by dipalmitoyl lecithin, downwards by dimyristoyl lecithin and was abolished

Table 2 Composition and activities of the partially substituted or fused complexes of rat liver plasma membranes with defined lecithins.

| Defined Lipid | Complex Type | % Defined Lipid in total Lipid Pool | Lipid:Protein Ratio (mg/mg) | Glucagon stimulated activity units/mg at 37 | Fluoride stimulated activity units/mg at 37°C |
|---|---|---|---|---|---|
| | Original Membranes | - | $0.55 \pm 0.04$ | 83.2 | 27.2 |
| DML | Subst. | $48 \pm 5$ | $0.33 \pm 0.03$ | 8.6 | 4.3 |
| DML | Fusion | $60 \pm 5$ | $1.35 \pm 0.33$ | 55 | 23 |
| DPL | Subst. | $55 \pm 5$ | $0.36 \pm 0.03$ | 8.6 | 3.8 |
| DOL | Subst. | $50 \pm 5$ | $0.40 \pm 0.03$ | 8.5 | 4.7 |
| DOL | Fusion | $60 \pm 5$ | $1.35 \pm 0.33$ | 39.8 | 22.4 |

<u>Table 3</u> Inflection temperatures and activation energies from Arrhenius plots of adenylate cyclase activities stimulated by various ligands in lipid-substituted complexes.

<table>
<tr><th rowspan="2">Complex</th><th rowspan="2">Ligand added</th><th rowspan="2">Break Point ($^{o}$C)</th><th colspan="2">Activation Energy (Kcal/mole)</th></tr>
<tr><th>Above break</th><th>Below Break</th></tr>
<tr><td rowspan="3">Dioleoyl lecithin substitution</td><td>Glucagon</td><td>-</td><td colspan="2">13$\pm$3</td></tr>
<tr><td>Fluoride</td><td>-</td><td colspan="2">22$\pm$2</td></tr>
<tr><td>des-his-glucagon + fluoride</td><td>-</td><td colspan="2">16$\pm$3</td></tr>
<tr><td rowspan="2">Dioleoyl lecithin fusion</td><td>Glucagon</td><td>-</td><td colspan="2">17$\pm$2</td></tr>
<tr><td>Fluoride</td><td>-</td><td colspan="2">20$\pm$2</td></tr>
<tr><td rowspan="5">Dimyristoyl lecithin substitution</td><td>Glucagon</td><td>22$\pm$</td><td>18$\pm$2</td><td>43$\pm$4</td></tr>
<tr><td>Fluoride</td><td>-</td><td colspan="2">23$\pm$1</td></tr>
<tr><td>des-his-glucagon + fluoride</td><td>22$\pm$1</td><td>18$\pm$3</td><td>40$\pm$6</td></tr>
<tr><td>GMP-P(NH)P</td><td>-</td><td colspan="2">25$\pm$3</td></tr>
<tr><td>GMP-P(NH)P + Glucagon</td><td>23$\pm$1</td><td>16$\pm$2</td><td>48$\pm$4</td></tr>
<tr><td rowspan="2">Dimyristoyl lecithin fusion</td><td>Glucagon</td><td>22$\pm$0.5</td><td>16$\pm$2</td><td>38$\pm$2</td></tr>
<tr><td>Fluoride</td><td>-</td><td colspan="2">21$\pm$2</td></tr>
<tr><td rowspan="2">Dipalmitoyl lecithin substitution</td><td>Glucagon</td><td>32.5$\pm$1</td><td>20$\pm$2</td><td>43$\pm$6</td></tr>
<tr><td>Fluoride</td><td>-</td><td colspan="2">23$\pm$2</td></tr>
</table>

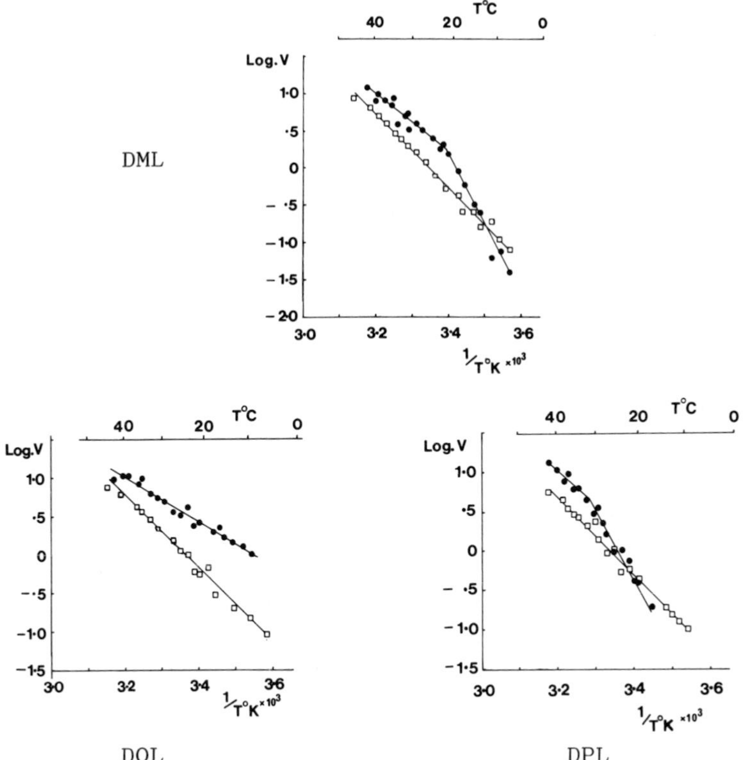


Fig. 3 Arrhenius plots of adenylate cyclase activities stimulated by various ligands in lipid-substituted complexes. ● Glucagon stimulation; □ fluoride stimulation

by dioleoyl lecithin. The break temperatures and activation energies for adenylate cyclase activity were the same in complexes prepared by fusion or substitution with a specific lecithin, in spite of the differences in the absolute activity obtained by the two techniques (Fig.3, Table 3). Very similar shifts in the break point were observed for stimulation by glucagon or des His glucagon in combination with fluoride or GMP-P(NH)P (Fig. 3, Table 3). This provides further evidence for a coupling interaction of the receptor with the catalytic unit mediated by des His glucagon.

These results can be rationalised within the hypothesis already described. The insensitivity of the fluoride stimulated response to the lipid environment may result from a negligible interaction of the uncoupled catalytic unit with the chain region of the bilayer, or it may be that no significant phase separations occur in the inner half of the bilayer, even when it is substantially modified by the presence of saturated lecithins. The changes in the biphasic Arrhenius plots of glucagon-stimulated activity are consistent with a lipid phase separation in the native membrane at 28.5$^{o}$C which is shifted by the insertion of defined lecithins into the lipid pool, depending on their phase transition temperatures. Dioleoyl lecithin has a phase transition at about -22$^{o}$C, and we suggest that the linear Arrhenius plot for the glucagon-stimulated activity in dioleoyl lecithin complexes is due to a large depression of the phase separation temperature in the mixed lipid pool to below 5$^{o}$C. Dimyristoyl lecithin, which has a transition temperature at 23.5$^{o}$C causes a much smaller depression of the phase separation temperature, whereas dipalmitoyl lecithin raises the temperature because its transition is at 41$^{o}$C. The large increase in activation energies observed with the saturated lecithins below the phase separation temperatures is consistent with this interpretation. A striking consequence of these very high activation energies is that the glucagon stimulated activities become lower than the fluoride stimulated activities at low temperatures (Fig. 3).

These experiments consistently indicate that the catalytic unit is rendered sensitive to the lipid environment of the receptor by coupling through the action of glucagon or des-His glucagon. However the results are consistent with both of the main models of the coupling mechanism, and in a further attempt to discriminate between the models, we have measured the apparent molecular weights of the two components in the coupled and uncoupled states.

## Apparent molecular weights of the glucagon receptor, adenylate cyclase and the coupled complex

We have used the technique of radiation inactivation to determine the apparent molecular weights of the receptor and the catalytic unit in the membrane. Briefly, the lyophilised membranes are irradiated in a 15 MeV electron beam which causes inactivation of the protein when absorption energy from an electron disrupts essential structural elements of the protein. The rate at which inactivation occurs depends on the radiation dose and the size of the target protein. By calibration with proteins of known molecular weights, the apparent molecular weight of a protein can be determined from the rate at which it is inactivated by an increasing radiation dose. The technique suffers from disadvantages similar to those of electron microscopy for membranes in that the samples are frozen and dehydrated by lyophilisation before irradiation and it is assumed that the organisation of the proteins remains similar to that in the functional membrane. However it provides the only available method of assessing the molecular size of identified proteins _in situ_, since the protein functions are assayed after irradiation and rehydration. The method has been used to provide apparent molecular weights for several membrane-bound proteins which are not incompatible with the available biochemical evidence (e.g. $Na^+$, $K^+$ATPase, mitochondrial $Mg^{2+}$ATPase, succinic dehydrogenase and the aggregated forms of acetyl cholinesterase).

Samples of lyophilised rat liver plasma membranes were irradiated without hormone, and after pretreatment with glucagon, or des His glucagon. At least 90% of adenylate cyclase activity for all ligands used was recovered on rehydration of samples which had not been irradiated. Typical first-order inactivation plots are given in Fig. 4 and the apparent molecular weights from all experiments are summarised in Table 4.

In the uncoupled state, either fluoride or GMP-P(NH)P which both activate the enzyme directly, yield an apparent molecular weight of about 160,000 whereas glucagon stimulated activity decays with an apparent molecular weight of about 390,000. However, after pretreatment with glucagon or des-His glucagon, the apparent molecular weight for both fluoride and glucagon stimulated activities is about 400,000 (±10%). Specific binding of $^{125}$I glucagon yielded an apparent molecular weight for the receptor of about 220,00 in the uncoupled state, and about 320,00 when coupled.

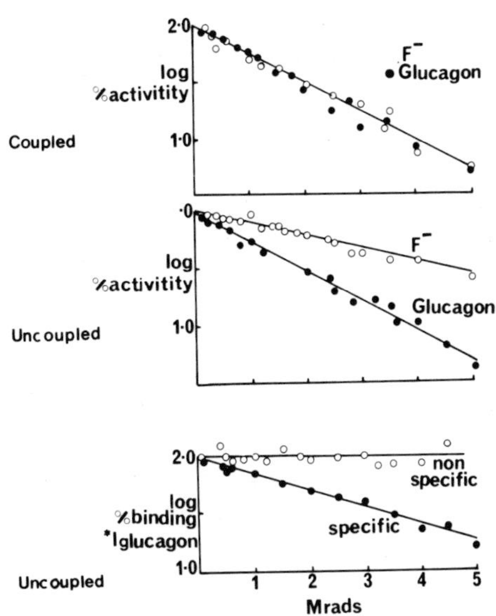


Fig. 4 First-order inactivation by irradiation of adenylate cyclase activity and specific glucagon binding in rat liver plasma membranes

Table 4

Apparent molecular weights of the glucagon receptor and adenylate cyclase, estimated by inactivation after irradiation of adenylate cyclase activity or specific binding of $I^{125}$ glucagon. The membranes were lyophilised without hormone (uncoupled), or after pretreatment with glucagon or des His glucagon (coupled). Values for membranes pretreated with des-His glucagon are bracketed.

| Assay | Adenylate cyclase - receptor complex Uncoupled | Coupled |
|---|---|---|
| Fluoride stimulation of catalytic unit | 160,000±23,000 | 379,000±65,000<br>(402,000±35,000) |
| GMP-P(NH)P stimulation of catalytic unit | 163,000±31,000 | Not determined |
| Glucagon stimulation through receptor | 389,000±51,000 | 427,000±47,000 |
| $I^{125}$ glucagon specific binding to receptor | 217,000±6,000 | 310,000±26,000<br>(328,000±40,000) |

These results suggest that the receptor and the catalytic unit are inactivated independently when the two components are uncoupled but in the coupled state the two components are so tightly associated that an electron hit in either the receptor or the catalytic unit causes inactivation of both components, and the apparent molecular weight is increased as observed. This energy transfer implies an intimate association between the coupled receptor and the catalytic unit, consistent with our previous conclusion that the catalytic unit responds to changes in the lipid environment of the receptor only in the coupled state. The data also imply that no effficient energy transfer occurs in the uncoupled state, showing that the two components are either independent, or associated much more weakly. The apparent molecular weights obtained after pretreatment with des His glucagon are very similar to those obtained with glucagon, and provide strong support for the biochemical evidence that des-His glucagon causes structural coupling of the receptor to the catalytic unit.

Although the results are incompatible with a number of suggested models of the coupling interaction (e.g. release of the catalytic unit from the receptor on the binding of hormone), they do not allow us to distinguish between the main classes of models outlined previously. The most significant result to be accounted for is the high molecular weight of about 390,000 for the complex stimulated by glucagon after irradiation in the uncoupled state. This is compatible with a permanent association between the two components, in which an electron hit only destroys one or other component of the couple, but the undamaged partner is unable to form a new functional couple because it remains associated with an inactive partner. Glucagon stimulated activity would then decay as observed with an apparent molecular weight equal to the sum of the components in the complex, because the unstimulated catalytic unit has negligible activity. For an independent diffusion model, we expect that if an electron hit inactivates both the active site and the coupling interface of an uncoupled catalytic unit, then the apparent molecular weight for glucagon stimulated activity would be expected to correspond to the larger component of the couple which is inactivated more rapidly (i.e. the receptor). This is not consistent with the results. However, if we assume that the coupling interface for the receptor on the catalytic unit can be preserved when the catalytic activity is inactivated, then the apparent molecular weight for glucagon stimulated activity will again correspond to the sum of the separate components as observed.

Although the data do not therefore finally distinguish the models it is likely that a biochemical analysis of the ability of the inactivated catalytic unit to couple to the receptor will resolve the problem.

References

1. Robison, G.A., Butcher, R.W. and Sutherland. E.W. (1967) Ann. N.Y. Acad. Sci. U.S.A. 139, 703-723

2. Birnbaumer, L. (1973) Biochim. Biophys. Acta. 300, 129-158

3. Helmreich, E.J.M. (1976) FEBS Lett. 61, 1-5

4. Bennett, V., O'Keefe, E. and Cuatrecasas, P. (1975) Proc. Nat. Acad. Sci. U.S.A. 72, 33-37

5. Levey, G.S., Fletcher, M.A., Klein, I. Ruize, E. and Schenk, A. (1974) J. Biol. Chem. 249, 2665-2673

6. Wisnieski, B.J., Parkes, T.G., Huang, Y.O. and Fox, C.F. (1974) Proc. Natl. Acad. Sci. U.S.A. 79, 1785-1789

7. Houslay, M.D., Metcalfe, J.C., Warren, G.B., Hesketh, T.R. and Smith, G.A. (1976) Biochim. Biophys. Acta. in press.

8. Kreiner, P.W., Keirns, J.J. and Bitensky, M.W. (1973) Proc. Natl. Acad. Sci. U.S.A. 71, 4381-4385

9. Houslay, M.D., Hesketh, T.R., Smith, G.A., Warren, G.B. and Metcalfe, J.C. (1976) Biochim. Biophys. Acta. in press

10. Warren, G.B., Toon, P.A., Birdsall, N.J.M. (1974) Proc. Natl. Acad. U.S.A. 71, 622-626

11. Houslay, M.D., Warren, G.B., Birdsall, N.J.M., and Metcalfe, J.C. (1975) FEBS Lett. 51, 146-151

12. Warren, G.B. Toon, P.A., Birdsall. N.J.M., Lee, A.G and Metcalfe, J.C. (1974) Biochim. 13, 5501-5507

# RECENT FINDINGS IN THE STRUCTURAL AND FUNCTIONAL ASPECTS OF THE PEPTIDE IONOPHORES

Yu.A.Ovchinnikov

Shemyakin Institute of Bioorganic Chemistry

USSR Academy of Sciences, Moscow, USSR

The last decade is witness to the birth and tempestuous development of a new field in bioorganic chemistry. Its objects are membrane–active complexones (see [1] and references therein), often called ionophores because of their ability to bind metal ions in solution and carry them along in one or the other stage of the transmembrane ion transporting process. It is this property which has won for the ionophores a secure place in the arsenal of modern biochemical and biophysical techniques as highly effective tools for studying ion movements across membranes.

The ionophores constitute a highly varied class of compounds, both natural and synthetic, cyclic and linear, differing greatly in molecular weights and possessing diverse functional groups. Respecting their mode of action they can be divided into two main classes; viz., carriers (sometimes called "cage carriers") and substances forming ion–conducting pores or "channels" (Fig.1). Among the carriers a pre–eminent place is occupied by macrocyclic depsipeptides of the groups of valinomycin and the enniatins. Of these, valinomycin

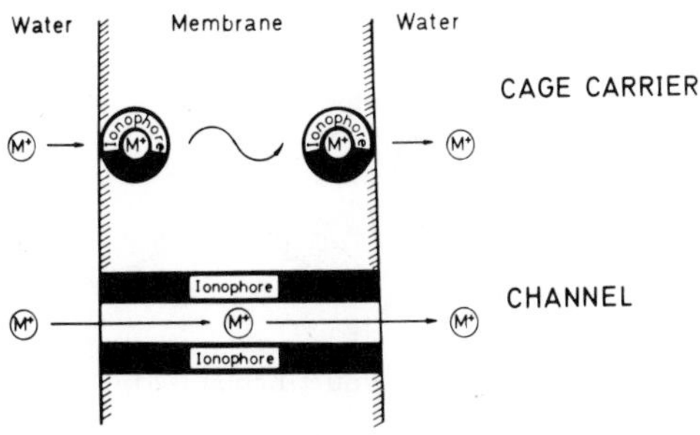


Fig.1. Schematic Illustration of the Mechanism of Cation Transport through Membranes.

⌊(D-Val-L-Lac-L-Val-D-Hylv)$_3$⌋ ⌊(L-MeVal-D-Hylv)$_3$⌋

valinomycin (1) enniatin B (2)

(Vm) itself is unsurpassed in its potassium ion transporting capacity and the K/Na selectivity of its membrane action. The enniatins on the other hand are mainly interesting because of their broad spectrum of action, members of this group being capable of binding and transporting ions of diverse sizes and valencies. The most representative of the "channel" type of ionophores are the 15–membered linear peptide, gramicidin A(Gr A) and its structurally closely allied gramicidins B and C.

L L D L D L D L D L D L D L
HCO-Val-Gly-Ala-Leu-Ala-Val-Val-Val-Trp-Leu-Trp-Leu-Trp-Leu-Trp-NH$(CH_2)_2$OH
gramicidin A (3)

For a number of years in the Shemyakin Institute of Bioorganic Chemistry of the USSR Academy of Sciences systematic synthesis of new ionophores and their subsequent structural, mechanistic and physicochemical study is being carried out.

The present paper presents new findings that have resulted from further study of the valinomycin, enniatin B (En) and gramicidin A groups of peptides.

## VALINOMYCIN, ENNIATIN B AND THEIR ANALOGS

All experience with membrane–affecting ionophores has shown that it is analysis of the spatial structure of both the free compounds and their complexes with metal ions which provides the key to comprehension of their biological properties. It was shown by a number of groups (see [1], pp. 118–140) that in solution valinomycin exists in the form of an equilibrium mixture of three different conformers A, B and C (Fig. 2). In form A which is predominant in heptane, $CCl_4$ or $CHCl_3$ all NH groups are intramolecularly hydrogen–bonded to the amide carbonyls, form B, the preferable form for solvents of medium polarity possesses three such bonds involving the NH groups of the D–Val residues; in form C the characteristic one for polar media, particularly

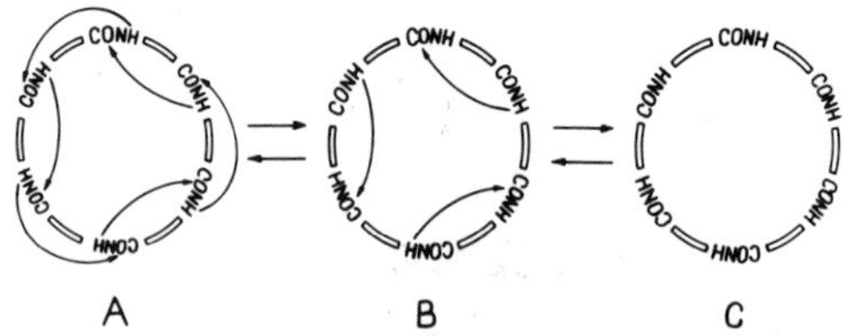


Fig. 2. The Conformational A⇄B⇄C Equilibrium Interconversion of Valinomycin

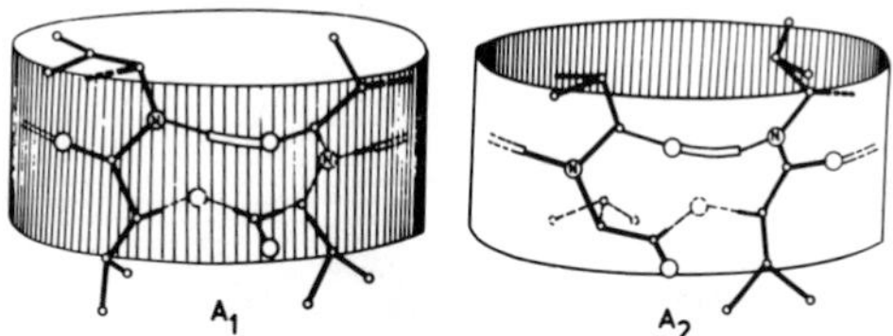


Fig.3. Schematic Representation of the $\underline{A}_1$ and $\underline{A}_2$ Structures of Valinomycin

at elevated temperatures, the NH groups are H–bonded to the solvent.

The hydrogen bonds of $\underline{A}$ in which each amide carbonyl is 4→1 H–bonded to the NH of the neighbouring amide group (in the direction of acylation) convert it into a system of six, fused 10–membered rings. Hence, in non–polar media the valinomycin molecule acquires a compact bracelet–like conformation with an internal diameter of ~8Å and height of ca. 4 Å. In principle, form $\underline{A}$ has a dual way of chain folding so that two forms $\underline{A}_1$ and $\underline{A}_2$ (Fig.3) with opposite ring chirality and differently oriented side chains are possible. Both the $\underline{A}_1$ and $\underline{A}_2$ forms can in turn exist in four conformations differing in the orientation of the ester carbonyls. From Fig.3 it can be seen that if the molecule is placed with the L–lactic acid residues in its upper part the acylation runs clockwise in the $\underline{A}_1$ and counterclockwise in $\underline{A}_2$.

In structure $\underline{A}_1$ the NH–CH protons are *cis,* whereas in $\underline{A}_2$ they are *gauche* ($\Theta \sim 0°$ and $\sim 120°$ respectively, Figs. 4 and 5). The high $^3J(H-NC^\alpha-H)$ values at first had led to the assumption that for non–polar solvents the preferable of the two structures was form $\underline{A}_1$. However, in the course of a systematic investigation in this institute into the possibility of using $^{13}C$ NMR spectroscopy for conformational purposes, resulting in establishment of an angular dependence of the $^3J(^{13}C'-NC^\alpha-H)$ coupling constant (Fig. 5) [2] , it turned out that in valinomycin this constant is small (~3.5 Hz) from which followed that, in conformity with the above angular dependence, the NH–CH protons must be *gauche* (see Fig. 5). Consequently in non–polar solvents valinomycin actually prefers the $\underline{A}_2$ conformation. A similar revision has been made of the $\Theta$ and $\Phi$

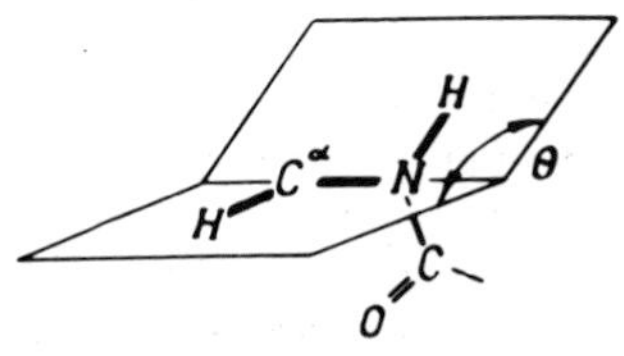


Fig.4. The $H-N-C^\alpha$ and $N-C^\alpha-H$ Dihedral Angles.

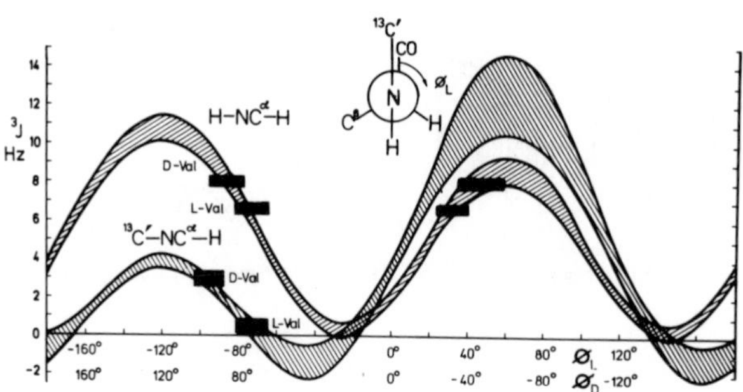


Fig. 5. The Dependence of the Vicinal $^3J(H-NC^{\alpha}-H)$ and $^3J(^{13}C'-NC^{\alpha}-H)$ on the Conformational Parameter Φ. Ordinates of the black rectangles refer to experimentally found vicinal coupling constants of Vm in cyclohexane; their abscissas – to corresponding intervals of possible dihedral angles Φ.

angles of the L–Val residues participating in the β–turns of the valinomycin propeller B, for which we now propose the conformation shown below in Fig. 6.

Hence the structural form of valinomycin in non–polar solvents turns out to be similar to that of the $K^+$–complex which in both the crystalline state [3,4] and solution is in form $A_2$ (see [1], pp. 118–140). The difference is in the orientation of the ester carbonyls which in the free compound point outward and in the complex, inward, and in greater stability of the intramolecular hydrogen bonding system of the latter. The conformational equilibrium of valinomycin in solution in the presence of potassium ions is shown in Fig.6. It is readily seen from the figure that in the A as well as the B conformation the free antibiotic is already capable of directly complexing the metal ion.

Further studies have also yielded a number of new findings concerning the spatial structures of the enniatin depsipeptides. On the one hand, X–ray analysis of crystalline enniatin B [5] has confirmed the earlier [6] conclusion that the so–called P–form with *trans*–methylamide and ester bonds and up–down alternating carbonyl groups (Fig. 7.) participates in the equilibrium interconversion.

The same holds for the enniatin B analog cyclo [–(L–Hylv–D–Hylv)$_3$–] (4) in which the N–methylamide groups have been replaced by ester groups (Fig.8) [7]. On the other hand a NMR study of a series of recently synthesized [8] other enniatin B analogs (5)–(12) in various media ($CDCl_3$, $CD_3OD$ etc.) has revealed [9] the presence of considerable amounts (30–50%) of conformers with *cis* N–methylamide groups, a phenomenon earlier observed in some of the diastereomers of this antibiotic [10], but hitherto unknown for its analogs with the natural LDLDLD set of amino and hydroxy acid configurations. The clearest defined expression of such structural niceties is in the N–methyl signals, as one can see, for instance, in Fig.9. It is hard to say whether the *cis–trans* isomerism of these amide groups plays any significant part in the membrane–affecting function of the enniatins, but one thing is clear, the ready formation of *cis*––methylamide bonds in solution is of prime importance in the equilibrating process of the cyclopeptides.

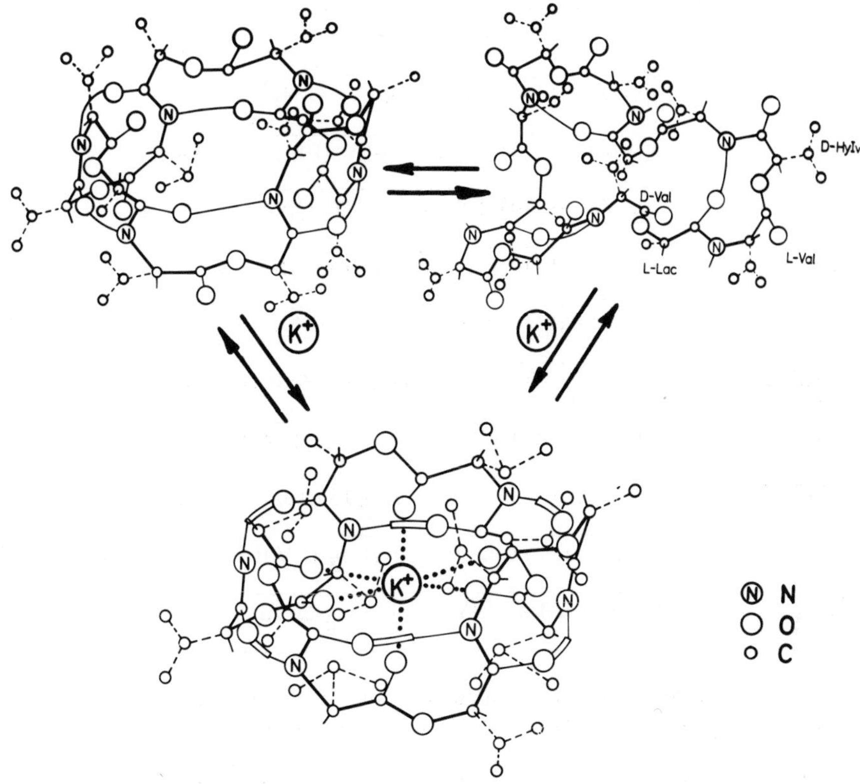


Fig.6. The Conformational Equilibrium of Valinomycin in Potassium Ion Containing Solutions.

Previous studies on the structure – membrane–affecting relations in the valinomycin and enniatin ionophores were centered on the effect of ring size, backbone polar groups, side chain bulkiness and asymmetric center configuration upon the stability of the alkali metal complexes, complexing kinetics, and lipophilicity and surface activity of both the free and complexed depsipeptides and upon the mode of their transmembrane ion transport [1]. This approach left unclear, however, the question of the effect on the ionophore properties of the side chain polarity and the electrical charge on the side chains. To shed light on this aspect of the problem a series of valinomycin and enniatin B analogs (13, 14) and (5, 7, 9 and 11), respectively, which contained either a carboxyl or amino function in the side chain were synthesized and the ability of these compounds to form complexes depending on the ionic state of the side chain functions and their behavior in artificial membrane systems was investigated.

The valinomycin analogs Vm(Lys) and Vm(Glu) differed from the parent compound in the replacement of a L–Val residue by a L–Lys or L–Glu

| No. | Abbr. | Formula |
|---|---|---|
| 5 | En(MeLys) | ⌊(L-MeVal-D-Hylv)$_2$-L-MeLys-D-Hylv⌋ |
| 6 | En(MeLysPht) | ⌊(L-MeVal-D-Hylv)$_2$-L-MeLys(Pht)-D-Hylv⌋ |
| 7 | En(Lys) | ⌊(L-MeVal-D-Hylv)$_2$-L-Lys-D-Hylv⌋ |
| 8 | En(LysPht) | ⌊(L-MeVal-D-Hylv)$_2$-L-Lys(Pht)-D-Hylv⌋ |
| 9 | En(MeGlu) | ⌊(L-MeVal-D-Hylv)$_2$-L-MeGlu-D-Hylv⌋ |
| 10 | En(MeGluBzlNO$_2$) | ⌊(L-MeVal-D-Hylv)$_2$-L-MeGlu(OBzlNO$_2$)-D-Hylv⌋ |
| 11 | En(Glu) | ⌊(L-MeVal-D-Hylv)$_2$-L-Glu-D-Hylv⌋ |
| 12 | En(GluBzlNO$_2$) | ⌊(L-MeVal-D-Hylv)$_2$-L-Glu(OBzlNO$_2$)-D-Hylv⌋ |
| 13 | Vm(Lys) | ⌊(D-Val-L-Lac-L-Val-D-Hylv)$_2$-D-Val-L-Lac-L-Lys-D-Hylv⌋ |
| 14 | Vm(Glu) | ⌊(D-Val-L-Lac-L-Val-D-Hylv)$_2$-D-Val-L-Lac-L-Glu-D-Hylv⌋ |

residue and the enniatin analogs, En(MeLys) and En(MeGlu) by the replacement of a L-MeVal residue by a L-MeGlu or N-MeLys residue.

In contrast to the uni–charged complexes Vm·$M^+$ and En·$M^+$ ($M^+$–stands for an alkali metal) the Vm(Lys)·$M^+$ and En(MeLys)·$M^+$ complexes with their ionized $\varepsilon$–$NH_2$ groups carry a charge of $2^+$ while the complexes Vm(Glu)·$M^+$ and En(MeGlu)·$M^+$ with an ionized $\gamma$–carboxyl of the glutamic acid residue

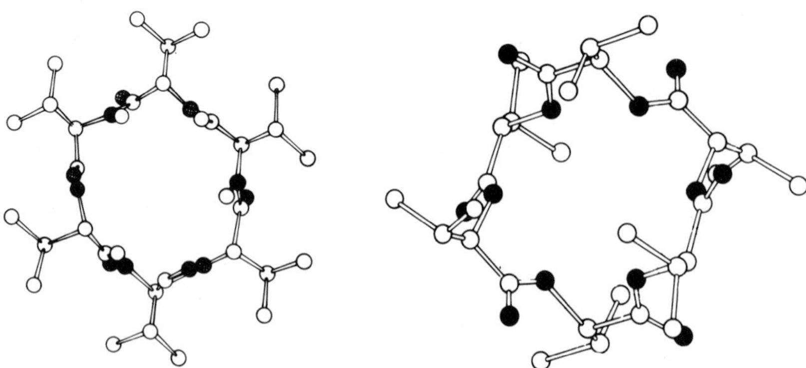

Fig.7. The Conformation of Enniatin B in the Crystalline State.

Fig.8. The Crystalline Conformation of Cyclo[-(L-Hylv-D-Hylv)$_3$-].

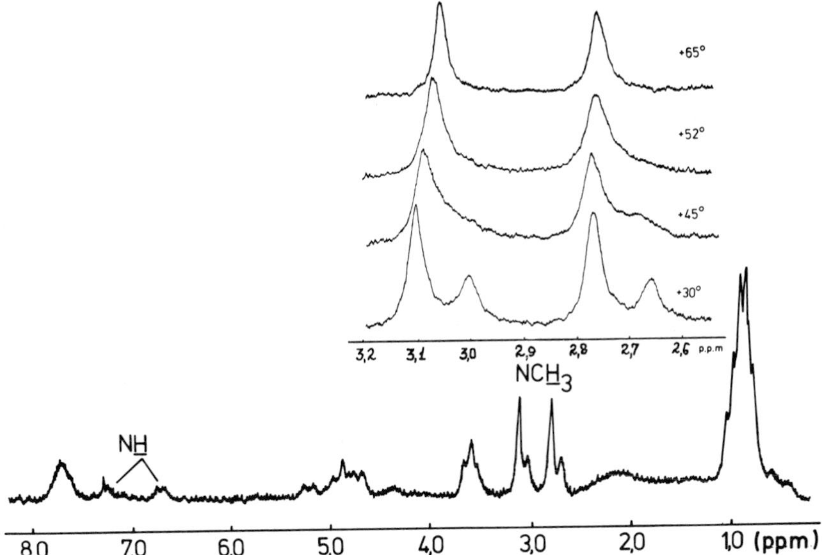


Fig. 9. The NMR-[1]H Spectrum of En(Lys) in $CCl_4$. In the Upper Part of the Figure is Shown the Expanded N-Methyl Region at Different Temperatures.

are electrically neutral. This circumstance should affect the stability of the metal ion complexes, the magnitude of the effect serving as measure of the screening of the metal ion in the molecular cavity from the environment, such that the larger the effect, the less the amount of screening. On the other hand the Vm(Glu) and En(MeGlu) should have properties in common with the nigericin antibiotics which are characterized by the presence of carboxylic groups and the ability to form electrically neutral complexes with metal ions (see [1], pp.198–210). Finally Vm(Lys) and En(MeLys) with their ionized amino groups should be interesting tools for analysis of the modes of action of ionophores in membranes, capable of revealing novel, hitherto inaccessible information.

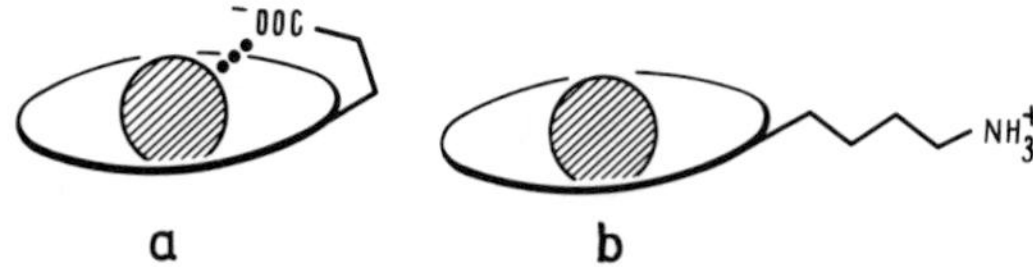


Fig.10. Schematic Representation of the Conformation of Equimolar Complexes of (a) En(MeGlu) and En(Glu) and (b) En(MeLys) and En(Lys) with Alkaline Metal Ions.

The metal–binding properties of compounds (5,7,9,11,13 and 14) were investigated with the side chains in both the neutral and charged states. For this purpose the CD–monitored salt titration of the cyclodepsipeptide was carried out in the presence of 1.2 equivalents of hydrogen chloride or triethylamine. The results obtained are summarized in Table 1. They show that the

Table 1. Stability Constants ($M^{-1}$) of $Na^+$ and $K^+$ Complexes of Valinomycin and Enniatin B Derivatives as Compared with the Parent Compounds

| Compound | Solvent | $Na^+$ | | $K^+$ | | $Na^+$ | $K^+$ | Ref. |
|---|---|---|---|---|---|---|---|---|
| | | HCl 1.2 eq. | $Et_3N$ 1.2 eq. | HCl 1.2 eq. | $Et_3N$ 1.2 eq. | | | |
| 1 Vm | | | | | | | $1.6 \cdot 10^4$ | [11] |
| 13 Vm(Lys) | 80% aqueous ethanol | 10 | 10 | 1900 | 3600 | | | [12] |
| 14 Vm(Glu) | | 10 | 10 | 3300 | 4300 | | | [12] |
| 2 En | | | | | | 340 | 2100 | [6] |
| 5 En(MeLys) | | 50 | 50 | 60 | 110 | | | [12] |
| 7 En(Lys) | 96% aqueous ethanol | 150 | 150 | 80 | 450 | | | [12] |
| 9 En(MeGlu) | | 370 | 360 | 300 | 3400 | | | [12] |
| 11 En(Glu) | | 270 | 600 | 130 | 1200 | | | [12] |

majority of the analogs both with charged or with electrically neutral side chains form complexes with potassium ions that in general are of somewhat lower stability than complexes of valinomycin or enniatin B. Contrary to the enniatin analogs, the valinomycin derivatives reveal no signs of complexing with $Na^+$ thereby retaining the high K/Na selectivity of the parent antibiotic. Noteworthy is the fact that the potassium complexes of the charged form of the enniatin analogs En(MeGlu) and En(Glu) (in the presence of 1.2 eq.$Et_3N$) are markedly more stable than the corresponding complexes in which the carboxyl group is non–ionized (i.e. in the presence of 1.2 eq.HCl). Most likely the ionized carboxyl complexes are additionally stabilized by electrostatic interaction of the bound cation with the carboxylate anion (Fig.10a). This effect confirms an earlier made conclusion that the bound metal ion in the enniatin complexes is not

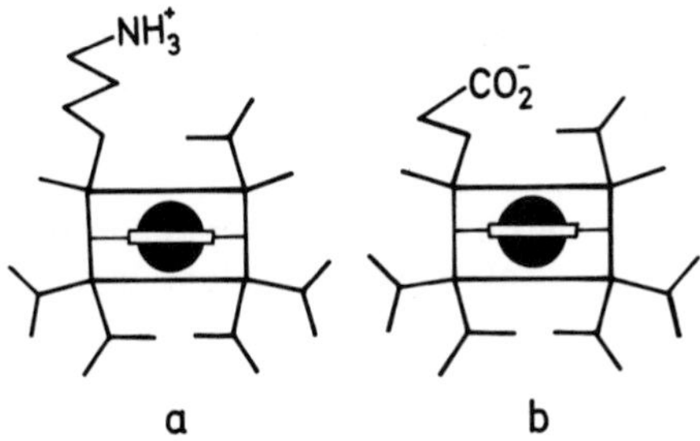


Fig. 11. Schematic Representation of Equimolar Complexes of (a) Vm(Lys) and (b) Vm(Glu).

completely screened from the anion or the solvent [6]. The En(MeLys) and En(Lys) lysine side chains are apparently streatched (Fig.10b), so that its ionization only weakly affects the complex stability. The side chain has also almost no effect on the $K^+$–complex stability of the valinomycin derivatives, in harmony with the effective screening of the centrally located cation from the counterion due to their well known rigid "bracelet" structure (Fig.11).

Judging from the stability of the complexes, these valinomycin and enniatin B derivatives have all prerequisites for ionophoric function. In fact, in the experiments with the lecithin liposomes, Vm(Lys), Vm(Glu), En(MeLys) and En(MeGlu) mediate potassium ion transport along the salt concentration gradient, the rate of potassium chloride efflux from the liposomes in the presence of these compounds as can be seen from Table 2, being commensurate with the corresponding parameters for valinomycin or enniatin B. The liposome experiments were carried out at pH 6.5, i.e. under conditions where one could expect ionization (at least in aqueous solution) of both the lysine $\varepsilon$–$NH_2$ group (pK$\approx$9) and the glutamic acid $\gamma$–COOH group(pK$\approx$3).

One of the characteristic properties of the ionophores is their ability to transfer the cations they bind from the aqueous into the organic phase (see [1], pp.102–109). On the example of Vm(Lys) and Vm(Glu) the effect of the ionic state of the side chains on this process was investigated by measuring by an extraction technique the ability of these compounds to mediate the transition of potassium ions from water into methylene chloride at pH 5, 7 and 9. Picrate ($Pi^-$) was selected as the lipophilic anion because of its high extinction coefficient and the possibility of determining spectrophotometrically its concentration in the organic phase. The $Pi^-$ in the organic phase is due to the complex cation $LM^+ \cdot Pi^-$ where L is the ionophore and is, therefore, equal to the amount of potassium transferred from the aqueous phase. Two series of experiments (A and B) were carried out. In series A conditions for the salt "saturation" of valinomycin, i.e. for the maximum entrance of the potassium ion into the organic phase were established by making up high potassium chloride concentrations in the aqueous phase; in series B, the concentrations and solution volumes were so selected that while there should be, as before, excess $K^+$ and $Pi^-$ with

Table 2. The Effect of Valinomycin, Enniatin B and Their Analogs on the Efflux of Potassium Ions from Liposomes

| No. | Compound | $K^+$ outflow 5 min. after ionophore addition (in % of total $K^+$ content in the liposomes) |
|---|---|---|
| 1 | Vm | 34 |
| 2 | En | 15 |
| 5 | En(MeLys) | 18 |
| 9 | En(MeGlu) | 9 |
| 13 | Vm(Lys) | 29 |
| 14 | Vm(Glu) | 31 |

respect to ionophore, their concentration should yet be insufficient for attaining complete complexation. The series A experiments yielded the stoichiometry of the process. Series B gave the relative capacity of the ionophore to transfer potassium ions into the organic phase. The results obtained are summarized in Table 3 from which it follows that for the very high excesses of salt (series A) the ionophore $K^+$–complexes in the methylene chloride turned out in every case to be equimolar; the pH of the medium had no effect on the valinomycin–mediated extraction of potassium picrate, being always one picrate anion per ionophore molecule. The series B data for Vm(Lys) at pH 7 and 9 and Vm(Glu) at pH 5 and 7 showed that the efficiency of the potassium salt extraction by valinomycin and its analogs diminishes when a polar group is incorporated into the side chain. The sharp increase in $Pi^-$ concentration of the extracted Vm(Lys) on passing from pH 7 to 5 is apparently due to protonation of its amino group and formation of a di–charged complex cation of the type (K+)∿∿$NH_3^+$ coupled with two equivalents of $Pi^-$.

With Vm(Glu) at pH 9, on the other hand, no signs of picrate ion transition into the organic phase were observed, but that a $K^+$–complex did form was established from its breakdown in the chloroform solution on shaking it with aqueous hydrochloric acid (pH 5) and determining the $K^+$ in the aqueous phase by flame photometry. From the data obtained it followed that at pH 9, i.e. with the carboxyl group in an ionized state, Vm(Glu) extracts $K^+$ from water into methylene chloride in the form of a 1:1 macrocycle:cation complex and that the extractable form of the complex is electrically neutral, since the positive charge of the potassium ion is compensated by the negative charge of the carboxylate anion.

Thus, under certain conditions Vm(Glu) acquires properties characteristic of nigericin, monensin and allied antibiotics, a large group of carboxyl–containing ionophores that carry out non–electronogenic cation transport in membrane systems according to the following scheme:

Table 3. Valinomycin and Analogs–mediated Potassium Picrate Extraction from Water into Methylene Chloride

| No. | Compound ($c = 10^{-4}$M) | pH | Concentration of picrate anion in methylene chloride, M($\times 10^4$) A | B | Extractable form of the complex (schematic) | Degree of complexation A | B |
|---|---|---|---|---|---|---|---|
| 1 | Vm | 5 | 1.01 | 0.75 | ⊕ | Quantitative formation of equimolar complex | 0.75 |
| | | 7 | 1.02 | 0.74 | | | |
| | | 9 | 0.99 | 0.76 | | | |
| 13 | Vm(Lys) | 5 | 1.98 | 0.82 | ⊕~~$NH_3^+$ | | 0.41 |
| | | 7 | 1.01 | 0.46 | } ⊕~~$NH_2$ | | |
| | | 9 | 1.00 | 0.45 | | | 0.45 |
| 4 | Vm(Glu) | 5 | 0.98 | 0.51 | } ⊕~~COOH | | 0.50 |
| | | 7 | 0.99 | 0.49 | | | |
| | | 9 | 0.01 | 0.01 | ⊕~~$COO^-$ | | 0.56 |

A typical test for the ability of nigericin ionophores to mediate $M^+ \rightleftarrows H^+$ exchange in artificial membrane systems is to carry out conductance experiments in a U–tube (see [1] , p. 276). In such a system neutral ionophores transport metal ions only in the presence of lipophilic anions while the nigericin group of antibiotics is capable of exchanging metal cations for hydrogen. In view of the above said with respect to Vm(Glu), we tested it by the U–tube method, covering the bottom with a solution of Vm(Glu) in methylene chloride; to this was added water in the left arm, and 1 M aqueous potassium chloride solution in the right arm. Runs were made at differing initial pH values of the aqueous phase (pH 5, 7 and 9). No changes in pH occurred in the course of a week at

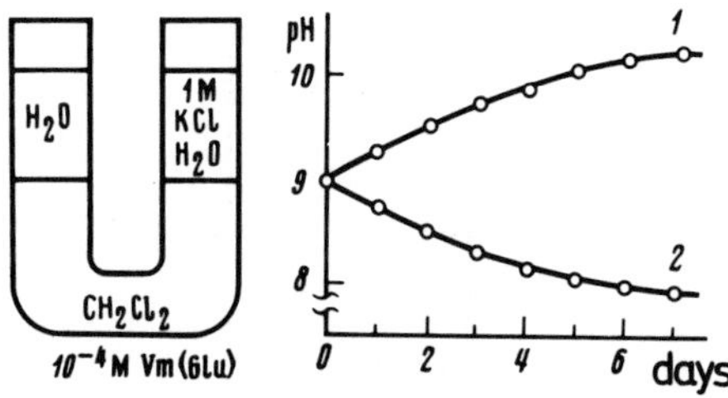


Fig.12. The Rate of $M^+ \rightarrow H^+$ Exchange across a Vm(Glu)–containing Liquid Membrane. Curve 1 Shows the Change in pH in the Left Arm and Curve 2 in the Right Arm of the U–Tube.

pH 5 and 7. At pH 9 the right hand side of the U–tube became acidified and the left hand side, alkaline (Fig.12), i.e. $K^+$ ions were replaced by $H^+$. Consequently the U–tube experiment provided additional proof of the observation that in neutral and acid solutions Vm(Glu) is a conventional neutral ionophore (of the type of valinomycin, the enniatins or the nactins) performing only electrogenic cation transport, whereas as the pH is increased non-electrogenic transport enters the scene and the analog starts behaving like an antibiotic of the nigericin group.

Thus, a unique type of membrane-active complexones has been synthesized and characterized. The compounds are in many respects , spatial structure, metal-binding properties, lipophilicity of the $K^+$–complexes etc., similar to the naturally occurring depsipeptide ionophores; but at given pH values they differ from the latter in being electrically non-neutral. As a result they have widened the area of application of the metal-binding complexones for studying ion transport processes across biological and artificial membranes.

Illustration of the possibilities uncovered here can serve the behavior of Vm(Lys) in a bilayer under ε–amino protonating conditions. In conformity with the theoretical model of Markin, Liberman et al. [13] one could have expected a negative slope in some region of the steady state current–voltage curve, because with increase in electric field intensity the ◯∿∿$NH_3^+$ cations are "pressed in" towards the cathodically polarizable boundary of the membrane and are thereby excluded from the potassium ion transport cycle (Fig. 13). Hence the intra-membrane flow of $K^+$–free ionophore molecules should diminish as the applied voltage is increased and this should decrease the rate of ion transport across the anodically-polarizable boundary of the bilayer. Actually in conventional membranes from bovine brain phospholipids one does not observe such "falling" characteristics perhaps due to the fact that the protonated molecules contribute little to the process owing to the high positive potential jump across the membrane boundary [12] , but they do appear when a "dipolar modifier" (1-methyltetrachlorobenzimidazole + $Cu^{2+}$ ) is added to the membrane so as to lower the positive potential difference between the membrane and the aqueous bathing solution (Fig. 14). Thus this derivative served as the means for experimental substantiation of the model.

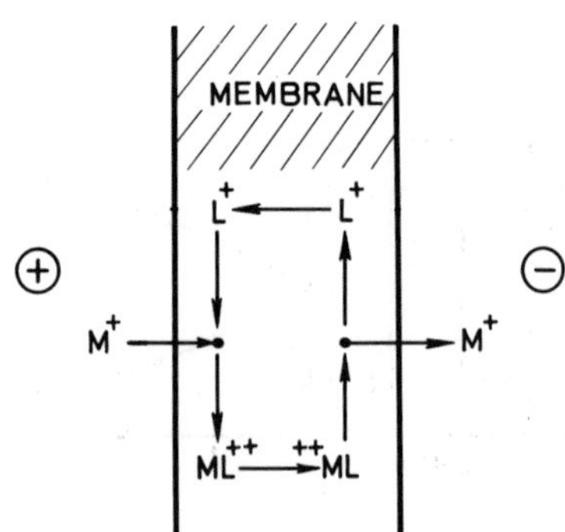


Fig.13. Model of Transmembrane Cation Transport by Di–charged Complexes.

The behavior of the enniatins and especially of valinomycin in various membrane systems has been extensively investigated. However, in these investigations use was made largely of electrochemical methods, which do not give direct answer to such mechanistically important questions as location of the ionophore in the membrane, its manner of interaction with the lipid and the effect on this interaction of cations, of the nature and state of the lipid, etc. Such questions could be resolved by spin or fluorescence labeling of the ionophores, thereby permitting a straightforward determination of their mobility and their immediate environment by the methods of electron spin resonance (ESR) or fluorescence spectroscopy. Of considerable consequence is that in this way high dilutions can be used and the experiments can be made with heterogeneous in particular biological systems over a wide range of experimental conditions. A convincing demonstration of the power of such an approach has been given by Veatch et al. [14] who prepared the dansyl derivative of gramicidin C and studied its behavior in bilayers.

With this in mind we synthesized spin and fluorescence labeled valinomycin derivatives (15 and 16) and enniatin B derivatives (17 - 20) issuing from analogs (5,7 and 13) containing a free amino group.

Experiments with liposomes showed that compounds (15 - 20) were on a par with valinomycin or enniatin B in their trans-membrane potassium ion transport capacities (Table 4), in other words the insertion of spin or fluorescence labels into the molecules does not interfere with their ionophoric activity.

Taking advantage of the spin labels in compounds (15, 17, 18) we used ESR spectroscopy to study their metal-binding properties. In all cases when potassium thiocyanate was added to a solution of the depsipeptide in 96% aqueous alcohol there was a fall in the signal intensity (J, see Fig. 15) and an increase in the correlation time ($\tau_c$) indicating that complexing lowers the mobility of the radical-con-

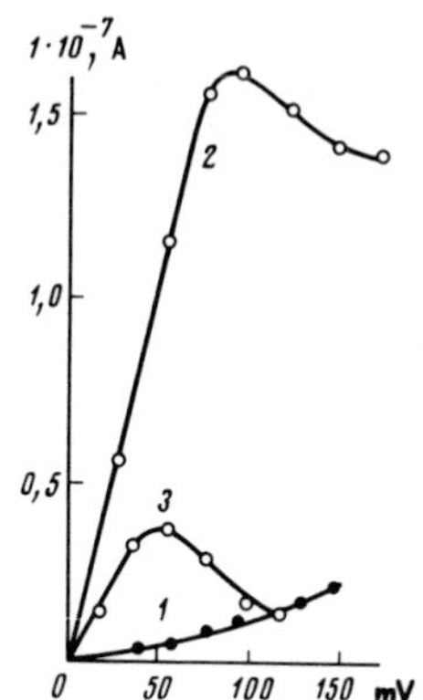


Fig.14. The Effect of 1-Methyltetrachlorobenzimidazole (MTB) on the Current--Voltage Characteristics of Membranes in Solutions Containing $1.3 \cdot 10^{-6}$ Vm(Lys), $1 \cdot 10^{-5}$M $CuSO_4$, $5 \cdot 10^{-5}$M Hydroquinone and Differing Amounts of KCl.(1) 1M KCl (Reference); (2) 1 M KCl + 1M MTB; (3) 3M KCl + $10^{-5}$M MTB; pH 3.5±0.2.

| No | Abbrev. | Formula |
|---|---|---|
| 15 | Vm(LysSL) | ⌊(D-Val-L-Lac-L-Val-D-Hylv)$_2$-D-Val-L-Lac-L-Lys(SL)-D-Hylv⌋ |
| 16 | Vm(LysDns) | ⌊(D-Val-L-Lac-L-Val-D-Hylv)$_2$-D-Val-L-Lac-L-Lys(Dns)-D-Hylv⌋ |
| 17 | En(MeLysSL) | ⌊(L-MeVal-D-Hylv)$_2$-L-MeLys(SL)-D-Hylv⌋ |
| 18 | En(LysSL) | ⌊(L-MeVal-D-Hylv)$_2$-L-Lys(SL)-D-Hylv⌋ |
| 19 | En(MeLysDns) | ⌊(L-MeVal-D-Hylv)$_2$-L-MeLys(Dns)-D-Hylv⌋ |
| 20 | En(LysDns) | ⌊(L-MeVal-D-Hylv)$_2$-L-Lys(Dns)-D-Hylv⌋ |

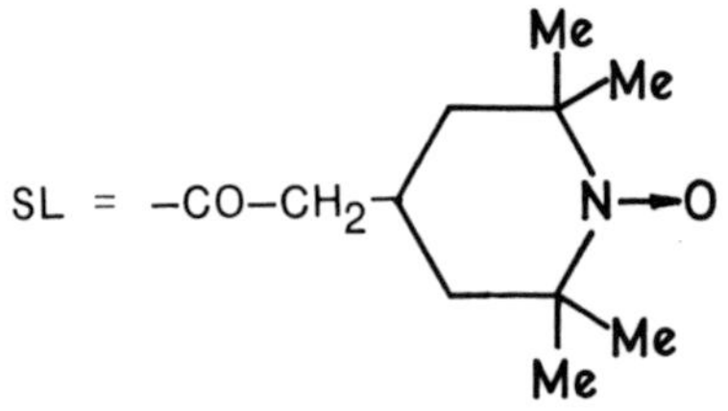

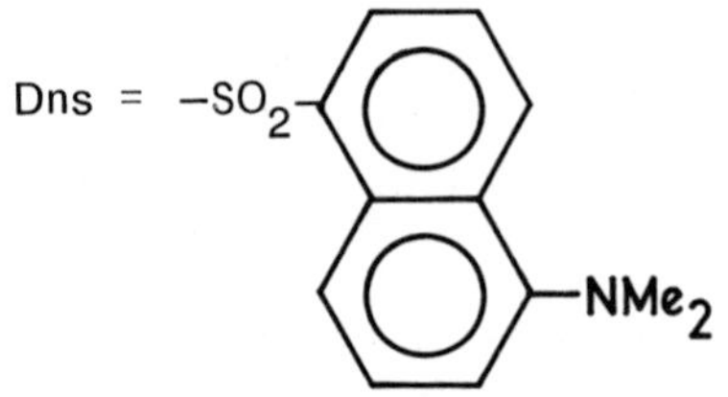

SL = $-CO-CH_2-$ (2,2,6,6-tetramethylpiperidine ring, N→O; Me, Me, Me, Me)

Dns = $-SO_2-$ (naphthalene ring, $-NMe_2$)

Table 4. The Effect of Valinomycin, Enniatin B and Their Spin and Fluorescence Labeled Analogs on the Efflux of Potassium Ions from Egg Lecithin Liposomes

| No. | Compound ($c = 10^{-4}$M) | Amount of released $K^+$ in % of initial KCl content after 5 min at 25° |
|---|---|---|
| 1 | Vm | 34 |
| 2 | En | 15 |
| 15 | Vm(LysSL) | 35 |
| 16 | Vm(LysDns) | 32 |
| 17 | En(MeLysSL) | 16 |
| 18 | En(LysSL) | 17 |
| 19 | En(MeLysDns) | 14 |
| 20 | En(LysDns) | 15 |

Table 5. Rotational Correlation Time ($\tau_c$) for Spin–labeled Cyclodepsipeptides and Their $K^+$–Complexes

| No. | Compound | $\tau_c \cdot 10^{11}$ sec | |
|---|---|---|---|
| | | Free cyclodepsipeptide | $K^+$–Complex |
| 15 | Vm(LysSL) | 2.5 | 3.0 |
| 17 | En(MeLysSL) | 2.0 | 4.0 |
| 18 | En(LysSL) | 2.3 | 4.5 |

taining fragment (Table 5). The stability constants determined from the titration curves for the complexes (Fig. 15b) are 18000 $M^{-1}$ for Vm(LysSL), 300 $M^{-1}$ for En(MeLysSL) and 1500 $M^{-1}$ for En(LysSL). Judging from the $\tau_c$ value, the nitroxyl radical is more immobilized in the En(MeLysSL) and particularly in the En(LysSL) complexes than in the Vm(LysSL) complex. In the case of En(MeLysSL) this result could be taken as indicating possible interaction of the weakly screened cation in the enniatin complexes with the nitroxyl O atom (by analogy with the En(MeGlu)·$K^+$–complex see Fig.10a).

In the Vm(LysSL)·$K^+$–complex the cation is much more effectively screened from the environment and so does not interact with the nitroxyl moiety. In En(LysSL) on the other hand accessibility of the nitroxyl radical to the cation is apparently additionally favored by the replacement of the N–methyl group by an NH group. This can also explain the higher stability of the En(LysSL) than the En(MeLysSL) complex.

The high metal binding and ionophoric activity of the spin–labeled derivatives combined with the high sensitivity of their ESR spectra to subtle changes of molecular structure demonstrate their considerable potentiality in the study of membrane systems.

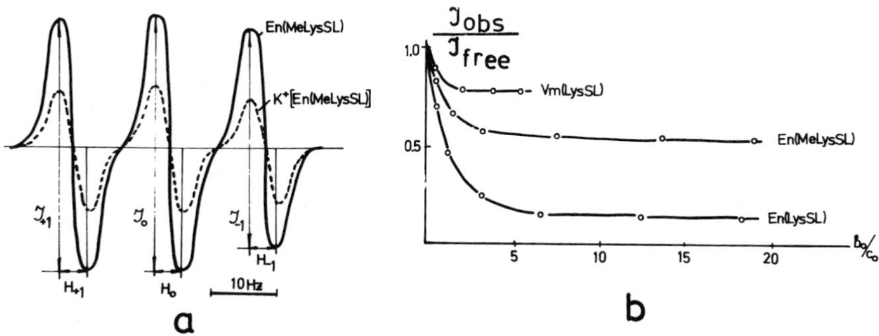


Fig.15. (a) ESR Spectra of Compound (17) (continuous line) and of its $K^+$–Complex (dotted line) in 96% Aqueous Ethanol and (b) Potassium Thiocyanate Titration of the Spin Labeled Analogs in 96% Aqueous Ethanol.

Table 6. Fluorescence spectral data for Vm(LysDns) and DnsLys ($\lambda_{exc.}$ = 340nm)

| Solvent* (solute concentration 1.3 $10^{-6}$M) | Vm(LysDns)* | | | DnsLeu | | |
|---|---|---|---|---|---|---|
| | $\lambda_{max}$ | Q | P | $\lambda_{max}$ | Q | P |
| Heptane–ethanol (99:1) | 460 | 0.23 | <1% | 460 | 0.22 | <1% |
| Ethanol | 500, 520 | 0.14 | 1.5% | 500, 520 | 0.18 | 1.5% |
| Water–ethanol (99:1) | 460 | 0.016 | 31–37% | 550 | 0.20 | 1.5% |
| Egg lecithin liposomes | 500, 520 | 0.13 | 10–15% | – | – | – |

* On adding 1M KCl solution to the aqueous Vm(LysDns) solution P=50%; in all other cases the salt addition did not affect the $\lambda_{max}$, Q and P values.

The dansyl derivative of valinomycin (16) was investigated by means of fluorescence spectroscopy. The position of $\lambda_{max}$ , the quantum yield (Q) and the fluorescence polarization (P) were measured in solvents of varying polarity for Vm(LysDns), both free and its $K^+$ -complex. The results were compared with those for the simple model, dansyl-L-leucine (DnsLeu) in the same solutions. They are summarized in Table 6.

As might have been expected, on passing from heptane to ethanol the fluorescence peak shifts towards longer wavelengths and the quantum yield of fluorescence diminishes to some extent. The behavior of Vm(LysDns) was always very similar to that of the model leucine derivative.

Further augmentation of solvent polarity (aqueous solutions) leads in the case of DnsLys to still larger shifts of $\lambda_{max}$ to longer wavelengths and to considerable decrease in the Q values. With Vm(LysDns), although the quantum yield was normal for water, the observed $\lambda_{max}$ was characteristic for heptane pointing out, we believe, to association of the depsipeptide under such conditions with the formation of a micellar solution. Support for such an assumption can be seen in the anomalously high polarization of fluorescence.

The Vm(LysDns) was further studied in an artificial membrane system, composed of egg lecithin liposomes. Its $\lambda_{max}$ and Q values indicated that the dansyl label under such conditions must be in a highly polar environment; i.e. it must be localized in the polar head region of the lipid, rather than in the interior of the membrane. Hence, it may be considered likely that the ionophore backbone must also be situated quite near the membrane surface. One such possible disposition is shown in Fig.16. Taking into account the similar ion-carrying properties of Vm(LysDns) and of the parent valinomycin it would be only natural to conclude that the latter should be similarly located in the membrane. This would be in harmony with the high surfactant properties of valinomycin observed earlier in this laboratory [15] and with the CD measurements on this compound in dimyristyllecithin liposomes [16] . The fluorescence spectra of Vm(LysDns) proved to be little sensitive to the $K^+$ complexing in ethanol (Table 6) indicating the invariability of the label's environment under

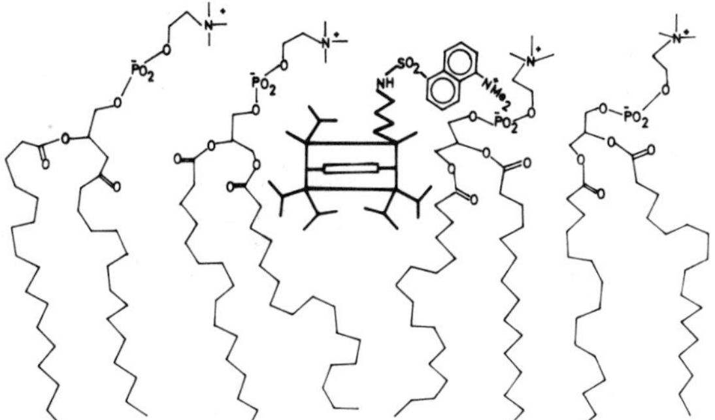

Fig.16. The Preferred Dislocation of Vm(LysDns) in Liposome Membranes (Schematically).

such conditions. In water the degree of polarization was found to increase, apparently manifesting further association of the depsipeptide and providing additional evidence of the high lipophilicity of the complex cation. As in alcohol no significant spectral changes were observed in the liposome suspensions on adding potassium chloride, i.e. under conditions of intensive transmembrane ion transport (see Table 4). Consequently the diffusion time of the complex ion of Vm(LysDns) within the membrane must be small compared with sojourn of the ion near the surface, a conclusion which must be borne in mind when analyzing the shapes and magnitudes of the energy barriers in the transmembrane transport process.

Recently it has been discovered in this laboratory that valinomycin can form 2 : 1 macrocycle : cation complexes in the form of "sandwiches", as well as the conventional 1 : 1 complex, the former much less stable than the latter. A hypothetical structure was suggested for these sandwiches (Fig.17) and it was conjectured that they may participate in the transmembrane ion transport process [17, 18]. It occurred to us that "sandwiching" might be facilitated if two depsipeptide rings were joined together by a sufficiently large chain that by folding could bring the rings into a position favorable for both to interact with a single ion, so to say intramolecular "sandwiching".

Accordingly, from the functional derivatives Vm(Lys) and Vm(Glu) we prepared the bis-valinomycin analog Vm(Lys)–Vm(Glu) (21) by acyl chloride condensation [9]. However, analysis of its potassium chloride titration curve in ethanol (controlled by CD spectroscopy) showed that the rings behave independently, each binding its own ion to form a di-ionic complex of low stability (370 ± 50 M ).

The same holds for extraction experiments. The analog (21) like the monomeric analog Vm(LysPht) (22), taken for comparison, mediated the stoichiometric (per ion binding center) extraction of potassium picrate from water into methylene chloride over a large range of concentrations. In other words both ion binding centers of bis-valinomycin are capable of solubilizing in a nonpolar medium potassium ions with an efficiency indistinguishable under the experimental conditions from the efficiency of the nearest analog with a single center (Table 7).

An explanation for this follows from an examination of molecular models from which it can be seen that arrangement of the two rings of the bis-analog into a conformation like that shown in Fig.17. Thus it seems to be impossible for the rings to assume such a position

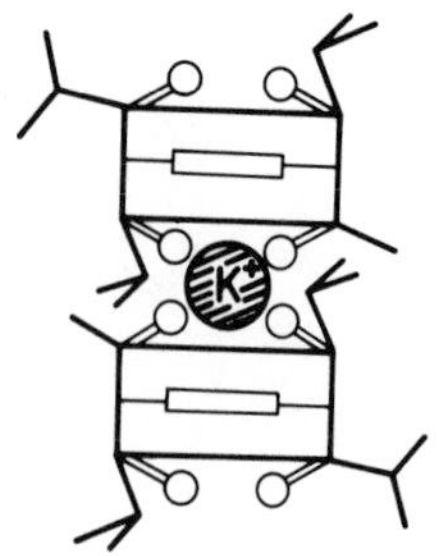


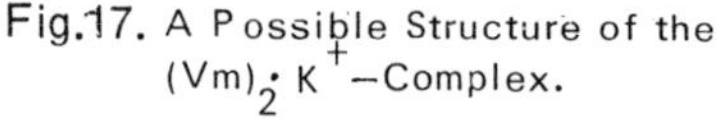
Fig.17. A Possible Structure of the $(Vm)_2 \cdot K^+$ –Complex.

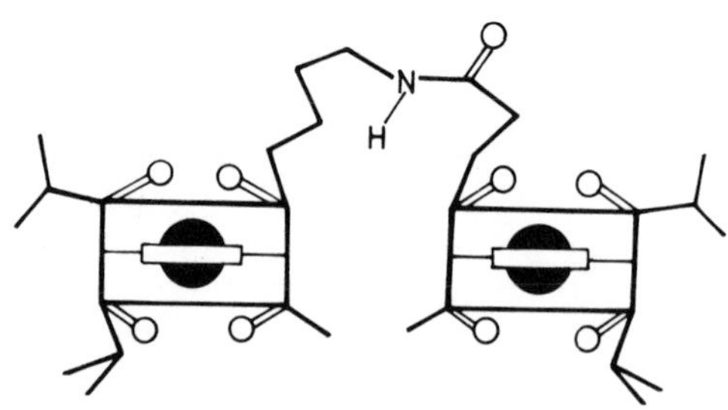


Fig.18. Proposed Structure of the $[Vm(Lys)-Vm(Glu)] \cdot (K^+)_2$ –Complex.

that the cation undergoing complexation could interact with the carbonyl groups on the sterically least hindered "lactyl" ends of the "cylinders" where the lactic acid residues are located (lower ends in Fig.18).

Just recently we have synthesized a new series of valinomycin analogs (23 – 26).

$\overline{\text{(D-Val-L-Pro-L-Val-D-Hylv)}_n}$

n = 3 (23)

n = 4 (24)

$\overline{\text{(D-Val-L-Lac-L-Val-D-Pro)}_n}$

n = 3 (25)

n = 4 (26)

Table 7. Extraction of Potassium Picrate from Water into Methylene Chloride Containing $10^{-4}$ M Ionophore

| No. | Compound | Picrate anion concentration in methylene chloride, $M \cdot 10^4$ | |
|---|---|---|---|
| | | 1M KCl | $5 \cdot 10^{-4}$ M KCl |
| 21 | Vm(Glu)–Vm(Lys) | 2.01 | 2.08 |
| 22 | Vm(LysPht) | 1.04 | 1.05 |
| 27 | En(MeGlu)–En(MeLys) | 0.99 | – |

The compounds were qualitatively shown to all form exceptionally stable 2:1 complexes with potassium and cesium ions in 96% ethanol. This is illustrated in Fig.19 by the results of titration of analog (25) with cesium chloride. Thus the formation of double complexes should be considered to be a common property

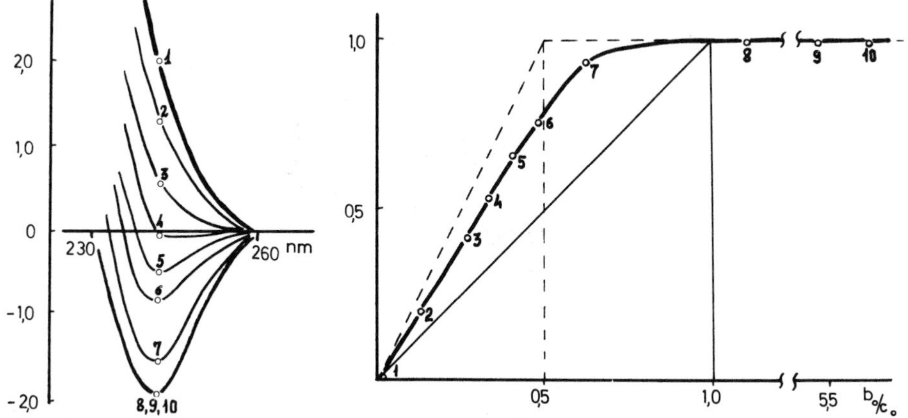


Fig.19. The Change in (a) the CD Curve and (b) the Titration Curve for Compound (25) on Titrating with Cesium Chloride in 96% Aqueous Alcohol. Initial Concentration $10^{-1}$M, b/c Cyclodepsipeptide/Salt Molar Ratios.

of the members of the valinomycin group of depsipeptides. We are now in the progress of a quantitative study of this phenomenon for different solutions. Its implications for the transmembrane transport of the ions as yet remain obscure.

Quite the opposite holds for the enniatin antibiotics, where there is strong evidence for the "non–stoichiometric" complex playing a major role in their mediation of transmembrane alkali ion transport [6,19,20] . With the advent of the dimeric complexes new support is now available for such mode of transport and they have moreover allowed more light to be shed on the conceptual aspect of the problem. The dimers we used in the enniatin series were En(MeLys) — En(MeGlu) (27) and the desmethyl derivative En(Lys) – En(Glu) (28) synthesized in this institute [9] . Because of the low lipophilicity of (28) extraction experiments were carried out with (27), whereas (28) was subjected to titration with potassium chloride; the small quantities of the more difficultly synthesized (27) precluded its use for titration, Both compounds, however, were of very similar properties, giving grounds for treating them together.

The stability constant of the 1 : 1 "intramolecular sandwich" complex of (28) (Fig.20) in 96% aqueous alcohol turned out to be $(1.4 \pm 0.3) \cdot 10^4$ $M^{-1}$ from the titration experiments showing the complex to be of exceptionally high stability. The same holds for (27), which even at very high potassium chloride concentrations carried over the potassium ions into the organic phase only in equimolar quantities, i.e. in the form of an "intramolecular sandwich".

Of considerable import is that under the same conditions no signs of sandwich formation were observed in the case of sodium ions. Now, it is well known that

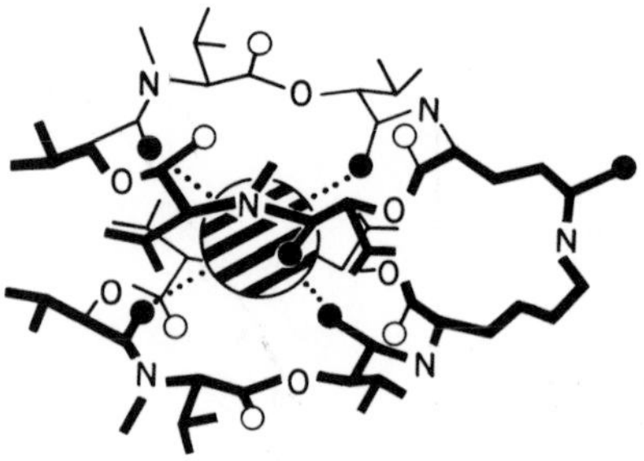

Fig.20. Conformation of the [En(Glu)—En(Lys)]·$K^+$ Complex in Solution.

in contrast to the enniatin complexing reaction with its practically imperceptible K/Na selectivity, enniatin-mediated transmembrane ion transport is markedly K/Na selective [19]. Since sodium ions are not sandwiched even by dimers, one might well conjecture that this ion is transported across membranes by enniatin B in the equimolar form.

The opposite is the case with potassium ions. The high stability of their 2 : 1 complexes with the enniatin dimers, is evidence of a much higher tendency for the potassium ions to form such complexes, a conclusion which should be valid also for the monomers. This, together with the marked K/Na selectivity of the enniatin-mediated transmembrane transport process, in contrast with the low selectivity of the monomer complexing reaction, points out to the high probability of potassium transport of the enniatins across membranes in the form of the double (2:1) complex.

## GRAMICIDIN A AND ITS ANALOGS

It was initially assumed that gramicidin A (Gr A) functions as an ionophore, but Haydon and Hladky [21] have shown that in membranes Gr A forms ion-conducting channels. A fluorescence study by Veatch et al. [14] of the membrane-bound Gr A analog, dansyl Gr C and of its mediated ion conductivity in the membrane led to the conclusion that two Gr A molecules are required for the formation of each "active" ion-conducting channel.

It seemed logical to find some relationship between this mechanism and the fact that Gr A exists in solution as a dimer chromatographically separable into four conformationally different species (1, 2. 3 and 4) [22]. It was, of course, clear that the membrane affecting properties of the antibiotic must be governed by its conformati However, the membrane conformation of Gr A has not been unequivocally established, at least two models, the $\pi_{(L,D)}$-helix [23] and the β–double helix [22,24] having been proposed for it (Fig.21). Dimerization of the $\pi_{(L,D)}$–helix was postulated as occurring by head to head association (although other possibilities were not excluded) [23] in the β–double helix, all hydrogen bonds were assumed to be intermolecular and the chains oriented either parallely or antiparallely.

Species (1) and (2) have very similar CD spectra (see Fig.22) [22], that of (4) is

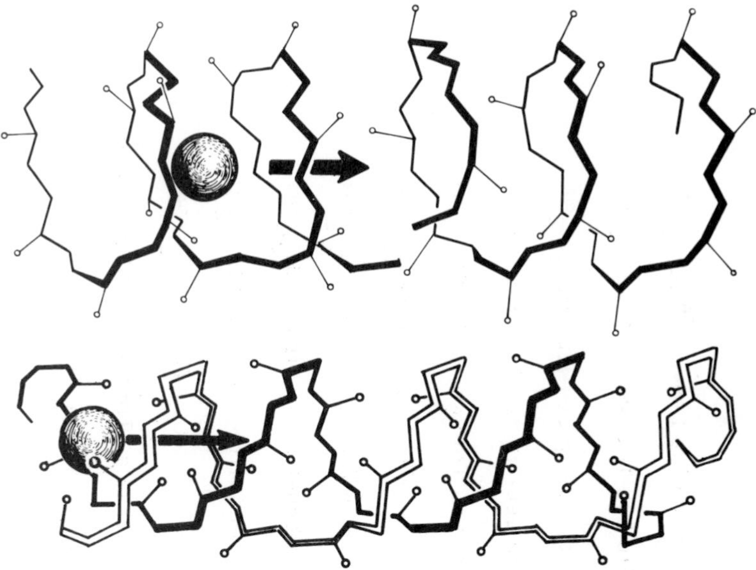

Fig.21. Schematic Representation of the $\pi_{(L,D)}$–Helix (Upper Structure) and the β–Double Helix of Gramicidin A.

approximately their mirror image, while species (3) displays a spectrum differing considerably from the other. All four exhibit a strong amide I band in the IR spectrum at 1633cm$^{-1}$ whereas the IR spectrum of (3) has an additionally resolved component at 1680cm$^{-1}$. Because the elementary cells of the $\pi_{(L,D)}$–helix and the β–double helix differ little in symmetry they should display similar IR–bands, a fact which made the authors retain both possible structures without choosing between them.

Other points which are obscure in the Gr A behavior are the relation between the primary structure of the antibiotic and its associative ability, conformational states, and membrane–affecting properties.

Thus, although much progress has been achieved in elucidation of the Gr A properties much yet awaits an answer. Considerable responsibility for the unclarities lies in the dearth of experimental data that could lend unequivocal support to or provide sound reason for rejecting the models proposed.

As contribution to filling in this gap, we have synthesized a number of Gr A analogs [25], differing from the parent antibiotic in chain length (compounds (29) – (35), (38) and (39)) in the L and D sequence of the amino acid residues (compounds (37) and (38)) and in some of the residues themselves (compound (36)). The analogs were then investigated with respect to their associative capacity by determining the concentrational dependence of the CD curves and the fluorescence polarization (FP) over the maximum possible concentration range ($10^{-2}$ – $10^{-5}$ M) in ethanol and dioxane (see Table 8).

| No. | Formula |
|---|---|
| 3 | HCO-Val-Gly-Ala-Leu-Ala-Val-Val-Val-Trp-Leu-Trp-Leu-Trp-Leu-Trp-NH$(CH_2)_2$OH |
| 29 | HCO-Val-Gly-Ala-Leu-Ala-Val-(— — )-Trp-Leu-Trp-Leu-Trp-Leu-Trp-NH$(CH_2)_2$OH |
| 30 | HCO-Val-Gly-Ala-Leu-Ala-Val-Val-Val-( — — )-Trp-Leu-Trp-Leu-Trp-NH$(CH_2)_2$OH |
| 31 | HCO-Val-Gly-(— — — — )-Val-Val-Trp-Leu-Trp-Leu-Trp-Leu-Trp-NH$(CH_2)_2$OH |
| 32 | HCO-Val-Gly-Ala-Leu-Ala-( — — — — )-Leu-Trp-Leu-Trp-Leu-Trp-NH$(CH_2)_2$OH |
| 33 | HCO-Val-Gly-Ala-Leu-Ala-Val-Val-Val-( — — — — )-Trp-Leu-Trp-NH$(CH_2)_2$OH |
| 34 | HCO-Val-Gly-(— — — — — —)-Trp-Leu-Trp-Leu-Trp -Leu-Trp-NH$(CH_2)_2$OH |
| 35 | HCO-Val-Gly-Ala-Leu-Ala-(— — — — — —)-Leu-Trp-Leu-Trp-NH$(CH_2)_2$OH |
| 36 | HCO-Val-Gly-Val-Leu-Val-Leu-Val-Leu-Trp-Leu-Trp-Leu-Trp-Leu-Trp-NH$(CH_2)_2$OH |
| 37 | HCO-Val-Gly-Ala-Leu-Ala-Val-Val-Val-Gly-Trp-Trp-Leu-Trp-Leu-Trp-NH$(CH_2)_2$OH |
| 38 | HCO-Val-Gly-Ala-Leu-Ala-Val-Val-Val-Gly-Trp-Leu-Trp-Leu-Trp-Leu-Trp-NH$(CH_2)_2$OH |
| 39 | HCO - Leu-Val-Gly-Ala-Leu-Ala-Val-Val-Val-Trp-Leu-Trp-Leu-Trp-Leu-Trp-NH$(CH_2)_2$OH |

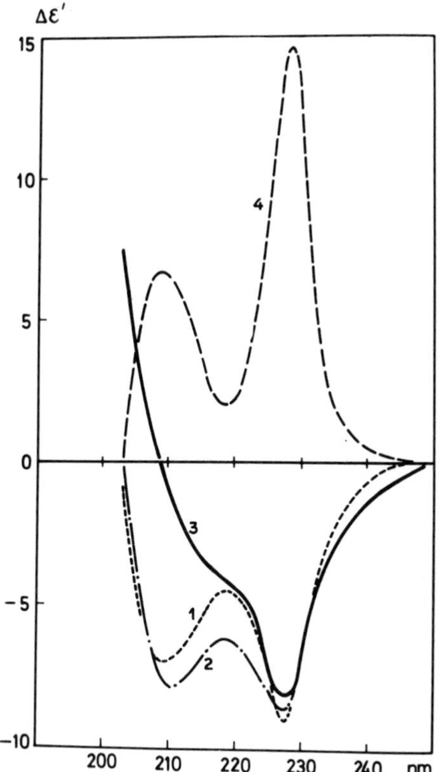


Fig.22. CD Curves of the Individual Gr A Dimeric Forms in Dioxane [22]. The Numbers on the Curves Designate the Corresponding Species.

The CD spectral dependence on concentration for some of the analogs are shown in Figs.23, 24, the spectra always being obtained under equilibrium conditions. The figures also show the differential dichroic absorption at $\lambda$ =228 nm vs. lg $c$, where $c$ is the overall concentration of the antibiotic.

In the majority of cases, an increase in the dimer concentration is accompanied by changes in the CD curves, thus providing the means for following the dimerization process.

The CD parameters also provide a clue to both the monomer and the dimer conformations. Generally speaking, a clear-cut relationship between the CD curves and the chain length or the nature of the displaced residue can be observed only for some, not all the analogs. A comparison of the CD spectra of the analog dimers (the most favorable dimerizing conditions being concentrated solutions in dioxane) with those of the separate Gr A dimeric species showed that in certain analogs the conformational equilibrium is shifted towards one or the other of these species. From this standpoint the analogs can be divided into four groups (Table 8). In one, consisting of analogs (30)

Table 8. Stability Constants ($K_{dimer}, M^{-1}$) of Gr A Analog Dimers

| | No. | Solvent | Dimerization Constant | |
|---|---|---|---|---|
| | | | CD | FP |
| species 3 | 30 | ethanol | $(3.0 \pm 0.4) \cdot 10^2$ | $(3.5 \pm 1) \cdot 10^2$ |
| | | dioxane | — | $(2.0 \pm 0.5) \cdot 10^3$ |
| | 33 | ethanol | $(1.4 \pm 0.4) \cdot 10^2$ | $(2 \pm 4) \cdot 10^2$ |
| | | dioxane | $(9.6 \pm 0.5) \cdot 10^4$ | $\sim 10^5$ |
| species 4 | 36 | ethanol | $(3 \pm 2) \cdot 10^4$ | $(9 \pm 4) \cdot 10^3$ |
| | | dioxane | $\sim 10^5$ | $(4.0 \pm 1.5) \cdot 10^4$ |
| | 31 | ethanol | $(3 \pm 1) \cdot 10^2$ | $\sim 10^3$ |
| | | dioxane | $\sim 10^4$ | $\sim 10^4$ |
| | 34 | ethanol | $(1 \pm 0.5) \cdot 10^2$ | $\sim 10^2$ |
| | | dioxane | $(3.5 \pm 2) \cdot 10^3$ | $(2.6 \pm 2) \cdot 10^3$ |
| see text | 29 | ethanol | $(2.5 \pm 0.5) \cdot 10^2$ | $(3 \pm 1) \cdot 10^2$ |
| | | dioxane | – | $(5.5 \pm 0.5) \cdot 10^2$ |
| | 39 | ethanol | $(3.2 \pm 0.7) \cdot 10^2$ | $(3.8 \pm 0.5) \cdot 10^2$ |
| | | dioxane | – | $(2 \pm 1) \cdot 10^2$ |
| see text | 38 | ethanol | – | $\lesssim 10^3$ |
| | | dioxane | – | – |
| | 32 | ethanol | $< 10^2$ | $\lesssim 10^2$ |
| | | dioxane | $\sim 10^3$ | $\sim 3 \cdot 10^3$ |
| | 35 | ethanol | $\lesssim 10^2$ | $\lesssim 10^2$ |
| | | dioxane | – | – |

and (33), the CD spectra are close to that of the Gr A species (3) (Fig.23). Of particular interest here is analog (33) since its CD spectrum at concentrations for complete dimerization in dioxane can be described in terms of a combination of all four individual Gr A species, with species (3) amounting to 65% of the total.

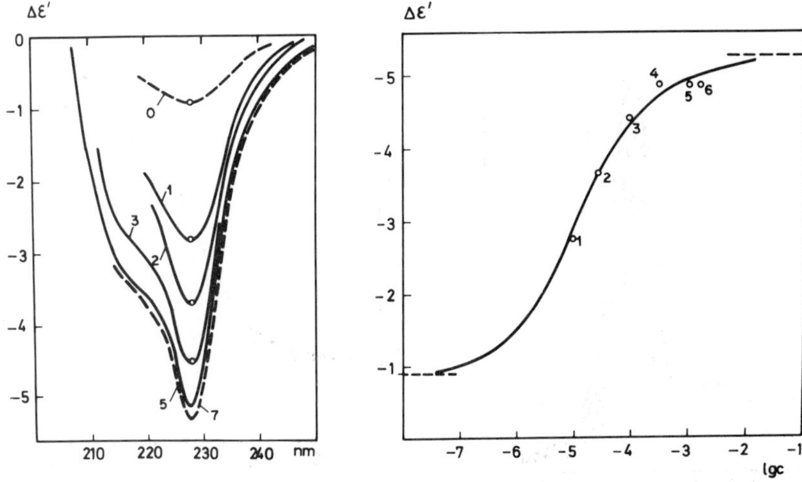


Fig.23. Concentrational Dependence of (a) – the CD Curves and (b)–Δε' for λ= 228 nm of Analog (33) in Dioxane. The Dashed Lines are for the Monomer (O) and for the Dimer (7).

A second group whose CD spectra closely resemble that of Gr A species (4) comprises analogs (31), (34) and (36), the most interesting being analog (36) which possesses the species (4) conformation to the extent of ca. 80%.

The two analogs (29) and (39) comprise a third group whose CD spectra resemble those of the Gr A species (1) and (2) but are about half their amplitude. These spectra cannot be described as a linear combination of the spectra of the four Gr A species.

The last group contains analogs (32), (38) and (35), characterized by the fact that the CD amplitudes are less than 0.2 those characteristic of the Gr A dimers (Fig.22).

A comparison of the shift in the analog dimer equilibrium with the structure of the monomers uncovers a number of interesting facts. Thus, decrease in the relative Trp content is paralleled by augmentation of the species (3) content. In group 2 (Table 8) with predominant species (4) conformations, on the contrary, the Trp content is higher than for the other analogs. Moreover, in analog (36) Ala has been replaced by Val and Val by Leu, i.e. the less bulky hydrophobic residues have been replaced by more bulky ones. This all evidences that an essential part in stabilization of species (4), is played not only by H-bonds, but also by hydrophobic interaction.

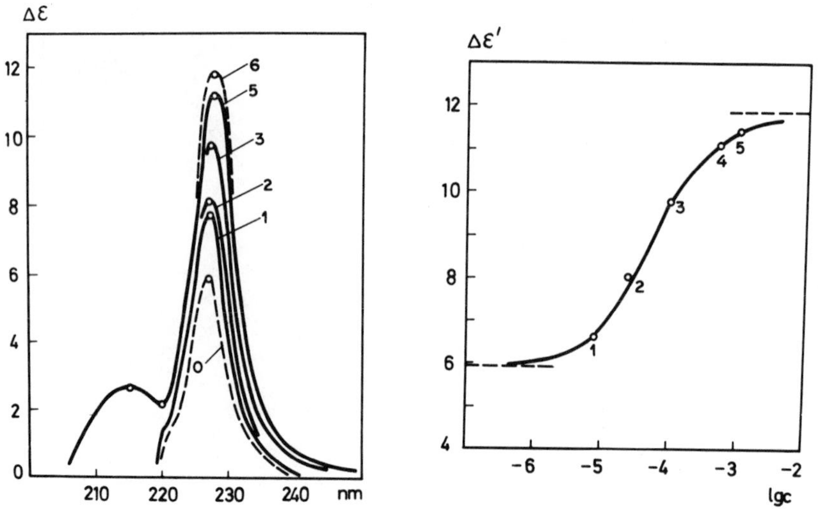


Fig.24. Concentrational Dependence of (a) – the CD Curves and (b) – Δε' at λ= 228 nm for Analog (31) in Dioxane. The Dashed Lines are for the Monomer (O) and the Dimer (6).

Let us now, in the light of the new data presented here, consider the effect of Gr A and its synthesized analogs on the ion conductance of lipid bilayers.

It is well known that in the presence of very small amounts of Gr A the membrane current at constant voltage undergoes jumps, indicating the "opening" and "closing" of the conductance channels. At the same time the channels have definite conductance values, of which some are realized more often than others. In egg lecithin membranes Gr A forms four basic types of channels differing in both conductance and dwell times. It would, therefore, seem quite logical to relate these types with the four dimer species we have discussed above. Grounds for this may be found in the fact that with the thirteen - membered analog (30) which in solution exists mainly in the form of a single type of dimer, we have observed mainly a single type of channel with a conductance of 7 pmho. At the same time, with another thirteen - membered analog (29) whose dimeric forms in solution differ from the Gr A species, the histogram characterizing the conductance distribution of the channel displays a wide range of values. The fifteen - membered analog (36) was also found to be active in bilayers but its individual channels are characterized by a much larger range of conductances than might have been expected since its solution contains predominantly a single dimer species. True, whereas there are several different conducting types of this analog in the bilayer, the differences between them are much less than in the case of the naturally occurring Gr A and we are inclined to ascribe them to small conformational differences that could have escaped detection in solution.

Other indirect support of the proposal advanced here may be seen in the fact that the CD curves of analog (38) which displays no conductance channels in bilayer membranes are of low intensity, evidence of its lack of preferable conformations in solution. The fact that in this analog the L,D alteration of the asymmetric centers is violated shows that such ordering must be required for the formation of the Gr A type channels.

Thus, the correlation which apparently exists between the number of different solution dimers and the number of different types of conductance channels serves as argument in favor of the concept that the conductance channels of the antibiotic are formed by means of its dimers and allows one to penetrate deeper into the conductance mechanism. Apparently, the dimer species does not seem to be crucial, but this requires further study.

## REFERENCES


1. Yu.A.Ovchinnikov, V.T.Ivanov, A.M.Shkrob. Membrane Active Complexones, Elsevier, Amsterdam, 1974.
2. V.F.Bystrov, Yu.D.Gavrilov, V.T.Ivanov, Yu.A.Ovchinnikov, in preparation.
3. M.Pinkerton, L.K.Steinrauf, P.Dawkins, Biochem.Biophys.Res.Communs, 35, 512 (1969).
4. K.Neupert–Laves, M.Dobler, Helv.Chim.Acta, 58, 432 (1975).
5. G.N.Tischenko, Z.Karimov, B.K.Vainshtein, A.V.Evstratov, V.T.Ivanov, Yu.A.Ovchinnikov. FEBS Letters, in press.
6. Yu.A.Ovchinnikov, V.T.Ivanov, A.V.Evstratov, I.I.Mikhaleva, V.F.Bystrov, S.L.Portnova, T.A.Balashova, E.A.Meshcheryakova, V.M.Tulchinsky, Int.J.Pept.Prot.Res., 6, 465 (1974)
7. T.G.Shishova, V.I.Simonov, V.T.Ivanov, I.I.Mikhaleva, T.A Balashova, Yu.A.Ovchinnikov, Bioorgan.Chim.(USSR), 1, 1689 (1975).
8. V.T.Ivanov, L.V.Sumskaya, I.I.Mikhaleva, I.A.Laine, I.D.Ryabova, Yu.A.Ovchinnikov, Chim.Prirodn.Soed. (USSR), 346 (1974).
9. L.V.Sumskaya, T.A.Balashova, T.S.Chumburidze, V.T.Ivanov, Yu.A.Ovchinnikov, Bioorgan.Chim. (USSR), 2 (in press).
10. V.T.Ivanov, A.V.Evstratov, T.A.Balashova, E.A.Meshcheryakova, S.L.Portnova, V.F.Bystrov, Yu.A.Ovchinnikov, Bioorgan. Chim. (USSR), 2 (in press).
11. I.M.Andreev, G.G.Malenkov, A.M.Shkrob, M.M.Shemyakin, Molec. Biol. (USSR), 5, 614 (1971).
12. L.V.Sumskaya, N.M.Chekhlayeva, L.I.Barsukov, O.P.Terekhov, V.V.Djomin, A.M.Shkrob, V.T.Ivanov, Yu.A.Ovchinnikov, Bioorgan. Chim. (USSR), 2, 351 (1976).
13. V.S.Markin, V.F.Pastushenko, L.I.Krishtalik, E.A.Liberman, V.P.Topaly, Biophysika (USSR), 14, 462 (1969).
14. W.R.Veatch, R.Mathies, M.Eisenberg, L.Stryer, J.Mol.Biol.,99, 75 (1975).
15. M.M.Shemyakin, Yu.A.Ovchinnikov, V.T.Ivanov, V.K.Antonov, E.I.Vinogradova, A.M. Shkrob, G.G.Malenkov, A.V.Evstratov, I. D.Ryabova, I.A.Laine, E.I.Melnik, J.Membr.Biol. 1, 402 (1969).

16. E.Grell, Th.Funck, F.Eggers, in Membranes (G.Eisenman, ed.), Marcel Dekker Inc., New-York-Basel, 1975, vol. 3, pp.2–216
17. V.T.Ivanov, Ann. N.Y.Acad.Sci., 264, 221 (1975).
18. V.T.Ivanov, L.A.Fonina, N.N.Uvarova, S.A.Koz'min, T.B.Kropotnitskaya, N.M.Chekhlayeva, T.A.Balashova, V.F.Bystrov, Yu.A.Ovchinnikov, Peptides Chemistry,Structure and Biology (R.Walter, J.Meienhofer, eds.), Ann. Arbor Sci. Publ., Ann Arbor (1975).
19. V.T.Ivanov, A.V.Evstratov, L.V.Sumskaya, E.I.Melnik, T.S.Chumburidze, S.L.Portnova, T.A.Balashova, Yu.A.Ovchinnikov, FEBS Letters, 36, 65 (1973).
20. Yu.A.Ovchinnikov, FEBS Letters, 44, 1 (1974).
21. D.A.Hayden, S.B.Hladky, Quart. Rev., Biophys., 5, 187 (1972).
22. W.R.Veatch, E.T.Fossel, E.R.Blout, Biochemistry, 13, 5249 (1974).
23. P.W.Urry, M.C.Goodall, J.D.Glickson, D.F.Mayers, Proc. Nat. Acad. Sci. US, 68, 1907 (1971).
24. E.T.Fossel, W.R.Veatch, Yu.A.Ovchinnikov, E.R.Blout, Biochemistry, 13, 5264 (1974).
25. E.N.Shepel, St.Iordanov, I.D.Ryabova, A.I.Miroshnikov, V.T.Ivanov, Yu.A.Ovchinnikov, Bioorgan. Chim. (USSR), 2, 581 (1976).

# INCORPORATION OF INTEGRAL MEMBRANE PROTEINS INTO LIPOSOMES

Gera D. Eytan[*], Gottfried Schatz[§] and Efraim Racker

Section of Biochemistry, Molecular & Cell Biology
Cornell University, Ithaca, New York 14853

## I. INTRODUCTION

One of the most intriguing features of biological membranes is their asymmetry. A variety of techniques has shown that most membrane components are not randomly distributed, but specifically oriented with respect to the two membrane surfaces (1,2). In the case of the mitochondrial inner membrane, cytochromes *c* and $c_1$ are localized on the outer side (C-side) and the mitochondrial ATPase $F_1$ on the inner or matrix side (M-side) of the membrane (2). Cytochrome *c* oxidase spans the membrane (3-5) so that some subunits of this oligomeric enzyme are situated on the outer side, some on the inner side and some in the interior of the membrane (4). We are beginning to understand how this asymmetric architecture determines the various vectorial functions of biological membranes but we know next to nothing about how this asymmetry arises *in vivo*. For example, it would be interesting to know to what extent the asymmetric transmembranous orientation of cytochrome oxidase is caused by an asymmetry in the enzyme itself, or by specific properties such as charge, curvature, phospholipid asymmetry or protein composition of the receptor membrane.

During the past few years we have approached these questions by inserting integral membrane proteins into liposomes and studying the asymmetry and the functional properties of the resulting proteoliposomes. The mitochondrial proton pump (6,7), the three sites of oxidative phosphorylation (8,9,10), the adenine nucleotide transporter (11), the proton pump of bacteriorhodopsin (12) and the $Ca^{2+}$- and $Na^{+}K^{+}$ pumps (13,14) are among the reconstituted systems that were most extensively characterized. The three reconstitution procedures which we have used preferentially are 1. the cholate dialysis method, 2. the sonication method and 3. the cholate dilution method (cf 15). All of these methods have two major disadvantages: first, the directionality of reconstitution cannot be controlled; some of the proteins are assembled in the same orientation as in natural membranes, some in the opposite orientation and some of them randomly. Second, these procedures do not allow stepwise reconstitution of integral proteins. These procedures are therefore unsuitable for investigating the sequential and unidirectional assembly of membrane proteins into membranes.

In this communication we describe experiments with a new reconstitution procedure in which membrane proteins are incorporated into preformed liposomes or proteoliposomes without sonication or detergents. This procedure does not share the disadvantages of the above-mentioned methods. It should now be possible to systematically study the factors which govern the asymmetric assembly of biological membranes.

## II. INCORPORATION OF INTEGRAL MEMBRANE PROTEINS INTO PREFORMED LIPOSOMES

The general procedure used for the preparation of unilamellar liposomes was as follows: A solution of phospholipids (25 μmoles) dissolved in chloroform-methanol (2:1) was dried in a small pyrex test tube under a stream of nitrogen, dissolved in ether and dried again. After addition of 1.0 ml of 50 mM $KP_i$ (pH 7.0) or 10 mM Hepes (pH 7.0) containing 40 mM KCl, the mixture was sonicated in a nitrogen atmosphere in a small bath-type sonicator (Model G1225P1, Laboratory Supplies Co., Hicksville, N.Y.) until it was clarified. Depending on the composition of phospholipids this may vary between 5 and 20 minutes.

The resulting liposomes were incubated with the integral membrane protein (1 mg) for 30 minutes at room temperature under conditions specified in the legends of the figures.

## III. INCORPORATION OF CYTOCHROME OXIDASE

As shown in Fig. 1 the incorporation of cytochrome oxidase into phosphatidylserine-containing liposomes described above is rapid. When cytochrome oxidase is incubated with preformed liposomes, most of the enzyme activity becomes "masked" within a few minutes and can be regained only by either uncouplers or detergents. Incorporation of cytochrome oxidase is measured by the developing respiratory control. We define the respiratory control ratio (RCR) as the ratio between the cytochrome *c*-dependent oxygen consumption in the presence of uncoupler and the corresponding activity in the absence of uncoupler. RCR of 5 for example means that at least

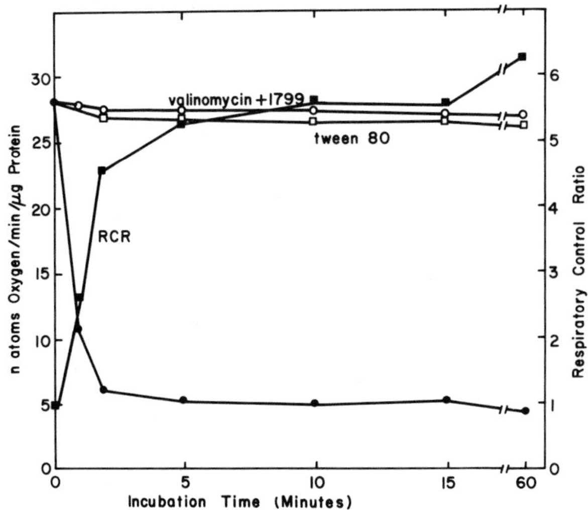


*Fig. 1. Time course of cytochrome oxidase incorporation.* Phospholipid vesicles were formed with phosphatidylethanolamine (13.2 µmoles), phosphatidylcholine (4.4 µmoles) and phosphatidylserine (7.4 µmoles) by sonication to clarity in 1 ml of 50 mM $KP_i$ buffer (pH 7.0) containing 10 mM $MgCl_2$. Purified cytochrome oxidase (1 mg) was added and the mixture was incubated at room temperature. Samples were withdrawn and assayed in the absence (close circles) or presence (open circles) of either Tween 80 (3%) or valinomycin (0.5 µg/ml) plus 1799 (20 µM).

80% of the total activity was incorporated into the liposomes. This is a minimum estimate since it assumes total proton impermeability of the liposomes.

This procedure does not require the addition of detergent. On the contrary, addition of either cholate or lysolecithin, even at low concentrations, inhibits incorporation. The detergents do not interfere with the assay as they have no effect on already reconstituted proteoliposomes. Nevertheless, it was important to show that the incorporation procedure was not dependent on the presence of large amounts of detergents bound to the isolated membrane proteins. It is well-known that cytochrome oxidase preparations isolated by standard procedures contain considerable amounts of detergents. Recently Yu *et al* (16) described a method for preparing lipid-depleted cytochrome oxidase with cholate as the only detergent. When we used this procedure with radioactive cholate, we found that the isolated enzyme after resuspension in detergent free medium still contained an equal weight of cholate. Repeated precipitations of the cytochrome oxidase with ammonium sulfate reduced the residual cholate to around 4% of the enzyme weight.

Further washes or even extraction with ether or hexane did not remove this residual cholate. The cholate-depleted enzyme was insoluble but could still be incorporated into liposomes containing acidic phospholipids (Fig. 2). In separate experiments it was shown that the amount of residual detergents did not increase the leakage of $^{86}Rb$ from the resulting proteoliposomes.

Phosphatidylserine could be replaced by phosphatidylinositol and even by synthetic dicetylphosphate. Cardiolipin was the most effective acidic phospholipid. Its optimal concentration was 10 to 20 mole per cent of the total phospholipid which is also its concentration in the mitochondrial inner membrane. The optimal concentrations of phosphatidylinositol or phosphatidylserine were 30 mole per cent. In the presence of cardiolipin, the only other phospholipid required for incorporation of cytochrome oxidase was

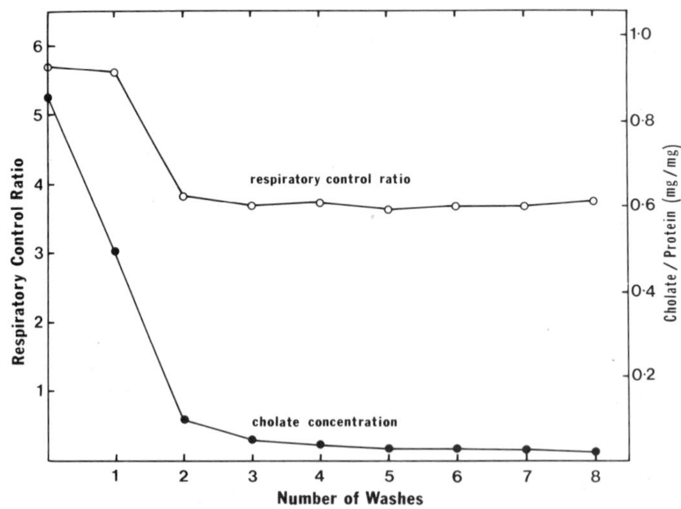


*Fig. 2. Removal of cholate from cytochrome oxidase does not impair its subsequent incorporation into liposomes.* Cytochrome oxidase (10 mg) was prepared according to Yu *et al* (16) except that $^{14}C$-cholate (40 μCi/mmole) was used. The purified enzyme was resuspended in 1 ml 50 mM $KP_i$ (pH 7.0) -0.4 M KCl and then precipitated twice at 35% ammonium sulfate saturation by centrifugation at 100,000 g for 20 min. After these washes the resuspended enzyme (10 mg/ml) was insoluble and further washes were carried out by pelleting the enzyme without ammonium sulfate. After every wash the pellet was resuspended in 1 ml of the above buffer and samples were withdrawn for radioactive determinations and incorporation into liposomes as described in the legend to Fig. 1.

phosphatidylcholine whereas in the presence of either phosphatidylserine or phosphatidylinositol, an excess of phosphatidylethanolamine over phosphatidylcholine was required for optimal incorporation. Metal ions such as calcium, magnesium or manganese accelerated the rate of incorporation but had little or no effect on the final extent of incorporation.

## IV. OPTIMAL PHOSPHOLIPID TO PROTEIN RATIO FOR INCORPORATION

As shown in Fig. 3, optimal incorporation of cytochrome oxidase required at least a 20-fold excess of phospholipids. However, analogous experiments with other membrane proteins yielded quite different results. When we incorporated reduced coenzyme Q ($QH_2$)-cytochrome *c* reductase (complex III) and again measured respiratory control as an assay for functional incorporation, near-maximal incorporation into preformed liposomes occurred at a lipid to protein ratio of 5 (Fig. 3). At a lipid to protein ratio of only 2 a respiratory control ratio of 4 was observed, whereas a lipid to protein ratio of 10 was required for a similar RCR of cytochrome

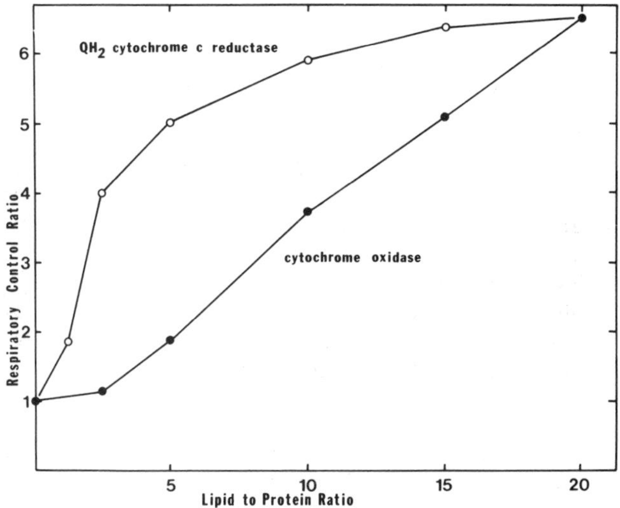


*Fig. 3. Effect of lipid to protein ratio on the reconstitution of cytochrome oxidase and $QH_2$ cytochrome c reductase.* Cytochrome oxidase (closed circles) and $QH_2$ cytochrome *c* reductase (open circles) vesicles were prepared by incorporation of 1 mg/ml protein into liposomes at the indicated lipid to protein ratios. The phosphatidylserine containing liposomes were prepared as described in the legend of Fig. 1.

oxidase. It is interesting that cytochrome oxidase could be successfully incorporated (RCR of four) into these preformed $QH_2$-cytochrome *c* reductase vesicles with a lipid to protein ratio of 2. Thus, cytochrome oxidase can be incorporated at a lipid to protein ratio approaching more closely that of a natural membrane provided another suitable membrane protein is present in the liposome.

## V. ORIENTATION OF INCORPORATED PROTEINS

As mentioned earlier, reconstitution in the presence of large amounts of detergents may yield mixed particle populations with respect to the orientation of incorporated membrane proteins. For example, roughly 40% of the activity of oligomycin-sensitive ATPase vesicles prepared by the cholate dilution procedure could be detected only in the presence of detergent (3% of Emasol 1130, Table I). In contrast, all the activity of oligomycin-sensitive ATPase reconstituted by the new incorporation procedure could be assayed even in the absence of detergent. This implies that most, if not all of the enzyme was oriented unidirectionally, with its hydrolytic site exposed to the suspending medium. On the other hand, the $^{32}P_i$-ATP exchange was actually higher in vesicles reconstituted by the cholate dilution procedure. The ATPase oriented in the opposite direction did not collapse the proton gradient because the liposomes are impermeable to ATP (13).

## VI. EFFECT OF FATTY ACYL GROUP SATURATION ON INCORPORATION

The experiments mentioned so far were carried out with phosphatidylethanolamine and phosphatidylcholine prepared from mitochondria. When the mitochondrial phosphatidylcholine was substituted by synthetic dioleylphosphatidylcholine, incorporation of cytochrome oxidase was not affected. On the other hand, no incorporation was observed with vesicles prepared with fully saturated synthetic phospholipids such as dipalmitoylcholine or dimyristoylcholine. An exception to this rule was only observed with saturated phospholipids containing short fatty acyl groups such as dilauroyl phosphatidylcholine (Table II).

In order to further study the interfering effect of dimyristoyl phosphatidylethanolamine on the incorporation of cytochrome oxidase, the enzyme was incubated with liposomes containing various amounts of this saturated phospholipid (Fig. 4). There was a sharp drop in the incorporation of cytochrome oxidase if the concentration of dimyristoyl phosphatidylethanolamine rose to 10 or 15%. Since cytochrome oxidase was still incorporated into vesicles which were completely devoid of any phosphatidylethanolamine (not shown), the saturated phospholipid was not merely unsuitable for incorporation but actively inhibited it. Interestingly, dimyristoyl phosphatidylethanolamine had no significant effect on the incorporation of $QH_2$-cytochrome *c* reductase (Fig. 4).

TABLE I

Orientation of oligomycin-sensitive ATPase in proteoliposomes prepared by different procedures

The oligomycin-sensitive ATPase (7) was reconstituted into liposomes either by the procedure described in this paper or by the cholate dilution procedure (17). In the latter case, reconstitution was carried out as in the incorporation procedure except that 0.8% cholate was included in the incubation medium. Emasol 1130 (3%) was added either before the ATPase assay (+ Emasol) or after it (- Emasol), but before the determination of the liberated $P_i$. Also standard $P_i$ curves must be performed in the presence of Emasol since the detergent seriously alters the colorimetric results. ATPase was assayed in the presence of both valinomycin (0.5 µg/ml) and 1799 (20 µM).

| Mode of reconstitution | ATPase activity (µmoles ATP/min/mg protein) | | $^{32}P_i$-ATP exchange (nmoles $^{32}$ATP formed/ min/mg protein) |
|---|---|---|---|
| | - Emasol | + Emasol | |
| Incorporation procedure | 4.2 | 4.3 | 45 |
| Cholate dilution | 2.6 | 4.4 | 72 |

TABLE II

Effect of synthetic phospholipids on reconstitution of cytochrome oxidase

Liposomes were prepared in 50 mM $KP_i$ pH 7.0 either with mitochondrial phosphatidylethanolamine:phosphatidylcholine:phosphatidylserine (1:1:1) or with one of these components replaced by the corresponding synthetic phospholipid mentioned in the Table. Cytochrome oxidase was incorporated at room temperature as described in this paper except that incubation was prolonged to 60 minutes.

| Phospholipid used | Nature of fatty acids | Respiratory control ratio |
|---|---|---|
| dilauroyl phosphatidylcholine | 12:0 | 4.0 |
| dimyristoyl phosphatidylcholine | 14:0 | 1.0 |
| dimyristoyl phosphatidylethanolamine | 14:0 | 1.0 |
| dipalmitoyl phosphatidylcholine | 16:0 | 1.0 |
| dipalmitoyl phosphatidylethanolamine | 16:0 | 1.0 |
| dioleyl phosphatidylcholine | 18:1 | 4.1 |
| mitochondrial phospholipids (cf. legend) | - | 4.5 |

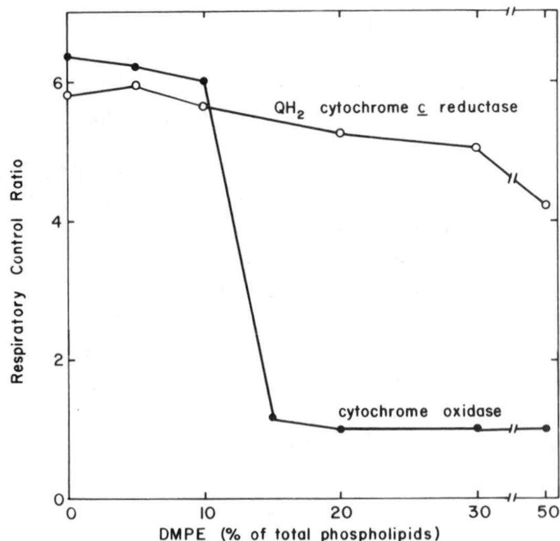


*Fig. 4. Effect of dimyristoyl phosphatidylethanolamine on the incorporation of cytochrome oxidase and $QH_2$ cytochrome c reductase into liposomes.* Cytochrome oxidase and $QH_2$ cytochrome *c* reductase were prepared as described in the legend of Fig. 3 except that the indicated amounts of dimyristoyl phosphatidylethanolamine were included during the preparation of the liposomes.

Is the inhibiting effect of dimyristoyl phosphatidylethanolamine on cytochrome oxidase incorporation related to the overall concentration of saturated fatty acyl groups in the liposome or to the concentration of fully saturated phospholipid molecules? We approached this question by checking the incorporation of cytochrome oxidase into vesicles containing varying amounts of dimyristoyl phosphatidylethanolamine and either egg phosphatidylcholine or mitochondrial phosphatidylcholine (Fig. 5). Egg phosphatidylcholine is much more saturated (0.35 double bond /acyl group) than mitochondrial phosphatidylcholine (2 double bonds/acyl group). It was found that only a slightly higher concentration of dimyristoyl phosphatidylethanolamine was required for inhibition of the incorporation of cytochrome oxidase into the vesicles with highly unsaturated phosphatidylcholine. This indicates that the inhibition of incorporation is caused by the presence of the fully saturated dimyristoyl phosphatidylethanolamine molecules and not by the overall concentration of saturated fatty acyl groups in the liposomes.

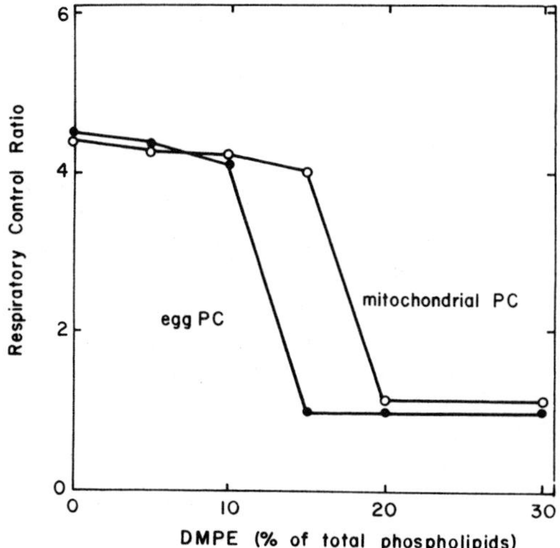


*Fig. 5. Effect of the degree of unsaturation of phosphatidylcholine on the incorporation of cytochrome oxidase.* Incorporation was performed as described in the legend of Fig. 3 except that in one series of experiments phosphatidylcholine from egg (full circles) and in another series mitochondrial phosphatidylcholine (open circles) was used.

Varying the temperature either during the incorporation of cytochrome oxidase and/or during its assay did not affect the inhibition by dimyristoyl phosphatidylethanolamine. This suggests that this inhibition does not simply reflect changes in the transition temperature of the liposomes. The inhibition is also not caused by an increased leakiness of the liposomes since control experiments with $^{86}$Rb-loaded liposomes revealed no effect of dimyristoyl phosphatidylethanolamine on liposome leakiness.

Does the saturated phospholipid inhibit the proper incorporation of cytochrome oxidase or the controlled function of the incorporated enzyme? In order to answer this question we made use of a recently developed procedure for fusing proteoliposomes (18). This procedure is based on the fact that liposomes containing phosphatidylserine, phosphatidylcholine and phosphatidylethanolamine fuse to form large vesicles in the presence of $Ca^{2+}$. Cytochrome oxidase was first incorporated into liposomes containing mitochondrial phosphatidylethanolamine, phosphatidylcholine and phosphatidylserine. The resulting proteoliposomes exhibited a respiratory control ratio of 5. They were then fused with protein-free liposomes of a similar phospholipid composition except that the phosphatidylethanolamine was fully saturated dimyristoyl phosphatidylethanolamine. The fused vesicles (which now contained a high percentage of saturated phosphatidylethanolamine) still exhibited a respiratory control ratio of 5, indicating that cytochrome oxidase was functioning normally. In the converse experiment, cytochrome oxidase was first incubated with liposomes containing only the saturated phosphatidylethanolamine and the incubated vesicles were then fused with liposomes containing only mitochondrial phospholipids. The fused vesicles lacked respiratory control even though their content of saturated phosphatidylethanolamine was no higher than that of the vesicles obtained in the reciprocal fusion experiment mentioned above. We conclude that the saturated phospholipid inhibits the proper incorporation of cytochrome oxidase rather than functioning of the already incorporated enzyme.

Which step of the incorporation process is inhibited by saturated phospholipids? Cytochrome oxidase was incubated with a mixture of two liposome populations; one population contained 30% dimyristoyl phosphatidylethanolamine and the other only mitochondrial phospholipids. An intermediate respiratory control ratio was obtained, as expected for random distribution of the enzyme between the "unsaturated" vesicles (respiratory control ratio = 5) and the largely saturated vesicles (respiratory control ratio = 1). Addition of calcium to the mixture of "saturated" and "unsaturated" liposomes before incubation with cytochrome oxidase abolished any respiratory control of the resulting proteoliposomes. In this case, the two liposome populations fused before the incorporation step and the resulting vesicles had 15 mole percent dimyristoyl phosphatidylethanolamine which completely inhibited the incorporation. The saturated phospholipid does thus not interfere with the binding of the enzyme to the liposome but must affect some later stage of the correct incorporation process.

We suspect that the initial binding of cytochrome oxidase to liposomes is dependent on the nature of the head-group of the phospholipids whereas the nature of the fatty acyl group is critical for the correct orientation of the protein in the membrane. Similar inhibition by saturated synthetic phospholipids was observed by Kagawa *et al* (7) with reconstituted vesicles catalyzing $^{32}P_i$-ATP exchange.

## VII. SEQUENTIAL INCORPORATION OF PROTEINS INTO LIPOSOMES

One of the advantages of the incorporation procedure is that it allows the sequential insertion of proteins into vesicles. The most striking effects were observed on incorporating cytochrome oxidase and the hydrophobic moiety of the oligomycin-sensitive ATPase into the same liposome. The presence of the two protein components in the same vesicle could be conveniently assayed since the hydrophobic protein fraction serves as a proton channel and abolishes respiratory control of the incorporated cytochrome oxidase (19). The determining factor in the effectiveness of the hydrophobic protein is not its absolute concentration but the ratio of this fraction to cytochrome oxidase vesicles. Thus, in the presence of increasing amounts of cytochrome oxidase vesicles, a given concentration of hydrophobic proteins became less effective for uncoupling (not shown). Does cytochrome oxidase, already incorporated into liposomes, affect the subsequent incorporation of hydrophobic proteins of mitochondrial ATPase? To answer this question, a fixed amount of hydrophobic protein fraction was incubated with liposome mixtures containing a fixed amount of cytochrome oxidase vesicles and increasing amounts of protein-free liposomes. If the hydrophobic proteins are randomly incorporated into both types of liposomes, the presence of excess protein-free liposomes should decrease the uncoupling (i.e. raise the respiratory control ratio) of the cytochrome oxidase liposomes. As shown in Fig. 6, this was not the case. All of the added hydrophobic protein fraction had therefore incorporated only into the cytochrome oxidase vesicles. This result is all the more surprising since both liposome populations had the same phospholipid composition. Essentially the same result was obtained when the cytochrome oxidase vesicles were prepared from an enzyme that had been almost totally depleted of lipids and detergents (cf above).

Are membrane proteins always incorporated more readily into proteoliposomes than into protein-free liposomes? The answer is no. Cytochrome oxidase was incubated with a series of liposome mixtures consisting of a fixed amount of liposomes with incorporated hydrophobic proteins and increasing amounts of protein-free liposomes. Preferential incorporation of the cytochrome oxidase into the liposomes containing hydrophobic protein fraction should have resulted in particles lacking respiratory control. Conversely, random incorporation of cytochrome oxidase into the two types of liposomes should have yielded respiratory control values that increased with increasing amounts of protein-free liposomes. As shown in Fig. 7, the observed respiratory control values were even higher than those expected for a random distribution of cytochrome oxidase. We conclude therefore that cytochrome oxidase incorporated preferentially into the protein-free liposomes. Depending on the system investigated, the presence of incorporated proteins in liposomes may therefore stimulate or inhibit the subsequent incorporation of a second protein. Further efforts will be required to learn whether these effects reflect direct protein-protein interaction or a modification of the liposomes by the incorporated proteins.

## VIII. CONCLUSION

The sequential *in vitro* incorporation of membrane proteins into preformed liposomes may give us access to many important and

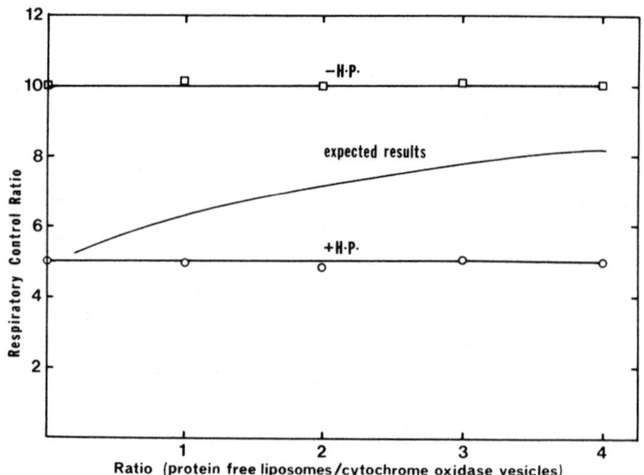


*Fig. 6. Uncoupling of cytochrome oxidase vesicles by hydrophobic proteins in the presence of excess protein-free liposomes.* Liposomes were prepared by sonication of phosphatidylethanolamine: phosphatidylcholine:phosphatidylserine (3:1:2, 25 mM) in 50 mM $KP_i$ (pH 7.0) -0.05 mM EDTA. Cytochrome oxidase (1 mg/ml) was incorporated into part of the liposomes by incubation at room temperature for 30 min. The rest of the liposomes were incubated under the same conditions with no protein. Cytochrome oxidase vesicles (2.5 mM lipids) and various amounts of protein-free liposomes were incubated with or without hydrophobic proteins (0.25 mg/ml) in the same buffer at room temperature for 30 min. Expected results for random distribution of the hydrophobic proteins among the protein-free liposomes and the cytochrome oxidase vesicles were calculated, assuming that in this case the uncoupling should be equal to that observed in the presence of cytochrome oxidase vesicles concentration equal to the combined concentrations of the liposomes and cytochrome oxidase vesicles.

hitherto unanswered questions of membrane biochemistry. Can one identify specific domains either on the protein or the receptor liposome, which directs the asymmetric insertion of a membrane protein? Which subunits of the oligomeric protein complexes studied in the report are involved in the attractive and repulsive phenomena observed during sequential incorporation? Why do saturated long chain phospholipids inhibit the correct incorporation of some membrane proteins but not that of others?

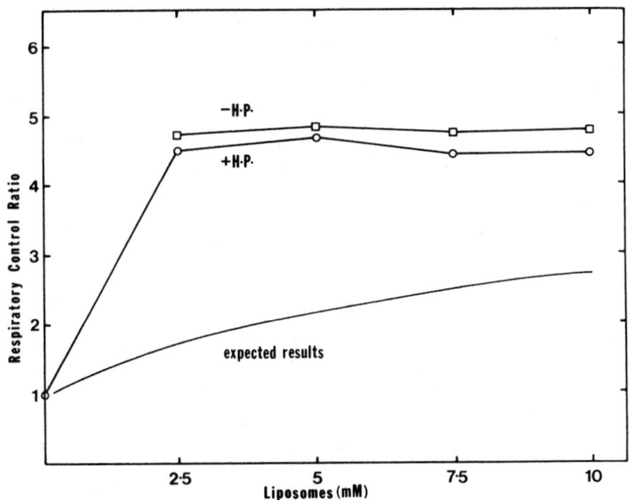


*Fig. 7. Incorporation of cytochrome oxidase into hydrophobic protein vesicles in the presence of excess liposomes.* Hydrophobic protein vesicles were prepared by sonicating until clarity phosphatidylethanolamine:phosphatidylcholine;phosphatidylserine (3:1:2, 25 mM) in 50 mM $KP_i$ (pH 7.0) 1% cholate, adding the hydrophobic proteins (2 mg/ml, final concentration) and dialyzing overnight at 4°C against 200 volumes 50 mM $KP_i$, pH 7.0. Protein-free liposomes were prepared according to a similar procedure without the hydrophobic proteins. Various amounts of liposomes and cytochrome oxidase (0.1 mg/ml) were incubated with or without hydrophobic protein vesicles (2.5 mM phospholipids) in 50 mM $KP_i$, pH 7.0, 0.5 mM EDTA. After 30 min at room temperature samples (0.05 ml) were assayed for respiratory control of cytochrome oxidase activity. Expected results for random distribution of the cytochrome oxidase among the liposomes and hydrophobic protein vesicles were calculated, assuming that cytochrome oxidase bound to hydrophobic protein vesicles is completely uncoupled, while the respiratory control ratio of cytochrome oxidase incorporated into the liposomes is equal to that observed upon incorporation of the enzyme into liposomes with no hydrophobic protein vesicles present. The actual results indicate that cytochrome oxidase is preferentially incorporated into the protein-free liposomes over the hydrophobic protein vesicles.

Any results derived from such *in vitro* studies will have to be weighed against the possibility that the system described here does not reflect the events occurring during the biogenesis of biological membranes. Indeed, recent studies on the formation of the mitochondrial inner membrane (20,21) have made it fairly clear that this membrane is not made by the incorporation of complete protein complexes such as cytochrome oxidase, oligomycin-sensitive ATPase or $QH_2$-cytochrome *c* reductase, but probably by insertion of individual or even nascent (22) chains into a preexisting membrane. On the other hand, it cannot be overlooked that some of the data presented here suggest intriguing parallels between the *in vitro* system and the situation in living cells. For example, it is often found that the mutational alteration or loss of a single membrane polypeptide results in secondary (pleiotropic) modifications of the membrane; for example, mutants specifically lacking cytochrome oxidase often show greatly depressed levels of mitochondrial ATPase (23). It may also be relevant that yeast cells growing anaerobically in the absence of added unsaturated fatty acids accumulate large amounts of phospholipids containing short-chain saturated acyl groups (24,25). It may be recalled that such phospholipids can mimic the action of unsaturated phospholipids in supporting incorporation of cytochrome oxidase into preformed liposomes. Finally, unpublished observations from our laboratory suggest that the affinity of hydrophobic ATPase proteins for cytochrome oxidase vesicles can be drastically lowered by seemingly minor modifications of cytochrome oxidase such as binding of cytochrome *c* or partial reduction. This might point to a possible mechanism for modulating the incorporation of membrane proteins. Biological examples of such delicate modulations are the regulation of mitochondrial biogenesis by oxygen (26,27), glucose (28) or heme (29) and the regulation of chloroplast membrane formation by the chlorophyll-synthesizing system (30). While we do not wish to imply that the experiments described here explain the mechanism of these regulations, we would like to raise the possibility that the oxidation state of cytochromes may control the incorporation of these and other proteins into preexisting membranes.

*Acknowledgment*: This investigation was supported by Grant No. PCM 73-01025A02 from the National Science Foundation.

*Abbreviations*: RCR, respiratory control ratio; PC, phosphatidylcholine; PE, phosphatidylethanolamine: DMPE, dimyristoyl phosphatidylethanolamine; $F_1$, mitochondrial ATPase (oligomycin-insensitive); H.P., hydrophobic protein fraction from mitochondria required for formation of the proton channel of the transmembranous ATPase complex; 1799, bis-(hexafluoroacetonyl)acetone.

*Present address: Department of Biology, Technion, Haifa, Israel

§Present address: Biozentrum der Universität Basel, Abteilung Biochemie, CH-4056 Basel, Switzerland

## REFERENCES

1. Singer, S.J. (1974) Ann. Rev. Biochem. 43, 805-834.
2. Racker, E. (1970) in Essays in Biochemistry (P.N. Campbell and F. Dickens, eds.) Academic Press, Vol. 6, p. 1-22.
3. Schneider, D.L., Kagawa, Y. and Racker, E. (1972) J. Biol. Chem. 247, 4074-4079.
4. Eytan, G.D., Carroll, R.C., Schatz, G. and Racker, E. (1975) J. Biol. Chem. 250, 8598-8603.
5. Hackenbrock, C.R. and Hammon, K.M. (1975) J. Biol. Chem. 250, 9185-9197.
6. Kagawa, Y. and Racker, E. (1971) J. Biol. Chem. 246, 5477-5487.
7. Kagawa, Y., Kandrach, A. and Racker, E. (1973) J. Biol. Chem. 248, 676-684.
8. Ragan, C.I. and Racker, E. (1973) J. Biol. Chem. 248, 2563-2569.
9. Leung, K.H. and Hinkle, P.C. (1975) J. Biol. Chem. 250, 8467-8471.
10. Racker, E. and Kandrach, A. (1971) J. Biol. Chem. 246, 7069-7071.
11. Shertzer, H.G. and Racker, E. (1974) J. Biol. Chem. 249, 1320-1321.
12. Racker, E. and Stoeckenius, W. (1974) J. Biol. Chem. 249, 662-663.
13. Knowles, A.F. and Racker, E. (1975) J. Biol. Chem. 250, 3538-3544.
14. Racker, E. and Fisher, L.W. (1975) Biochem. Biophys. Res. Commun. 67, 1144-1150.
15. Racker, E. (1975) in Proceedings of the Tenth FEBS Meeting (J. Montreuil and P. Mandel, eds.) North-Holland/American Elsevier, Vol. 41, pp. 25-34.
16. Yu, C-A., Yu, L. and King, T.E. (1975) J. Biol. Chem. 250, 1383-1392.
17. Racker, E., Chien, T-F. and Kandrach, A. (1975) FEBS Letters, 57, 14-18.
18. Miller, C. and Racker, E. (1976), J. Membrane Biology, in press.
19. Racker, E. (1972) J. Membrane Biol. 10, 221-235.
20. Schatz, G. and Mason, T.L. (1974) Ann. Rev. Biochem. 43, 51-87.
21. Tzagoloff, A., Rubin, M.S. and Sierra, M.F. (1973) Biochim. Biophys. Acta 301, 71-104.
22. Kellems, R.E. and Butow, R.A. (1972) J. Biol. Chem. 247, 8043-8050.
23. Ebner, E., Mason, T.L. and Schatz, G. (1973) J. Biol. Chem. 248, 5360-5368.
24. Jollow, D., Kellerman, G.M. and Linnane, A.W. (1968) J. Cell Biol. 37, 221.
25. Paltauf, F. and Schatz, G. (1969) Biochemistry 8, 335-339.
26. Slonimski, P.P. (1953) La formation des enzymes respiratoires chez la levure, Masson, Paris.
27. Criddle, R.C. and Schatz, G. (1969) Biochemistry 8, 322-334.
28. Mahler, H.R. (1973) CRC Crit. Rev. Biochem. 1, 381-460.
29. Gollub, E.G., Trocha, P., Liu, P.K. and Sprinson, D.B. (1974) Biochem. Biophys. Res. Comm. 56, 471-477.
30. Ohad, I. (1975) in Membrane Biogenesis (Tzagoloff, A., ed.) Plenum Press, New York, pp. 279-350.

# LIPID ASYMMETRY, CLUSTERING AND MOLECULAR MOTION IN BIOLOGICAL MEMBRANES AND THEIR MODELS

Peter R. Cullis, Ben De Kruijff, Alister E. McGrath,
Christopher G. Morgan and George K. Radda

Dept. of Biochemistry, University of Oxford
South Parks Road, Oxford

The term 'biological membrane' is often used in a sense that implies considerable uniformity among different types of membranes. Such uniformity of course, can only refer to some common principles in organisation (structure) but not in function. There is after all no reason to suppose that say carrier mediated transport, protein synthesis, energy coupling or signal transmission all operate by a similar mechanism. It is perhaps the large diversity of membrane functions based on structural similarities that makes research in this area of such current interest.

In that all membranes are made up of lipids and proteins they are similar. (Carbohydrates, also constituents, are often considered in a separate class, being on the membrane surface). Yet the differences in fine detail (lipid-protein ratio, lipid and protein composition and structure) must be responsible for the functional variations. Combining spectroscopic studies (providing some of the fine structural detail) with biochemical ones (describing functions) we can determine some structure-function relations. Present spectroscopic techniques alone cannot provide an overall structural view while diffraction and electron microscopic measurements are likely to produce sufficient detail only in specialised circumstances (1). We must therefore combine the results of different types of measurements. In this paper we shall examine some of the properties of the chromaffin-granule membrane and describe observations on model membranes that are relevant to our understanding of the former system.

## I. THE BIOCHEMISTRY OF CHROMAFFIN GRANULES

Chromaffin granules are membrane limited vesicles of about 2000 Å diameter contained within the adrenal medullary chromaffin cells. They are the major storage vesicles for catecholamines which they concentrate to 0.55M together with ATP (0.125M) inside the vesicles. Their two major biochemical functions are the release and uptake of catecholamines. The former process is thought to occur by excocytosis, i.e. the fusion of the membrane of the granule with that of the cell resulting in the release of the total content of the storage system. The trigger for this event is $Ca^{2+}$(2), although the exact form of the triggering is not known. The required energy for amine uptake is derived from the hydrolysis of ATP. The ATP-ase seems to be located on the outside of the vesicles and in a $Mg^{2+}$ dependent reaction drives the accummulation of adrenaline via a proton-linked mechanism (3,4,5). It is likely that the granule membrane contains a catecholamine carrier since uptake can be competitively inhibited by reserpine (6) and also shows saturation kinetics.

The role of the other membrane proteins (notably an NADH-oxidase, cytochrome b and DOPAmine hydroxylase) is not well understood.

The composition of the membrane lipids is unusual (Table 1) with perhaps the two most emphasized features being the relatively high proportion of lysolecithin and cholesterol.

TABLE 1. PHOSPHOLIPID COMPOSITION OF BOVINE CHROMAFFIN GRANULES

| Phospholipids | |
|---|---|
| Lysolecithin | 16.8 |
| Sphingomyelin | 12.8 |
| Phosphatidylcholine | 27.5 |
| Phosphatidylinositol | 8.2 |
| Phosphatidylserine | 2.5 |
| Phosphatidylethanolamine | 31.8 |
| Phosphatidic acid<br>Cardiolipin | 1.03 |

Taken from Winkler, H., Schneider, N. Ziegler, E. Naunyn-Schmiedebergs Arch.Exp.Path.Pharmak. 1967,256, 407-415.

As in most studies on biological membranes so far we can most readily ask questions about the structure and motions of the phospholipid components and to attempt to define their role in modulating biological activities. We shall therefore first summarise our work on phospholipid model membranes and then examine the functional relevance of our observations.

## II. MODEL MEMBRANE SYSTEMS

Two model membrane systems are commonly employed to study the bilayer properties of phospholipids. Unsonicated aqueous dispersions (liposomes) of many phospholipids consist of layers of concentric bilayers arranged in an onion skin configuration. The much smaller (250Å) diameter) "vesicles" obtained on sonication of liposomes consist of a single bilayer separating inner and outer aqueous phases. It should be noted that vesicles are "good" model systems as the lipid packing and the local motion available to the phospholipid do not appear to be unduly perturbed by the high curvature of the vesicle entity. In this regard it has been shown that the gel-liquid crystalline phase transition for dipalmitoyl lecithin occurs at the same temperature and has the same heat content in both vesicle and liposome preparations (7), strongly suggesting that the lipid packing is not significantly perturbed in the vesicle membrane. Further, $^{31}$P NMR results show that the local motion in the phosphate region of the polar headgroup is similar in both vesicles and liposomes (8).

Both systems have advantages in certain situations for studying motional and structural details of the constituent phospholipids. In particular the small vesicles tumble rapidly, resulting in high resolution NMR spectra, anionic shift reagents such as ferricyanide may then be used to obtain details of the outside-inside distribution of choline containing lipids such as sphingomyelin and phosphatidylcholine, thus giving information on the vesicle size. In the case of $^{31}$P NMR, different classes of phospholipid present in the vesicle may be resolved separately. Cationic broadening and shift reagents may then be employed to obtain information on both the vesicle size and possible assymetric distributions of the constituent phospholipids across the vesicle membrane. These points will be elaborated in the next section.

The much larger liposomes on the other hand do not have such rapid tumbling rates as vesicles. Thus the $^{31}$P NMR signals observed in such systems have a "solid state" lineshape which only reflects the local anisotropic motion available to the phospholipid in a bilayer configuration. Biological membrane preparations obtained by osmotic lysis also consist of relatively large membrane fragments. The observation of similar solid state $^{31}$P NMR signals in such biological membrane preparations thus strongly indicates regions of bilayer phospholipid structure. In the case of the chromaffin granule membrane a large percentage of the constituent phospholipids must experience a relatively fluid bilayer environment as indicated by the observation of $^{31}P$ NMR spectra characteristic of liquid crystalline phospholipids in a bilayer configuration (9).

## 1. Asymmetry Of Lipids In Vesicles

It is well known that in biological membranes not only the protein components but also phospholipids are asymmetrically distributed across the membrane.

Recently on the basis of chemical studies (10), nuclear magnetic resonance (NMR) measurements (11,12) and theoretical considerations (13) asymmetric distribution of phospholipids across vesicle membranes has been reported.

We have examined in detail the factors that are important in determining such asymmetric lipid distributions. In this we have used three methods: (i) The proton NMR signals from the choline head groups of lecithin or sphingomyelin located on the outside of a vesicle can be shifted by the addition of ferricyanide (14,15). (ii) most classes of phospholipids can be observed separately by $^{31}P$ NMR. Here we have used a non-penetrating broadening reagent ($Co^{2+}$) to measure lipid distributions (15). (iii) Finally, the $^{31}P$ NMR spectra of different lipids are easily resolved into outside/inside components by the addition of a non-permeating shift reagent like $Nd^{3+}$ (7). Since the technical details of these measurements have been described elsewhere (7,12,15) only the conclusion will be summarised here.

The ratio of the molecules in the outer and inner monolayers (Ro/i) in phosphatidylcholine vesicles clearly depends on the sizes of the vesicles. This can be demonstrated experimentally in two ways. Sonicated egg lecithin vesicles can be fractionated by gel-filtration according to their size (15), or the size can be varied in pure phosphatidylcholines by changing the nature of hydrocarbon chains (7). In both cases the Ro/i approaches unity for the larger vesicles and increases above 2.0 for the smaller ones (Table 2). (We should mention here that recently we have demonstrated that the sizes of phospholipid vesicles can be calculated from the widths of the $^{31}P$ NMR lines as a result of the fact that the linewidths are dependent on the tumbling rate of the vesicles, which depends on the vesicle size through Stokes law (16)).

In vesicles containing mixtures of phospholipids with different head groups both charge and the packing properties of the head group are important in determining phospholipid distribution (12,15). Thus in mixtures of lecithin with phosphatidylethanolamine, phosphatidic acid, phosphatidylserine and phosphatidylinositol, phosphatidyl

Table 2. DISTRIBUTION OF PHOSPHATIDYLCHOLINE (PC) ON THE OUTSIDE AND INSIDE LAYERS OF PHOSPHATIDYLCHOLINE VESICLES

$$R_{o/i} = \frac{\text{amount of phosphatidylcholine outside monolayer}}{\text{amount of phosphatidylcholine inside monolayer}}$$

All $R_{o/i}$ measurements at 30°C, except for the 16:0/16:0-phosphatidylcholine and 18:0/18:0-phosphatidylcholine vesicles where $R_{o/i}$ was measured at 50 and 60°C, respectively. Error in $R_{o/i}$ is 0.05. (from ref.7)

| Vesicle composition | $R_{o/i}$ | Membrane thickness (Å) | Calculated vesicle outer radius (Å) |
|---|---|---|---|
| 14:0/14:0-PC | 2.65 | 32 | 84 |
| 16:0/16:0-PC | 2.2 | 37 | 112 |
| 18:0/18:0-PC | 1.7 | 42 | 180 |
| 16:1c/16:1c-PC | 1.8 | 28 | 110 |
| 18:1c/18:1c-PC | 1.75 | 32 | 131 |
| 18:1t/18:1t-PC | 2.0 | 36 | 123 |
| Egg PC | 2.0 | 35 | 120 |

Table 3. DISTRIBUTION OF PHOSPHOLIPIDS IN MIXED VESICLES

| Lipid mixture (1:1) | lipid component | $R_{o/i}$ |
|---|---|---|
| PC-PS(pH 7.2) | total | 2.3 |
| | PC | 2.45 |
| | PS | 2.06 |
| PC-PS(pH 4.9) | total | 2.3 |
| | PC | 3.7 |
| | PS | 1.15 |
| PC-PE(pH 7.2) | total | 1.41 |
| | PC | 1.76 |
| | PE | 1.17 |
| PC-PA(pH6) | PC | 2.8 |
| 14:0/14:0-PC-cholesterol | PC | 1.9 |
| | cholesterol | |

Continuation of Table 3.

| Lipid mixture (1:1) | Lipid component | $R_{o/i}$ |
|---|---|---|
| 18:1c/18:c-PC-cholesterol | PC | 2.95 |
| | cholesterol | 0.46 |

PC: phosphatidylcholine; PS: phosphatidylserine
PE: phosphatidylethanolamine; PA: phosphatidic acid

(From refs. 7,15,16)

choline prefer to be located at the outside of the bilayer (Table 3). The last four lipids all have a smaller head group than lecithin and this is clearly an important factor in determining the observed asymmetry. In contrast the slight preference of sphingomyelin for the outside layer in mixtures with lecithin (Table 3) may well be due to a difference in geometry of the headgroup compared to that of lecithin, requiring a slightly larger area.

The effect of charge on phospholipid asymmetry has been demonstrated by Michaelson _et al_. (11) and by our work on the effect of the distribution of phosphatidylserine in mixed vesicles (15). As expected (13) the effect of increased charge is to decrease the preference of the charged phospholipid for the inside of the bilayer. In the case of phosphatidylserine this effect is in competition with the opposite preference due to head group size (15).

The presence of cholesterol in membranes raises special problems. The outside/inside ratio ($R_{o/i}$) of both saturated and unsaturated phosphatidylcholine species is not much affected by the incorporation of up to 30mol% cholesterol.

Above this level of cholesterol the outside /inside ratio of the phospholipid is markedly increased for phosphatidylcholines with cis-unsaturated fatty acid chains. In contrast this effect was either absent or in the opposite direction when the fatty acid chains were fully saturated or contained trans-unsaturation (16). However, since in all instances above 30mol% cholesterol the vesicle sizes also increased (but not below 30%) the distribution of the lipid across the bilayer became asymmetric with a disproportionately larger amount of cholesterol on the inside. We have suggested that under these conditions cholesterol-cholesterol (as opposed to cholesterol-phospholipid) interactions are mainly responsible for the increased vesicle sizes and also for the preferential location of often over 50mol% cholesterol on the inside of the bilayer. The dynamic shape of the cholesterol frame apparently favours placing of adjacent molecules on the inner monolayer, i.e. by locating the hydroxyl group on the concave surface.

It is now appropriate to enquire as to the biological significance, if any, of these observations on small lipid vesicles.

Phospholipids and cholesterol have been shown to be asymmetrically distributed in a variety of membranes. If this asymmetry is entirely biosynthetic in origin one may argue that studies on small vesicles bear no relevance to the real situation. If this is so at worst we have learned how to produce asymmetric vesicles which may allow one to determine how such lipid distributions affect transport and coulombic interaction with some proteins in situations that are analogous to the known lipid orientations in biological membranes. (The fact that in erythrocytes lecithin and sphingomyelin are largely on the outside while phosphatidylserine and phosphatidylethanolamine are mainly present in the inner layer (17) may then be no more than fortunate.)

The distribution experiments on vesicles also give us some insight into the packing requirements of different phospholipid head groups.

At the other extreme, one may wish to consider the possibility that biological membranes *in vivo* do possess large areas of high curvature and that the packing of the lipids in these regions are governed by the same kinds of interactions that we have discussed. Or an interesting possibility might be that during the biosynthesis of the membrane the asymmetric phospholipid distribution is initially set up in such curved regions. Transporting epithelial cells certainly have highly folded cell surfaces (18) and it has been suggested that the membranes at the tips and bases of microvilli and the deeper parts of basal infolds could be sites of high water permeability. Other examples include mitochondrial cristae and the ends of cisternae of endoplasmic reticulum. Intuitively, regions of high curvature would be expected to occur during processes like exocytosis and pinocytosis. From the functional point of view such regions therefore could have special significance.

## 2. Phase Behaviour of Phospholipids

In recent years many reports have dealt with existence of phospholipid "phase transitions" in biological membranes. Generally non-linear Arrhenius plots for membrane linked functions and "discontinuities" in probe (i.e. spin and fluorescent labels) behaviour have been used is support of such conclusions. It is unlikely that temperature dependent phase changes have any direct biological relevance but their observation provides a valid and valuable method or understanding the way the physical state of the membrane lipids modulates biochemical functions.

In pure phospholipid bilayers (or their mixtures) two types of well documented phase changes may take place: (i) gel-liquid crystalline transitions and (ii) lateral phase separations.

(i) Co-operative gel-liquid crystalline transitions are a result of the tight crystal like packing of the hydrocarbon chains of phospholipids being expanded to produce a less ordered and expanded structure.

Here we only wish to emphasize two sets of observations.

(a) The first is that $^{31}P$ NMR can be used to follow such changes in vesicles (7), liposomes (8) and in some special biological membranes (19). This is because below the phase transition temperature the rotational motion of phospholipids is hindered (9) resulting in large NMR line widths, as the temperature is raised towards the transition temperature the spectra undergo motional narrowing owing to the onset of rapid axial rotation of the phospholipid. Perhaps an important feature of such observations is that line narrowing takes place over a relatively broad temperature range ($\sim 15^{\circ}C$) and is essentially complete several degrees below the calorimetrically observed phase transition temperatures.

(b) The second set of observations relies on an entirely different type of measurement. Here we rely on the introduction of probe molecules (in our case various fluorescent molecules (20))into the phospholipid or membrane system. The apparent rotational motion of such molecules (derived from measurements of the polarisation and life-time of fluorescence) reflects the expected phase changes in single lipid systems (21). In those measurements too the observed temperature range for the transition is somewhat broader than that obtained from calorimetric data. The most likely reason for this is the progressive exclusion of the probe molecule from the crystalline regions of the phospholipids (see below and (22)).

From the biological point of view, where one often relies on correlations between functions and physical (structural) measurements, it is therefore important to understand the precise contributions to the observed changes in the physical measurements if the observations are to be interpreted in terms of the effect of structure and motion on the functional behaviour.

(ii) When the difference between the transition temperatures of two lipids is too large to allow co-crystallization of the fatty acid chains in their mixture <u>lateral phase separation</u> takes place.

Thus in an equimolar mixture of 18:1c/18:1c-phosphatidylcholine (dioleoyl lecithin) and 16:0/16:0-phosphatidylcholine (dipalmitoyl lecithin) two phase transitions are observed. The phase transition of 16:0/16:0-phosphatidylcholine, in the mixture, is broadened and shifted to lower temperatures because of interactions with 18:1c/18 :1c - phosphatidylcholine which remains in the liquid-crystalline state down to -20°C (7). Effects corresponding to lateral phase separation in vesicles composed of mixed phospholipid species may also be observed with $^{31}P$ NMR . Fig. 1 shows the $^{31}P$ NMR spectra

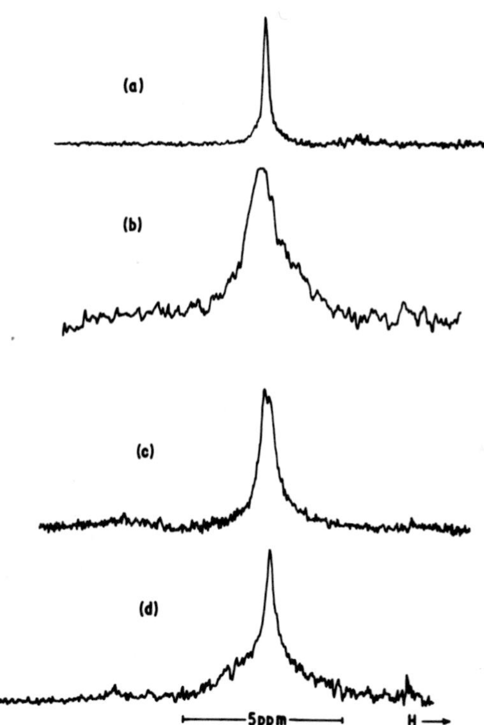


Fig.1. 36.4 MHz $^{31}P$ NMR spectra at 10°C of (a) 18:1c/18:1c-phosphatidylcholine vesicles,(b) 16:0/16:0-phosphatidylcholine vesicles, (c) 16:0/16:0-phosphatidylcholine-18:1c/18:1c-phosphatidylcholine (1:1) vesicles and (d) 16:0/16:0-phosphatidylcholine-18:1c/18:1c-phosphatidylcholine vesicles in the presence of 6mM $Co^{2+}$.

of 18:1c/18:1c-phosphatidylcholine (Fig.1a) 16:0/16:0 phosphatidylcholine (Fig.1b) and 16:0/16:0-phosphatidylcholine-18:1c/18:1c-phosphatidylcholine (1:1) vesicles (Fig.1c) at 10°C. The spectrum of the 16:0/16:0 phosphatidylcholine-18:1c/18:1c-phosphatidylcholine (1:1) vesicles is composed of a broad line, which is ascribed to 16:0/16:0-phosphatidylcholine molecules in the gel state and two narrower components, presumably due to the liquid crystalline 18:1c/18:1c-phosphatidylcholine molecules on the inside and outside of the vesicle. The outside resonances can be broadened beyond detection by the addition of $Co^{2+}$. The resultant spectrum of the inside resonances at 10°C, as shown in Fig. 1a is composed of a broad and narrow line, indicating the occurrence of phase separation in the inside monolayer of this vesicle. Fig.1 also demonstrates that the chemical shift difference between the outside and inside resonances of the narrow component is much larger than the chemical shift difference between the outside and inside resonances of 18:1c/18:1c-phosphatidylcholine vesicles.

In a similar type of experiment in mixtures of dipalmitoyl and dilauroyl phosphatidylcholines lateral phase separation occurs between the two endothermic phase transitions at 8 and 38°C (21).

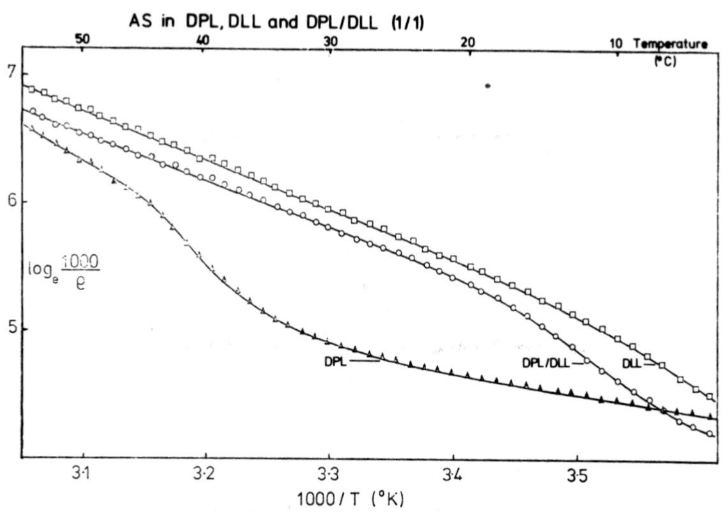


Fig.2 Arrhenius plots of the rate of rotation of 12-(9-anthroyloxy)-stearate bound to dipalmitoyl and dilauroyl phosphatidylcholines and in equimolar mixture of the two. Probe concentration 1μM, lipid concentration 0.5mM. Excitation at 385nm, emission measured above 410nm. 12-AS in dipalmitoyl phosphatidylcholine (▲), in dilauroyl phosphatidylcholine (□), in equimolar mixture (○).

When the fluorescent probe 12-(9-anthroyloxy-stearic acid is introduced in this mixed system the probe mobility is insensitive to the higher of the two transitions although as mentioned before it detects the phase transition in pure dipalmitoyl lecithin (Fig.2). This leads to the conclusion that in regions of lateral phase separation the probe is almost exclusively localised in the fluid phase. This observation has important implications when fluorescent probes (and possibly other labels) are used in estimating the fluidity ("microviscosity") of biological membranes. The likelihood is that the fluorescent molecule, being an impurity, is always excluded from the regions of the membrane containing lipids in a more ordered form and hence will only measure the motion of the probe in the less structured domains of the membrane.

It is relevant and interesting to note that in lipid-protein complexes (23) and biological membranes (24) proteins have a preference for the disordered lipid regions when phase separation presents a choice to the "protein impurity". Because of this it is important to have a method for measuring the presence of gel and fluid phases (and their amounts) in biological membranes. It is evident that both probe methods and to a lesser extent NMR tend to detect preferentially the more mobile components in mixtures of phospholipids.

Recently we have devised a new method that involves the introduction of positrons, positively charged anti-electrons, into the lipid matrix and the determination of their decay times. The properties of positrons are currently investigated in experimental physics and theoretical chemistry (25,27). Since, however, the concepts and measurements are not familiar to most biologists we shall briefly summarise them here.

Positrons are positively charged particles with the mass of an electron. There are several sources of such particles, but the most conveniently and commonly used source is the isotope $^{22}$Na which decays to an isotope of neon by positron emission. This emission is accompanied by a gamma photon from the excited state of neon, and for most purposes the gamma emission and positron emission can be regarded as simultaneous. Once formed, the positron must dissipate most of its energy before annihilation with an electron is possible. The positron may then either remain free, in which case it has a natural lifetime of the order of tens of picoseconds before annihilation, or else may bind an electron without being immediately annihilated. The bound state is known as 'positronium' and may be formed in either the singlet or triplet state with a yield weighted by the multiplicity. The singlet state, p-positronium, has a lifetime of the order of 125 picoseconds in free space before selfannihilation with the emission of two simultaneous 0.511 MeV gamma photons at 180°. The triplet, o-positronium, is much longer lived with a free space lifetime of about 140 ns, whereafter it decays by the emission of three simultaneous gamma photons at 120° to conserve momentum. Since the triplet lifetime is comparatively long, the species is susceptible to various 'quenching processes' which shorten the measured decay time by providing additional routes for annihilation. One such process is intersystem crossing with subsequent fast annihilation of the singlet positronium, while another of chemical interest is the oxidation leading to a 'bare' positronium. One other process leading to premature annihilation of the triplet state positronium is 'pick-off' annihilation. In this process, the o-positronium, after reaching thermal energy, interacts with one of the electrons from an outer orbital of a surrounding atom. The o-positronium is thus annihilated in a two-photon event. In molecular materials this 'pick-off' annihilation is the predominant route for loss of o-positronium in most circumstances. The pick-off annihilation rate is sensitive to the physical state of the material, and subsequently it is possible to distinguish different phases of a given substance by measurement of this rate.

One difficulty in the measurement is that the large amount of water in biological materials results in a significant contribution to the annihilation pick-off rate. Previously an external source of $^{22}Na$ sandwiched within two thin mica films and placed between the scintillation detectors of the measuring apparatus was used. We have taken advantage of the presence of water in biological samples by incorporating the $^{22}Na^{+}$ directly into the solution. In this approach a relatively small amount of isotope is needed and the effect of pick-off annihilation within the walls of the containing vessel is minimised.

Positron decay times were measured using conventional fast-slow coincidence circuitry (28). The operational time resolution was 0.7 nsec,or better,using $^{60}Co$ as calibration source.

Typical decay time spectra obtained for aqueous dispersions of dipalmitoyl lecithin are shown in Fig.3. The 'tail' of the decay is a result of pick-off annihilation in the lipid matrix: the component due to water contributes to the earlier part of the decay (see Table 4.) The lifetime component resulting from pick-off annihilation in the lipid remains constant within the accuracy of the analysis at temperatures below the gel-liquid crystalline transition temperature. Above this temperature, the decay time becomes longer and again remains constant at high temperatures. Decay times of 2.8 and 3.3ns are found for the frozen and fluid phases respectively. For dioleoyl and dilauroyl phosphatidylcholine, which are fluid at room temperature, pick-off decay times of 3.3ns are found. In a 1:1 mixture of dioleoyl and dipalmitoyl phosphatidylcholine at a temperature where phase separation occurs, the positron pick-off

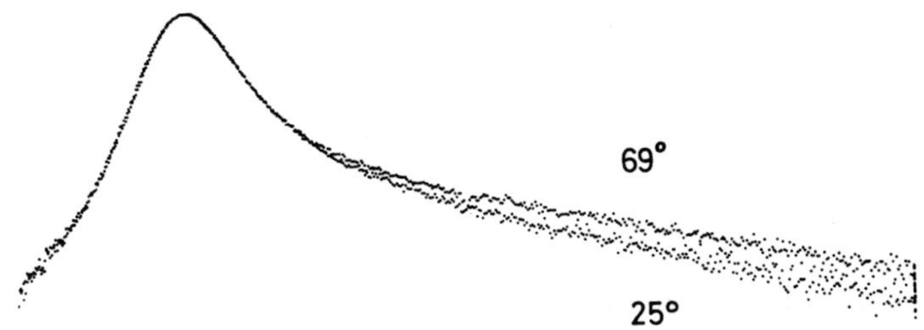


Fig.3. Decay time spectra for positrons in dipalmitoyl phosphatidylcholine at 69° and 25°. Log (intensity) is plotted against time, and least squares analysis of the decay 'tails' gives lifetimes of 2.8 and 3.3 ns at 25 and 69° respectively. Gel-liquid crystalline transition temperature for this lipid is 41.5°.

Table 4. PICK-OFF DECAY TIMES FOR O-POSITRONIUM IN PHOSPHOLIPIDS

| System | Temperature (°C) | Positronium lifetime (ns) | State of liquid |
|---|---|---|---|
| Water | 18 | 1.9 | ---- |
| 18:1c/18:1c-PC | 18 | 3.3 | liquid crystal |
| 12:0/12:0-PC | 18 | 3.3 | liquid crystal |
| egg PC | 18 | 3.25 | liquid crystal |
| 16:0/16:0-PC | 69 | 3.3 | liquid crystal |
| 16:0/16:0-PC | 25 | 2.8 | gel |
| 18:1c/18:1c-PC 16:0/16:0-PC (1:1) | 18 | 3.05 | gel and liquid phases |

Samples contained 25-50μCi of $^{22}Na^+$ (as NaCl) in 20mg of lipid dispersed in 0.5 ml buffer (10 mM trishydroxymethylaminomethane, pH 8.5).

annihilation rate would be expected to resolve into components characteristic of annihilation in fluid and frozen phases. Since these rates are similar (decay times of 3.3 and 2.8ns respectively), with equal weighting of components the resultant decay would be difficult to analyse, but would visually resemble a single component of intermediate lifetime. The measured positron pick-off decay time for the mixed lipid system was 3.05nsec (Table 4).

Clearly the decay times of positrons are sensitive to the fluidity of phospholipid-water systems. Positron lifetimes characteristic of fluid and frozen lipids have been established. In a system where fluid and frozen phases co-exist at room temperature on account of lateral phase separation, an intermediate apparent life-time is found. Careful analysis of the decay times is thus likely to provide a method for estimating the extent of phase separation in biological membranes.

### 3. Lateral Diffusion and Cluster Formation

Rapid lateral diffusion of phospholipids and of some membrane proteins has been demonstrated by several methods in recent years. It is true to say that we do not know if such observations have any direct biological relevance. Nevertheless, their demonstration undoubtedly contributes to the dynamic view of membrane structures (29), first proposed in the 'fluid mosaic model' by Singer and Nicholson (30).

We have recently introduced a new fluorescence method for diffusion studies on lipid and membrane systems (22). The fluorescent molecule 12-(9-anthroyloxy)-stearic acid that has been widely used as a lipid like probe (20) dimerises on irradiation with light of 366nm wavelength.

This dimerization proceeds by a diffusion limited second order mechanism in many solvents and in homogeneous fluid lipid dispersions and vesicles. The 'apparent diffusion coefficients' (for details see ref.22) for this probe in a variety of systems are shown in Table 5. In oriented lipid multilayers these diffusion coefficients are similar to those found by other techniques. It is, however, significant that the photodimerisation rates for the probe in fluid lipid <u>vesicles</u> are greater than those found for the rates in oriented multilayers of the same lipids.

There are two effects which may contribute towards the higher rates of diffusion of 12-(9-anthroyloxy)-stearic acid in vesicles. Firstly, the curvature of vesicles may cause structural differences permitting more rapid diffusion than in the multilayers. Secondly, the Brownian motion of the <u>bulk aqueous phase</u>, experienced by the probe in vesicles but <u>not</u> in multilayers, may increase the rate of lateral diffusion (31).

To determine which of these effects is responsible, the rates of diffusion of a probe which is located exclusively in the hydrocarbon region of phospholipids, and thus experiencing the Brownian motion of the bulk aqueous phase in neither vesicle nor oriented multilayers were investigated using excimer formation by pyrene derivatives as a means of comparing diffusion rates in the two model systems. Pyrene is located in the hydrocarbon interior of model membranes, while pyrene-butyric acid is accessible to the bulk aqueous phase in vesicles on account of its amphiphilic character. For a given lipid: probe ratio the amounts of excimer formed in vesicles and oriented multilayers were comparable for pyrene. This is consistent with the data of Callis and Vanderkooi, who calculated a diffusion coefficient of $3.0 \times 10^{-8}$ $cm^2 sec^{-1}$ for this probe in vesicles (32). For pyrene butyric acid excimer formation was considerably greater in vesicles than in oriented multilayers for a given lipid:probe ratio.

Table 5. INITIAL RATES OF BLEACHING OF 12-(9-ANTHROYLOXY)-STEARIC ACID IN VARIOUS MEDIA

| Medium | $k_{bleaching}$ | Probe Distribution |
|---|---|---|
| n-Butanol | $2.9 \cdot 10^{-7}$ | isotropic |
| Water | $2.0 \cdot 10^{-6}$ | clustered |
| Dodecane | $3.7 \cdot 10^{-6}$ | partly clustered |
| Lauric acid | $2.0 \cdot 10^{-6}$ | clustered |
| Dilauroyl-PC | | |
| vesicles | $2.8 \cdot 10^{-7}$ | isotropic |
| multilayers | $2.0 \cdot 10^{-8}$ | " |
| Dielaidoyl-PC | | " |
| vesicles | $2.3 \cdot 10^{-7}$ | |
| multilayers | $2.5 \cdot 10^{-8}$ | " |
| Dioleoyl-PC | | |
| vesicles | $1.9 \cdot 10^{-7}$ | " |
| multilayers | $2.0 \cdot 10^{-8}$ | " |
| Dipalmitoyl-PC | | |
| vesicles | $2.0 \cdot 10^{-6}$ | clustered |

Bleaching rates are in arbitrary units at 20°C. Diffusion coefficients can be calculated from these as in ref. 22. These calculations involve certain assumptions, but the ratio of rates in different lipid systems gives the ratio of diffusion coefficient directly.

These observations suggest that molecules which are accessible to the bulk aqueous phase will have significantly greater diffusion coefficients in phospholipid vesicles than in oriented multilayers on account of reduction of translational drag by the surrounding water (31).

Finally we should mention that anomalously high rates of photodimerization of 12-(9-anthroyloxy)-stearic acid are seen in water, in heterogenous hydrocarbons like liquid paraffin, phospholipids below their phase transition temperatures and in heterogeneous but <u>fluid</u> lipids like egg lecithin (Table 5.) In all these cases we have attributed the observed anomaly to the formation of localised high concentrations (<u>clusters</u>) of the fluorescent molecule. While it is not surprising that below the gel-liquid crystalline transition temperature of the phospholipid the probe tends to be excluded from the gel phase of the lipid matrix (see in lateral phase separation above) clustering in fluid lipids (and in hydrocarbon mixtures) is not necessarily obvious. We have suggested before on the basis of fluorescence polarisation experiments that even in liquid systems 'short range order' or 'liquid clustering' takes place (21). This would also account for the unusually high photodimerization rates of the fluorescent molecule.

As far as the biological problems are concerned 'correlated fluid motion' (which in a sense is equivalent to the formation of short range order above the melting temperature) could have a role in the rapid opening and closing of 'channels' in the membrane that could be responsible for passive diffusion processes across the lipid bilayer.

## III. THE RELATION BETWEEN THE PROPERTIES OF PHOSPHOLIPIDS AND BIOCHEMICAL FUNCTIONS OF THE CHROMAFFIN GRANULE MEMBRANE

We now return to the problem of the chromaffin-granule membrane and some of its special properties that may be important in the process of exocytosis (membrane fusion). We record that both the ATPase and NADH: acceptor oxido-reductase activities of the chromaffin-granule membrane have discontinuous Arrhenius temperature versus activity relationships with 'transitions' at 33°C (33). The transition for the NADH : acceptor oxido-reductase is removed by treatment with the detergent triton X-100. It is important to emphasize that such anomalies in Arrhenius behaviour in themselves may have no special significance (a fact that is often overlooked in membrane studies). What is significant in the chromaffin granule system is that the fluorescence properties of four different fluorescent probes (33,34) and the behaviour of five other types of spin labels (35) all indicate that some ordering of the membrane lipids occur below 33°C. What makes this observation particularly attractive is that this transition temperature is so close to the physiological operating range of the system. It is almost as if this membrane is poised to undergo some transition (responding to the triggering signal) as part of the fusion process. The cholesterol content of the membrane which is close to the region of 30% cholesterol per total lipid, where cholesterol segregation (see above) is just beginning to become a possibility, may well be significant to this. Although at present it is clear that the biological system is still too complex to draw definite conclusions, the model studies have allowed us to point to possibilities and defined the limitations and advantages of the various methods that hopefully will lead to a solution.

We thank the Science Research Council for financial support. C.G. Morgan is a S.R.C. Post-Doctoral Fellow, P.R. Cullis is a Medical Research Council (Canada) Post-Doctoral Fellow and B. De Kruijff was a recipient of a stipend of the Netherlands Organisation for the Advancement of Pure Research (Z.W.O.)

1. Henderson, R., and Unwin, P.N.T. (1975) Nature, 257, 28-32.
2. Douglas, W.W., (1974) Biochem.Soc.Symp. 39, 1-28.
3. Bashford, C.L., Radda, G.K., Ritchie, G.A. (1975) FEBS Lett. 50 21-24.
4. Bashford, C.L., Casey, R.P., Radda, G.K., Ritchie, G.A. (1975) Biochem.J. 148, 153-155.
5. Bashford, C.L., Casey, R.P., Radda, G.K., Ritchie, G.A. Neuroscience, 1976 (in press).
6. Kirschner, N. (1962). J.Biol.Chem. 237, 2311-2317.
7. De Kruijff, B., Cullis, P.R., Radda, G.K. (1975), Biochim. Biophys.Acta. 406, 6-20.
8. Cullis, P.R., De Kruijff, B., Richards, R.E., (1976), Biochim. Biophys.Acta. 426, 433-446.
9. McLaughlin, A.C., Cullis, P.R., Hemminga, M.A., Hoult, D.I., Radda, G.K., Ritchie, G.A., Seeley, P.J., Richards, R.E., (1975) FEBS Lett. 57,213-218.
10. Litman, B.J., (1973), Biochemistry, 12, 2545-2554.
11. Michaelson, D.M., Horwitz, A.F., Klein, M.P. (1973), Biochemistry, 12, 2637-2645.
12. Barker, R.W., Barrett-Bee, K., Berden, J.A., McCall, C.E., Radda, G.K., BBA Library Vo.13, 321-335,
13. Israelachvili, J.N., (1973), Biochim.Biophys.Acta 323, 659-663.
14. Kostelnik, R.J., Castellano, S.M., (1972), J.Magn.Res. 7, 219-223.
15. Berden, J.A., Barker, R.W., Radda, G.K., (1975), Biochim.Biophys. Acta., 375, 186-208.
16. De Kruijff, B., Cullis, P.R., Radda, G.K., (1976), Biochim. Biophys.Acta. in press.
17. Zwaal, R.F.A., Roelofsen, B., Colley, C.M. (1973), Biochim. Biophys.Acta. 300, 159-182.
18. Oschman, J.L., Wall, B.J., Gupta, B.L. (1974), 28, Symp.Soc. Exptl.Biol., 305-350.
19. De Kruijff, B., Cullis, P.R., Radda, G.K., Richards, R.E.,(1976) Biochim.Biophys.Acta., 419, 411-424.
20. Radda, G.K., Vanderkooi,J., (1972). Biochim. Biophys.Acta. 265, 509-549.
21. Bashford, C.L., Morgan, C.G., Radda, G.K. (1976), Biochim. Biophys.Acta. 426, 157-172.
22. McGrath, A.E., Morgan, C.G., Radda, G.K., (1976), Biochim. Biophys. Acta. 426, 173-185.
23. Grant, C.W.M., Hong-Wei Wu, S., McConnell, H.M., (1974), Biochim. Biophys.Acta. 363, 151-158.
24. Shechter, E., Letellier, L., Gulik-Krzywicki, T. in "Molecular Aspects of Membrane Phenomena", Kaback, H.R., Neurath, H., Radda, G.K., Schwyzer, R. & Wiley, W.R. eds. 1975. Springer Verlag, 39-63.
25. Brandt, W., (1974), Appl.Phys. 5, 1-7
26. Tao, S.J., (1974), Appl.Phys. 3, 1-23.
27. Walker, W.W., Kline, D.C., (1974), J.Chem.Phys. 60, 4990-3.

28. Williams, T.L., Ache, H.J., (1969), J.Chem.Phys. 50, 4493-4501.
29. Radda, G.K., (1975), Phil.Trans.R.Soc.Lond. 272, 159-171.
30. Singer, S.J., Nicholson, G.L., (1972) Science, 175, 720-731.
31. Saffman, P.G., Delbrück, M., (1975), Proc.Natl.Acad.Sci.72, 3111-3113.
32. Vanderkooi, J.M., Callis, J.B., (1974), Biochemistry, 13, 4000-4006.
33. Radda, G.K., (1975), Phil.Trans.R.Soc.Lond. 270, 539-549.
34. Bashford, C.L., Johnson, L.N., Radda, G.K., Ritchie, G.A., (1976) Europ.J.Biochem. in press.
35. Ritchie, G.A., (1975) D.Phil.Thesis, Oxford.

# OLIGOSACCHARIDES OF THE MEMBRANE GLYCOPROTEINS OF SEMLIKI FOREST VIRUS

Ossi Renkonen, Marja Pesonen and Kari Mattila

Dept. of Biochemistry, University of Helsinki
Haartmaninkatu 3, 00290 Helsinki 29, Finland

"All cells come with a sugar coating" is a brief and important summary of a very large body of biological experience. Most of this coating is in the form of glycoproteins. During the evolution of life the materials located on cell surfaces have assumed important functions in a variety of intercellular interactions. Hughes (1975) has recently reviewed some of these functions, including the role of glycoproteins as antigenic determinants, surface receptors and as elements of intercellular recognition and adhesion.

In an attempt to simplify the study of the cellular plasma membrane our group has selected the envelope of Semliki Forest virus (a small animal virus of the toga virus group) as a model membrane.

This membrane is, in cellular dimensions, a very small structure consisting of only a few hundred protein molecules and of about 30.000 lipid molecules (Laine et al., 1973). Moreover, when Semliki Forest virus (SF virus) infects a suitable host cell, the synthesis of the cellular macromolecules comes to a full stop. The infected cell offers thus an ideal system for observing the synthesis and the assembly of the viral structures.

The present paper describes some work with the glycoproteins of SF virus. Many other viruses, too, are currently being studied with similar methods and with similar motivations (Klenk, 1974). In particular the work with Sindbis virus, a close relative of SF virus, carried out by Burge and Strauss (1970), Sefton and Burge (1973), Sefton and Keegstra (1974), Keegstra et al. (1975), Sefton (1976) and Schlesinger et al. (1976) has inspired our work.

## Composition and Structure of SF virus

SF virus is composed of a nucleocapsid surrounded by a membrane. The nucleocapsid contains the viral RNA and one lysine-rich protein species (Simons and Kääriäinen, 1970). The nucleocapsid is assembled in the cytoplasm of the host cell and it acquires its surrounding membrane by a budding process during maturation (Acheson and Tamm, 1967). The viral membrane contains lipid and protein. The lipids are arranged into a bilayer structure (Harrison et al., 1971), and they are apparently derived from the host cell plasma membrane (Renkonen et al., 1971).

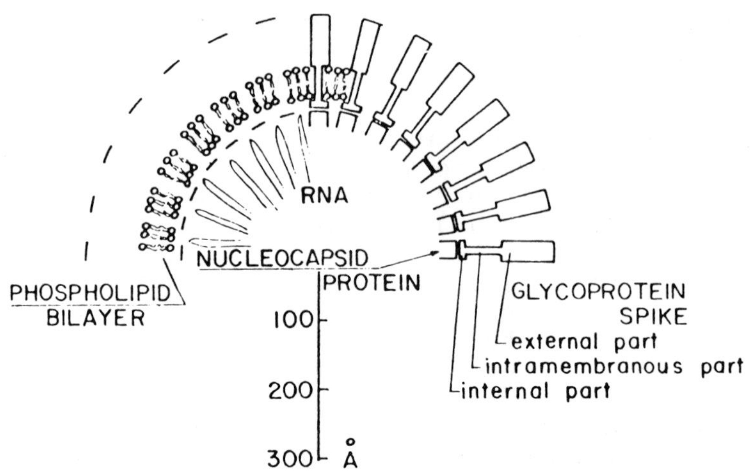


Fig. 1. Proposed structure of Semliki Forest virus (taken from the doctoral thesis of H. Garoff, Helsinki, 1974)

Three membrane proteins are found in SF virus (Garoff et al., 1974); they form spikes on the external surface of the membrane. All three membrane proteins contain covalently bound carbohydrates (Garoff et al., 1974). Two of these glycoproteins, $E_1$ (Mw 49000) and $E_2$ (Mw 52000), contain hydrophobic segments which "anchor" them into the lipid bilayer (Utermann and Simons, 1974). In contrast to these intrinsic membrane proteins, $E_3$ (Mw 10000) is an extrinsic membrane glycoprotein, believed to have no hydrophobic interaction with the lipids.

The nucleocapsid protein of SF virus is not glycosylated (Ranki et al., 1972; Kennedy and Burke, 1972).

The three membrane proteins and the nucleocapsid protein appear to be present in about equimolar ratios in the purified virion (Garoff et al., 1974). The results of chemical analysis suggest that a few hundred copies of each are present (Laine et al., 1973). This agrees well with the findings of v. Bonsdorff and Harrison (1975), who have recently demonstrated 240 subunits on the surface of Sindbis virus.

It is likely that the nucleocapsid proteins interact with one or both of the spike glycoproteins $E_1$ and $E_2$ which seem to extend through the lipid bilayer into a close contact with the nucleocapsid (Garoff and Simons, 1974). A model of the structure of Semliki Forest virus is shown in Fig. 1.

Table 1. Carbohydrate Composition of SFV glycoproteins

| | (moles carbohydrate/mole protein) | | |
|---|---|---|---|
| | E1 | E2 | E3 |
| N-acetylglucosamine | 7 | 8 | 9 |
| Mannose | 5 | 12 | 4 |
| Galactose | 3 | 3 | 4 |
| Fucose | 1 | 1 | 2 |
| Sialic Acid | 2 | 4 | 3 |
| Total carbohydrate % by weight | 7.5 % | 11.5 % | 45.1 % |

## Protein-bound Oligosaccharides of SF virus

The three membrane glycoproteins have been separated from each other on a preparative scale and their carbohydrates analyzed (Table 1) (Garoff et al., 1974). All three glycoproteins appeared to contain galactose, mannose and fucose in addition to N-acetyl glucosamine and sialic acid. The carbohydrate compositions suggest that most of the oligosaccharide units are linked through N-glycosidic bonds to amide groups of asparagine residues in the polypeptide chains. The relative amounts of the different monosaccharides indicate that the oligosaccharides of $E_1$ and $E_3$ may be mostly of "A-type" (Johnson and Clamp, 1971), whereas the high mannose content of $E_2$ suggests that it may contain "B-type" oligosaccharides.

Closer analysis of the oligosaccharides was carried out by labeling the virus in vivo with radioactive monosaccharides in the medium of the host cells. The viral proteins were then separated with discontinuous polyacrylamide gel electroforesis in the presence of sodium dodecyl sulfate (Fig. 2). The three membrane glycoproteins were eluted from the gels and subjected to pronase digestion, which cleaves off most of the polypeptide and leaves the oligosaccharide attached to an amino acid or a small peptide. These pronase-glycopeptides were then analyzed by gel filtration.

The results obtained in these experiments (Mattila et al., 1976) confirmed that $E_1$ and $E_3$ contain oligosaccharides of the A-type with mannose, galactose, fucose and N-acetyl glucosamine as constituent monosaccharides. $E_2$ revealed B-type oligosaccharide with only mannose and N-acetyl glucosamine. In addition, $E_2$ revealed material which labeled heavily from [$^3$H]-galactose, but only marginally from labeled glucosamine, fucose and mannose.

The apparent molecular weights of the A-type glycopeptides measured by gel filtration were 3400 daltons in $E_1$ and 4000 daltons in $E_3$; the B-type unit of $E_2$ was of 2000 daltons, and the galactose-rich oligosaccharide of $E_2$ was of 3100 daltons.

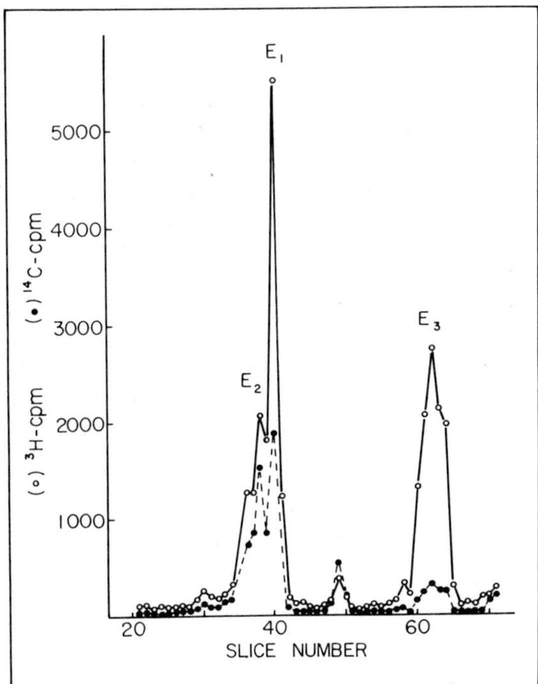


Fig. 2. Gel electrophoresis of SF virus proteins labeled with [$^{14}$C]-mannose ( • ) and [$^{3}$H]-galactose ( ◦ ). (taken from Mattila et al., 1976).

We hoped that the number of oligosaccharide chains in each protein could now be estimated by combining the data of the carbohydrate content of the proteins (Table 1) and the molecular weights of the glycopeptides. There was, however, a problem here; the size of the peptide part in the glycopeptides is not known. In some instances no extra amino acids are found in the pronase glycopeptide (Arima et al., 1972), in others the peptide appears to be rather large (Baenziger and Kornfeld, 1974).

To overcome this difficulty we have hydrazinolyzed the pronase glycopeptides of SF virus. This reaction, which has been developed into a potentially valuable tool of glycoprotein analysis in Montreuil's laboratory (Bayard and Roux, 1975), cleaves amide and ester linkages, but not glycosidic bonds. Therefore it should reduce conveniently the peptide structure attached to the oligosaccharides in the pronase glycopeptides. From N-glycosidic glycopeptides the reaction should yield aminoglycosides of the oligosaccharides, in de-N-acetylated form (Fig 3), and from O-glycosidic glycopeptides it should give oligosaccharides bound an amino acyl hydrazide.

The hydrazinolysates of the SF virus glycopeptides were re-N-acetylated and the molecular weights of the oligosaccharides were estimated by gel filtration. Our preliminary results suggest that the mixed "amino acid-free" A-type oligosaccharides have molecular weights around 3000 daltons. Analyzed in the same way, the molecular weight of the "amino acid-free" B-type oligosaccharide of $E_2$-protein appears to be about 1250 daltons (Rasilo et al., unpublished).

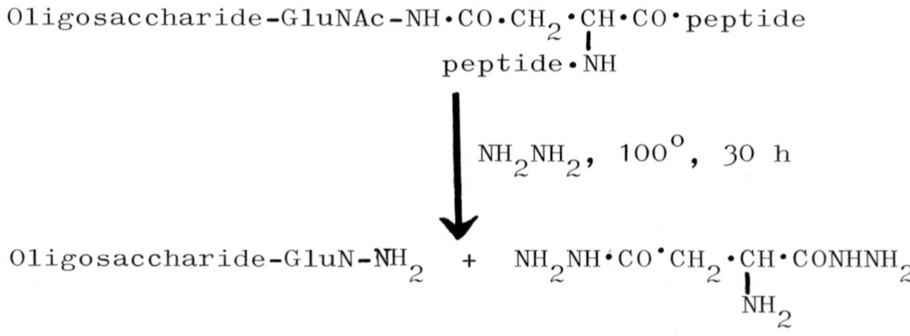


Fig. 3. Hydrazinolysis of a N-glycosidic glycopeptide

The results of chemical analysis shown in Table 1 suggest that $E_1$ may contain an oligosaccharide of about 3400 daltons, and $E_3$ an oligosaccharide of about 4200 daltons. When these data are combined with the results of hydrazinolysis, it appears that there may be a little more carbohydrate in $E_1$, and particularly in $E_3$, than is required for one mole of oligosaccharide. However, on the basis of results obtained in sequencing the A-oligosaccharides (see below), we believe that the best estimate is one mole oligosaccharide in both proteins. The uncertainties inherent in the chemical analysis (Table 1), and in the estimation of the molecular weights of both the proteins and the oligosaccharides, are so great that the number of oligosaccharide chains obviously must be studied also with the aid of tryptic peptides.

In $E_2$-protein the number of glucosamine and mannose residues (Table 1), which form the "1250 dalton oligosaccharides", suggests that there can be as many as three B-type oligosaccharides. Even this estimation is quite uncertain; $E_2$ isolated on the hydroxylapatite column is likely to be contaminated by $E_1$ and $E_3$, and it is not quite certain that all glucosamine and mannose units of $E_2$ belong to the B-type oligosaccharides (Mattila et al., 1976).

## The Structure of the A-type Oligosaccharides

When the SF virus was labeled in vivo with 1-$[^3H]$-mannose and its all three glycoproteins, without any previous fractionation, were digested with pronase, a gel filtration pattern like that in Fig 4 was obtained. The material eluting at the void volume (fraction no. 23) was probably undegraded proteins, the peak at fraction no. 35 consisted of the A-type glycopeptides of $E_1$ ($E_1A$) and $E_3$ ($E_3A$). The peak at fraction no. 40 contained the B-glycopeptides of $E_2$ ($E_2B$). The galactose rich glycopeptide of $E_2$ ($E_2X$) was not labeled in this experiment; it probably occupied a position between the A and B peaks.

The A-type glycopeptides from this and similar experiments were subjected to a series of stepwise degradations with pure exo-hydrolases. The reduction in the molecular weight and the cleavage of radioactive monosaccharides were monitored with gel filtration (Pesonen and Renkonen, 1976). The sequential degradations were carried out with neuraminidase followed by

L-α-fucosidase, D-β-galactosidase, D-β-N-acetyl-glucosaminidase, D-α-mannosidase, D-β-mannosidase and D-β-N-acetyl-glucosaminidase.

The molecular weight of the glycopeptide was reduced by about 1000 daltons after the neuraminidase treatment, which suggests the presence of 3.4 moles of distal L-α-sialic acid.

Incubation of the native glycopeptide with L-α-fucosidase released all fucose label and decreased the apparent molecular weight of the glycopeptide with 100 daltons.

Treatment of the asialo-glycopeptides with D-β-galactosidase of Jack bean reduced the molecular weight by 500 daltons, equivalent with three galactose units. All galactose label was liberated.

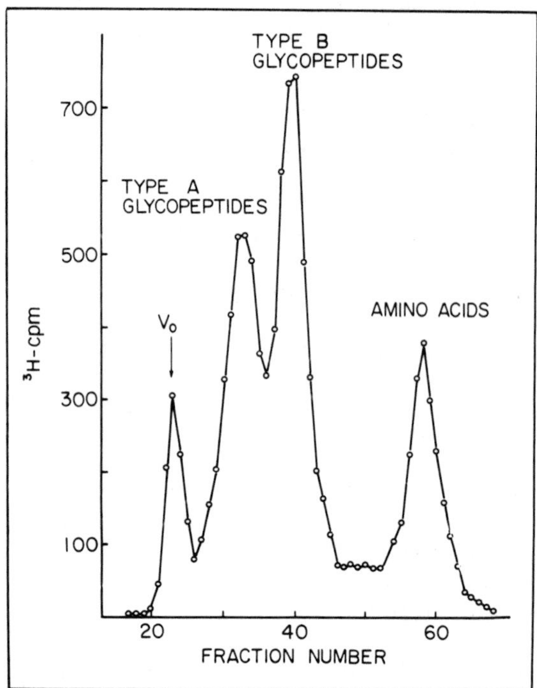


Fig. 4. Gel filtration of $[^3H]$-mannose-labeled pronase glycopeptides of SF virus on Biogel P6 (taken from Mattila et al., 1976).

Incubation of the resulting glycopeptide with N-acetyl-β-glucosaminidase resulted in the liberation of 58 % of the glucosamine label as monosaccharide, and in the reduction of the molecular weight of the glycopeptide from 2200 to 1360 daltons. This drop is equivalent to 4.2 moles of N-acetyl glucosamine. In this case the not-liberated glucosamine label (42 %) is equivalent to 3.2 moles. However, it is possible that there are only two moles of the "not-liberated" glucosamine (see below). This would implicate that three moles of N-acetyl glucosamine were liberated in the present experiment, and that the drop in the apparent molecular weight as indicated by gel filtration was too large.

Subsequent treatment of the glycopeptide with α-mannosidase released 70 % of the mannose label, and 25 % of the mannose label was found in a glycopeptide of 1200 daltons, together with glucosamine radioactivity. The not-released mannose proved to be due to one mole of β-mannose units (see below), which implies that the present incubation released 2.8 moles of α-mannose units. The drop in the molecular weight of the glycopeptide as indicated by gel filtration was too small in this case.

The residual 25 % of mannose label were completely liberated by treatment with pure β-mannosidase obtained from professor Y-T. Li. The reduction in the apparent molecular weight was, this time, reasonable, 140 daltons, corresponding to 0.9 moles of mannose.

The glycopeptide from all mannose units which had been stripped off was incubated with β-N-acetyl glucosaminidase. In our first experiment, two thirds of the glucosamine label was released as monosaccharide; from our second sample a more prolonged incubation released only one third of the counts. Averaging, we conclude that half of the glucosamine label present after demannosylation is releasable with β-N-acetyl glucosaminidase, and half is not. This last half is believed to represent glucosamine units attached to asparagine.

The apparent molecular weight of the N-acetyl glucosaminyl peptide was 720 daltons, which implies that the apopeptide probably was of about 500 daltons.

These findings are compatible with the basic structure shown in Fig. 5. To this general structure are attached about one mole of fucose and, perhaps, one mole of mannose. The proposed structure should be regarded as a tentative suggestion only, since it represents average data obtained with a mixture of $E_1A$ and $E_3A$. These may, or may not be identical; for instance the number and the length of the distal chains attached to the $\alpha$-mannose units, and the number of fucose residues, may be different in $E_1A$ and $E_3A$. The structure of Fig 5 is familiar from a number of animal systems, and is reportedly present also in BHK cells (Ogata et al., 1976), which were used as host cells in our experiments. It has been generally thought that the viral genome is too small to carry the genes for all sugar transferases needed for the oligosaccharide synthesis. It has seemed possible, however, that the virus could specify one or two enzymes, which, together with cellular enzymes, would carry out the synthesis of an unique viral oligosaccharide. Even this, obviously, is not the case in type A oligosaccharides of SF virus; the whole structure appears to be cell-specific.

## Glycosylation of SF virus Membrane Proteins

The structural data presented above seem to justify the use of SF virus as a probe for the processes of glycosylation of membrane proteins within the host cells.

In addition to the structural proteins of SF virus, $E_1$, $E_2$ and $E_3$, virus specific nonstructural proteins are present in infected cells. A nonstructural protein NSP62 can be detected in relatively large amounts (Hay et al., 1968; Ranki et al., 1972; Morser et al., 1973). It is believed to be a precursor protein of the membrane glycoproteins $E_2$ and $E_3$ (Ranki et al., 1972; Simons et al., 1973; Garoff et al., 1974; Lachmi et al., 1975).

In smaller amounts a still larger nonstructural glycoprotein NSP97, thought to be a precursor of NSP62 and $E_1$ can be detected (Morser and Burke, 1974). All the structural proteins of the virus are translated as one precursor polyprotein having an apparent molecular weight of about 130 000 daltons (Keränen and Kääriäinen, 1975; Kääriäinen, 1976). This has been shown mainly by using temperature sensitive mutants of the virus.

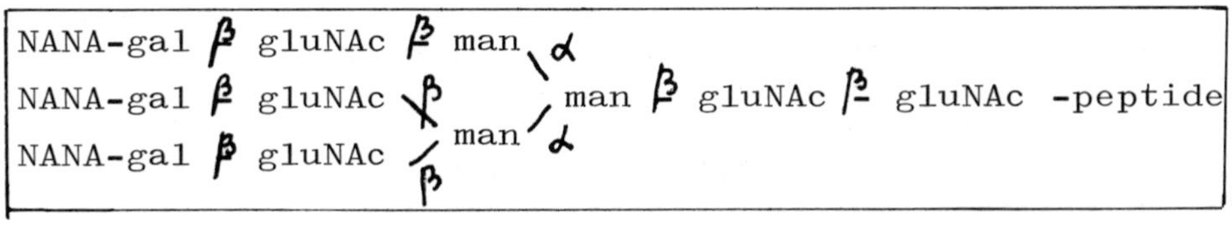

```
NANA-gal β gluNAc β man \ α
NANA-gal β gluNAc β       man β gluNAc β gluNAc -peptide
NANA-gal β gluNAc / man / α
                  β
```

Fig. 5. Proposed general structure of the A-type oligosaccharides in SF virus (taken from Pesonen and Renkonen, 1976).

The presence of the glycosylated precursor proteins in SF virus infected cells offers an attractive possibility to study how the proteolytic cleavage of these precursors into the final structural proteins is related in time and space to the stewise process of protein glycosylation.

At present the mechanisms operating in the assembly of A-type oligosaccharides are known to some extent e.g. for the plasma glycoproteins secreted by liver (Molnar, 1975) and for the immunoglobulin light chains secreted by myeloma cells of mice (Melchers, 1973). The A-oligosaccharides of these proteins are glycosylated stepwise at several sites within the cells. Roughly, the evidence suggests that the first N-acetyl glucosamine residues may become attached to the peptide while this is still in the rough endoplasmic reticulum, attached to the ribosomes. The core region consisting of N-acetyl glucosamine and mannose residues appears to be assembled at the endoplasmic reticulum. On the other hand, the Golgi apparatus appears to be responsible for the addition of the distal N-acetyl glucosamine, galactose, and sialic acid units (Hughes, 1975). Fucosyl transferase, too, is assumed to be present in the distal compartments of the cell, in Golgi and plasma membrane (Molnar, 1975; Riordan et al., 1974).

As the A-type oligosaccharides of SF virus resemble those found in plasma proteins and immunoglobulins (Baenziger and Kornfeld, 1974) it seems possible that their assembly, too, is a stepwise process which can be traced, more clearly, perhaps, than is usual, with the aid of the precursor-product relationship found in the virus specific proteins of SF virus infected cells.

Kaluza (1975) has shown that after a pulse of 90 min with [$^3$H]-mannose, NSP62 has a higher mannose label than $E_1 + E_2$. Similar pulses with $^3H$ -fucose, $[^3H]$-glucosamine and $^3H$ -galactose indicated that NSP62 obtains the label from all these tritiated sugars, but to a smaller extent than $E_1 + E_2$. The glycosylation can be impaired with 2-deoxy-D-glucose or D-glucosamine (Kaluza et al., 1973). These antimetabolites have no effect on the synthesis of carbohydrate-free protein, but they inhibit the multiplication of SF virus (Kaluza et al., 1972).

When the glycosylation of SF virus proteins is impaired by omission of sugars in the culture medium, or with 2-deoxy-glucose or glucosamine, aberrant forms of glycoproteins are formed (Kaluza, 1975). They have lower molecular weights, and contain less carbohydrates, especially mannose, than the normal glycoprotein molecules. Pulse-chase experiments indicate that the aberrant precursor glycoproteins are cleaved very slowly if at all into (aberrant) final products. If, however, the impairment is caused by omission of sugars in the culture medium, the radioactivity can be chased from the aberrant NSP62 into the normal $E_1 + E_2$ (Kaluza, 1975). This suggests a completion of the carbohydrate chains under these conditions.

Our group has been interested in the oligosaccharides which can be found on NSP62. We have observed earlier that the NSP62, which accumulates in cells treated with canavanine (this arginine analogue blocks the cleavage of NSP62 into $E_3$ and $E_2$), lacks fucose (Ranki et al., 1972). In our more recent experiments (Stenvall et al., unpublished) we have analyzed, after a 7h-pulse with tritiated sugars, the normal NSP62 by subjecting it to pronase digestion and subsequent gel filtration. Our preliminary results indicate that the galactose containing oligosaccharide $E_2X$ appears to have gained its full molecular size already on NSP62; it appears to be rather heterogeneous, though. This material contains galactose label, but no label from mannose or glucosamine. All glucosamine and mannose label in NSP62 appears in a single, sharp glycopeptide peak of 2000 daltons, which does not contain galactose label.

In contrast, the A-type glycopeptides obtained by pronase digestion of the $E_3$ found in the cells appear to have the full size of about 4000 daltons; also the B-type and the X-type oligosaccharides are full-sized in the $E_2$-protein isolated from the cells.

The 2000 dalton glycopeptide of NSP62 can be either $E_2B$ or an unfinished form of $E_3A$ (pre-$E_3A$), or both. The full-sized $E_2B$ is of 2000 daltons and contains only glucosamine and mannose (see above). The galactose-free "pre-A" obtained by treatment of the A-type oligosaccharides with sialidase and galactosidase has a Mw around 2200 daltons (see above), and that form of "pre-A" which lacks the distal glucosamine units, has a Mw around 1400 daltons (see above). If, indeed, pre-$E_3A$ is present in NSP62, it may contain the proximal glucosamines, the mannoses and even some of the distal glucosamines, but not the galactose or sialic acid units.

If the presence of "pre-$E_3A$" in NSP62 can be substantiated these data would imply that NSP62 is probably cleaved into $E_2$ and $E_3$ before it reaches a site where the A-type oligosaccharides are finished and obtain their full size and their most distal monosaccharide units. This site is quite likely the Golgi apparatus.

The data on $E_2X$ on NSP62 are interesting as they reveal no glucosamine or mannose label and help to distinguish $E_2X$ clearly from the A-type oligosaccharides. Moreover, the incorporation of galactose label into $E_2X$ of NSP62 before any galactose is incorporated into the "pre-$E_3A$" carried on the same protein may imply two galactosylation sites of the viral proteins in BHK cells.

Tunicamycin, an inhibitor of biosynthesis of membrane glycoproteins (Takatsuki and Tamura, 1971), inhibits the incorporation of glucosamine in acid insoluble products, and, specifically, the formation of polyisoprenyl N-acetyl glucosaminyl pyrophosphate (Tkacz and Lampen, 1975).

Following these findings which seemed to suggest that polyisoprenyl sugar pyrophosphates play a role in the glycosylation of membrane proteins in eucaryotic systems, we attempted to extract lipid soluble, labeled glucosamine derivatives from BHK cells infected with SF virus. Much to our disappointment we were unable to identify labeled polyisoprenyl sugar pyrophosphates in our system (Somerharju et al., unpublished).

## The Outlook

$E_1$ protein forms the hemagglutinating apparatus of SF virus (Helenius and Simons, 1976). The sialic acid units can be removed from the intact virus without affecting its hemagglutinating properties. But when the virus is completely deglycosylated all of its hemagglutinating activity is lost (Kennedy, 1974). When the virus is grown in a mutant cell line deficient in a specific N-acetyl glucosamine transferase the distal sialic acid-galactose-N-acetyl glucosamine chains of the viral oligosaccharides seem to be missing, and the hemagglutinating activity of the virus is clearly reduced (Schlesinger et al., 1976).

The stewise degradation, which we have carried out with the oligosaccharides $E_1A$ and $E_3A$, should, perhaps, be possible also with the intact $E_1$ protein. It should clarify which is the necessary oligosaccharide structure(s) making $E_1$ a hemagglutinin.

The possibility of obtaining $E_1$ oligosaccharides in the form of pronase glycopeptides and hydrazinolysates should make it possible to see, with hemagglutination-inhibition type of experiments, whether the "hemagglutination-oligosaccharide" also plays a role in the adsorption of the virus in the process of infection of BHK cells, and the other hosts used for SF virus, chick embryo fibroblasts and cultured mosquito cells.

The small SF virion particle should make an ideal model for "topographical" studies of the structure of the glycocalyx, e.g. with galactose oxidase (Gahmberg, 1976), and other enzymes.

The possibility of chemical labeling of the oligosaccharides after hydrazinolysis, e.g. by re-N-acetylation with labeled acetic anhydride, should be quite helpful in their structural analysis. Complete sequencing of large oligosaccharides are now possible with pure tracer techniques, as shown by Pesonen and Renkonen (1976).

It remains to be seen whether glycosylation of the viral proteins in some way may label them for transport across cell membranes. - It is striking that the G-glycoprotein of Vesicular stomatitis virus, which is situated on the outer surface of the virion, is synthetized on membrane bound ribosomes, whereas the M-protein, another membrane protein of the virus situated on the inner side of the membrane and carrying no carbohydrates, is synthetized at least partly on cytoplasmic ribosomes (Morrison and Lodish, 1975). This line of investigation has gained new momentum, and it has become more confused, after the recent findings of Dallner's group suggesting that even the cytoplasmic surface of rough endoplasmic reticulum may contain glycosylated proteins (Depier and Dallner, 1975).

Anyways, the unusually large number of different oligosaccharides of Semliki Forest virus, and the large number of different host cells where this virus may be grown, together with the curious way its proteins are synthetized, should make it a very useful probe for studies on the glycosylation and transport of membrane proteins within eucaryotic cells.

## Acknowledgments

We thank Drs L. Kääriäinen, C.G. Gahmberg, K. Simons and A. Helenius for many helpful and stimulating discussions. Grants from Sigrid Jusélius Foundation, Wihuri Foundation, Suomen Kulttuurirahasto and the Finnish Academy of Science are acknowledged. Many of the enzymes were gifts from Dr. Y-T. Li.

## References

Acheson, N.H. and Tamm, F. (1967), Virology 32, 128.

Arima, T., Spiro, M.J. and Spiro, R.G. (1972), J. Biol. Chem. 247, 1836.

Baenziger, J. and Kornfeld, S. (1974), J. Biol. Chem. 249, 1889.

Bayard, B. and Roux, D. (1975), FEBS Lett. 55, 206.

v. Bonsdorff, C.H. and Harrison, S.C. (1975), J. Virology 16, 141.

Burge, B.W. and Strauss, Jr., J.H. (1970), J. Mol. Biol. 47, 449.

Depierre, J.W. and Dallner, G. (1975), Biochim. Biophys. Acta 415, 411.

Gahmberg, C.G. (1976), J. Biol. Chem. 251, 51.

Garoff, H., Simons, K. and Renkonen, O. (1974), Virology 61, 493.

Garoff, H. and Simons, K. (1974), Proc. Nat. Acad. Sci., U.S. 71, 3988.

Harrison, S.C., David, A., Jumblatt, J. and Darnell, J.E. (1971), J. Mol. Biol. 60, 523.

Hay, A.J., Skehel, J.J. and Burke, D.C. (1968), J. Gen. Virol. 3, 175.

Helenius, A. and Simons, K. (1976), submitted for publication.

Hughes, R.C. (1975) in "Essays in Biochemistry", vol. 11, p. 1, eds P.N. Campbell and W.N. Aldridge, Academic Press.

Hughes, R.C. (1976) in "Membrane Glycoproteins", Butterworths.

Johnson, J. and Clamp, J.R. (1971), Biochem. J. 123, 739.

Kaluza, G., Scholtissek, C. and Rott, R. (1972), J. Gen. Virol. 14, 251.

Kaluza, G., Schmidt, M.F.G. and Scholtissek, C. (1973), Virology 54, 179.

Kaluza, G. (1975), J. Virol. 16, 602.

Keegstra, K., Sefton, B. and Burke, D. (1975), J. Virol. 16, 613.

Kennedy, S.I.T. (1974). J. Gen. Virol. 23, 129.

Kennedy, S.I.T. and Burke, D.C. (1972), J. Gen. Virol. 14, 87.

Keränen, S. and Kääriäinen, L. (1975), J. Virol. 16, 388.

Klenk, H-D. (1974), in "Current Topics in Microbiology and Immunology", vol. 16, p. 29, Springer-Verlag.

Kääriäinen, L. (1976) in "Membrane Assembly and Turnover", vol. 4, eds. G. Poste and G.L. Nicolson (in press), Elsevier.

Lachmi, B., Glanville, N., Keränen, S. and Kääriäinen, L. (1975), J. Virol. 16, 1615.

Laine, R., Söderlund, H. and Renkonen, O. (1973), Intervirology 1, 110.

Mattila, K., Luukkonen, A. and Renkonen, O. (1976), Biochim. Biophys. Acta 419, 435.

Melchers, F. (1973) in "Membrane Mediated Information", vol. 2, p. 39, ed. P.W. Kent, Medical and Technical Publishing Company, Lancaster, England.

Molnar, J. (1975), Molecular and Cellular Biochemistry 6, 3.

Morrison, T.G. and Lodish, H.F. (1975), J. Biol. Chem. 250, 6955.

Morser, M.J., Kennedy, S.I.T. and Burke, D.C. (1973), J. Gen. Virol. 21, 19.

Morser, M.J. and Burke, D.C. (1974), J. Gen. Virol. 22, 395.

Ogata, S-I., Muramatsu, T. and Kobata, A. (1976), Nature 259, 580.

Pesonen, M. and Renkonen, O. (1976), submitted for publication.

Ranki, M., Kääriäinen, L. and Renkonen, O. (1972), Acta path. microbiol. Scand. Section B, 80, 760.

Rasilo, M.L., Pesonen, M. and Renkonen, O., unpublished experiments.

Renkonen, O., Kääriäinen, L., Simons, K. and Gahmberg, C.G. (1971), Virology 46, 318.

Riordan, J.R., Mitranic, M., Slavic, M. and Moscarello, M.A. (1974), FEBS Lett. 47, 248.

Schlesinger, S., Gottlieb, C., Gelb, U. and Kornfeld, S. (1976), J. Virol. 17, 239.

Sefton, B.M. and Burge, B.W. (1973), J. Virol. 12, 1366.

Sefton, B.M. and Keegstra, K. (1974), J. Virol. 14, 522.

Sefton, B. (1976), J. Virol. 17, 85.

Simons, K. and Kääriäinen, L. (1970), Biochim. Biophys. Res. Commun. 38, 981.

Simons, K., Keränen, S. and Kääriäinen, L. (1973), FEBS Lett. 29, 87.

Somerharju, P., Kates, M. and Renkonen, O., unpublished experiments.

Stenvall, H., Haahtela, K. and Renkonen, O., unpublished experiments.

Takatsuki, A. and Tamura, G. (1971), J. Antibiotics 24, 785.

Tkacz, J.S. and Lampen, J.O. (1975), Biochim. Biophys. Res. Commun. 65, 248.

Utermann, G. and Simons, K. (1974), J. Mol. Biol. 85, 569.

# PHOSPHOLIPASES AS STRUCTURAL AND FUNCTIONAL PROBES FOR CIRCULATING LIPOPROTEINS

Angelo M. Scanu

Departments of Medicine and Biochemistry, University of Chicago, and Franklin McLean Memorial Research Institute,*

Chicago, Illinois 60637, USA

## A. INTRODUCTION

In spite of extensive investigations by a variety of physical and chemical techniques, the structural organization of the protein and lipid constituents of the major classes of serum lipoproteins, very low- (VLDL), low- (LDL) and high-density (HDL), remains to be determined (1,2). In the case of phospholipids, their location in the outer polar surface region of the lipoproteins appears to be accepted; yet their mode of interaction with proteins and with the other lipid constituents is essentially unknown. Some theories propose that the polar head groups are involved in electrostatic interactions with particular regions of the polypeptide chains (3), others favor a predominance of hydrophobic interactions between the hydrocarbon chains of phospholipids and the non-polar face of the apolipoproteins (4-6). Even within the framework of current theories on serum lipoprotein structure the role played by the individual phospholipids (phosphatidylcholine, sphingomyelin, phosphatidylethanolamine, etc.) is totally unexplored.

The use of lipolytic enzymes as probes for the structure of serum lipoproteins is not new and a large number of observations have shown that lipoprotein lipids are susceptible to enzymatic hydrolysis. Yet, the structural information obtained by this method has not been conclusive, largely because the purity of the enzyme

---

*Operated by the University of Chicago [Contract E(11-1)-69] for the United States Energy Research and Development Administration.

preparations was not rigorously determined, the kinetics of the reactions were not carefully analyzed, and the products were often not characterized.

From the functional standpoint it is established that the intravascular process of VLDL-LDL degradation is attended by a progressive loss of lipids, presumably through the action of lipolytic enzymes (7). In such a process, a VLDL particle containing an average of 6,000 moles of phospholipids per mole is transformed into a particle LDL, which has just 650 moles of phospholipids per mole, while the hydrolysis of triglycerides takes place presumably through the concerted action of lipoprotein lipase and lecithin-cholesterol acyl transferase (LCAT). Although the mechanism of phospholipid hydrolysis in this degradation is undetermined, very recently a phospholipase $A_1$, released in circulation by heparin, has been implicated in the process (8). Thus, it is becoming apparent that phospholipases may be intimately involved in the regulation of transport and degradation of circulating phospholipids and, indirectly, in the structure and metabolism of serum lipoproteins.

As an example of the role that phospholipases may play in lipoprotein metabolism, I wish to discuss studies carried out in this laboratory utilizing the enzyme phospholipase $A_2$ (EC 3.1.1.4, phosphatide acyl-hydrolase) from either *Crotalus atrox* or *Crotalus adamanteus* to investigate the location and reactivity of phospholipids in human $LDL_2$ and $HDL_3$. Detailed reports on these studies have appeared elsewhere (9,10).

Phospholipase $A_2$ was chosen as the probe enzyme for several reasons. First, the reaction catalyzed by phospholipase $A_2$,

$$\begin{array}{c} CH_2O\text{-}\overset{O}{\overset{\|}{C}}\text{-}R_1 \\ | \\ R_2\overset{O}{\overset{\|}{C}}\text{-}O \blacktriangleright C \blacktriangleleft H \\ | \\ CH_2\text{-}O\text{-}\underset{\underset{O}{|}}{\overset{O}{\overset{\|}{P}}}\text{-}O\text{-}X \end{array} + H_2O \xrightarrow{\text{phospholipase } A_2} \begin{array}{c} CH_2\text{-}O\text{-}\overset{O}{\overset{\|}{C}}\text{-}R_1 \\ | \\ HO \blacktriangleright C \blacktriangleleft H \\ | \\ CH_2\text{-}O\text{-}\underset{\underset{O}{|}}{\overset{O}{\overset{\|}{P}}}\text{-}O\text{-}X \end{array} + R_2\overset{O}{\overset{\|}{C}}\text{-}OH$$

X = choline, serine or ethanolamine

is highly specific for phosphatidylcholine, phosphatidylethanolamine, and phosphatidylserine, with no detectable activity toward sphingomyelin. The substrates under consideration, $LDL_2$ and $HDL_3$, have well-defined amounts of hydrolyzable and non-hydrolyzable phospholipids. Second, potential structural changes induced by phospholipase $A_2$ hydrolysis may be detected readily since both $LDL_2$ and $HDL_3$ have been characterized extensively. Third, the action of this enzyme results in a single and mild chemical modification of the lipoprotein, free of side reactions, which is well suited for precise quantitative kinetic analysis. Finally, the *in vitro* hydrolysis of $LDL_2$ and $HDL_3$ by phospholipase $A_2$ is likely to represent a model of physiological importance since such enzymatic modification may occur *in vivo* and may be involved in processes of lipoprotein-cell interactions.

## B. RESULTS WITH LOW-DENSITY LIPOPROTEIN, SUB-CLASS $LDL_2$

The physical properties and the chemical composition of $LDL_2$ are presented in Table 1. From the specificity of phospholipase $A_2$ toward phospholipids *in vitro*, it can be predicted that a mole of $LDL_2$ has about 430 moles of hydrolyzable phospholipids and 175 moles of non-hydrolyzable sphingomyelin.

In the absence of defatted bovine serum albumin (BSA), a known fatty acid acceptor, the hydrolysis of $LDL_2$ by phospholipase $A_2$ was attended by a fast release of protons which gradually decreased and then stopped (Fig. 1). At this time, all of the hydro-

Table 1

Properties of Human Serum $LDL_2$

| Physical parameter | | Value |
|---|---|---|
| Flotation rate | | 6.4 $S_{f(1.063)}$ |
| Density of medium | | 1.019-1.063 g/ml |
| Hydrated density | | 1.030 g/ml |
| Molecular weight | | $2.3 \times 10^6$ |
| Diameter | | 170-230 Å |
| Electrophoretic mobility | | β |
| **Chemical composition** | **% weight** | |
| Amino acids | 21 | |
| Phospholipids | 22 | Moles/mole LDL* |
| Phosphatidylcholine | | 400 |
| Sphingomyelin | | 175 |
| Phosphatidylethanolamine | | 19 |
| Phosphatidylserine | | 9 |
| Lysophosphatidylcholine | | 18 |
| Cholesterol | 8 | |
| Cholesteryl esters | 37 | |
| Glycerides | 11 | |
| Non-esterified fatty acids | 1 | |

*Assuming MW = $2.3 \times 10^6$.

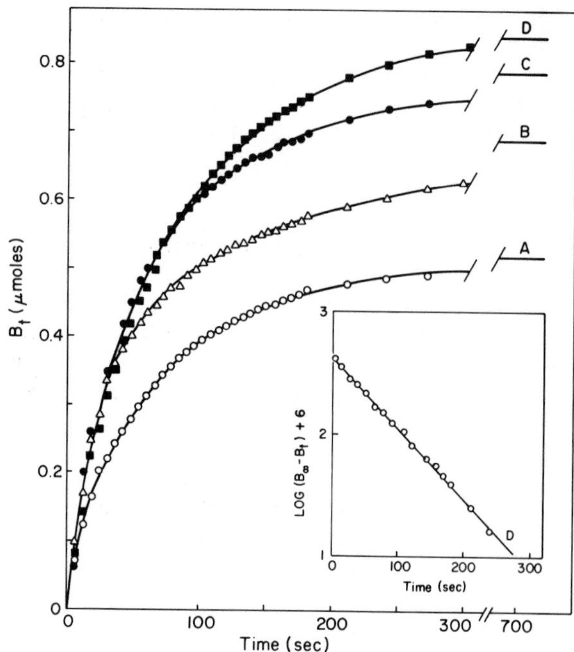


Fig. 1. Time course of the hydrolysis of $LDL_2$ phospholipids by phospholipase $A_2$ in the absence (A) and presence of (B) $0.34\times10^{-4}$M, (C) 0.68 $10^{-4}$M, and (D) 1.34 $10^{-4}$M defatted bovine serum albumin. $B_t$ is the quantity of base added at time t. The inset is the corresponding first-order plot for reaction D. The reactions were carried out at 25° in 150 mM NaCl, 1 mM borate, pH 7.4, in a total volume of 2 ml, with 10 mM NaOH used as the titrant. $E_0=1\times10^{-7}$M, $S_0=4.1\times10^{-4}$M, $[CaCl_2]$=7 mM. The theoretical yield is 0.875 μmoles.

lyzable phospholipids had been transformed into their corresponding lyso-derivatives, whereas there were no measurable changes in the content of sphingomyelin, cholesterol, or cholesteryl esters. In spite of the rather extensive hydrolysis, the lipoprotein particle remained stable and, upon re-isolation in the ultracentrifuge, retained properties very similar to those of $LDL_2$ before hydrolysis (Table 2), except for an increased electrophoretic mobility due to the carboxylate ion of bound fatty acids produced by the enzymatic reaction. The initial rate of phospholipid hydrolysis, as measured by the rate of proton release, was proportional to the enzyme concentration and to the concentration of $LDL_2$. However, the time course of the lecithin hydrolysis did not obey simple kinetic laws and was characterized by a non-quantitative release of protons: at the end of the reaction only 61% of the base corresponding to the total proton concentration had been consumed. This together with

Table 2

Properties of $LDL_2$ Following Complete Hydrolysis by Phospholipase $A_2$

| Physical parameters | Without BSA | With BSA |
|---|---|---|
| $S_{f(1.063)}$ | 5.2 | 4.6 |
| Diameter | 260 ± 22 Å | 260 ± 22 Å |
| Circular dichroism | Same as normal | Same as normal |
| Small-angle x-ray scattering | Same scattering curve as normal | |
| Reactivity to anti-LDL | Same as normal | |
| Electrophoretic mobility | Increased anodic migration | Anodic migration between that of native $LDL_2$ and $LDL_2$ cleaved in the absence of BSA |

| Chemical composition | % weight distribution | | | |
|---|---|---|---|---|
| Proteins | 21.0 | | 22 | |
| Total phospholipids | 16.0 | % | 14.9 | % |
| Phosphatidylcholine | | | | |
| Phosphatidylethanolamine | | 0 | | 0 |
| Phosphatidylserine | | | | |
| Lysolecithin | | 65 | | 50 |
| Sphingomyelin | | 35 | | 50 |
| Cholesterol | 8.0 | | 9.4 | |
| Cholesterol esters | 38.0 | | 42.0 | |
| Triglycerides | 7.10 | | 7.13 | |

the increase in negative charge on $LDL_2$ shown by electrophoresis, suggested that the electrostatic effects due to the accumulation of the negative charges resulted in incomplete ionization of the fatty acid products even at pH = 7.4. Support for this hypothesis was obtained from the hydrolysis of $LDL_2$ in the presence of BSA. At 1% concentration of albumin (stoichiometric amount required to bind all fatty acids produced during the course of the reaction) the proton yield was increased to approximately 95% of the theoretical value. This was true whether the albumin was added before or at the end of the reaction (Fig. 1). Albumin also removed 30% of the lysolecithin formed in the hydrolysis. In spite of this relatively large loss of lipid, the $LDL_2$ particles, when they had been re-isolated by ultracentrifugation, were indistinguishable from those hydrolyzed in the absence of albumin (Table 2). Furthermore, in the presence of albumin the proton release approached theoretical yields; therefore, the pH-stat reaction curve shown in Fig. 1(B) directly measures the conversion of substrate to product. If P is the molarity of the product formed and $P_\infty$ that of the product formed at the end of the reaction, a plot of log $(P_\infty - P)$ versus time, t, yields straight lines for the entire course of the reaction (inset of Fig. 1). From the slope of these lines, the apparent first-order constant, $k_{exp}$, can be calculated. The value of $k_{exp}$ was found to be directly proportional to the enzyme concentration over a 15-fold range and independent of the initial $LDL_2$ concentration over a 10-fold range. Thus, in the presence of an amount of albumin sufficient to accept the fatty acid products, the enzymatic digestion of $LDL_2$ obeys the simple kinetic law: $dP/dt = kE_0(P_\infty - P)$, where k is the true second-order rate constant of the enzymatic reaction and $E_0$, the initial enzyme concentration. Dividing $k_{exp}$ by the enzyme concentration, one obtains the value $k = 1.3 \times 10^5\ M^{-1}\ sec^{-1}$. In the absence of added albumin, the initial rate of the reaction remained unchanged, but the later phases of the reaction were markedly slower than one would have expected if a first-order kinetic law was obeyed. The deviation from first-order kinetics was analyzed as being due to competitive product inhibition by the fatty acids, this inhibition being relieved by serum albumin.

## C. RESULTS WITH HIGH-DENSITY LIPOPROTEIN, SUB-CLASS $HDL_3$

The properties of the $HDL_3$ particle are summarized in Table 3.

Table 3

Properties of Human Serum $HDL_3$

| Physical parameter | Value |
|---|---|
| Flotation rate | 2.05 $S_{f(1.21)}$ (S) |
| Density of medium | 1.12-1.21 g/ml |
| Hydrated density | 1.13 g/ml |
| Molecular weight | $1.75 \times 10^5$ |
| Diameter | 50-90 Å |
| Electrophoretic mobility | $\alpha_1$ |

| Chemical composition | % weight | Moles/mole HDL* | |
|---|---|---|---|
| Amino acids | 56 | | |
| Phospholipids | 23 | | |
| Phosphatidylcholine | | 38 | |
| Sphingomyelin | | 5 | |
| Phosphatidylethanolamine | | 1 | 49 |
| Phosphatidylserine | | 1 | |
| Lysophosphatidylcholine | | 4 | |
| Cholesterol | 3 | | |
| Cholesteryl esters | 12 | | |
| Glycerides | 4 | | |
| Non-esterified fatty acids | 2 | | |

*Assuming MW 175,000.

From the specificity of phospholipase $A_2$ *in vitro*, it can be predicted that a mole of $HDL_3$ contains 40 moles of hydrolyzable phosphoslipids and 5 moles of non-hydrolyzable sphingomyelin. In the absence of BSA, phospholipase $A_2$ from *C. adamanteus* quantitatively hydrolyzed all of the hydrolyzable phospholipids and left sphingomyelin intact (Table 4), yielding a stable, water-soluble complex which could be re-isolated by ultracentrifugation. As observed with $LDL_2$, the number of protons liberated was only about 85% of that expected from the quantity of phospholipids hydrolyzed (Table 3), even though the chemical analysis showed complete hydrolysis. The modified lipoproteins, analyzed following re-isolation in the ultracentrifuge, contained no phosphatidylcholine, phosphatidylethanolamine, or phosphatidylserine, but instead contained their lyso-derivatives and fatty acid products (Table 4). Neither the physical

Table 4

Properties of $HDL_3$ Following Complete Hydrolysis by Phospholipase $A_2$

| Physical parameters | Without BSA | | With BSA (1%) | |
|---|---|---|---|---|
| Flotation rate, $S_{f(1\cdot21)}^{\circ}$ (S) | 1.55 | | 0.97 | |
| Sedimentation velocity $S_{20,w}^{\circ}$ (S) | 4.13 | | 4.61 | |
| $D_{20,w}^{\circ} \times 10^7$ cm$^2$/sec | 4.05 | | 4.35 | |
| $[\theta] \times 10^{-4}$ deg cm$^2$/dmole, 208 nm | 2.31 | | 1.96 | |
| $[\theta] \times 10^{-4}$ deg cm$^2$/dmole, 222 nm | 2.18 | | 1.89 | |
| Small angle x-ray scattering radius of gyration and electro-density distribution | No marked changes from native | | | |
| Reactivity to anti-HDL | No changes from native | | | |
| Electrophoretic mobility (relative to normal = 1) | 1.6 | | 1.2 | |
| **Chemical composition** | **% weight** | | **% weight** | |
| Protein* | 56 | | 60 | |
| Total phospholipids | 16 | % | 13 | % |
| Phosphatidylcholine | | | | |
| Phosphatidylethanolamine | | 0 | | 0 |
| Phosphatidylserine | | | | |
| Lysophosphatidylcholine | | 87 | | 80 |
| Sphingomyelin | | 13 | | 20 |
| Total cholesterol | 14 | | 15 | |
| Glycerides | 4 | | 4 | |
| Non-esterified fatty acids | 10 | | 8 | |

*The distribution in A-I, A-II, and C-polypeptides was the same as in native $HDL_3$.

parameters nor the protein-lipid distribution of the digested particle was changed substantially from those of the native $HDL_3$. The only significant change observed was the increased electrophoretic mobility of the digested particles, due to the increase of their negative charge (Table 1). An analysis of the time course of the reaction (Fig. 2) indicated that the initial velocity, $v_0$, was independent of the initial concentration of hydrolyzable phospholipids, $S_0$, over a 40-fold range ($9.5$ to $400 \times 10^{-4}$ M), and was proportional to the enzyme concentration, $E_0$, in the range of 2 to $75 \times 10^{-8}$ M. For all values of $S_0$ and $E_0$, the time course of the reaction was first-order with respect to the concentration of unreacted hydrolyzable phospholipids, S, as shown by the linearity of the $\ln[P_\infty - P_t]$ vs. time plots over the entire reaction. The experimental rate constants, $k_{exp}$, were directly proportional to the enzyme concentrations and inversely proportional to the initial substrate concentration. The rate equation had the following form: $v = k[E_0S/S_0]$.

A series of hydrolytic reactions was carried out with constant enzyme and substrate concentrations at temperatures between 15° and 47°. At all temperatures examined, the reactions were found to fol-

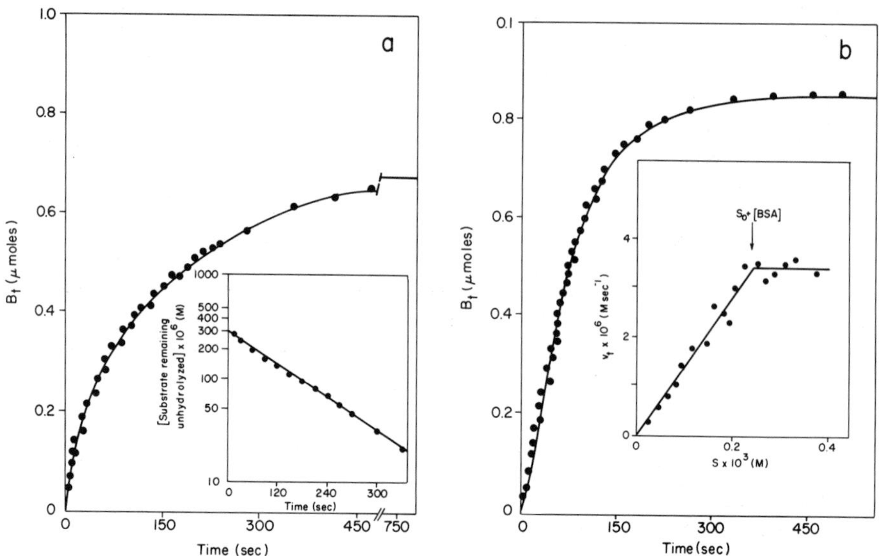


Fig. 2. Time course of the hydrolysis of $HDL_3$ phospholipids by phospholipase $A_2$ (a) in the absence and (b) in the presence of $1.4 \times 10^{-4}$ M defatted albumin (BSA). Insets are the corresponding first-order semi-logarithmic plot (a) and $v_t$ vs. S plot (b). The reaction was carried out at 25° in a total volume of 2 ml, with 0.01 M NaOH used as the titrant, 1 mM borate NaCl buffer, pH 7.4, $E_0 = 4.2 \times 10^{-8}$ M, $S_0 = 3.9 \times 10^{-4}$ M, $[CaCl_2] = 7$ mM.

low first-order kinetics. The Arrhenius plot for this reaction ($\log k_{exp}$ against $1/T$) was linear. The apparent activation energy for the reaction, calculated from the slope of the curve, was 15.2 Kcal/mole.

In the presence of optimal concentrations of albumin (i.e., 0.4 molar equivalent to hydrolyzable phospholipids), the proton release was quantitative, indicating again that BSA acts as an acceptor for lysophospholipids and fatty acid products. The quantity of hydrolytic products released varied with the albumin concentration. At a 0.13 molar equivalent of substrate phospholipids, still about 90% of the lysophospholipids and fatty-acid remained in $HDL_3$. The modified particle was similar in physical properties to that digested in the absence of BSA, except for the changes in flotation rate and electrophoretic mobility (Table 4) expected from the loss of lipids to albumin. The time course of the reaction showed a considerable zero-order portion, i.e., the rate of the reaction, $v_t$, was independent of the remaining substrate concentration, S, over a wide range. The zero-order portion appeared to be proportional to the albumin concentration. The final portion of the reaction, however, still followed first-order kinetics as observed in the absence of albumin.

The results of the overall analysis of the kinetic data in the presence and absence of albumin led to the conclusion (9) that hydrolysis of $HDL_3$ by phospholipase $A_2$ follows Michaelis-Menten type kinetics with $K_m \simeq 1 \times 10^{-4}$ M. The apparent first-order time course in the absence of albumin is due to competitive product inhibition which is relieved by albumin without any alteration of the kinetic parameters:

$$\left(\frac{k_{cat}}{K_m} = 7 \times 10^5\ M^{-1}sec^{-1} \text{ and } K_1\text{(product inhibition)} = 1 \times 10^{-4}\ M\right)$$

The kinetics of phospholipid hydrolysis can also be measured by $^{31}$P-NMR spectroscopy (11), in which use is made of the resonance differences between the hydrolyzable phospholipids and their lysoderivatives. When the action of phospholipase $A_2$ on HDL was measured by this technique again a first-order rate law was observed, thus confirming the results obtained titrametrically (10).

## D. CONSIDERATIONS ON THE STRUCTURE OF $LDL_2$ AND $HDL_3$

The results for $LDL_2$ as presented in Section B and detailed elsewhere (10) permit us to draw several conclusions regarding the arrangement of phospholipids in this particle. The observation that phosphatidylcholine, phosphatidylserine, and phosphatidylethanolamine are accessible to digestion by phospholipase $A_2$ indicates that these phospholipids are probably located at the surface of the particle. Moreover, the fact that at least 85% of the phospholipids are kinetically equivalent excludes the existence of slowly interconvertible pools. This conclusion is also supported by the $^{31}$P-NMR results (10). The phospholipids, although possibly interacting among themselves or with proteins, still have sufficient freedom of motion to permit enzyme cleavage at the C-2 ester linkage. Also, the adherence of the reaction to simple kinetic laws suggests that the polar heads of phospholipids have a loose or "fluid-like" packing similar to that observed in liquid-expanded lipid monolayers. Finally, the fact that, following hydrolysis, $LDL_2$ can be isolated as a stable particle, with only minor variations in properties as compared to the native lipoprotein, leads us to suggest that lecithins may have a rather limited stabilizing role in the native particle.

The above considerations apply as well to $HDL_3$ digested by phospholipase $A_2$. The observed rate constants for the two particles were of the same order of magnitude: $k_{cat}/K_m = 7 \times 10^5 M^{-1}sec^{-1}$ for $HDL_3$ and $1.3 \times 10^5$ $M^{-1}sec^{-1}$ for $LDL_2$: the respective product inhibition constants were $K_I = 1 \times 10^{-4}$ M ($HDL_3$) and $3 \times 10^{-4}$ M ($LDL_2$), and inhibition is suppressed by serum albumin in both cases. There were significant dissimilarities between the two particles however, in regard to the kinetic parameters of hydrolysis by phospholipase $A_2$. Although the apparent competitive inhibition constant, $K_I$, was of the same order of magnitude for the fatty acids on the surface of both lipoproteins, the experimentally measured values of $K_m$ were very different for the two lipoproteins ($K_m \simeq 10^{-4}$ M for $HDL_3$; $K_m >$ $10^{-3}$ M for $LDL_2$), resulting in zero-order kinetics for $HDL_3$ and first-order kinetics for $LDL_2$ under identical experimental conditions. This great difference in the Michaelis constants indicates that the phospholipids on the surfaces of the two particles do not have the same environment and that the enzyme is capable of recognizing these differences. Surface differences between $LDL_2$ and $HDL_3$ have been suggested by the results of recent small-angle x-ray scattering (11) and nuclear magnetic resonance studies (13). The conclusions which we derived by physical methods are supported by our current kinetic analyses which provide the first chemical demonstration of the structural uniqueness of these two lipoproteins. The difference between $LDL_2$ and $HDL_3$ is also documented by the observation that defatted albumin, present in optimal amounts, extracted from $LDL_2$ all of the fatty acids produced during enzymatic hydrolysis, but failed to remove up to 30% of the total fatty acids produced in the reaction involving $HDL_3$. Furthermore, albumin extracted up to 70% of the lysolecithin produced in the phospholipase $A_2$-digested $HDL_3$, but only 30% from hydrolyzed $LDL_2$.

In spite of the above surface differences, both $LDL_2$ and $HDL_3$ remained remarkably stable in aqueous solutions at both low- and high salt concentrations, after all of the lecithins had been converted into lysolecithins and fatty acids, and even after total or partial loss of these reaction products to albumin. This observation leads us to suggest that lecithins are not an essential structural component in either $LDL_2$ or $HDL_3$, and that lecithins may be replaced by a much lesser amount of fatty acids and/or lysolecithin without disruption of the particle structure. How can these facts be reconciled with the existing models of lipoprotein structure? In the case of $LDL_2$, the above interpretation is compatible with the recent NMR data of Brasure *et al.* (11), the small-angle x-ray scattering studies of Tardieu *et al.*(12), and the x-ray and thermocalorimetric data of Dekelbaum *et al.* (14). It is at variance, however, with the trilayer model proposed by Finer *et al.* (5) from NMR studies and also with the conclusions from early x-ray studies (15). For $HDL_3$, the results are compatible with the models proposed on the basis of small-angle x-ray scattering studies (16) and the recent NMR work by Henderson *et al.* (13), but they do not support the view that there would be strong electrostatic interactions between the polar head groups of lecithins and the "ion-pairs" of HDL polypeptides (3). Furthermore, the proposal that the lecithins do not play an essential structural role in $LDL_2$ and $HDL_3$ appears at variance with previous proposals suggesting that a protein-phospholipid complex represents the primary association in the process of lipoprotein assembly (16). It should be stressed, however, that the current studies provide no information concerning the interaction between the apoproteins and sphingomyelin, since the latter is not cleaved by phospholipase $A_2$. It is also possible that, in the early steps of lipoprotein assembly, protein-phospholipid interactions are important or even essential, but that they become comparatively less significant once the lipoprotein particle reaches its final stable structure. In this context, it would be of interest to compare the reactivity toward phospholipase $A_2$ of native particles and particles obtained during the various phases of reconstitution, and to examine the digested particles also in terms of their core components, with particular reference to cholesteryl esters. The techniques of nuclear magnetic resonance, small-angle x-ray scattering, and thermocalorimetry may prove useful in this regard.

## E. FUNCTIONAL CONSIDERATIONS

The participation of various lipolytic enzymes in the processes attending the catabolism and interconversion of the circulating lipoproteins is well established (1,2). Enzymes which have been studied thus far include lipoprotein lipase, lecithin-cholesterol acyl-transferase, and phospholipase $A_1$, the last, particularly after intravenous administration of heparin. Although the details on their molecular structure and mechanism of action are still limited, the key role of these enzymes in lipid metabolism rests on solid experimental data derived both from animal experimentation and from the study of human genetic variants. No information is currently available on the potential role of phospholipase $A_2$ in lipoprotein metabolism. The studies which have been conducted in this laboratory have made use of enzyme preparations obtained from snake venom; this condition is far from physiological. On the other hand, the recent work by Etienne *et al.* (17) and Law *et al.* (personal communication) has shown that there is in circulation a precursor form of a phospholipase $A_2$ which can be converted into its active form by the action of platelet extracts or various proteolytic enzymes. Isolation and characterization of the pro- and active enzymes are in progress. Once the pure enzyme is obtained, studies of the kind outlined in this report for the venom phospholipase $A_2$ ought to be carried out on serum lipoproteins so that the specificity of action, and, if appropriate, the properties of the hydrolyzed products can be determined. An alternative way of proving the physiological role of phospholipase $A_2$-like enzymes is to inject intravenously into animals well characterized phospholipase $A_2$-digested lipoproteins and to compare their intravascular distribution and catabolism with those of untreated lipoproteins. In this context, it would be of interest to determine the cellular localization of the particles taken up from the circulation, for example, by the use of appropriate radioactive labels.

The action of phospholipase $A_2$ on serum lipoproteins may also be of physiological importance in their interaction with cells, brought about by the changes in lipoprotein surface charge consequent to enrichment in lysolecithin and fatty acids. Based on the documentation that serum lipoproteins are capable of regulating the metabolism of intracellular cholesterol, we have initiated a study of the interaction of phospholipase $A_2$-digested HDL with human blood granulocytes (Ritter, M. and Scanu, A.M., to be published). In these studies, the granulocytes are separated from the other blood components by centrifugation and are then incubated with lipoprotein-free sera supplemented with $HDL_3$ before and after phospholipase $A_2$-digestion. Of the various parameters examined, the most significant was the cholesterol egress, which was found to be about three times higher in the cells exposed to the phospholipase $A_2$-digested $HDL_3$. While the molecular basis of this phenomenon is still under investigation, the results already clearly indicate that appropriate surface modifications in lipoproteins may play a role in the regulation of cell metabolism, at least *in vitro*. Obviously, before a physiological role is considered, it is necessary to demonstrate that these effects are specific for phospholipase $A_2$-treated lipoproteins, and that human granulocytes do not represent a unique case among mammalian cells. The physiological role of phospholipase $A_2$ is also suggested by the recent observation made in this laboratory (N.Pattnaik, to be published) that VLDL, contrary to $LDL_2$ and $HDL_3$, is not cleaved by phospholipase $A_2$ in the presence of whole serum, but it is hydrolyzed when the serum is removed. The search for a potential phospholipase $A_2$ inhibitor in the serum is now under way. Also under consideration are experiments where the phospholipase $A_2$ activity will be examined in post-heparin plasma and compared with that of phospholipase $A_1$, an enzyme whose purification has recently been achieved (18) and its activity against VLDL reported (8).

Finally, another important role of phospholipase $A_2$ may be that of reversing the LCAT reaction. Preliminary experiments in this laboratory have shown (N. Pattnaik, to be published) that the reversal does indeed occur *in vitro*. Thus the action of phospholipase $A_2$ *in vivo* could help in maintaining the appropriate stoichiometric balance between circulating lecithins and free and esterified cholesterol. The facile transfer of fatty acids from phospholipase-digested lipoproteins to albumin also suggests that circulating lipoproteins may play a role in the transport by exchange of unsaturated fatty acids from lecithins and cholesteryl esters. It is also most likely that a dynamic equilibrium of fatty acid distribution exists between albumin and all circulating lipoprotein species.

## F. CONCLUDING REMARKS

The present studies have shown that enzymatic probes can be applied successfully to the study of the structure and function of circulating lipoproteins. Phospholipase $A_2$ permitted highly specific modifications in both low- and high-density lipoproteins, under mild reaction conditions, with no evidence of side reactions. Furthermore, it was possible by the use of this enzyme, to make precise kinetic analyses, and to characterize the substrate particles reisolated following enzymatic digestion. The present studies have also shown that the hydrolyzable phospholipids which are constituents of circulating lipoproteins, are a natural substrate for the enzyme--an observation which suggests that this reaction may have physiological importance in lipoprotein metabolism. The wealth of information derived from the current work ought to encourage precise kinetic analyses of other lipolytic enzymes known to cleave lipid constituents of circulating lipoproteins. The intensive work now being directed at the purification of these enzymes should make possible their use as structural probes to provide an understanding of their physiological role in fat metabolism.

It is important, to realize, however, that the use of enzymes as probes is attended by certain limitations. As we mentioned, enzyme purity is an absolute necessity which is not always easily achieved. Also, some enzymes are bulky reagents which are sensitive to small changes in reaction conditions, thus limiting their effectiveness to narrow experimental conditions. Finally, enzyme reactivity may depend on gross steric features and on surface effects to which small molecules like lipids might be insensitive. Thus, it is evident that the choice of the probe enzyme must be made carefully if the modifications which are specifically desired are to be achieved. If these limitations are recognized, however, the probing of lipoproteins by lipolytic enzymes ought to yield fruitful results.

ACKNOWLEDGMENTS

The work cited in this report was carried out in collaboration with Drs. Lawrence Aggerbeck, Nikhil Pattnaik and Ferenc Kézdy. Dr. Kézdy also provided valuable comments and suggestions during the preparation of this manuscript. The author wishes to thank Mrs. E. Lanzl for critically editing the manuscript and Mrs. Joey Czerwonka and Mrs. Gen LaPinska for the careful typing.

The work was supported by funds from the United States Public Health Service (NIH Grant HL-08727), the Chicago Heart Association (Grant C75-15) and the United States Energy Research and Development Administration [Franklin McLean Memorial Research Institute--operated by the University of Chicago under Contract E(11-1)-69].

REFERENCES

1. Scanu, A.M., Edelstein, C. and Keim, P.: Serum lipoproteins. In The Plasma Proteins, 2d Ed., Vol. I. F. Putnam, Editor. New York: Academic Press, Inc., 1975, pp. 317-391.
2. Morrisett, J.D., Jackson, R.L. and Gotto, A.M., Jr.: Lipoprotein structure and function. Annu. Rev. Biochem. 44: 183-207, 1975.
3. Segrest, J.P., Jackson, R.L., Morrisett, J.D. and Gotto, A.M., Jr.: A molecular theory of lipid-protein interactions in plasma lipoproteins. FEBS Lett. 38: 247-253, 1974.
4. Assmann, G., Sokoloski,E.A.and Brewer, H.B., Jr.: $^{31}$P-Nuclear magnetic resonance spectroscopy of native and recombined lipoproteins. Proc. Natl. Acad. Sci. USA 71: 549-553, 1974.
5. Finer, E.C., Henry, R., Leslie, R.B. and Robertson, R.N.: NMR studies of pig low- and high-density serum lipoproteins:molecular motion and morphology. Biochim. Biophys. Acta 380: 320-337, 1975.
6. Stoffel, W., Zieremberg, O., Tunggal, B. and Schreiber, E.: $^{13}$C-Nuclear magnetic resonance spectroscopy: evidence for hydrophobic lipid-protein interaction in human high density lipoproteins. Proc. Natl. Acad. Sci. USA 71: 3696-3700, 1974.
7. Eisenberg, S. and Levy, R.I.: Lipoprotein metabolism. Adv. Lipid Res. 13: 1-89, 1975.
8. Eisenberg, S.: Phospholipid hydrolysis during degradation of rat plasma very low-density lipoproteins. Circulation 52: (Suppl. II) 17, 1975.
9. Pattnaik, N.M., Kézdy, F.J. and Scanu, A.M.: Kinetic study of the action of snake venom phospholipase $A_2$ on human serum high density lipoproteins. J. Biol. Chem., 251: 1984-1990, 1976.
10. Aggerbeck, L., Kézdy, F.J. and Scanu, A.M.: Enzymatic probes of lipoprotein structure: hydrolysis of human serum low-density lipoprotein-2 by phospholipase $A_2$. J. Biol. Chem., in press, 1976.

11. Brasure, E.B., Henderson, T.O., Gloneck, T., Pattnaik, N.M. and Scanu, A.M.: Kinetic analysis of phospholipase $A_2$-modified human serum high density lipoprotein-3 by $^{31}P$ nuclear magnetic resonance spectroscopy. Fed. Proc., in press, 1976.
12. Tardieu, A., Mateu, L., Sardet, C., Weiss, B., Luzzati, V., Aggerbeck, L. and Scanu, A.M.: Structure of human serum lipoproteins in solution. II. Small-angle x-ray scattering study of $HDL_3$ and LDL. J. Mol. Biol. 101: 129-153, 1976.
13. Henderson, T.O., Kruski, A. W., Davis, L.G., Gloneck, T. and Scanu, A.M.: $^{31}P$-Nuclear magnetic resonance studies on serum low- and high-density lipoproteins: effect of paramagnetic ion. Biochemistry 14: 1915-1920, 1975.
14. Dekelbaum, R.J., Shipley, G.G. and Small, D.M.: Thermal transitions in human plasma low-density lipoproteins. Science 190: 392-394, 1975.
15. Mateu, L., Tardieu, A., Luzzati, V., Aggerbeck, L. and Scanu, A.M.: On the structure of human serum low-density lipoproteins. J. Mol. Biol. 70: 105-116, 1972.
16. Scanu, A.M.: Structural studies on serum lipoproteins. Biochim. Biophys. Acta 265: 471-509, 1972.
17. Etienne, J., Gruber, A. and Polonovski, J.: Occurrence of phospholipase activity from erythrocytes during coagulation in rat serum: role of platelets. Paroi Arterielle 1: 119-123, 1973.
18. Enholm, C., Shaw, W., Greten, H. and Brown, W.V.: Purification from human plasma of a heparin-released lipase with activity against triglycerides and phospholipids. J. Biol. Chem. 250: 6756-6761, 1975.

# THE FLUID MOSAIC MODEL OF MEMBRANE STRUCTURE

S. J. Singer

Department of Biology, University of California at San Diego
La Jolla, California 92093, USA

## ABSTRACT

The fluid mosaic model had its origins in an analysis of the equilibrium thermodynamics of membrane systems, which led to the suggestion that the integral proteins of membranes are amphipathic molecules. This thermodynamic analysis continues to have considerable explanatory and predictive power for problems of membrane structure and function. In this paper, the analysis is applied to explain or predict the relatively hydrophobic amino acid composition of integral proteins, their large content of α-helical secondary structure, the characteristics of the short-range interactions of lipids and integral proteins, the molecular asymmetry of the integral proteins and phospholipids of membranes, the possible biogenesis of such asymmetry, and the mechanisms of transport of small ionic and polar molecules through membranes.

## 1. INTRODUCTION

The fluid mosaic model of membrane structure (Singer, 1971; Singer and Nicolson, 1972) is now generally accepted as a working model for the molecular organization of the integral proteins and the ionic and polar lipids of membranes. The general perception of this model (Fig. 1) is a morphological or topological one, depicting the membrane as a two-dimensional solution of globular integral proteins in a fluid solvent of the lipid bilayer. However, the model was originally derived largely from a consideration of the equilibrium thermodynamics of membrane components in an aqueous environment. The general acceptance of the topological features of the fluid mosaic model has not led to a corresponding general appreciation of the thermodynamic reasoning that led to it. We suggest

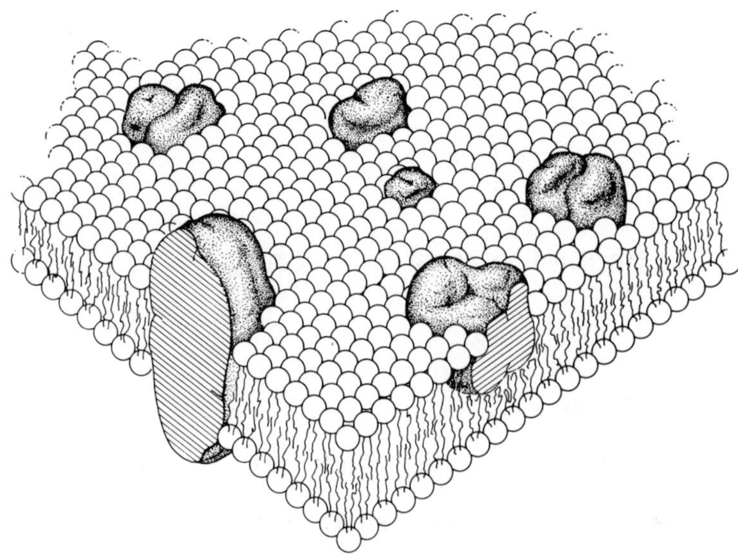

Figure 1 The fluid mosaic model of membrane structure, in schematic three-dimensional and cross-sectional views. The solid bodies with stippled surfaces represent the amphipathic globular integral proteins, embedded in a fluid bilayer of lipid. The circles represent the ionic and polar head groups of the phospholipid molecules; the wavy lines represent the fatty acid chains. From Singer and Nicolson (1972).

that these thermodynamic considerations have a great deal to contribute to an understanding of membrane structure and function, as we attempt to show in this brief article.

## 2. THERMODYNAMIC CONSIDERATIONS

We have extensively discussed the thermodynamics of membrane systems elsewhere (Singer, 1971), and will only summarize here. It is first assumed that at least the local structure of a membrane, consisting of an individual protein molecule, its surrounding lipid molecules, and the adjacent aqueous medium, are all in equilibrium, and that equilibrium thermodynamics therefore applies to that element of structure. For our present purposes, only four major types of molecular interactions occurring in an aqueous environment need be considered: A) hydrophobic; B) hydrophilic; C) hydrogen bonding; and D) electrostatic.

A) Hydrophobic Interactions. These interactions are well-appreciated in molecular biology, since they were first discussed in detail by Kauzmann (1959). They are responsible for the sequestering of non-polar groups away from contact with water. The free energy changes associated with hydrophobic interactions can be roughly estimated from the free energy required to transfer a mole of methane from benzene to water solution, which is + 2.6 kcal/mole at 25°C. Terms of this magnitude, summed over the methylene groups of the fatty acid chains of the ionic and polar lipids of membranes, must play an important role in determining the stability of the bilayer structure of phospholipids in water, a structure in which the fatty acid chains are sequestered from contact with water. Similarly, however, the non-polar amino acid residues of the integral proteins of membranes must also be largely sequestered from contact with water at equilibrium.

B) Hydrophilic Interactions. While the significance of hydrophobic interactions is widely recognized, the same cannot be said of hydrophilic interactions. We have previously attempted to emphasize them as a major factor in membrane biology (Singer, 1971). By hydrophilic interactions is meant the thermodynamic tendency for ionic and highly polar groups to remain in immediate contact with water if offered a choice between an aqueous and a non-polar environment. To estimate the free energy changes involved in hydrophilic interactions, let us first consider how ionic groups may be taken out of an aqueous environment and placed in molecular contact with a nonpolar solvent that is in equilibrium with the aqueous. To bury an isolated electric point charge in a low dielectric constant medium requires a very large expenditure of free energy, and need not be considered. More likely mechanisms involve i) burying charged groups as ion pairs, consisting of one negatively and one positively charged ion, one of which may be a counterion; ii) if the ionic group is a weak acid or base, the binding or dissociation of a proton at pH 7.0 removes the ionic charge, and the discharged group may now be buried; or iii) the ionic group may form a coordination complex with a suitable hydrophobic ligand, and the resultant complex may be buried in the non-polar environment.

A simple model system with which one can estimate the free energy change in transferring an ion pair from water to a nonpolar solvent involves the transfer of the zwitterion glycine from water to a non-aqueous solvent (Singer, 1971) (the contribution of the $CH_2$ group may be neglected for our purposes). $\Delta G$ for this process ($\Delta G_{WS}$) may be determined approximately from the solubilities of glycine in water and other solvents as $RT \ln (\chi_W/\chi_S)$, where $\chi_W$ and $\chi_S$ are the mole fractions of glycine in the saturated solutions in water and in the solvent, respectively, at a given temperature. The data in Table I indicate the $\Delta G_{WS}$ is a large positive number for all non-aqueous solvents and generally increases as the solvent becomes less polar. The solubility of glycine is already very low in acetone,

Table 1

Solubility and Free Energy of Glycine Transfer in Various Solvents at 25°C[a]

| Solvent | Solubility (mole/liter) | Log $\chi$[b] | $\Delta G_{WS}$ (kcal/mole) | D[d] |
|---|---|---|---|---|
| Water | 2.886 | -1.247 | | 78.5 |
| Formamide | 0.0838 | -2.476 | 1.68 | 109.5 |
| Methanol | 0.00426 | -3.762 | 3.43 | 32.6 |
| Ethanol | 0.00039 | -4.638 | 4.63 | 24.3 |
| Butanol | 0.0000959 | -5.055 | 5.19 | 17.1 |
| Acetone | 0.0000305 | -5.648 | 6.00 | 20.7 |

[a]From Singer (1971). Data from Cohn and Edsall (1943).

[b]The symbol $\chi$ is the mole fraction of glycine.

[c]The term $\Delta G_{WS}$ is the approximate unitary free energy of transfer a mole of glycine from water to the solvent in question.

[d]Dielectric constant.

although it is still a fairly polar solvent. To get a rough estimate for the free energy of transfer of glycine from water to a hydrocarbon solvent of dielectric constant 2.0, we extrapolate from the data of Table I as follows: Water appears here as in other instances to be an anomalous solvent; formamide is therefore used as a reference solvent. The free energy of transfer of glycine from formamide to solvent S, $\Delta G_{FS}$, is taken to be porportional to $\Delta = (1/D_S)-(1/D_F)$ (Edsall and Wyman, 1958) where $D_S$ and $D_F$ are the dielectric constants of solvent S and of formamide, respectively. If a straight line is drawn through a plot of $\Delta G_{FS}$ against $\Delta$, a long extrapolation to $D_S = 2$ suggests that $\Delta G_{FS}$ ($\cong \Delta G_{WS}$) is greater than + 50 kcal/mole for a hydrocarbon solvent, an extraordinarily large number. (In fact, it is very likely that it would cost less free energy to transfer the proton from the $-NH_3^+$ to the $-COO^-$ group of the glycine molecule and for the resultant neutral species, rather than the zwitterion, to be dissolved in the hydrocarbon solvent.)

The point made in the previous paragraph is that it costs a very large amount of free energy to transfer an ion pair from a water solution to a hydrocarbon solvent in equilibrium with it. This conclusion is of particular relevance to the problem of phospholipid flip rates in membranes, as discussed below.

The second mechanism for burying an ionizable residue mentioned above is to discharge it and then internalize it as a polar rather than an ionic group. A <u>part</u> of the cost in free energy to accomplish this is the free energy change, $\Delta G_d$, required to discharge at pH 7.0 a group whose pK is different from 7.0; $\Delta G_d = 2.3\ RT|(pH-pK)|$. For example, for a carboxyl group of pK 4.5, $\Delta G_d = +\ 3.3$ kcal/mole. To bury the now discharged polar group will cost some additional free energy to transfer it from water to a hydrocarbon-like environment.[1]

The third mechanism for burying a charged group usually involves metal ions, which can form specific coordination complexes with molecules bearing a fixed number of electron-donating groups in appropriate sterochemical positions. The free energy of formation of the coordinate covalent bonds in the complex supplies the free energy required to bury the ionic charge within the complex, and in addition the charge may be delocalized and spread over a much larger volume. This mechanism, which applies for example to metal-porphyrin complexes in proteins, and to metal ion-ionophore complexes, is not of direct interest in this discussion because the major charged groups of proteins and phospholipids do not form such complexes with one another.

The overall conclusion about hydrophilic interactions is that it is generally very costly in free energy to sequester ionic residues of proteins and phospholipids away from contact with water at equilibrium. A similar conclusion applies to the saccharide residues of membrane glycoproteins and glycolipids (Singer, 1971).

C) Hydrogen Bonding. With hydrogen bond donor and acceptor groups (such as the $\rangle C=O$ and $\rangle NH$ groups of the peptide bond), there is not much of a difference in free energy whether those groups are hydrogen bonded to water molecules or to each other. In other words, $\Delta G$ for the following reaction is not very different from zero (Klotz and Farnham, 1968):

$$\rangle C=O\cdots H\text{-}OH + H_2O\cdots H\text{-}N\langle \rightleftharpoons \rangle C=O\cdots H\text{-}N\langle + H_2O\cdots H_2O$$

However, if groups capable of forming hydrogen bonds are sequestered from contact with water, it costs free energy ($\sim$ 4.5 kcal/mole) if such groups do not hydrogen bond to one another within the nonpolar medium. As a result, interpeptide hydrogen bonds are maximized within a nonpolar medium.

D) Electrostatic Interactions. For the reasons given in section 2B), charged groups will likely be in contact with water and will not be buried in the nonpolar region of a membrane at equilibrium. In an aqueous environment, however, electrostatic interactions generally appear to be relatively weak (Tanford, 1954) and it is therefore only when charged groups are brought into unusual proximity to one another (as is likely to be true at the membrane surface for the zwitterionic polar head groups of the membrane phospholipids)

[1] Histidyl residues, with pK $\sim$ 7, can be buried with less cost in free energy than other ionic amino acid residue.

that such interactions might be important in determining membrane structure and stability.

Before proceeding from this brief catalogue of molecular interactions to the problem of membrane structure, we may note the influence of hydrophobic and hydrophilic interactions on two well-studied macromolecular systems: phospholipid bilayers in water and water-soluble proteins. The phospholipid bilayer maximizes hydrophobic interactions by sequestering the nonpolar fatty acid chains away from water into the interior of the bilayer, and simultaneously maximizes hydrophilic interactions by exposing the ionic head groups to the aqueous phase. In the case of water-soluble proteins, X-ray crystallographic structure analysis has shown that the interiors of the globular protein molecules contain the major fraction of the hydrophobic amino acid residues of the proteins, thereby maximizing hydrophobic interactions. Of further great interest, however, is the fact that essentially all of the amino acid residues that are normally ionized near neutral pH are on the surfaces of the protein molecules that are in contact with water. This is no doubt an expression of the significance of hydrophilic interactions.

## 3. THERMODYNAMICS AND MEMBRANE STRUCTURE

As remarked above, the bilayer is the equilibrium structure adopted by phospholipids to accommodate to hydrophobic and hydrophilic interactions. To apply these thermodynamic constraints to the membrane-associated proteins, we have first distinguished two major categories of such proteins, peripheral and integral (Singer, 1971; Singer, 1974), on the basis of their relative ease of dissociation from the membrane and their solubility characteristics in aqueous solutions once dissociated from the membrane. In our view, peripheral membrane proteins (cytochrome C of mitochondrial membranes is a good example) and ordinary soluble proteins are likely to have very similar amino acid compositions and structures. The integral proteins of membranes, however, are presumed to interact intimately with the lipids of the membrane, and it is these proteins whose structures we need to consider.

A. The Structures of Integral Proteins. We suggested some time ago (Lenard and Singer, 1966), as did Wallach and Zahler (1966) independently, that all integral proteins of membranes were more-or-less globular molecules partially protruding from the membrane and also partially embedded in it. On those portions protruding from the membrane into the aqueous phase were concentrated essentially all of the ionic amino acid residues of the protein (and highly polar residues such as saccharides) in contact with water, whereas those portions of the protein embedded in the membrane interior in contact with the nonpolar fatty acid chains, were devoid of such ionic and highly polar residues. If the integral protein spanned the thickness of the membrane, it would have two hydrophilic regions, one exposed on each side of the membrane with a hydrophobic segment in between embedded in the membrane.

Such amphipathic molecules were hypothesized in order to maximize both hydrophobic and hydrophilic interactions, and to achieve a structure of minimum free energy in an environment containing phospholipids and water. This concept has become a commonplace only recently as a result of amino acid sequence and other structural studies carried out with a number of integral proteins that have been isolated from membranes, including cytochrome $b_5$ (Strittmatter et al., 1972), cytochrome $b_5$ reductase (Spatz and Strittmatter, 1973) glycophorin (Segrest et al., 1973) and several others, which show that these molecules clearly have amphipathic structures as postulated. In the case of glycophorin, it has been shown that it spans the erythrocyte membrane with a hydrophilic segment protruding from each surface and a hydrophobic segment in between.

In several of these cases, the further interesting feature is that the three-dimensional amphipathic structure is determined by a <u>linear</u> amphipathy of the primary amino acid sequence. Cytochrome $b_5$ is an example of such a structure. Of its total of 152 amino acid residues, the amino-terminal 104 residues form a globular structure that protrudes from the membrane into the aqueous phase, while the carboxy-terminal 48 residues are largely embedded in the membrane. The amino acid composition and three-dimensional structure (Matthews et al., 1971) of the amino-terminal portion are not significantly different from that of water-soluble proteins, but the composition of the carboxy-terminal portion is much more hydrophobic. The two portions, however, appear to fold independently of one another into distinct domains connected by a stretch of polypeptide chain that is highly susceptible to proteolytic cleavage. In other words, the molecule <u>in situ</u> in the membrane is probably a monomeric dumb-bell shaped entity such as is schematically represented in Fig. 2A.

We have suggested (Singer, 1971, 1974), furthermore, that in addition to structures similar to Fig. 2A, integral membrane proteins may exist as subunit aggregates, conforming to the requirements imposed by hydrophobic and hydrophilic interactions. Of particular interest would be aggregates, such as depicted in Fig. 2B, which would span the thickness of the membrane and which would form a continuous water-filled channel down the central axis of the aggregate. Such a channel could be lined with ionic groups, since these would remain in contact with water. On the surfaces of the subunits that were exposed to the aqueous environment on either face of the membrane, would be situated the remaining ionic and highly polar groups, while the embedded portions that were in direct contact with the lipid interior of the membrane would be devoid of ionic groups. We have proposed (section 5) that all transport proteins have subunit aggregate structures similar to Fig. 2B.

MONODISPERSE

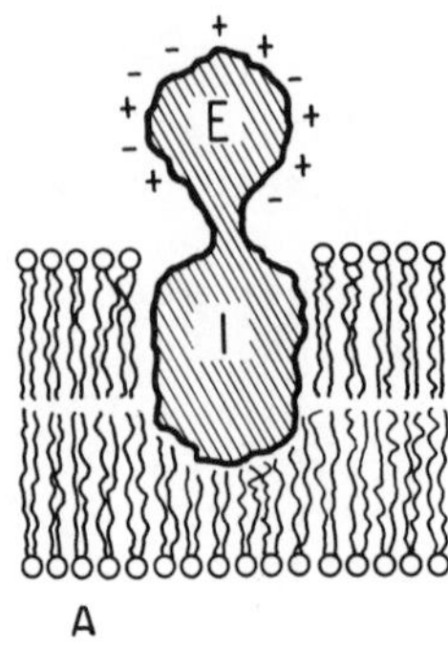


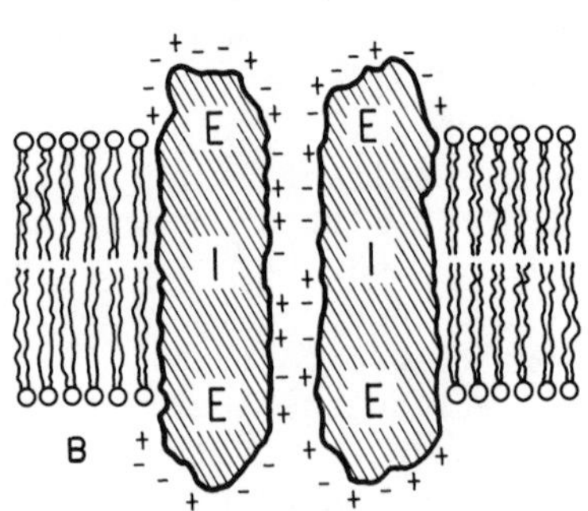


Figure 2 Two of the thermodynamically possible structures of integral proteins, depicted schematically. Whether such a protein is monodisperse or exists as a subunit aggregate is determined by its amino acid sequence, if equilibrium conditions prevail. The subunit aggregates, if they span the membrane, can generate water-filled pores through the membrane. E and I refer to exterior (exposed) and interior regions of the protein, respectively. From Singer (1974).

An important corollary is that when amino acid sequence data are obtained for integral proteins that exist as subunit aggregates in the membrane, they may not exhibit the simple linear amphipathic structure found with monomeric proteins like cytochrome $b_5$; the linear arrangement of the hydrophilic segments of the molecule along each polypeptide chain may be more complex.

B. The Amino Acid Composition of Integral Proteins. From such considerations of integral protein structure, one can qualitatively explore the question: how might the amino acid compositions of integral proteins differ, not only from those of water soluble proteins, but also amongst themselves? Capaldi and Vanderkooi (1972) collected data on the amino acid compositions of a number of proteins that are integral to membranes and showed that they were on the average relatively enriched in hydrophobic residues compared to water-soluble proteins. According to the thermodynamic restrictions discussed earlier, the empirical findings of Capaldi and Vanderkooi can be explained. We propose that the ionic amino acid residues are confined to the surfaces of protein molecules that are in contact with water. Let us make the further important assumption that the surface charge density produced by such residues is about the same on all such water-exposed surfaces, whether they are present on integral membrane proteins or on water-soluble proteins. For protein molecules of the same molecular weight and shape, it follows that the fraction of amino acid residues that is ionic would be smaller for the amphipathic integral membrane protein than for the water-soluble protein, because the former would have less surface exposed to water. One could further expect that polar non-ionic residues might be relatively more concentrated in the surfaces exposed to water than in the other portions of a protein molecule, and that the surface density of such residues was about the same for all such surfaces in contact with water. Again, it follows that for an integral membrane protein and a water-soluble protein of the same molecular weights and shapes, the former would have a relatively smaller fraction of polar, non-ionic residues than the latter.

The distribution of ionic and polar residues under these assumptions would therefore lead to the result that an integral protein should be relatively enriched in hydrophobic residues compared to a water-soluble protein of the same size and shape. For different integral membrane proteins, any structural features that decreased the ratio of their surface area in contact with water to the volume of the whole molecule should result in an enrichment of their relative contents of hydrophobic amino resides. Thus: i) the larger the molecular weight; ii) the more nearly spherical the shape; and iii) <u>the larger the volume fraction of the protein that was embedded</u> in the membrane interior, the greater should be the relative content of hydrophobic residues in the integral protein.

A more quantitative treatment incorporating these considerations could be developed, but would not be warranted given our present lack of information about the structures of integral proteins. Such a quantitative treatment could in principle lead to estimates of the degree to which individual integral proteins are embedded in membranes, and the degree to which they protrude.

We have made the point elsewhere (Singer, 1974) that since even the most hydrophobic integral proteins known still contain appreciable numbers of ionic residues, it is highly unlikely that any integral protein molecules are entirely embedded within the membrane, since their ionic residues would as a result have to be sequestered from contact with water.

C. <u>The Helicity of Integral Proteins</u>. The discussion of hydrogen bonding in section 2C, together with the structural features developed in section 3A, led us to the following predictions (Singer, 1971). Those globular portions of an integral protein molecule that protrude into the aqueous medium should exhibit the same wide range of α-helix, β-pleated sheets, and randomly folded secondary structures that is exhibited by water-soluble proteins. This is because hydrogen bonding by the $\rangle$NH and $\rangle$C = O groups of the polypeptide chain should occur equally well to water molecules or to one another. The particular mix of secondary structures exhibited in such protruding hydrophilic portions of integral proteins would then not be greatly constrained by hydrogen bonding. For those hydrophobic portions of the molecules embedded in the membrane, however, hydrogen bonding to water is ruled out, and $>NH\cdots O=C<$ interpeptide hydrogen bonding has to be extensive or the structure would be unstable. Since the α-helix is more efficient than 2- or 3-chain β-pleated sheets in allowing the formation of the maximum possible number of peptide hydrogen bonds, it follows that the embedded portions of integral protein molecules should be relatively rich in α-helix and, to a lesser extent, β-structures. This is almost certainly why simple proteins dissolved in nonaqueous solvents become extensively helical (Singer, 1971). These thermodynamic considerations bear on the fact that the proteins in a variety of intact membrane preparations have been found (Lenard and Singer, 1966, Wallach and Zahler, 1966) to exhibit a significant amount of α-helical structure (see Figs. 3 and 4). Even more interesting, however, is the recent demonstration (Henderson and Unwin, 1975) that the integral protein, bacterial rhodopsin, molecular weight 26,000, has about 80% of its <u>total</u> polypeptide chain in the α-helical conformation, and much of the chain is embedded in the bacterial membrane. These helical portions are folded back and forth across the membrane to give seven discrete helical segments running more-or-less parallel to one another and perpendicular to the plane of the membrane. It has also been suggested that that portion of the glycophorin molecule which is presumably embedded in the erythrocyte membrane is in the α-helical conformation (Segrest et al., 1974).

An important corollary of these considerations concerns the short-range interactions of lipids and integral proteins in the membrane. If the embedded portions of integral proteins are indeed extensively α-helical, consisting of bundles of parallel contiguous α-helices, then the <u>exterior surfaces</u> of the embedded portions, where the protein comes in contact with the fatty acid chains of the phospholipids, might quite generally be relatively smooth undulating cylinders lying roughly perpendicular to the plane of the membrane. This clearly bears on the properties of the fatty acid chains of the lipids immediately surrounding the embedded portion of an integral protein. In the few cases where the properties of such lipid-protein interactions have been examined by physico-chemical methods (Jost et al., 1973; Warren et al., 1975) it appears that there is a shell of one or at most two molecular thicknesses of lipid bound tightly to the protein. The lipid in this shell is relatively rigid, but the lipid in the adjacent shell already exhibits the translational mobility and the fatty acid chain fluidity characteristic of the bulk bilayer lipid. Such results would not be expected if, on the contrary, embedded portions of integral proteins had highly irregular surfaces, for in that event the steric accommodation of the fatty acid chains of the lipids immediately surrounding the protein might perturb and fluidize the lipid bilayer structure for some considerable distance away from the protein.

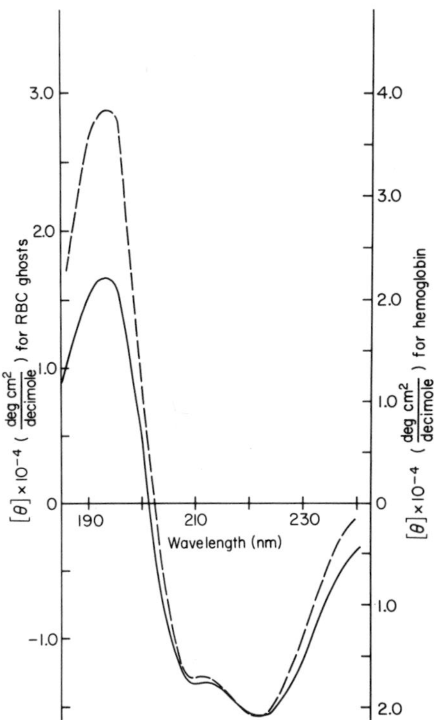


Figure 3 A comparison of the circular dichroism spectra of human hemoglobin (- - -) and intact human erythrocyte membranes (——) in 7mM phosphate buffer, pH 7.4, with respective ellipticy (θ) scales adjusted to give the same minimum near 222 nm. The double minimum near 208 nm and 222 nm is associated with polypeptide chains in the α-helical conformation, and in the case of hemoglobin, reflects the fact that about 75% of the protein is α-helix. Correspondingly, the spectrum indicates that the average helicity of the erythrocyte membrane protein is about 40%. From Singer (1971).

## 4. THE MOLECULAR ASYMMETRY OF MEMBRANES

There is much evidence that the integral proteins of membranes are asymmetrically positioned across the membrane, i.e., that they are predominantly, or exclusively, oriented in one or the other direction perpendicular to the plane of the membrane. It has been known for a long time, for example, that many membrane-bound enzymes express their activities on only one surface of their membranes, i.e., are so oriented as to have their active sites exposed only at that surface. In a related context, the oligosaccharide chains

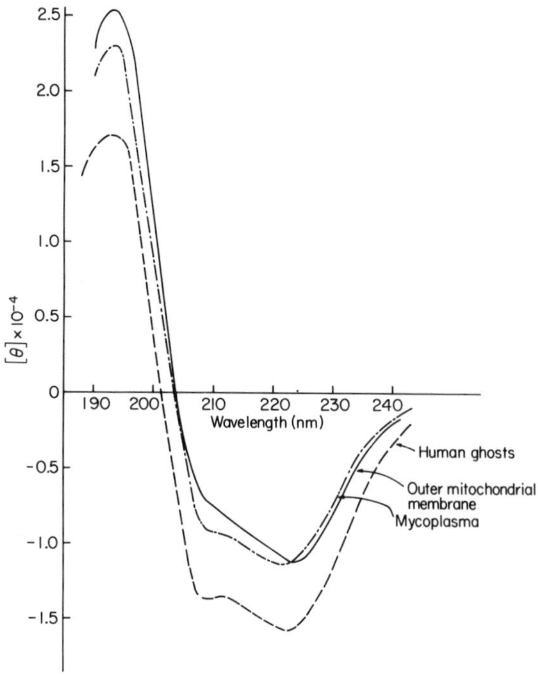


Figure 4 The circular dichroism spectra in neutral buffers of intact human erythrocyte membranes, of the outer membranes of rat liver mitochondria, and of the membranes of Mycoplasma laidlawii. All three spectra show the double minimum characteristic of partially α-helical conformation, averaged over the proteins in the respective membranes. From Singer (1971).

of glycoproteins bound to the plasma membranes of animal cells, are exclusively exposed at the exterior surfaces of those membranes (Nicolson and Singer, 1971; 1974) and on the corresponding single surfaces of intracellular membranes as well (Hirano et al., 1972). More recently, it has been shown that the phospholipids and glycolipids of some membranes are also asymmetrically distributed in the two half layers of the lipid bilayer. After the initial suggestion and inconclusive experiments of Bretscher (1972) that this was the case with human erythrocyte membranes, more significant enzymatic (Zwaal et al., 1973; Verkleij et al., 1973) and chemical experiments (Gordesky et al., 1975) indeed demonstrated such asymmetry, and similar conclusions have been derived with other membranes as well (Tsai and Lenard, 1975; Rothman et al., 1976).

It has been our view for some time (Singer and Nicolson, 1972) that the molecular asymmetry of the proteins and of the polar lipids of membranes is due to an initial asymmetrical insertion or synthesis of these components in the membrane, and that these asymmetries are maintained by the vanishingly small rates of flipping of the integral protein molecules and polar lipids from one surface of the membrane to the other. These slow flip rates are attributed to the very large free energies of activation required to transfer the ionic groups of the amphipathic integral proteins and the ionic and polar head groups of the lipids through the nonpolar interior of the membrane (section 2B). This conclusion has been strongly supported by recent measurements of phospholipid flip rates in bilayers (Roseman et al., 1975). After earlier reports suggesting that such flip rates were of the order of several hours (Kornberg and McConnell, 1971), Roseman et al., have shown that the half-time for the flipping of phosphatidyl-ethanolamine in synthetic bilayer vesicles cannot be less than 80 days at 22°C. As there are several artifacts which might result in apparently more rapid flip rates, this very long minimal half-time is probably nearer the true figure than earlier values. Since the lipid was in a fluid state in the experiments of Roseman et al., the slow flip rates can only be attributed to the large free energy of transfer of the zwitterionic head group of phosphatidyl-ethanolamine from water to the hydrocarbon interior of the membrane, as discussed in section 2B, resulting in a large free energy of activation for the flipping process.

Since integral membrane proteins, by the arguments of section 3A, are likely to possess hydrophilic segments exposed to the aqueous environment with many charged groups per segment, the free energy of activation for the flipping of such a molecule across the membrane should be even much larger than for an individual phospholipid molecule; i.e., the rate of flipping should be vanishingly small.

It has been claimed by several investigators (McNamee and McConnell, 1973; Renooij et al., 1976, Bloj and Zilversmit, 1976) that in real membranes, as opposed to synthetic lipid vesicles, the flip rates of phospholipids are much more rapid, of the order of minutes to hours. We believe that in each of these experiments, the apparently more rapid flip rates have some other explanation, but it would take too much space to discuss each experiment in detail here. It is also possible that under non-physiological conditions, flip rates of phospholipids and proteins may appear to be increased. For example, in a membrane containing substantial amounts of detergent (such as lysolecithin) large micellar domains produced by the detergent, and containing proteins and lipids, may flip from one surface of the membrane to the other. Barring such structural rearrangements of the membrane, however, we think that hydrophilic interactions act to keep phospholipid flip-rates in real membranes negligibly slow, certainly by comparison to phospholipid turn-over times in the membranes.

If the thermodynamic and structural considerations presented in this paper are accepted, it follows that the mechanisms involved in the biosynthetic origins of the molecular asymmetry of membranes need to be considered. How are the integral proteins inserted into membranes? For those molecules that are embedded only part way through the membrane (Fig. 2), how are those embedded exclusively on one side discriminated from those that are embedded exclusively on the other? And how are those proteins that span the thickness of the membrane, with exposed hydrophilic segments on either side, inserted with one orientation into the membrane? Are phospholipids with different polar head groups synthesized by enzymes that are themselves asymmetrically oriented in the membrane, thereby generating the phospholipid asymmetry? If so, what about a phospholipid which is present in unequal amounts on both sides of a membrane, such as phosphatidylcholine in the human erythrocyte membrane (Verkleij et al., 1973)? Is it possible that there are two different biosynthetic pathways involved in the generation of that lipid on the two different membrane surfaces? These intriguing questions have no satisfactory answers at present.

With regard to an integral protein that is embedded only part way through a membrane, but does not span it, it may attach to the membrane spontaneously directly from its site of synthesis. The hydrophobic half of the bi-partite amphipathic molecule may fold up and intercalate spontaneously into the nonpolar lipid interior of the membrane, leaving its hydrophilic half protruding from the membrane. Such spontaneous intercalation into a membrane seems to occur with isolated cytochrome $b_5$ (Strittmatter et al., 1972). On the other hand, it is difficult to envision a thermodynamically sound mechanism that would enable a single polypeptide chain that spans a membrane, and that has two hydrophilic ends and a hydrophobic middle portion, to intercalate into a membrane. Such a process would appear to require <u>one</u> of the hydrophilic ends (and <u>only</u> that one, if the protein is to be asymmetrically oriented in the membrane) to penetrate through the hydrophobic interior of the membrane, a process that would be highly unfavorable thermodynamically. An alternative possibility for the incorporation of such membrane-spanning polypeptide chains is that they also are originally synthesized as bipartite amphipathic molecules with one large hydrophilic half and one large hydrophobic half; but upon spontaneous intercalation of the hydrophobic half into one surface of the membrane, it may nearly span the membrane and come close to the opposite surface without protruding from it. At that stage, the nearly-protruding portion may be modified to acquire hydrophilic properties (as for example, by proteolytic processing, by phosphorylation or glycosylation, by deamidation of glutamine or asparagine residues, etc.). Such modifications, by placing ionic or highly polar groups on limited portions of the originally hydrophobic segment of the protein, may now change the protein conformation to come to a new equilibrium state. In that conformation, a second hydrophilic region might now protrude from that membrane surface opposite to the surface occupied by the original hydrophilic segment of the molecule. Such hydrophilic processing and resultant conformational changes might also lead to the conversion of monomeric integral proteins into sub-unit aggregates with water-filled channels (Fig. 2B).

## 5. THE MECHANISM OF TRANSLOCATION IN THE TRANSPORT OF SMALL HYDROPHILIC LIGANDS THROUGH MEMBRANES

The negligibly slow flip rates of integral proteins in membranes bear on the problem of transport through membranes. One of the two classes of mechanisms for the translocation event in transport that has been entertained in recent years is the "rotating carrier" mechanism, by which a specific transport protein in a membrane, with an active site to which the hydrophilic ligand becomes bound at one membrane surface, rotates or diffuses across the membrane and releases the ligand at the other surface. If transport proteins are amphipathic integral proteins, this mechanism is rendered highly unlikely by the thermodynamic considerations discussed earlier. Direct evidence to this effect has been lacking, however. Recently, it has been shown in our laboratory (Kyte, 1974; Dutton et al., 1976) and by others (Martinosi and Fortier, 1974) that the attachment of an antibody molecule (150,000 molecular weight) to the specific transport protein in the intact membrane does not affect the enzyme activity of that protein or the rate of transport which it mediates. These findings are difficult to reconcile with a rotating carrier mechanism in these cases, and we suggest that this mechanism is no longer tenable in general. Some time ago, on the basis of our thermodynamic analysis, we (Singer, 1971) proposed a different mechanism, unaware that it had in essence been earlier suggested by Jardetzky for other reasons (1966). We may call it the "aggregate rearrangement" mechanism (Fig. 5). All transport proteins are proposed to consist of sub-unit aggregates (Fig. 2B) spanning the thickness of a membrane with a continuous water-filled channel running down the axis of the aggregate. If the aggregate is a dimer of two identical chains, it would have a single two-fold axis of rotation perpendicular to the plane of the membrane, with a molecular structure similar, for example, to that of cytoplasmic malate dehydrogenase (Hill et al., 1972). The binding of the ligand to an active site situated within the channel, coupled to some energy-yielding step in the case of active transport, would then cause a quaternary rearrangement of the aggregate which translocated the ligand to the other side of the membrane (Fig. 5). With such a mechanism, the attachment of an antibody molecule to the transport protein might not affect the rate of transport if the antibody did not interfere with the quaternary rearrangement that was involved. The mechanism is also consistant with the properties of many soluble proteins and allosteric enzymes

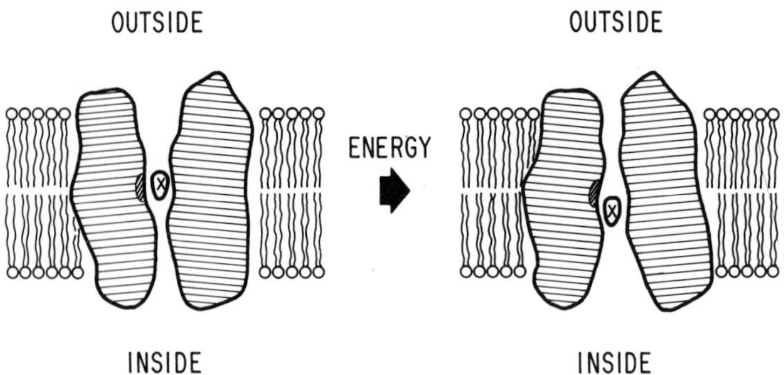


Figure 5 A schematic mechanism for the translocation event in active transport. A specific site for a small hydrophilic ligand X exists on the surface of the pore formed by a particular subunit aggregate (see Fig. 2B). (The aggregate is depicted as a dimer of non-identical units, but it could be a dimer, trimer, or tetramer species of identical or non-identical chains). Some energy yielding process is then converted into a quaternary rearrangement of the subunits, which translocates the binding site and X from one side of the membrane to the other. From Singer (1974).

which exist in solution as small subunit aggregates.

There is as yet not much experimental evidence regarding this hypothesis. However, two transport proteins that have been most extensively studied so far, the $Na^+$, $K^+$-ATPase that is the ubiquitous Na pump (Skou, 1960), and the anion transport protein in erythrocyte membranes (Cabantchik and Rothstein, 1974) have both been shown (Ruoho and Kyte, 1974; Bretscher, 1973) to span the width of the membrane, and in the case of the Na pump, to be present in the membrane as a noncovalently bound dimer of the membrane-spanning polypeptide chain (Kyte, 1975). Much more information is required, however, about the equilibrium and dynamic structures of transport proteins in membranes before any mechanism can be validated.

## ACKNOWLEDGMENT

The studies that are reviewed in this paper were supported by U. S. Public Health Service grants AI-06659 and GM-15971.

## REFERENCES

Bloj, B. and D. B. Zilversmit, Asymmetry and transposition rate of phosphatidylcholine in rat erythrocyte membranes, Biochemistry 15, 1277, 1976.

Bretscher, M. S., Phosphatidyl-ethanolamine: differential labelling in intact cells and cell ghosts of human erythrocytes by a membrane-impermeable reagent, J. Mol. Biol. 71, 523, 1972.

Bretscher, M. S., Membrane Structure: Some general principles, Science 181, 622, 1973.

Cabantchik, Z. I. and A. Rothstein, Membrane proteins related to anion permeability of human red blood cells. I. Localization of disulfonic stilbene binding sites in proteins involved in permeation, J. Membrane Biol. 15, 207, 1974.

Capaldi, R. A. and G. Vanderkooi, The low polarity of many membrane proteins, Proc. Nat. Acad. Sci. U.S.A. 69, 930, 1972.

Cohn, E. J. and J. T. Edsall, Proteins, Amino Acids and Peptides, p. 201, Reinhold Publ. Co., New York, 1943.

Dutton, A., E. D. Rees and S. J. Singer, An experiment eliminating the rotating carrier mechanism for the active transport of Ca ion in sarcoplasmic reticulum membranes, Proc. Nat. Acad. Sci. U.S.A. 73, 0000, 1976.

Edsall, J. T. and J. Wyman, Biophysical Chemistry, p. 258, Academic Press, New York, 1958.

Gordesky, S. E., G. V. Marinetti and R. J. Love, The reaction of chemical probes with the erythrocyte membrane, J. Membrane Biol. 20, 111, 1975.

Henderson, R. and P. N. T. Unwin, Three-dimensional model of purple membrane obtained by electron microscopy, Nature (Lond.) 257, 28, 1975.

Hill, E., D. Tsernoglou, L. Webb and L. J. Banaszak, Polypeptide conformation of cytoplasmic malate dehydrogenase from an electron density map at 3.0 Å resolution, J. Mol. Biol. 72, 577, 1972.

Hirano, H., B. Parkhouse, G. L. Nicolson, E. S. Lennox and S. J. Singer, Distribution of saccharide residues on membrane fragments from a myeloma-cell homogenate: its implications for membrane biogenesis, Proc. Nat. Acad. Sci. U.S.A. 69, 2945, 1972.

Jardetzky, O., Simple allosteric model for membrane pumps. Nature (Lond.) 211, 969, 1966.

Jost, P. C., O. H. Griffith, R. A. Capaldi and G. Vanderkooi, Evidence for boundary lipid in membranes, Proc. Nat. Acad. Sci. U.S.A. 70, 480, 1973.

Kauzmann, W., Some factors in the interpretation of protein denaturation, Advances Protein Chem. 14, 1, 1959.

Klotz, I. M. and S. B. Farnham, Stability of an amide-hydrogen bond in an apolar environment, Biochemistry 7, 3879, 1968.

Klotz, I. M. and J. S. Franzen, Hydrogen bonds between model peptide groups in solution, J. Amer. Chem. Soc. 84, 3461, 1962.

Kornberg, R. D. and H. M. McConnell, Inside-outside transitions of phospholipids in vesicle membranes, Biochemistry 10, 1111, 1971.

Kyte, J., The reactions of sodium and potassium ion-activated adenosine triphosphatase with specific antibodies. Implications for the mechanism of active transport, J. Biol. Chem. 249, 3652, 1974.

Kyte, J., Structural studies of sodium and potassium ion-activated adenosine triphosphatase. The relationship between molecular structure and the mechanism of active transport. J. Biol. Chem. 250, 7443, 1975.

Lenard, J. and S. J. Singer, Protein conformation in cell membrane preparations as studied by optical rotatory dispersion and circular dichroism, Proc. Nat. Acad. Sci. U.S.A. 56, 1828, 1966.

Martinosi, A. and F. Fortier, The effect of anti-ATPase antibodies upon the $Ca^{++}$ transport of sarcoplasmic reticulum, Biochem. Biophys. Res. Commun. 60, 382, 1974.

Matthews, F. S., P. Argos and M. Levine, The structure of cytochrome $b_5$ at 2.0 Å resolution, Cold Spring Harbor Symp. Quant. Biol. 36, 387, 1972.

McNamee, M. G. and H. M. McConnell, Transmembrane potentials and phospholipid flip-flop in excitable membrane vesicles, Biochemistry 12, 2951, 1973.

Nicolson, G. L. and S. J. Singer, Ferritin-conjugated plant agglutinins as specific saccharide stains for electron microscopy: application to saccharides bound to cell membranes. Proc. Nat. Acad. Sci. U.S.A. 68, 942, 1971.

Nicolson, G. L. and S. J. Singer, The distribution and asymmetry of mammalian cell surface saccharides utilizing ferritin-conjugated plant agglutinins as specific saccharide stains, J. Cell Biol. 60, 236, 1974.

Renooij, W., L. M. Van Golde, R. F. A. Zwaal and L. L. M. Van Deenan, Topological asymmetry of phospholipid metabolism in rat erythrocyte membranes, Eur. J. Biochem. 61, 53, 1976.

Roseman, M., B. J. Litman and T. E. Thompson, Transbilayer exchange of phosphatidylethanolamine for phosphatidylcholine and N-acetimidoylphosphatidylethanolamine in single-walled bilayer vesicles, Biochemistry 14, 4826, 1975.

Rothman, J. E., D. K. Tsai, E. A. Dawidowicz and J. Lenard, Transbilayer phospholipid asymmetry and its maintenance in the membrane of influenza virus, Biochemistry, in press, 1976.

Ruoho, A. and J. Kyte, Photoaffinity labeling of the ouabain-binding site on $(Na^+ + K^+)$ adenosinetriphosphatase, Proc. Nat. Acad. Sci. U.S.A. 71, 2352, 1974.

Segrest, J. P., I. Kahane, R. L. Jackson and V. T. Marchesi, Major glycoprotein of the human erythrocyte membrane. Evidence for an amphipathic molecular structure, Arch. Biochem. Biophys. 155, 167, 1973.

Segrest, J. P., T. Gulik-Krzywicki and C. Sardet, Association of the membrane-penetrating polypeptide segment of the human erythrocyte MN-glycoprotein with phospholipid bilayers, Proc. Nat. Acad. Sci. U.S.A. 71, 3294, 1974.

Singer, S. J., The molecular organization of biological membranes, in Structure and Function of Biological Membranes, pp. 145-222, edited by L. I. Rothman, Academic Press, New York, 1971.

Singer, S. J., The molecular organization of membranes, Ann. Rev. Biochem. 43, 805, 1974.

Singer, S. J. and G. L. Nicolson, The fluid mosaic model of the structure of cell membranes, Science 175, 720, 1972.

Skou, J. C., Further investigations on a $Mg^{++}$ + $Na^{+}$-activated adenosinetriphosphatase, possibly related to the active, linked transport of $Na^{+}$ and $K^{+}$ across the nerve membrane, Biochim. Biophys. Acta 42, 6, 1960.

Spatz, L. and P. Strittmatter, A form of reduced nicotinamide adenine dinucleotide-cytochrome $b_5$ reductase containing both the catalytic site and an additional hydrophobic membrane-binding segment, J. Biol. Chem. 248, 793, 1973.

Strittmatter, P., M. J. Rogers and L. Spatz, The binding of cytochrome $b_5$ to liver microsomes, J. Biol. Chem. 247, 7188, 1972.

Tanford, C., The association of acetate with ammonium and guanidinuim ions., J. Amer. Chem. Soc. 76, 945, 1954.

Tsai, K.-H. and J. Lenard, Asymmetry of influenza virus membrane bilayer demonstrated with phospholipase-C, Nature (Lond.) 253, 554, 1975.

Verkleij, A. J., R. F. A. Zwaal, B. Roelofsen, P. Comfurius, D. Kastelijn and L. L. M. Van Deenan, The asymmetric distribution of phospholipids in the human red cell membrane. A combined study using phospholipases and freeze-etch electron microscopy, Biochim. Biophys. Acta 323, 178, 1973.

Wallach, D. F. H. and P. H. Zahler, Protein conformations in cellular membranes, Proc. Nat. Acad. Sci. U.S.A. 56, 1552, 1966.

Warren, G. B., M. D. Houslay, J. C. Metcalfe and N. J. M. Birdsall, Cholesterol is excluded from the phospholipid annulus surrounding an active calcium transport protein, Nature (Lond.) 255, 684, 1975.

Zwaal, R. F. A., B. Roelofsen and C. M. Colley, Localization of red cell membrane constituents, Biochim. Biophys. Acta 300, 159, 1973.

# COUPLING OF CHEMICAL REACTION TO TRANSPORT OF SODIUM AND POTASSIUM

J. C. Skou

University of Aarhus, Institute of Physiology

DK-8000 Aarhus C, Denmark

## I. INTRODUCTION

The membrane bound ($Na^{+}$+$K^{+}$)-activated ATP hydrolyzing enzyme system ($Na^{+}$+$K^{+}$ ATPase in the following) couples the flux of sodium and of potassium along their electrochemical gradients across the cell membrane - the passive flux - to a flux of the same ions against their electrochemical gradients - the active transport (see Skou, 1975). The coupling is tight enough to give a steady-state concentration of sodium and of potassium in the cell which is away from thermodynamic equilibrium of the two cations.

The passive flux of sodium into the cell is coupled to the passive flux of potassium out by the membrane potential which again is set by the different permeabilities of the membrane for the two cations and of their concentration gradients.

But how is the passive flux coupled to the active transport? This can be divided into at least three problems.

1) How is the information about a change in the free energy of the cation gradients transferred to the coupling system - the $Na^{+}$+$K^{+}$ ATPase?

2) How is this information transformed into an active flux, i.e. what is the link between the exergonic chemical re-action - hydrolysis of ATP - and the active flux?

3) How is the active flux of sodium coupled to the active flux of potassium?

## II. COUPLING BETWEEN ACTIVE AND PASSIVE FLUX

It is characteristic for the $Na^{+}+K^{+}$ ATPase that the exergonic chemical reaction, the hydrolysis of ATP, coupled to a transport of sodium out and potassium into the cell is activated by a combined effect of sodium on a site on the system which is facing the internal solution in the intact membrane, the i-site in the following, and of potassium on a site facing the external solution, o-site, i.e. a ${}^{o}K_{m}/{}^{i}Na_{n}$ form of the system (i for inside, o for outside, m and n are numbers); each of the sites accepts a number of cations, probably three sodium or two potassium ions, and on each site there is competition between sodium and potassium (see Skou, 1975). It means that saturation of each of the sites with the one or the other of the cations is a function both of the absolute concentrations of sodium and potassium and of the ratio between the concentrations of the two cations in the internal, respective external solution.

If it is correct that the i-site and the o-site on the system exist simultaneously (see Skou, 1975) and assuming that each site does not exist in a hybrid form, the enzyme molecules will be divided between four different forms which are in equilibrium.

With saturating concentrations of ATP and magnesium the sodium-potassium affinity ratio on the i-site is about 3:1 while on the o-site it is about 1:100, meaning that with normal intra- and extracellular concentrations of sodium and of potassium the ratio between the different forms of the enzyme molecules will be about as shown in Figure 1, and the sodium-potassium dependent rate of hydrolysis of ATP (the ${}^{o}K_{m}/{}^{i}Na_{n}$ form) will then be about 35% of maximum (see Skou, 1975).

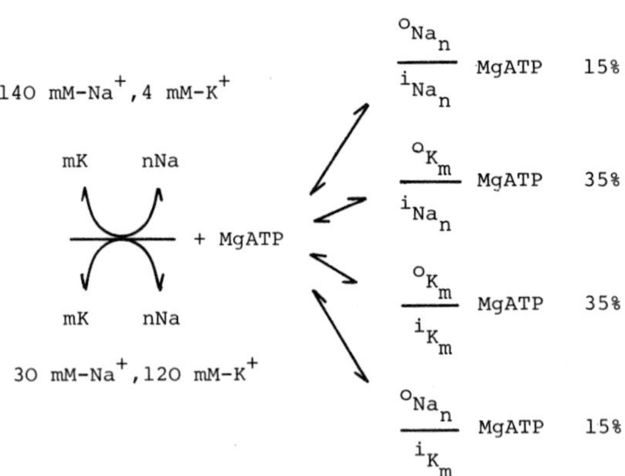


Fig. 1. For explanation see text.

It seems thus to be the ratio between sodium and potassium in the internal as well as in the external solution which determines the activity of the system and by this the rate of active transport of the two cations, meaning that a change in the passive flux via the change in sodium and potassium concentrations directly is transformed into a change in the active flux.

The rate of the chemical reaction, the hydrolysis of ATP, will as in any enzymatic reaction be dependent on the concentration of the substrate, ATP (and magnesium); for the sodium plus potassium activated reaction the apparent $K_m$ for MgATP is about 0,3 mM (see Post et al., 1965).

However, ATP also influences the apparent affinity for sodium relative to potassium both on the i- and on the o-site. On the i-site where sodium activates, ATP increases the apparent affinity for sodium relative to potassium from about 0.4:1 without ATP to about 3:1 with saturating concentrations of ATP (Skou, 1974a, b).

It means that at a given sodium:potassium ratio in the medium the number of enzyme molecules with a sodium saturated i-site increases with the ATP concentration, and by this the number of enzyme molecules on the active $^{o}K_m/^{i}Na_n$ form. The effect is due to ATP as such before ATP is hydrolyzed (Skou, 1974a). The isolated system has a high affinity for ATP ($K_{diss}$ about 0.2 μM) and this affinity is not changed or slightly increased by sodium; potassium on the other hand decreases the affinity for ATP and vice versa (Nørby and Jensen, 1971; Hegyvary and Post, 1971).

As this seems to be an effect of potassium on the i-site this could explain that ATP at a given sodium:potassium ratio shifts the equilibrium towards the sodium form of the i-site (Skou, 1974a); ATP seems to increase the rate by which potassium is released from the system (Post et al., 1972).

ATP has, however, also an effect on the apparent affinity for potassium relative to sodium on the o-site. This seems to be an effect of MgATP (Robinson, 1967) and may be related to the rate of hydrolysis of ATP. MgATP decreases the apparent affinity for potassium relative to sodium and with saturating concentrations of MgATP it is about 100:1.

With saturating concentrations of MgATP the normal 4 meq/l of potassium in the external medium give in the presence of 140 meq/l sodium about 70-75% potassium saturation of the o-site. With such non-saturating concentrations of potassium an increase in the potassium affinity when the MgATP concentration is decreased will tend to compensate for the decrease in activity which is due to the substrate effect of MgATP. But the substrate effect is much more pronounced than the effect of the increase in the apparent affinity for potassium.

The information about a change in the passive flux which gives a change in the concentrations of sodium and potassium is thus transferred to the coupling system via a direct effect of the cations on the activity of the coupling system. The effect is modulated by the ATP (and magnesium) concentration.

## III. COUPLING BETWEEN CHEMICAL REACTION AND ION FLUX

The active transport system can accomplish two types of transport of sodium and potassium (Glynn and Karlish, 1974). One in which there is a net transport of the cations involved, another in which there is no net transport but an exchange of alike cations.

The net transport can either be an exchange of sodium from inside for potassium from outside, the ${}^{o}K_m/{}^{i}Na_n$ form; it requires ATP in the internal medium and is coupled to the exergonic chemical reaction, the hydrolysis of ATP. Or it can be a low net efflux of sodium to an external medium which contains cholin but no sodium or potassium (Lew et al., 1973; Garrahan and Glynn, 1967a; Karlish and Glynn, 1974); it requires ATP in the internal medium (Lew et al., 1973). Experiments on the isolated system show that there is an absolute requirement for sodium on the i-site for activation of the exergonic reaction, while potassium on the o-site can be replaced by Rb, Cs, Li or $NH_4$ (Skou, 1960). Considering this lack of specificity for monovalent cations for activation on the o-site, it is tempting to suggest that cholin may also activate but with much lower Vmax than potassium and that the sodium efflux to the cholin medium is coupled to a cholin influx.

Finally, the net transport can be an exchange of sodium from outside for potassium from inside, the ${}^{o}Na_n/{}^{i}K_m$ form; this reguires ATP and Pi in the internal medium and is coupled to the endergonic reaction, the synthesis of ATP, a reversal of the pump (Garrahan and Glynn, 1967d; Glynn et al., 1970; Lant et al., 1970).

The exchange reaction with no net transport is either an exchange of sodium from inside for sodium from outside, the ${}^{o}Na_n/{}^{i}Na_n$ form, or an exchange of potassium for potassium, the ${}^{o}K_m/{}^{i}K_m$ form (Glynn and Karlish, 1974).

The Na:Na exchange requires ATP (Garrahan and Glynn, 1967b) and ADP (Glynn and Hoffman, 1971; Baker et al., 1971) in the internal medium. There is no or a very low net hydrolysis of ATP but probably a phosphorylation of the system from ATP followed by a reversal of the reaction with formation of ATP by a reaction of the phosphoenzyme with ADP - an ADP-ATP exchange reaction.

The K:K exchange requires ATP (Glynn et al., 1971) and Pi (Glynn et al., 1970) in the internal medium, but the chemical reaction underlying the exchange is not known.

The nature and the direction of the transport is thus set by the nature of the cations on the o- and the i-sites, respectively, which again is determined by the ratio of the cations in the external and the internal medium and of the affinities of the sites for the cations. It is, however, the chemical potential for the ATP-ADP-Pi system which determines whether the reaction with a given combination of cations on the two sites will procede or not and the rate of the reaction.

In the exergonic reaction three sodium ions are transported out and two potassium ions in for each ATP molecule hydrolyzed (see Glynn, 1968). The 3:2 ratio is not compensated for by a flow of an anion together with sodium or of another cation, a proton, with potassium, the transport is electrogenic (Thomas, 1972). It suggests that anions and protons cannot pass the membrane along the same route along which sodium and potassium are transported.

The tight coupling between the chemical reaction and the flow of the cations suggests that the cations cannot flow across the membrane through the system unless there is a reaction with the chemical substrate ATP-ADP-Pi, i.e. the system is "closed". Sodium and potassium can, however, be exchanged between the i-site and the surrounding medium without the presence of the chemical substrate (Skou, 1972a). It suggests that the system without the chemical substrate is closed somewhere between the i-site and the outside, probably between the i-site and the o-site (see Figure 2 and 3).

The number of sodium ions moved per ATP hydrolyzed seems to be the same whether the transport is with or against an electrochemical gradient (Sen and Post, 1964; Whittam and Ager, 1965; Garrahan and Glynn, 1967c). It suggests that the chemical reaction is not directly coupled to the movement of the cations but to a change of the system in such a way that the cations will move along gradients without being influenced by the electrochemical gradient across the membrane. This would require that the system is "closed" towards the surroundings and "open" for flow of the cations in between the o- and the i-site.

The reaction with the chemical substrate thus seems to change the system from a state in which it can exchange cations with the surrounding medium but not through the system to a state in which the cations can be exchanged through the system but not with the surrounding medium (see Figure 2 and 3); and this seems to be synchronized to creation of a driving force for sodium from the i- to the o-site and for potassium in the opposite direction.

SDS polyacrylamide gel electrophoresis of purified enzyme after digestion with trypsin in the presence of potassium show a polypeptide pattern which is different from that seen after digestion in the presence of sodium; it indicates a conformation of the system in the presence of potassium which is different from that seen in the presence of sodium (Jørgensen, 1975). This agrees with experiments in which it is seen that the system has a reactivity towards the SH blocking agent n-ethylmaleimide (NEM) in the presence of potassium which is different from that seen in the presence of sodium (Skou, 1972a).

In the NEM experiments it was shown that it is sodium and potassium on the i-site which have a different effect while the cations on the o-site have no influence. The experiments suggest a conformation of the ${}^{o}K_m/{}^{i}K_m$ form which is different from the ${}^{o}K_m/{}^{i}Na_n$ form and that this form has a conformation which is the same as the ${}^{o}Na_n/{}^{i}Na_n$ form. However, the ${}^{o}K_m/{}^{i}Na_n$ form has a much lower affinity for MgATP than the ${}^{o}Na_n/{}^{i}Na_n$ form (Post et al., 1965) indicating that there must also be a difference between these two forms of the system.

There seems thus to be a different conformation of the three different forms of the system and this is set by the combination of the cations on the two sites (Figure 2 and 3). As discussed above, it is also the combination of the cations on the two sites which determines whether the chemical reaction is turned into a reaction where work is done on the cations - net transport of sodium out and of potassium in - or work is done on the system - net flow of sodium in and potassium out - or no work is done - exchange of sodium for sodium or potassium for potassium. It seems therefore likely that it is the conformation set by the combination of the cations which so to say programs the system for the direction of the change in chemical potential when ATP-ADP-Pi react with the system and that the chemical potential of the ATP-ADP-Pi components determines whether the reaction can procede or not.

ATP has an effect on the conformation which goes beyond that of the cations and the effect depends on the cation combination on the sites. With potassium but no sodium, i.e. the ${}^{o}K_m/{}^{i}K_m$ form,

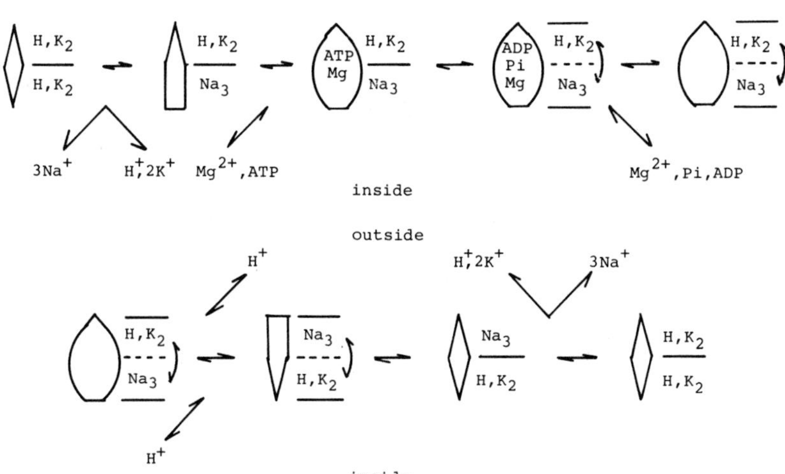


Fig. 2. Potassium affinity: outside ∧ , inside ∨
Sodium " : " ⊓ , " ⊔
Energized state : ( )
For further explanation see text.

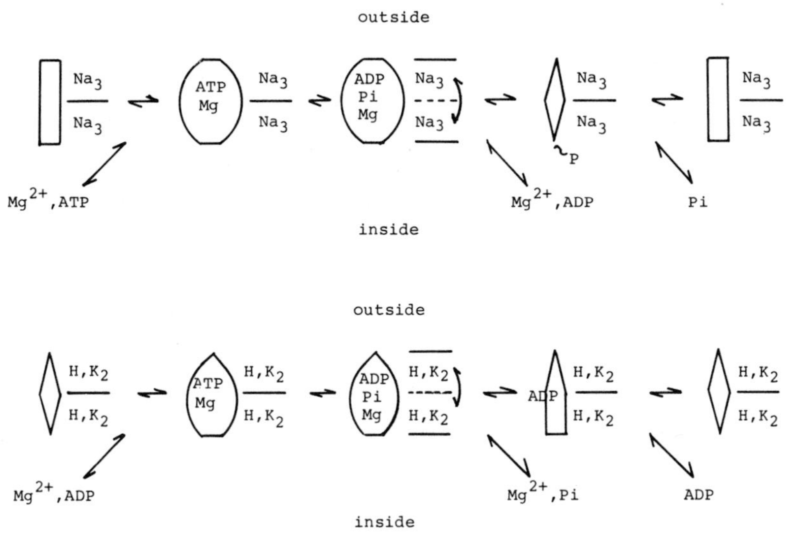

outside

Na3 / Na3 ⇌ ATP Mg Na3 / Na3 ⇌ ADP Pi Mg Na3 / Na3 ⇌ Na3 / Na3 ~P ⇌ Na3 / Na3

Mg2+,ATP Mg2+,ADP Pi

inside

outside

H,K2 / H,K2 ⇌ ATP Mg H,K2 / H,K2 ⇌ ADP Pi Mg H,K2 / H,K2 ⇌ ADP H,K2 / H,K2 ⇌ H,K2 / H,K2

Mg2+,ADP Mg2+,Pi ADP

inside

Fig. 3. For explanation see Fig. 2 and text.

ATP changes the sensitivity towards trypsin from a potassium to sodium sensitivity (Jørgensen, 1975), while there is no effect of ATP on the reactivity towards NEM (Skou, 1972a). On the other hand, with sodium ATP has no effect on the sensitivity towards trypsin, while ATP decreases the reactivity towards NEM both of the ${}^{o}K_m/{}^{i}Na_n$ and the ${}^{o}Na_n/{}^{i}Na_n$ form of the system.

There seems thus to be an effect of ATP on the conformation of all 3 forms, but the effect differs from the ${}^{o}K_m/{}^{i}K_m$ form on the one side and the ${}^{o}K_m/{}^{i}Na_n$ and the ${}^{o}Na_n/{}^{i}Na_n$ form on the other side. However, as discussed above, the different affinity of the ${}^{o}K_m/{}^{i}Na_n$ and the ${}^{o}Na_n/{}^{i}Na_n$ form for MgATP may suggest that these two forms also differs in their conformation.

It seems thus as if the nature of the cations on the two sites influences the conformation of the system and that the addition of ATP gives a further change in conformation, see Figure 2 and 3.

A change in conformation when ATP reacts with the system (in the presence of the cations) suggests that the ΔG change when ATP is bound, is transformed into "conformational" energy in the complex between ATP and the system, i.e. the complex represents an energized system.

The conformation set by the cations programs the system for the reaction which follows. On the ${}^{o}Na_n/{}^{i}Na_n$ form the reaction with ATP leads to a phosphorylation of the system (Post et al., 1965; Skou, 1965) and in the presence of ADP to a Na:Na exchange with no or a low hydrolysis of ATP (Glynn and Hoffman, 1971; Baker et al., 1971), i.e. no work is done on the cations. Potassium added to the phosphorylated system gives a dephosphorylation (Post et al., 1965) showing that it is due to lack of potassium that the ${}^{o}Na_n/{}^{i}Na_n$ form cannot release its phosphate. The ${}^{o}K_m/{}^{i}K_m$ form gives a K:K exchange and this requires ATP and Pi (Glynn et al., 1970; Glynn et al., 1971). By analogy one may suggest that the ${}^{o}K_m/{}^{i}K_m$ form can release Pi but not ADP and that it is the combined effect of potassium and sodium which is necessary to get a release of ADP as well as Pi from the system. It suggests that a proper release of ADP and Pi is crucial in the reaction which allows the system to do work on the cations, i.e. on the ${}^{o}K_m/{}^{i}Na_n$ form. On this form the release of ADP and Pi after hydrolysis of ATP will allow the ATP energized system to deenergize - relax - along a route which is different from the route along which it has been energized.

Thereby the deenergization may be used to energize the flow of the cations, i.e. create a driving force for sodium from the i- to the o-site and of potassium in the opposite direction in a system which due to the reaction is "closed" towards the surroundings but "open" for flow of cations in between the two sites.

## IV. COUPLING OF FLOW OF CATIONS

If it is correct that the i- and the o-site exist simultaneously, it is unlikely that the exchange of the cations between the i- and the o-site through the system can be coupled via a circulating carrier.

As discussed above, both the i- and the o-site change their affinities for the cations when the system reacts with ATP. The apparent affinity of the i-site is increased from a potassium higher than sodium to a sodium higher than potassium affinity when the system reacts with ATP (Skou, 1972a, b), meaning that the affinity must fall back to a potassium higher than sodium affinity when ATP is hydrolyzed and the hydrolysis products are removed from the system (ADP also increases the affinity for sodium, Jensen and Nørby, 1971). On the o-site MgATP increases the apparent affinity for sodium relative to potassium (Robinson, 1967) and this change in affinity seems to be related to the hydrolysis of ATP.

There seems thus to be an affinity change on the two sites related to the turnover cycle and in such a way that when ATP is hydrolyzed, the i-site goes from a sodium higher than potassium to a potassium higher than sodium affinity and at the same time the o-site seems to increase its affinity for sodium relative to potassium.

Provided that the i- and the o-site cannot exchange their cations with the surrounding medium and that the system is "open" for flow of cations in between the two sites, such a change in affinity could give a vectorial coupled flow of sodium from the i- to the o-site and of potassium the opposite way, Figure 2.

If the "opening" inside the system is transient, then the following "closing" inside, "opening" of the o-site towards the surroundings and the return of the affinity of the o-site towards a potassium much higher than sodium affinity, will allow potassium from outside by competition to displace sodium from the o-site, Figure 2.

Such an ion exchange based on an affinity change could give both a driving force for the ions and a coupling between the flow.

There are, however, two problems, one is that even if the apparent affinity for sodium on the o-site is increased, it never becomes higher than the affinity for potassium as one would expect necessary to give the exchange; it is always much lower than the apparent affinity for potassium (see Skou, 1975). The explanation could be that the o-site is "closed" towards the surrounding medium while the affinity is changed and that the affinity change can therefore not be seen.

The other problem is the 3:2 coupling between sodium and potassium. A possibility could be that the empty place in the presence of potassium could be occupied by a proton, i.e. that the site could bind 1 $H^+$ and 2 $K^+$ or 3 $Na^+$. As the transfer of the cations is electrogenic, the proton will not pass the membrane with the potassium ions but it may flow out from the o-site to the outside medium when the ions are exchanged and simultaneously there may be a flow of a proton in on the i-site from the inside, see Figure 2. This could be the proton released by the ATP hydrolysis which directly is transferred to the i-site.

An alternative way of creating a driving force for flow of the cations could be a transient potential difference set up inside the anion impermeable system which is "open" in between the sites but "closed" towards the surrounding medium and for example with the potential negative in the direction of the o-site. This could give a flow of sodium from the i-site towards the o-site and when the potential disappears, a flow of potassium from the o-site towards the i-site provided the permeability for potassium inside the system is higher than for sodium - a kind of a reversed action potential mechanism. A proton flowing out and in but not across the system could as in the exchange mechanism compensate for the 3:2 ratio.

It is tempting in this context to ask why is there a 3:2 coupling? Is it of functional significance that the transport is electrogenic or is it a byproduct of a way of constructing the pump; could for example a pK change underlying the suggested flow of protons be a necessary step when the reaction is programmed to do work on the cations?

A change in conformation means a change in distribution of electrons on the system. The link between conformational changes and changes in affinities or potentials set up locally inside the system may then be a change in distribution of electrons on the system accompanying the energization-deenergization process.

## Na:Na Exchange

In the Na:Na exchange there is no net flow of sodium and no work is done. However, ATP is necessary (Garrahan and Glynn, 1967b) and so is ADP (Glynn and Hoffman, 1971; Baker et al., 1971). There is no or only a very low net hydrolysis of ATP. With the isolated system in the test tube the reaction in the presence of magnesium, sodium and ATP leads to formation of a phosphoenzyme with a high rate, while the rate of dephosphorylation is low (Post el al., 1965; Skou, 1965).

The ATP induced change in conformation seems to disappear when the system becomes phosphorylated from ATP and in spite of having sodium but no potassium in the medium, it acquires a potassium sensitivity towards trypsin (Jørgensen, 1975).

Potassium added to the phosphoenzyme increases the rate of dephosphorylation, indicating that it is due to lack of potassium that the phosphate is not released from the ${}^{o}Na_n/{}^{i}Na_n$ form of the system when ATP is hydrolyzed but due to sodium that ADP is released. The system can therefore not deenergize as in the presence of sodium and potassium (the ${}^{o}K_m/{}^{i}Na_n$ form) where both ADP and Pi leave the system when ATP is hydrolyzed; instead the conformational energy is transformed into a high energy phosphate bond when ADP is released, Figure 3. ADP in the medium which is required for the Na:Na exchange will tend to decrease the rate of the forward reaction which leads to formation of the phosphoenzyme and thereby ADP increases the possibility of the system to deenergize along the route by which it is formed. As the hydrolysis of ATP leads to a closing of the system towards the surroundings and an opening in between the o- and the i-site, this reaction and its reversal can give a Na:Na exchange and this will chemically be connected to an ADP-ATP exchange reaction.

Oligomycin inhibits the Na:Na exchange seen in the intact membrane (Garrahan and Glynn, 1967c), but increases the ADP-ATP exchange reaction seen in the test tube (Blostein, 1970). It may suggest that oligomycin in some way uncouples the ADP-ATP exchange from the Na:Na exchange, leading to an inhibition of the Na:Na exchange and an increase in the ADP-ATP exchange.

## K:K Exchange

The K:K exchange gives no net transport of potassium. It requires ATP (Glynn et al., 1971) and Pi (Glynn et al., 1970) but not ADP (Simons, 1974) in the internal medium. And by analogy to the Na:Na exchange the requirement for Pi may suggest that it is ADP which cannot be released in a proper way, but Pi; the system can therefore not deenergize - relax - in a forward direction. Pi in the medium increases the possibility of the energized system to relax along the route by which it is formed and due to the opening and closing of the system, this may give, as in the Na:Na exchange, a K:K exchange, Figure 3.

In the discussed model the chemical reaction is linked to the flow of the cations via a conformational change of the system. The crucial point is the relaxation of the energized system which for a forward reaction and thereby for the coupling to an energy requiring process requires a proper release of Pi and ADP and this is dependent on the presence of potassium on the o-site and of sodium on the i-site of the system. The formation of a phosphoenzyme seen with the sytem on the ${}^{o}Na_n/{}^{i}Na_n$ form is due to an improper release of the hydrolysis products and thereby to a relaxation which leads to conservation of the energy in the phosphate bond and to no work done on the cations.

An alternative to this is a reaction in which the coupling between the chemical reaction and the transport is via a phosphorylation-dephosphorylation of the system (Post et al., 1965).

A phosphorylation-dephosphorylation could underlie the reaction in a two site system; on the other hand, as the system is phosphorylated in the presence of sodium but no potassium and the phosphoenzyme is dephosphorylated by potassium, it seems unlikely that a phosphorylation-dephosphorylation should not underlie the consequtive reaction of a one-site system. And even if there is a certain evidence that the system is a two-site system, the evidence is not conclusive. Therefore and for sake of simplicity the phosphorylation-dephosphorylation principle is exemplified in a one-site system, Figure 4.

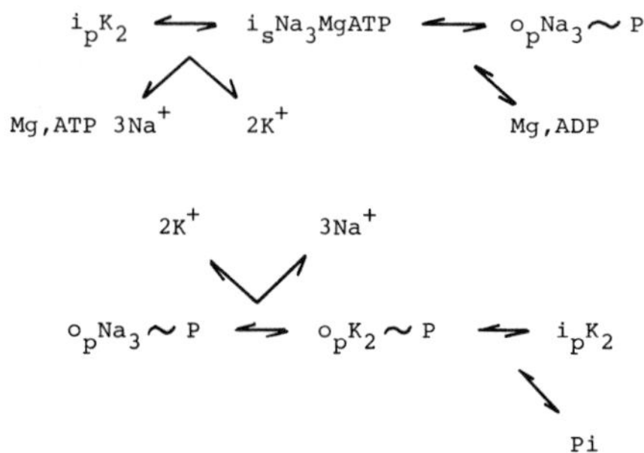


Fig. 4. For explanation see text.

## V. ONE-SITE SYSTEM

One site and a tight coupling between the chemical reaction and the transport suggests that the system alternates between a state in which it can exchange cations with the internal but not the external solution and another in which it can exchange cations with the external but not the internal solution. If this change in state is controlled by the chemical reaction and is synchronized to a change in affinity of the site, such a mechanism could give a vectorial coupled flow of the cations, Figure 4 (i for the site when facing the internal solution, o the external, p for the site having a potassium higher than sodium affinity, s a sodium higher than potassium affinity).

As discussed above, ATP increases the apparent affinity for sodium relative to potassium on the i-site (Skou, 1972a, b), and the trypsin experiments suggest that even in the presence of potassium the system is turned into the sodium form by ATP (Jørgensen, 1975). In the test tube, the system with sodium, magnesium and ATP becomes phosphorylated (Post et al., 1965; Skou, 1965); as addition of potassium leads to a dephosphorylation of the system in concentrations which are very low compared to the sodium concentrations, this effect of potassium must be on a site with an affinity for potassium much higher than sodium - the o-site. If the cation binding site is either an i- or an o-site, this must mean that the i-site by the phosphorylation has been turned into an o-site and at the same time its affinity has been changed from a sodium higher than potassium to a potassium higher than sodium affinity. Due to this, sodium on the o-site is exchanged for potassium from the outside and this leads to a dephosphorylation and a conversion of the site from an o- to an i-site. In the previous discussed two-site model, this effect of potassium is on the ${}^{o}Na_n/{}^{i}Na_n$ phosphorylated form of the system and it will give a dephosphorylation which is not connected to transport.

It is obvious that conformational changes are also of importance for the reaction in the discussed one-site model but it is the phosphorylation which links the chemical reaction to transport. In the discussed two-site model it is the formation of the enzyme - ATP complex which energize the system (change in conformation) and it is the following hydrolysis with removal of ATP and Pi which allows the system to deenergize (return of conformation) along another route than that by which it is energized and thereby couples the chemical reaction to the transport of the cations.

A comparison between the two models will show that it is very difficult experimentally to distinguish between the two principles and we have so far no experimental evidence which allows to decide whether the one or the other - or a third - principle underlies the coupling of the chemical reaction to the transport of the cations.

As discussed above, a one-site system will with all likelyhood depend on a phosphorylation-dephosphorylation reaction for the coupling between the chemical reaction and the transport of the cations; it is therefore of importance to get an answer to what should seem to be a simple but so far unsolved problem - is there one alternating site on the system or are there both an i- and an o-site at the same time?

## References

Baker, P. F., Foster, R. F., Gilbert, D. S. and T. I. Shaw, Sodium transport by perfused giant axons of Loligo, J. Physiol., Lond., 219, 487, 1971.

Blostein, R., Sodium-activated adenosine triphosphatase activity of the erythrocyte membrane, J. biol. Chem., 245, 270, 1970.

Garrahan, P. J. and I. M. Glynn, The sensitivity of the sodium pump to external sodium, J. Physiol., Lond., 192, 175, 1967a.

Garrahan, P. J. and I. M. Glynn, Factors affecting the relative magnitudes of the sodium:potassium and sodium:sodium exchanges catalysed by the sodium pump, J. Physiol., Lond., 192, 189, 1967b.

Garrahan, P. J. and I. M. Glynn, The stoicheiometry of the sodium pump, J. Physiol., Lond., 192, 217, 1967c.

Garrahan, P. J. and I. M. Glynn, The incorporation of inorganic phosphate into adenosine triphosphate by reversal of the sodium pump, J. Physiol., Lond., 192, 237, 1967d.

Glynn, I. M., Membrane ATPase and cation transport, Br. med. Bull., 24, 165, 1968.

Glynn, I. M. and J. F. Hoffman, Nucleotide requirements for sodium-sodium exchange catalysed by the sodium pump in human red cells, J. Physiol., Lond., 218, 239, 1971.

Glynn, I. M., Hoffman, J. F. and V. L. Lew, Some "partial reactions" of the sodium pump, Phil. Trans. R. Soc. B, 262, 91, 1971.

Glynn, I. M. and S. J. D. Karlish, The association of biochemical events and cation movements in (Na:K) dependent adenosine triphosphatase activity, in Membrane Adenosine Triphosphatase and Transport Processes, edited by R. Bronk, pp. 145-158, Biochem. Soc. Spec. Publ., 4, London, 1974.

Glynn, I. M., Lew, V. L. and U. Lüthi, Reversal of the potassium entry mechanism in red cells, with and without reversal of the entire pump cycle, J. Physiol., Lond., 207, 371, 1970.

Hegyvary, C. and R. L. Post, Binding of ATP to Na,K-ATPase, J. biol. Chem., 246, 5234, 1971.

Jensen, J. and J. G. Nørby, On the specificity of the ATP-binding site of ($Na^{+}$+$K^{+}$)-activated ATPase from brain microsomes, Biochim. biophys. Acta, 233, 395, 1971.

Jørgensen, P. L., Purification and characterization of ($Na^{+}$,$K^{+}$)-ATPase. V. Conformational changes in the enzyme. Transitions between the Na-form and the K-form studied with tryptic digestion as a tool, Biochim. biophys. Acta, 401, 399, 1975.

Karlish, S. J. D. and I. M. Glynn, An uncoupled efflux of Na from human red cells probably associated with Na dependent ATPase activity, Ann. N. Y. Acad. Sci., 242, 461, 1974.

Lant, A. F., Priestland, R. N. and R. Whittam, The coupling of downhill ion movements associated with reversal of the sodium pump in human red cells, J. Physiol., Lond., 207, 291. 1970.

Lew, V. L., Hardy, M. A. and J. C. Ellory, The uncoupled extrusion of $Na^{+}$ through the $Na^{+}$ pump, Biochim. biophys. Acta, 323, 251, 1973.

Nørby, J. G. and J. Jensen, Binding of ATP to brain microsomal ATPase. Determination of the ATP-binding capacity and the dissociation constant of the enzyme-ATP complex as a function of $K^+$-concentration, Biochim. biophys. Acta, 233, 104, 1971.

Post, R. L., Hegyvary, C. and S. Kume, Activation by adenosine triphosphate in the phosphorylation kinetics of sodium and potassium ion transport adenosine triphosphatase, J. biol. Chem., 247, 6530, 1972.

Post, R. L., Sen, A. K. and A. S. Rosenthal, A phosphorylated intermediate in adenosine triphosphate-dependent sodium and potassium transport across kidney membranes, J. biol. Chem., 240, 1437, 1965.

Robinson, J. D., Kinetic studies on a brain microsomal adenosine triphosphatase. Evidence suggesting conformational changes. Biochemistry, N. Y., 6, 3250, 1967.

Sen, A. K. and R. L. Post, Stoicheiometry and localization of adenosine triphosphate-dependent sodium and potassium transport in the erythrocyte, J. biol. Chem., 239, 345, 1964.

Simons, T. J. B., Potassium:potassium exchange catalysed by the sodium pump in human red cells, J. Physiol., Lond., 237, 123, 1974.

Skou, J. C., Further investigations on a $Mg^{++}$+$Na^+$-activated adenosintriphosphatase, possibly related to the active, linked transport of $Na^+$ and $K^+$ across the nerve membrane, Biochim. biophys. Acta, 42, 6, 1960.

Skou, J. C., Enzymatic basis for active transport of $Na^+$ and $K^+$ across cell membrane, Phys. Rev., 45, 596, 1965.

Skou, J. C., Effect of ATP on the intermediary steps of the reaction of the ($Na^+$+$K^+$)-dependent enzyme system. I. Studied by the use of N-ethylmaleimide inhibition as a tool, Biochim. biophys. Acta, 339, 234, 1974a.

Skou, J. C., Effect of ATP on the intermediary steps of the reaction of the ($Na^+$+$K^+$)-dependent enzyme system. II. Effect of a variation in the ATP/$Mg^{2+}$ ratio, Biochim. biophys. Acta, 339, 246, 1974b.

Skou, J. C., The ($Na^+$+$K^+$) activated enzyme system and its relationship to transport of sodium and potassium, Quart. Rev. Biophys., 7, 401, 1975.

Thomas, R. C., Electrogenic sodium pump in nerve and muscle cells, Phys. Rev., 52, 563, 1972.

Whittam, R. and M. E. Ager, The connection between active cation transport and metabolism in erythrocytes, Biochem. J., 97, 214, 1965.

# PROTON TRANSLOCATION BY BACTERIORHODOPSIN IN MODEL SYSTEMS

Walther Stoeckenius, San-Bao Hwang and Juan Korenbrot

Cardiovascular Research Institute
University of California, San Francisco
San Francisco, California 94143

## INTRODUCTION

Bacteriorhodopsin is a rhodopsin-like pigment found in the cell membrane of halobacteria. It occurs in discrete patches with a planar hexagonal lattice structure known as the purple membrane (Oesterhelt and Stoeckenius, 1971; Blaurock and Stoeckenius, 1971). When bacteriorhodopsin absorbs light, it undergoes a rapid cyclic photoreaction during which it translocates a proton across the membrane (Lozier et al, 1975; Lozier et al, 1976). In continuous light it acts as a light-driven proton pump generating a proton gradient and membrane potential. The cells can use the energy stored in the electrochemical gradient to synthesize ATP (Oesterhelt and Stoeckenius, 1973; Bogomolni et al, 1976; Danon and Stoeckenius, 1974). In intact cells it is difficult to measure parameters such as rapid absorption changes of the pigment, intracellular ion concentrations, and membrane potential, which are necessary to quantitate the light energy conversion in this system. Moreover, the energy metabolism of halobacteria has not been investigated in detail and contributions from other energy sources are difficult to evaluate. Reconstitution of the bacteriorhodopsin function in a well-characterized model system avoids most of the difficulties encountered in work with intact cells.

Purple membrane has been isolated in pure form (Stoeckenius and Kunau, 1968; Oesterhelt and Stoeckenius, 1974) and its functional reconstitution demonstrated in lipid vesicles (Racker, 1973; Racker and Stoeckenius, 1974; Racker and Hinkle, 1974; Kayushin and Skulachev, 1974; Knowles et al, 1975; Yoshida et al, 1975) and planar films (Drachev et al, 1974a; Drachev et al, 1974b; Yaguzhinsky et al, 1976). Light-induced potential or pH changes have been used as criteria for successful reconstitution. In some of the model systems, the light-generated chemiosmotic proton gradient has been coupled successfully to the synthesis of ATP by mitochondrial or bacterial ATPase (Racker and Stoeckenius, 1974; Yoshida et al, 1975; Yaguzhinsky et al, 1976). Little is known about the topology of the reconstituted systems. Vesicle systems preferentially translocate protons from the medium to the vesicle interior rather than in the opposite direction as intact cells (Oesterhelt and Stoeckenius, 1973; Bogomolni et al, 1976) or isolated cell envelopes do (MacDonald and Lanyi, 1975) implying that bacteriorhodopsin in the vesicles is oriented inside-out. We compare here the functional and morphological characteristics for a number of different lipid vesicle preparations and describe a new technique for preparing oriented interfacial films of purple membrane, which generate an electrical potential in the light.

## TECHNIQUES

Most techniques for the preparation of lipid vesicles involve adding dispersions of lipids to detergent-treated or sonicated purple membrane and removing the detergent by dialysis; the simplest technique is the sonication of purple membrane in the presence of added lipid (Racker, 1973). We have used this technique as well as the sonication in the presence of cholate (Racker and Stoeckenius, 1974) and shall refer to them as the sonication technique and the cholate technique. In addition, we have used Triton X-100 (Triton technique) as described in detail elsewhere (S.-B. Hwang and W. Stoeckenius, in preparation). This involves dissolving the purple membrane in 1.0% Triton X-100 at a detergent-to-protein weight ratio of 2. The dissolution at room temperature requires 48 hours. Subsequently, the dissolved membrane preparation is dialyzed in the cold for several weeks. Finally, we have pretreated purple membrane with 10% deoxycholate for 48 hours at room temperature, re-isolated the membrane on a sucrose density gradient and then added cholate-dispersed lipids. More than 90% of the cholate was removed by dialysis and the purple membrane vesicles were separated from the excess lipid by density gradient centrifugation (S.-B. Hwang and W. Stoeckenius, in preparation). This DOC/cholate technique gave the largest light responses and most consistent results.

The preparation of planar films has been described elsewhere (S.-B. Hwang, Ph.D. Thesis, in preparation; Hwang et al, 1976). Briefly, it involves suspension of purple membrane in hexane containing additional phospholipid. The suspension is spread at an air-water interface and the resulting mosaic monolayer covered with a layer of decane. Surface potential is measured with an ionization electrode.

## RESULTS AND DISCUSSION

Lipid Vesicles. Vesicle formation is indicated when the purple membrane-lipid preparations show sustained pH changes upon illumination. Action spectra demonstrate that this pH response is mediated by purple membrane. Isolated purple membrane sheets also show pH changes upon illumination, due to changes in protonation during the reaction cycle (Lozier et al, 1975); however, these changes are too small to be detected under the conditions used in the vesicle experiments. The pH changes in the vesicle suspensions are usually much larger and develop more slowly. Only when the reaction cycle kinetics of isolated purple membrane are modified, for instance by the presence of organic solvents so that deprotonated reaction cycle intermediate accumulates, does it become possible to detect pH changes in the suspension with a glass electrode and at the light intensities used with the vesicle experiments. Moreover, in most cases the vesicle preparations show a light-induced alkalinization of the medium, whereas purple membrane sheets always show an acidification.

While a large sustained pH response clearly indicates vesicle formation, its absence does not necessarily mean that vesicles have not formed. Neglecting the so far hypothetical case of vesicles too permeable to sustain a proton gradient, there remains the possibility that vesicles have formed but that the orientation of the bacteriorhodopsin in the vesicle wall is random. Such a topology would result in a translocation of protons across the vesicle wall occurring in both directions at the same rate and no change in pH would be observed. We have verified by electron microscopy that this random orientation does indeed occur in some preparations.

Freeze-fracturing reveals intramembrane proteins as particles on the fracture face (Branton, 1969). Bacteriorhodopsin is a typical intramembrane protein and in freeze-fracture preparations appears as a regular array of particles on the cytoplasmic membrane leaflet (Blaurock and Stoeckenius, 1971). The outer membrane leaflet shows a smooth fracture face. In intact cells, therefore, all convex fracture faces show particles in the hexagonal array of the bacteriorhodopsin lattice, whereas the concave fracture faces appear smooth. This feature allows us to detect the orientation of bacteriorhodopsin in the model systems, assuming that the fracture behavior remains the same as found in the native membrane which, fortunately, appears to be the case. Figure 1 shows a freeze-fracture electron micrograph of a DOC/cholate preparation reconstituted with soybean lecithin. Most of the particles are associated with the concave fracture faces and relatively few are present on the convex faces. Apparently, the preferential orientation of bacteriorhodopsin in the DOC/cholate vesicles is the opposite of the orientation found in the native membrane, and this agrees with the direction of the light-generated proton gradient, which is

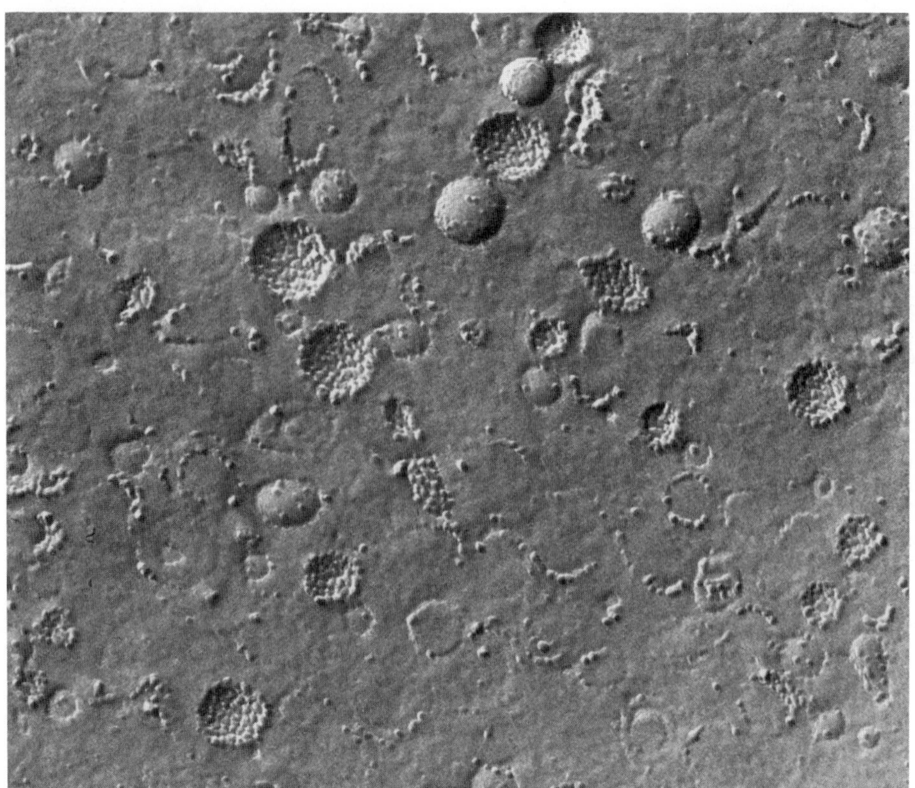

Figure 1 Freeze-fracture electron micrograph of DOC/cholate vesicles, prepared with purified soybean lecithin. The vesicles contain 35% lipid by weight as compared to 25% in the native purple membrane. All concave fracture faces are densely packed with particles, whereas few are seen on the convex faces. Magnification 127,000 X.

alkaline in the medium. The corresponding cholate vesicles show the same general appearance, but the preferential orientation is much less selective; the size variation of the vesicles is larger and many multilamellar structures are present. The same more random orientation is found in the vesicles prepared simply by sonication of purple membrane in the presence of added lecithin, and their size variation is still larger.

The appearance of DOC/cholate vesicles reconstituted with halobacteria phospholipids or vesicles prepared with the Triton technique have different morphologies. The Triton preparations

Figure 2 Freeze-etch electron micrograph of Triton vesicles. Most of them are very large and show particulate and smooth domains on the same fracture face. The particulate domains have a lattice structure very similar to the appearance of purple membrane in intact cells. Magnification 58,320 X.

contain large sheets and very large vesicles several microns in diameter. Both sheets and vesicles show particulate and smooth areas on the same fracture face (Figure 2). Similarly, no preferential orientation is visible on the fracture faces of DOC/cholate vesicles reconstituted with halobacterium lipids. These vesicles are much smaller than the Triton vesicles but show many flat sides and straight edges (Figure 3). The hexagonal lattice of bacteriorhodopsin is well preserved in the halobacteria lipid vesicles as well as in the Triton vesicles. Vesicle formation in both cases involves the native lipids and this apparently promotes retention or reformation of the planar bacteriorhodopsin lattice. The presence of the protein lattice in these vesicles is also confirmed by X-ray and electron diffraction data (Unwin and Henderson, 1975;

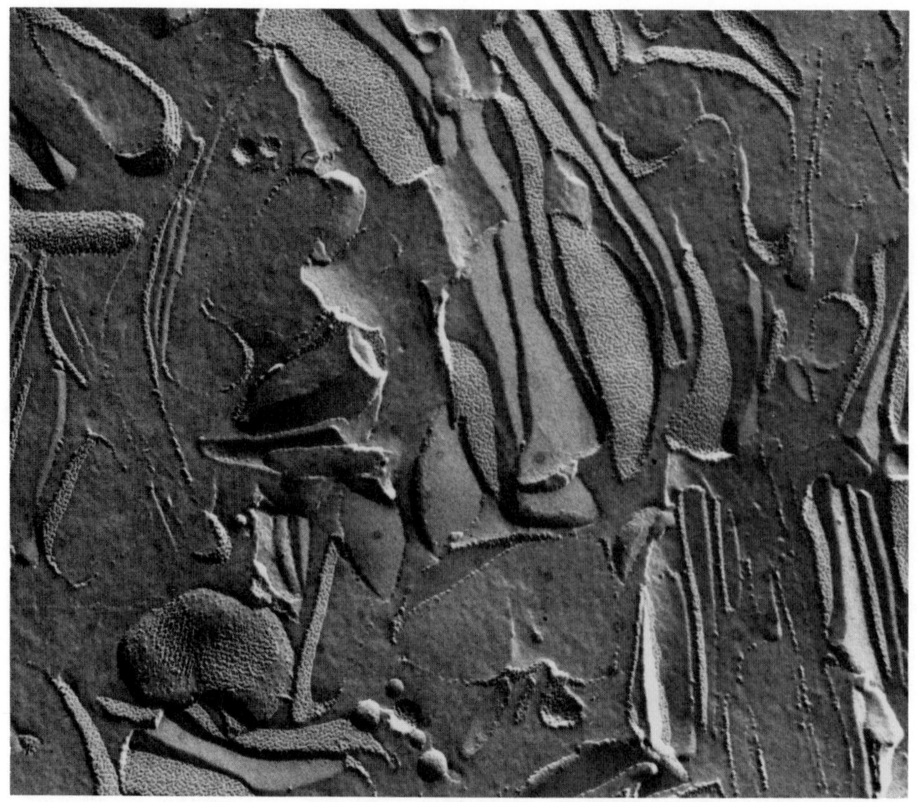

Figure 3 DOC/cholate vesicles prepared with polar lipids from *H. halobium*. Particles are seen on convex and concave surfaces. Most of the vesicles, however, have flat sides. No preferential orientation of the bacteriorhodopsin with respect to the vesicle interior is apparent. Magnification 54,450 X.

G. I. King, S.-B. Hwang and W. Stoeckenius, unpublished) and by CD spectra (Y.-W. Tseng, S.-B. Hwang and W. Stoeckenius, unpublished). A hexagonal lattice cannot occupy a curved surface without being distorted. This accounts for the flat sides and straight edges of the smaller vesicles. In the case of the large Triton vesicles, the curvature is probably negligible and may also be preferentially localized along the edges of the inverted domains. The observed pH response is compatible with the non-preferential orientation of the lattice domains; it is very small or absent altogether in both types of vesicles.

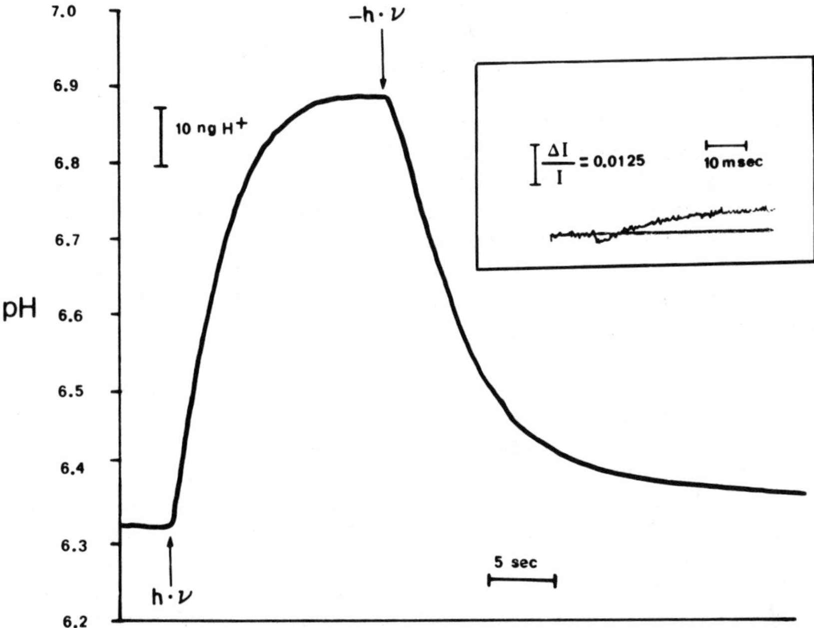


Figure 4 pH response of a DOC/cholate vesicle preparation reconstituted with purified soybean lecithin. The protein concentration is 43 μg $ml^{-1}$, the light intensity $10^6$ ergs $sec^{-1}$ $cm^{-2}$ from a 250-watt quartz iodine lamp measured at the sample through an orange cut-off filter (Schott OG5) and heat filter. The inset shows the transmission change of a pH indicator added to a similar preparation in response to a 1 msec light flash (from Lozier et al, 1976). The initial decrease in transmission indicates a rapid acidification of the medium followed by a larger and sustained alkalinization.

A typical pH response for a DOC/cholate preparation using purified soybean lecithin is shown in Figure 4. The response is far larger than the maximal response obtained with any of the other preparations. The difference in pH between dark and light conditions in the steady state is 34 protons per molecule of bacteriorhodopsin. The rate of proton translocation in the steepest part of the pH rise corresponds to 170 ng ion $H^+$ $sec^{-1}$ $mg^{-1}$ protein. The kinetics of the response appear complex. A small inflection is visible at the beginning of the pH rise. It is not an artifact of the recording system but is probably due to the fast release of protons from the small amount of bacteriorhodopsin which, as seen in the electron micrographs, is oriented right-side out. During the photoreaction cycle, protons are first released on the outer surface of the purple membrane before they are taken up on the inner surface (Lozier et al, 1976). With a faster recording system this initial release of protons can be observed as a small transient acidification of the medium, preceding the sustained alkalinization (Figure 4-inset). Similarly, the beginning acidification of the medium - when the light is turned off - is retarded because the inside-out oriented bacteriorhodopsin molecules completing their photoreaction cycle in the dark take up protons from the medium. This initial retardation is obvious from a closer inspection of the pH decay part of the curve in Figure 4. At least two more kinetic components can be extracted by "curve peeling" from both the rise and decay of the pH response.

While rational explanations for the complex kinetics can be advanced, it is difficult if not impossible to prove any of them. Even though the size distribution of the DOC/cholate vesicles is more uniform than that of any of the other preparations, it is still relatively broad as Figure 5 shows. Obviously, other things being equal, the few larger vesicles will dominate the response because of their larger contribution to the total internal volume. It is not known how the orientation and kinetics of the bacteriorhodopsin may vary with the size of the vesicle, and size-dependent differences in the passive permeability cannot be excluded. Moreover, the stability of bacteriorhodopsin in the vesicles without a planar lattice is reduced - especially at alkaline pH - and the photoreaction cycle kinetics differ somewhat from those observed in purple membrane sheets. All these factors complicate the quantitative analysis of the pH response to a degree that solid data cannot be obtained with a reasonable effort, and it appears better to improve the preparation techniques first. Unfortunately, we have so far been unable to obtain uniform large vesicles in which the hexagonal lattice is intact. Also, while it can be shown that a membrane potential is generated by light in the vesicles (Racker and Hinkle, 1974; Kayushin and Skulachev, 1974), the techniques used for measuring the potential are either slow or controversial and difficult to calibrate; a model system based on a planar film separating two macroscopic compartments seems to be much better suited for that purpose. Nevertheless, the vesicle system has distinct advantages - especially for combined spectroscopic and permeability studies - and should be further developed.

Planar Films. Mueller-Rudin type black lipid films have been used either with purple membrane added to the organic solvent (Drachev et al, 1974b) or with purple membrane-lipid vesicles added to the aqueous phase on one side of the film (Drachev et al, 1974a). Alternatively, purple membrane has been adsorbed from the aqueous phase to an octane-water interface (Boguslavsky et al, 1975; Yaguzhinsky et al, 1976). With these systems, photopotentials have been recorded which are attributed to light-driven proton translocation by bacteriorhodopsin. In the water-octane system it is

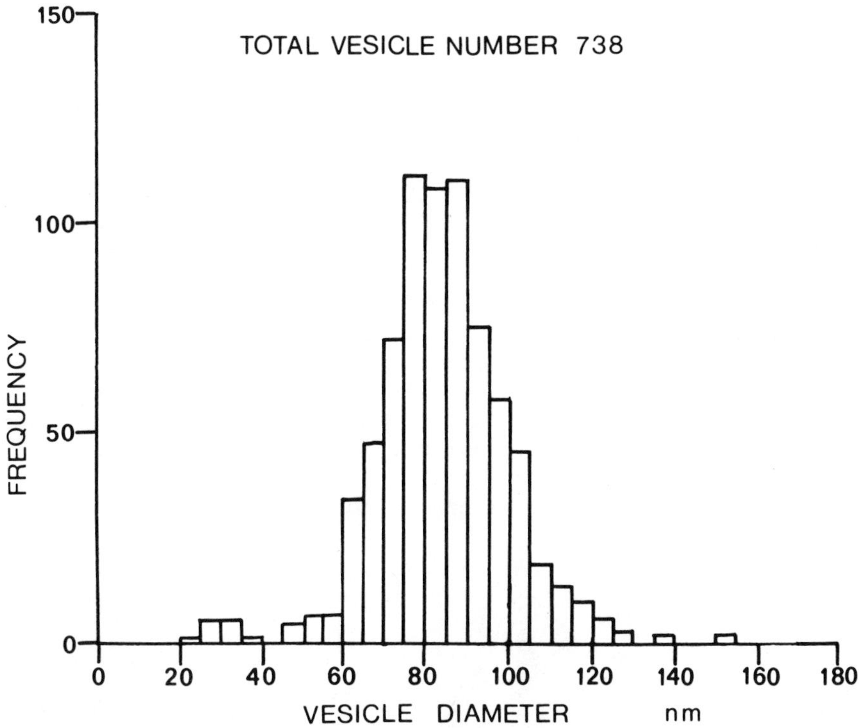


Figure 5 Size distribution of DOC/cholate vesicles prepared with soybean lecithin. The vesicle size was measured in electron micrographs of shadowed preparations fixed with $OsO_4$ and dried with the critical point technique.

necessary to add a proton acceptor to the organic phase, presumably to reduce the Born charging energy required to move a proton from a high to a low dielectric constant medium (Parsegian, 1969). In all cases, the amount and orientation of purple membrane present in the film or the interface are unknown. We have developed a system that overcomes this difficulty (Hwang et al, 1976).

We assumed that small differences in surface charge on both sides of the membrane and/or the dipole moment of purple membrane would result in a preferential orientation of the sheets in the high electric field strength at an air-water interface. We therefore dispersed the purple membrane in an organic solvent - hexane - and spread this suspension on a clean water surface. The purple membrane must be sonicated before spreading to disaggregate and reduce the size of the fragments and additional phospholipid must be added to the suspension to obtain satisfactory films. This technique results after the hexane has evaporated in a monolayer of lipid containing purple membrane fragments dispersed randomly in the lipid film. This "mosaic" film can be transferred to a support and observed in the electron microscope (Figure 6). The purple membrane pieces are randomly distributed and occupy ∿ 35% of the surface. After a second phospholipid monolayer has been layered on top of the mosaic film, the preparation can be freeze-fractured. This reveals that, wherever the purple membrane has been fractured, more than 80% of the pieces are oriented with their cytoplasmic side towards the water (Figure 7). We have also prepared Langmuir-Blodgett multilayers and studied their optical properties. The absorption spectrum is unchanged in the multilayers; the light-dark adaptation reaction (Lozier et al, 1975) and the fast cyclic photoreaction are still demonstrable (Figure 8). Values for the amount of purple membrane present in the film calculated from the absorbance of the multilayers at 570 nm or from the proportion of the total area covered by purple membrane pieces in the electron micrographs are

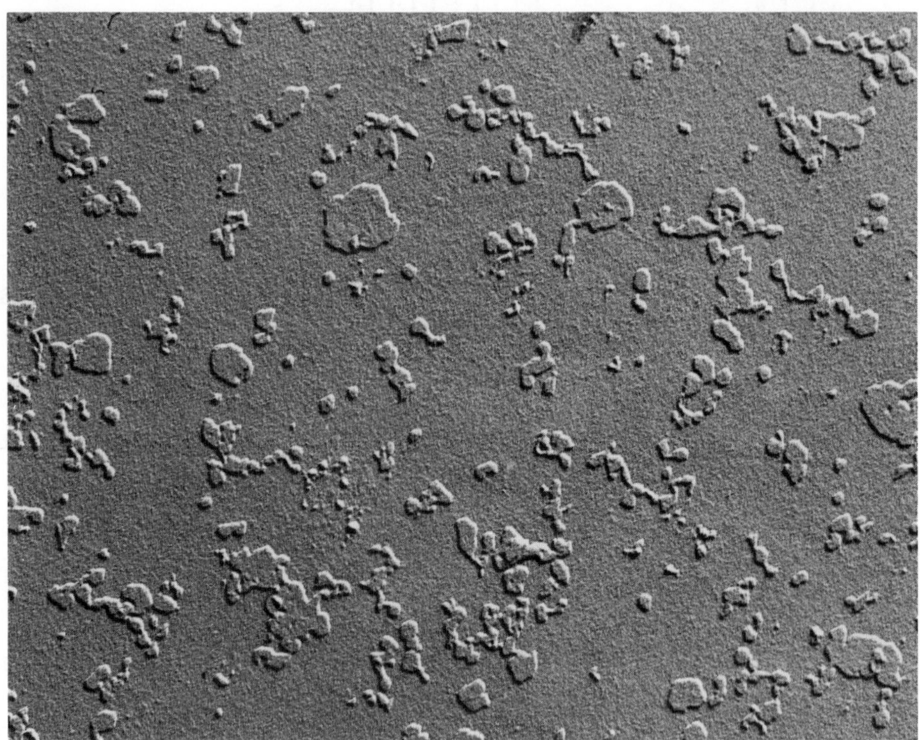

Figure 6 Surface film of purple membrane and phospholipid has been transferred from the water surface to a glass support and shadowed. The randomly distributed purple membrane pieces are reduced in size by a brief sonication before spreading. Magnification 74,250 X.

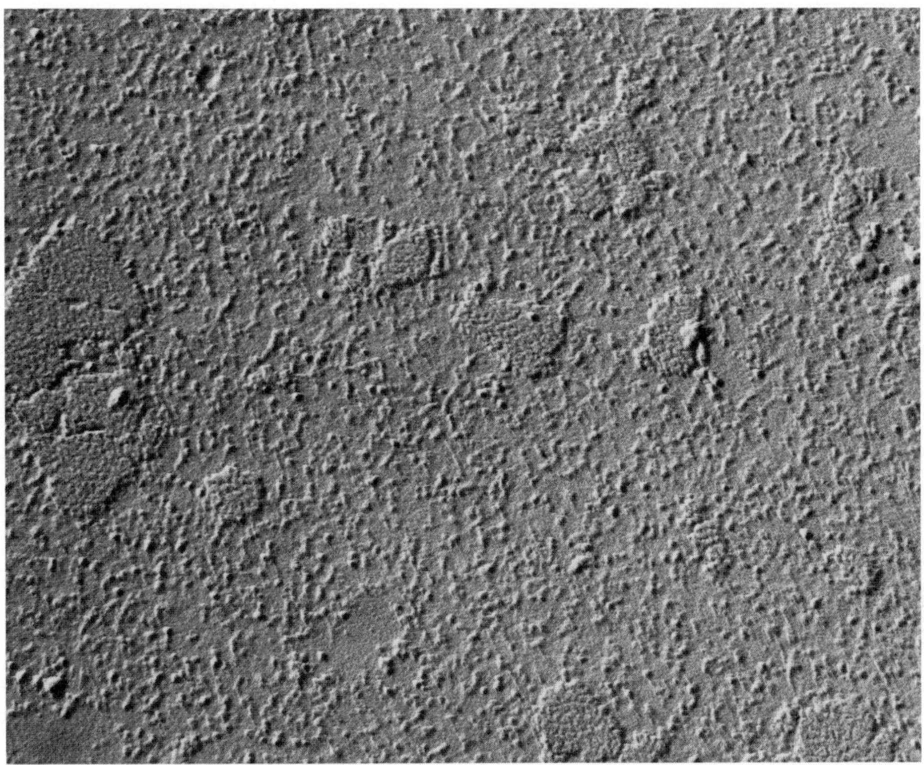

Figure 7 Freeze-etch electron micrograph of a mosaic film on a glass support. The purple membrane pieces show the particle and lattice structure characteristic for fracture face of the cytoplasmic leaflet. The film was transferred to the support so that the surface facing the water now faces the supporting glass slide. Magnification 74,250 X.

in good agreement. The amount present is a linear function of the membrane to phospholipid ratio applied, up to weight ratios of 10 (Figure 9). At higher purple membrane concentrations the membrane pieces in the film begin to overlap. The best and most consistent results were obtained with weight ratios of 7, where 35% of the area is occupied by purple membrane. Very small changes in surface potential are seen when such a film is illuminated at the air-water interface or after it has been overlaid with a layer of decane. However, after dinitrophenol (DNP) or p-trifluoromethoxyphenylhydrazone (FCCP) has been added, illumination elicits large increases in surface potential. The action spectrum for the effect corresponds to the absorption spectrum of the purple membrane (Figure 10). The maximal photovoltage is maintained constant as

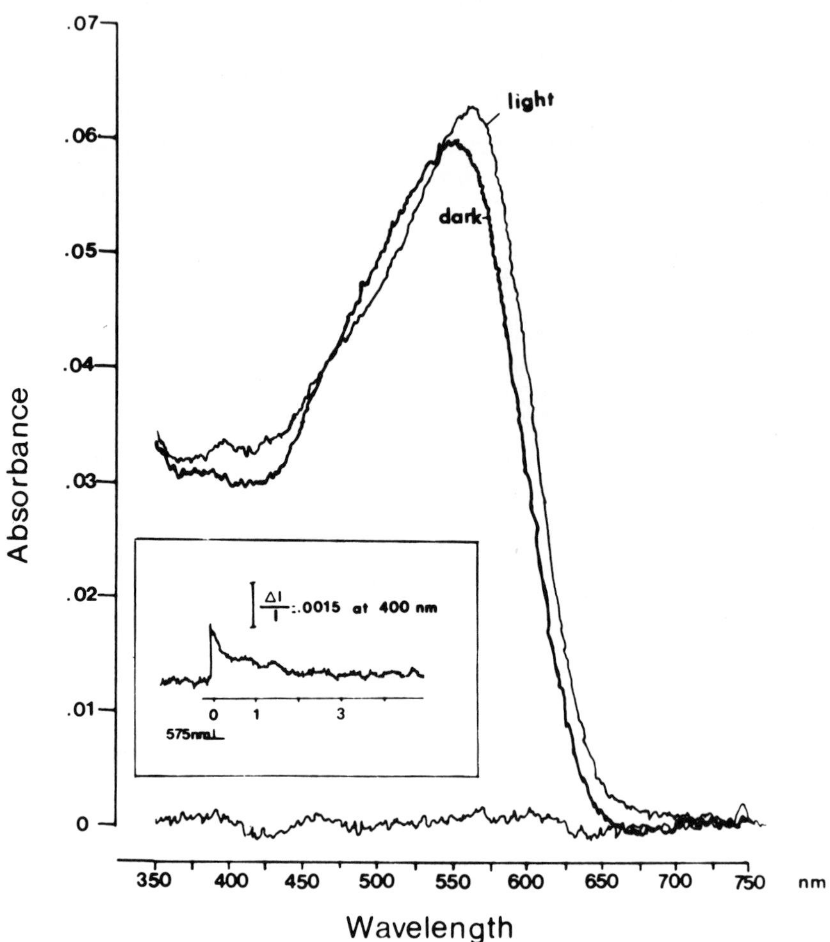


Figure 8 Absorption spectrum of a Langmuir-Blodgett film consisting of 80 mosaic film monolayers. The purple membrane absorption spectrum is unchanged and shows the typical small shift and absorbance changes of the light-dark adaptation reaction. Inset: Transient absorbance increase of multilayer at 400 nm after a 1 µsec flash of 575 nm light. The rise and decay of the 412 nm intermediate are recorded. The decay kinetics are slower than in purple membrane suspensions presumably caused by drying of the membrane.

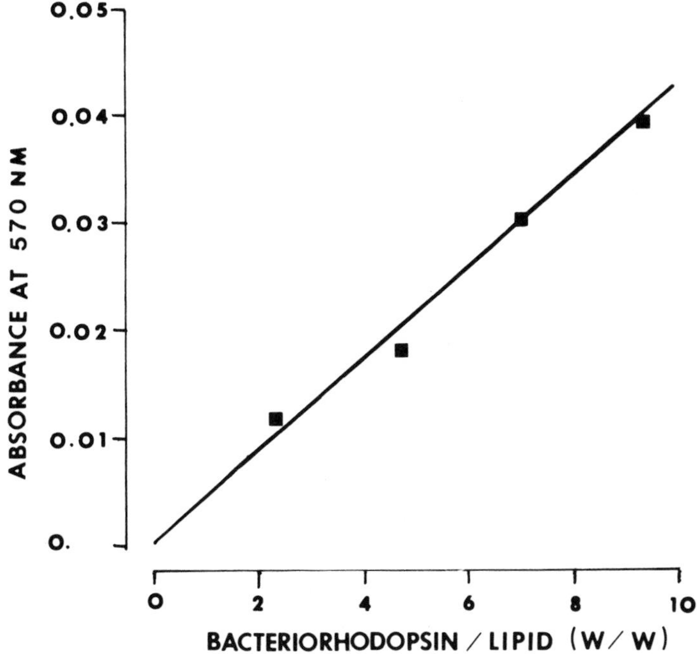


Figure 9 Absorbance of 40 monolayers vs. the protein to lipid weight ratio applied to the surface. The total amount of lipid was kept constant.

long as the light is on and is proportional to the light intensity; so is the rate of the voltage rise, whereas the decay half-time is independent of light intensity (Figure 11). Varying the proton acceptor concentration and keeping the light intensity constant yields first an increase in the photovoltage, followed by a decrease at higher concentrations. This decrease is apparently due to facilitation of charge transfer back through the interface, because the half time for the potential decay after the light has been turned off stays constant at low acceptor concentrations and begins to increase after the photovoltage has passed through its maximum.

We have developed the following tentative and qualitative explanation for the observed phenomena. The proton acceptor AH equilibrates between the water and decane and is dissociated to a substantial extent in the water. Due to the lipid solubility of the anion ($A^-$), its concentration in the decane is higher than that of its counterion ($H^+$). These assumptions are consistent with the

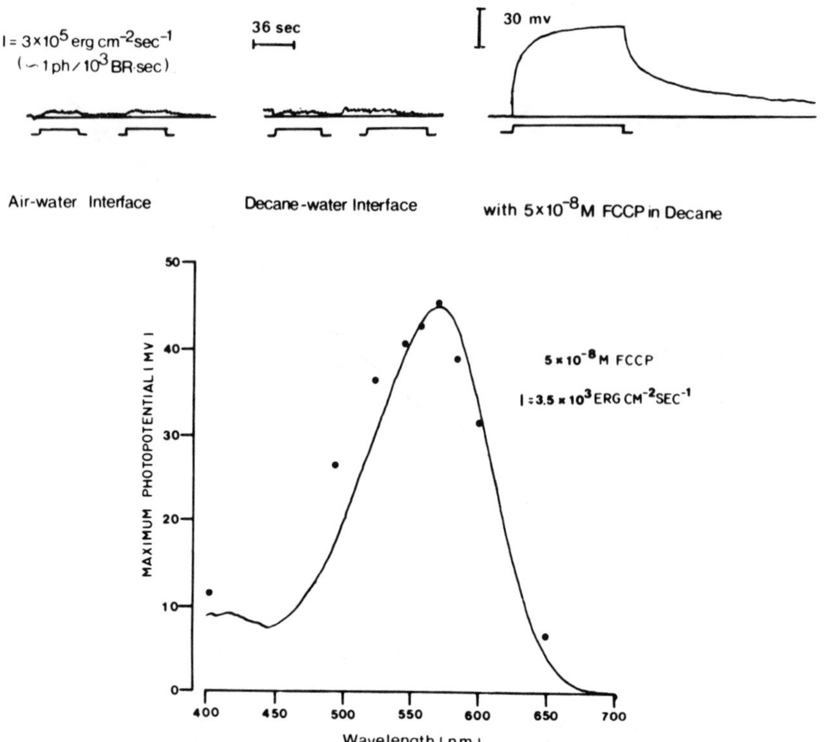


Figure 10 Photopotentials at the air-water and decane-water interface. Illumination periods are indicated below the voltage traces. In the lower figure the points indicate the maximal photovoltage obtained at a given wavelength, the line the absorption spectrum of the purple membrane scaled to give the best fit to the points. The light intensity of 3 X $10^3$ erg $cm^{-2}$ $sec^{-1}$ corresponds to approximately 1 photon per $10^3$ bacteriorhodopsin molecules per second.

observation that addition of the proton acceptor decreases the surface potential in the dark and that it does not matter whether the proton acceptor is added to the water or decane phase, but time must be allowed for equilibration. When light causes a transfer of protons from the water to the decane, they are accepted by $A^-$ to form AH and probably other larger complexes. Thus a photovoltage is generated. The increase in the concentration of AH and/or the larger complexes will lead to back diffusion of protons to re-establish equilibrium. With increasing concentration of added AH, back-diffusion begins to dominate and the photopotential decreases.

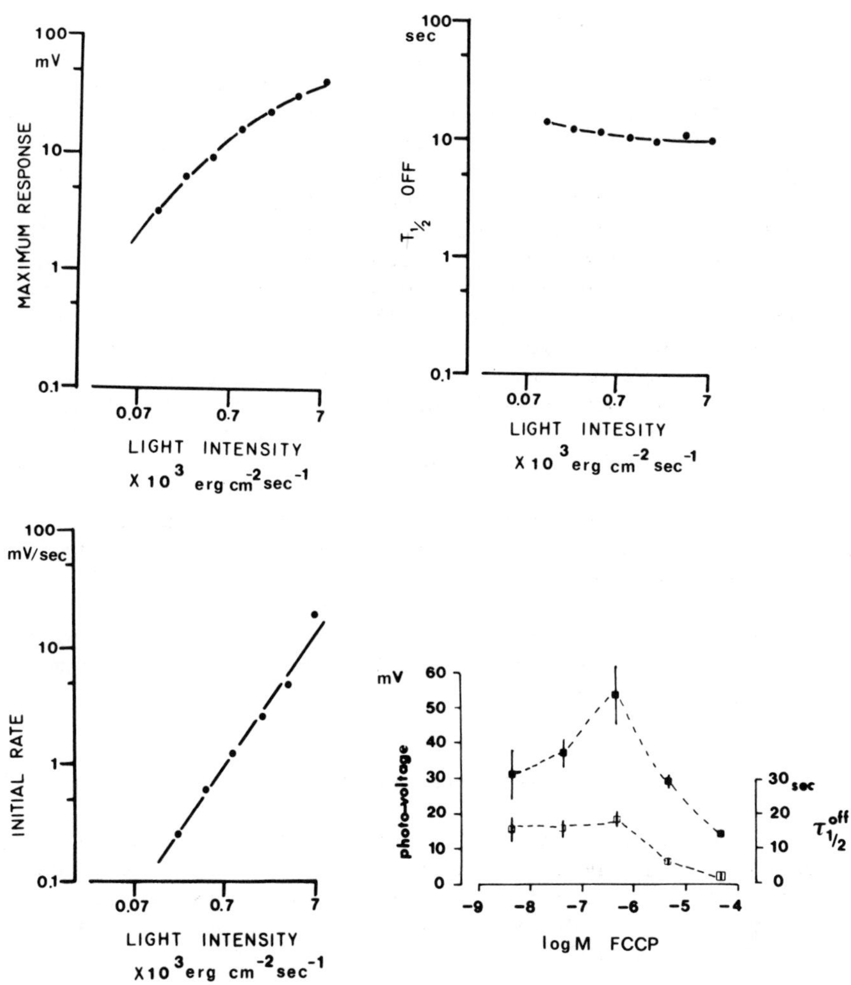


Figure 11 The initial rate of the photovoltage rise and the maximal voltage obtained are linearly proportional to the light intensity at the low intensities used here. It is not clear whether the small bending of the curve for light intensity vs. maximum photovoltage is significant. The decay rate of the photovoltage is essentially independent of light intensity, but increases at higher FCCP concentrations, while the photovoltage goes through a maximum near $10^{-6}$ M FCCP. The light intensity in these experiments was $7 \times 10^3$ erg $cm^{-2}$ $sec^{-1}$ in a broad band around 570 nm, which corresponds to ∿ 1 photon/bacteriorhodopsin molecule and second.

The rate of proton translocation into the decane is limited by the light intensity in our experiments.

As is the case with the vesicle system, the kinetics of the light response in the planar films are complex and we are limited in our measurements by the response time of the measuring system. Again, it appears desirable to improve the experimental conditions before attempting a detailed analysis. However, a comparison of our planar films with the system used by Boguslavsky et al (1975) reveals an interesting difference which should be pursued. They find that increasing the acceptor concentration in the octane phase above the optimal value does not reduce the photovoltage. The only difference between the two systems that is likely to be significant is the absence of a monolayer of lipid between the purple membrane pieces in their system, unless such a layer is formed by lipids extracted from the purple membrane present. It would be interesting to see whether this monolayer is necessary for the increase in back-flow at high acceptor concentrations.

Both model systems strongly support the postulated function of the purple membrane and the evidence is much less ambiguous than that obtained from intact cells. However, they must be further improved for valid quantitative data to be obtained. Our results also clearly demonstrate that no conclusions should be drawn from the observed pH response unless the topology of the model system is known.

## ACKNOWLEDGMENT

We thank Knute A. Fisher for advice and help with the freeze-fracturing of the films, Richard H. Lozier for the flash spectroscopy and Roberto A. Bogomolni for discussion. The work was supported by Program Project grant HL-06285 and grant EY-01586 from the U.S.P.H.S. and NASA grant NSG-7151.

## REFERENCES


Blaurock, A. E. and W. Stoeckenius, Structure of the purple membrane, Nature New Biol. 233, 152, 1971.

Bogomolni, R. A., R. A. Baker, R. H. Lozier and W. Stoeckenius, Light-driven proton translocations in *Halobacterium halobium*, Biochim. Biophys. Acta, in press, 1976.

Boguslavsky, L. I., A. A. Kondrashin, I. A. Kozlov, S. T. Metelsky, V. P. Skulachev and A. G. Volkov, Charge transfer between water and octane phases by soluble mitochondrial ATPase ($F_1$), bacteriorhodopsin and respiratory chain enzymes, FEBS Letters 50, 223, 1975.

Branton, D., Membrane structure, Ann. Rev. Plant Physiol. 20, 209, 1969.

Danon, A. and W. Stoeckenius, Photophosphorylation in *Halobacterium halobium*, Proc. Nat. Acad. Sci. USA 71, 1234, 1974.

Drachev, L. A., A. A. Jasaitis, A. D. Kaulen, A. A. Kondrashin, E. A. Liberman, I. B. Nemecek, S. A. Ostroumov, A. Yu. Semenov and V. P. Skulachev, Direct measurement of electric current generation by cytochrome oxidase, $H^+$-ATPase and bacteriorhodopsin, Nature 249, 321, 1974a.

Drachev, L. A., A. D. Kaulen, S. A. Ostroumov and V. P. Skulachev, Electrogenesis by bacteriorhodopsin incorporated in a planar phospholipid membrane, FEBS Letters 39, 43, 1974b.

Hwang, S.-B., J. I. Korenbrot and W. Stoeckenius, Light-dependent proton transport by bacteriorhodopsin incorporated in an interface film, J. Supramolecular Structure, in press, 1976.

Kayushin, L. P. and V. P. Skulachev, Bacteriorhodopsin as an electrogenic proton pump: Reconstitution of bacteriorhodopsin proteoliposomes generating $\Delta\psi$ and $\Delta$pH, FEBS Letters 39, 39, 1974.

Knowles, A. F., A. Kandrach, E. Racker and H. G. Khorana, Acetyl phosphatidylethanolamine in the reconstitution of ion pumps, J. Biol. Chem. 250, 1809, 1975.

Lozier, R. H., R. A. Bogomolni and W. Stoeckenius, Bacteriorhodopsin: A light-driven proton pump in *Halobacterium halobium*, Biophys. J. 15, 955, 1975.

Lozier, R. H., W. Niederberger, R. A. Bogomolni, S.-B. Hwang and W. Stoeckenius, Kinetics and stoichiometry of light-induced proton release and uptake from purple membrane fragments, *Halobacterium halobium* cell envelopes, and phospholipid vesicles containing oriented purple membrane, Biochim. Biophys. Acta, in press, 1976.

MacDonald, R. E. and J. K. Lanyi, Light-induced leucine transport in *Halobacterium halobium* envelope vesicles: A chemiosmotic system, Biochemistry 14, 2882, 1975.

Oesterhelt, D. and W. Stoeckenius, Rhodopsin-like protein from the purple membrane of *Halobacterium halobium*, Nature New Biol. 233, 149, 1971.

Oesterhelt, D. and W. Stoeckenius, Functions of a new photoreceptor membrane, Proc. Nat. Acad. Sci. USA 70, 2853, 1973.

Oesterhelt, D. and W. Stoeckenius, Isolation of the cell membrane of *Halobacterium halobium* and its fractionation into red and purple membrane, in Methods in Enzymology Volume XXXI, Biomembranes Part A, ed. S. Fleischer and L. Packer, pp. 667-678, Academic Press, New York-San Francisco-London, 1974.

Parsegian, A., Energy of an ion crossing a low dielectric membrane: Solutions to four relevant electrostatic problems, Nature 221, 844, 1969.

Racker, E., A new procedure for the reconstitution of biologically active phospholipid vesicles, Biochem. Biophys. Res. Commun. 55, 224, 1973.

Racker, E. and P. C. Hinkle, Effect of temperature on the function of a proton pump, J. Membrane Biol. 17, 181, 1974.

Racker, E. and W. Stoeckenius, Reconstitution of purple membrane vesicles catalyzing light-driven proton uptake and adenosine triphosphate formation, J. Biol. Chem. 249, 662, 1974.

Stoeckenius, W. and W. H. Kunau, Further characterization of particulate fractions from lysed cell envelopes of *Halobacterium halobium* and isolation of gas vacuole membranes, J. Cell Biol. 38, 337, 1968.

Unwin, P. N. T. and R. Henderson, Molecular structure determination by electron microscopy of unstained crystalline specimens, J. Mol. Biol. 94, 425, 1975.

Yaguzhinsky, L. S., L. I. Boguslavsky, A. G. Volkov and A. B. Rakhmaninova, Synthesis of ATP coupled with action of membrane protonic pumps at the octane-water interface, Nature 259, 494, 1976.

Yoshida, M., N. Sone, H. Hirata, Y. Kagawa, Y. Takeuchi and K. Ohno, ATP synthesis catalyzed by purified DCCD-sensitive ATPase incorporated into reconstituted purple membrane vesicles, Biochem. Biophys. Res. Commun. 67, 1295, 1975.

# STATE OF ASSOCIATION OF MEMBRANE PROTEINS

Charles Tanford

Department of Biochemistry, Duke University Medical Ctr.

Durham, North Carolina 27710, USA

## ABSTRACT

The native environment of a membrane protein can be closely simulated by the small micelles formed by appropriate detergents. This permits solubilization of membrane proteins with retention of their native structure and biological activity. Molecular weights and polypeptide chain compositions can then be determined by well-established methods of solution physical chemistry. We have found that some membrane proteins (e.g., cytochrome $b_5$) are monomeric, whereas others are oligomeric. The $Ca^{++}$-stimulated ATPase from sarcoplasmic reticulum appears to be a trimer (possibly tetramer) of identical polypeptide chains. Each chain has two very similar halves, which suggests that the transmembrane portion of this protein consists of six (possibly eight) symmetrically arranged elements, which perhaps create an aqueous channel for the passage of $Ca^{++}$ ions through the membrane.

## INTRODUCTION

Membrane proteins, like water-soluble proteins, may frequently be oligomeric, i.e.composed of two or more like or unlike polypeptide chains. Where such an associated state exists, it is likely to be intimately related to the functional properties of the protein, and the experimental determination of the state of association is thus a vital step in the characterization process.

For water-soluble proteins the problem of determining subunit composition has become a routine exercise: polypeptide chain compositions of 300 different proteins are listed in a recent review that covers the literature up to 1972 (Klotz et al., 1975).

The procedure employed has been essentially the same for all of them, and consists of three steps: (1) purification of the protein in soluble form; (2) denaturation and dissociation to the constituent polypeptides, and determination of how many different kinds there are, their relative abundance, and their molecular weights; (3) measurement of the molecular weight of the intact native protein, usually by use of the analytical ultracentrifuge. This three-step procedure is direct and virtually error-proof. Moreover, there is no simple alternative. If unambiguous results are to be obtained, the same procedure has to be used for membrane proteins. The major problem in doing so has been in the first step, in obtaining the protein in solution, without denaturation and in a state of high purity. This problem is rapidly being overcome by the use of detergents for solubilization. The subsequent problem of measuring molecular weights in the presence of detergent is not severe, as I shall show later in this paper, and accuracy comparable to measurements in the absence of detergent can be achieved.

## USE OF DETERGENTS

Membrane proteins exist in nature at the interface between a phospholipid bilayer and the adjacent aqueous medium. A substantial number of them penetrate into the liquid hydrocarbon core of the bilayer, and some traverse the entire width of the bilayer, and in either situation part of the protein molecular surface must be predominantly hydrophobic and in particular must be devoid of ionic amino acid side chains. If one hopes to solubilize such a protein with retention of its native structure, one must simulate the bilayer environment in solution, and there are sound theoretical reasons for believing that detergents fulfil this requirement. Detergent molecules in which the hydrophobic moiety is a long aliphatic hydrocarbon chain form small oblate ellipsoidal (disk-like) micelles with a fluid hydrocarbon core (Tanford, 1974). The thickness of the core, like the thickness of the core of a phospholipid bilayer, is determined by the length of the hydrocarbon chains: if a detergent with appropriate alkyl chains is used the middle of the micelle will almost exactly reproduce the dimensions of the hydrophilic-hydrophobic interface of the bilayer, as is shown in Figure 1. Exact simlation may actually not always be necessary, because the micelle dimensions can be distorted by contact with protein to fit the requirements of the protein's hydrophobic surface. Triton X-100, which has a mixed aromatic-aliphatic chain quite unlike the acyl chains of phospholipids, has been used successfully for the solubilization of several membrane proteins without loss of biological activity (Helenius and Simons, 1975).

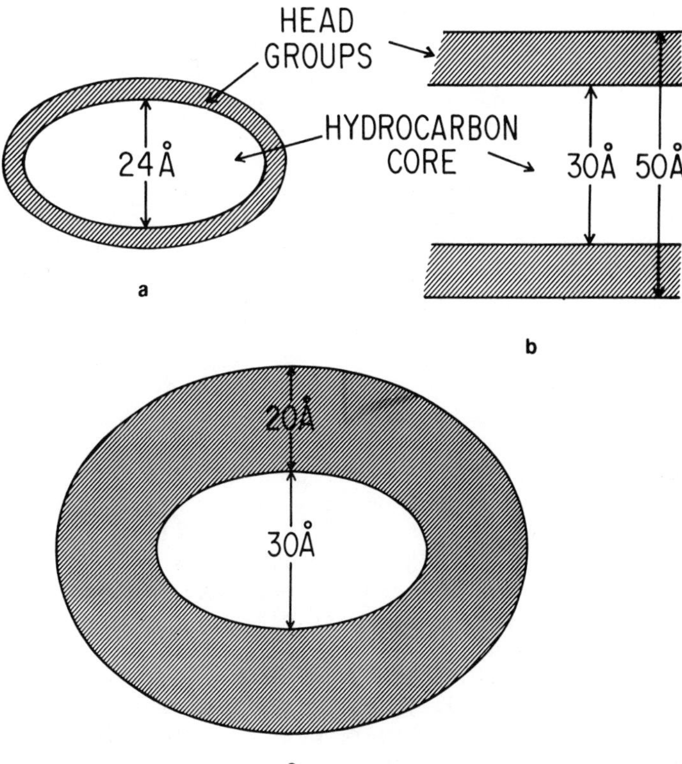


Figure 1. Approximate dimensions of typical disk-like detergent micelles, in comparison with typical membrane phospholipid bilayer dimensions: (A) sodium dodecyl sulfate; (B) phospholipid bilayer; (C) Lubrol WX. The unshaded portion of each diagram represents the fluid hydrocarbon core of the structure, its thickness being determined by the length of the hydrocarbon chains of the lipid or detergent molecules. The shaded portion represents the solvent-permeated head group region.

As for the polar head group of the detergent, it should be nonionic because nonionic detergents do not ordinarily denature proteins. (Ionic detergents can be expected to preserve the hydrophobic portion of a membrane protein in its native conformation, but would tend to denature the parts of the protein that lie on the aqueous side of the interface in the intact membrane). If phospholipid head groups per se are required to maintain the native conformation of the protein, lysophospholipids can be used: they are "detergents" as the term is here used and form small micelles similar to those formed by other detergents. Alternatively, diacyl phospholipids essential for maintenance of the native structure of a membrane protein may tend to remain associated with the protein and can be retained when the protein is dissolved in a detergent micelle (Le Maire et al., 1976).

It should be noted that detergents suitable for maintaining a protein in its native state may not be ideal for the initial disruption and solubilization of the membrane from which the protein is derived (Helenius and Simons, 1975). This does not present an experimental problem since there is no difficulty in replacing one detergent by another.

## PHOSPHOLIPID VESICLES

The ideal solubilizing medium from the point of view of simulation of the native environment would be small soluble phospholipid vesicles. Even the smallest vesicles of this type, as originally described by Huang (1969), have a molecular weight of about $2.1 \times 10^6$, and this is much too large for our purposes. Ultracentrifugal studies require particles that do not contain more than one independent (unassociated) protein molecule. The lipid/protein weight ratio in the vesicle would thus usually be in the range of 10:1 to 100:1, and this would severely stretch the ability of the experiment to reveal the properties of the protein constituent alone. Moreover, a very high overall lipid concentration would be needed to provide a sufficient excess of vesicles over protein molecules to assure the actual absence of multiple copies of protein molecules in a single vesicle.

On the other hand, phospholipid vesicles may be necessary in some instances to test for retention of native functional properties. An example is provided by ion-activated ATPase that also serve as ion pumps: ATPase activity can be assayed in detergent solution, but the ability to pump ions across a bilayer can only be tested when the protein is incorporated into phospholipid vesicles. In this situation one can gain assurance that the polypeptide chains comprising the enzymatically active protein in detergent solution are also sufficient for translocation, simply by demonstrating that active transport is retained when the protein is transferred from the detergent solution to a vesicular state. However, there is no guarantee that a change in state of association may not have occurred in the transfer, though the likelihood of such change may be small under appropriate circumstances, e.g., if the detergent-solubilized enzyme contains bound phospholipid. Evidence as to the identity of the structures in the detergent-solubilized and vesicular states can be obtained by a variety of measurements that can be made in both states: they include spectroscopic measurements, studies of the effects of chemical cross-linking reagents, and examination of fragments obtained by mild proteolysis.

## CONSTITUENT POLYPEPTIDES AND THEIR MOLECULAR WEIGHTS

Among water-soluble oligomeric proteins, 85% are oligomers of identical polypeptide chains, whereas 15% contain polypeptides of more than one kind (Klotz et al., 1975). Similar information for membrane proteins is readily obtained by carrying out fractionation (density gradient centrifugation, chromatography of various kinds) in detergent solution. A molar ratio of two detergent micelles per polypeptide chain should be sufficient to avoid having two or more polypeptides in the same micelle unless they actually belong together.

Measurement of the molecular weights of constitutent polypeptides may be a more laborious task for membrane proteins than for water-soluble proteins. Polyacrylamide gel electrophoresis in sodium dodecyl sulfate (SDS), though a very sensitive analytical tool for separating polypeptides of different molecular size, does not always provide reliable numerical estimates of molecular weight. The probability of inaccurate estimates is especially high for membrane polypeptides possessing hydrophobic regions, which can combine with whole SDS micelles, so that the overall interaction with SDS can be quite different from that of common water-soluble proteins used as molecular weight standards in the determination. A more serious problem is that guanidine hydrochloride, which disrupts both inter-chain and intra-chain interactions in virtually all water-soluble proteins, producing randomly coiled and unassociated polypeptides, has proved to be ineffective for most membrane proteins studied in our laboratory: polypeptides containing long sequences of predominantly hydrophobic amino acid residues are usually not dissociated to monomeric form. To obtain reliable polypeptide chain molecular weights for membrane polypeptides it is thus necessary to use sedimentation equilibrium measurements after the addition of high concentrations of SDS (and reduction of disulfide bonds) to promote complete dissociation (Grefrath and Reynolds, 1974). The measurement is not difficult in principle, but the auxiliary determination of SDS binding is very time-consuming.

## MOLECULAR WEIGHT OF THE NATIVE PROTEIN

The final step in our procedure is to determine the molecular weight of the protein in solution in a detergent (sometimes with residual bound lipid), under conditions where it has been shown that all measurable functional and structural properties of the native state are retained. The measurement can be made rigorously by the method of sedimentation equilibrium. As mentioned before, a two-fold molar excess of micelles over protein molecules should suffice to insure that the number of micelles containing more than one copy of the protein molecule is negligibly small. Use of an analytical ultracentrifuge equipped with photoelectric scanner allows determination of the equilibrium distribution of protein-containing micelles without interference from protein-free micelles.

The simplest theoretical treatment of the equilibrium distribution is in terms of the thermodynamics of multiple component systems (Casassa and Eisenberg, 1964), in which the quantity that is measured in a sedimentation equilibrium experiment is formally designated as $M_P(1-\phi'\rho)$. Here $M_P$ is the molecular weight of the protein moiety of the particle, excluding bound detergent, lipid or solvent, $\rho$ is solvent density, and $\phi'$ is the effective partial specific volume of the protein moiety, which includes the contributions of bound detergent and lipid. In a thermodynamically ideal solution we can rigorously express $M_P(1-\phi'\rho)$ as (Tanford et al., 1974)

$$M_P(1-\phi'\rho) = M_P[(1-\bar{v}_P\rho) + \delta_D(1-\bar{v}_D\rho) + \delta_L(1-\bar{v}_L\rho)] \qquad (1)$$

where $\delta_D$ is the amount of bound detergent in g per g of protein, such as one would measure by equilibrium dialysis or equilibration on a chromatographic column (Makino et al., 1973), $\delta_L$ is the equivalent quantity for bound lipid; and $\bar{v}_P$, $\bar{v}_D$ and $\bar{v}_L$ are the partial specific volumes of the protein moiety, of bound detergent, and of bound lipid, respectively. One assumes that $\bar{v}_D$ and $\bar{v}_L$ are the same as they would be in a pure detergent micelle or lipid bilayer: it is highly improbably that this can lead to a significant error in the final result (Tanford et al, 1974).

The amount of bound detergent is easily measured if the detergent is available in radioactive form. If phospholipid is retained, the amount bound is readily determined by lipid phosphorus analysis. Its contribution to $1-\phi'\rho$ is generally very small because the buoyant density of phospholipid is close to that of the solvent. With both $\delta_D$ and $\delta_L$ known, $M_P$ is obtained directly from the sedimentation equilibrium plot by using equation 1.

Nonionic detergents are generally not available in radioactive form and analysis for them by chemical means is cumbersome. The values of $\bar{v}_D$ for these detergents are however relatively large (often between 0.90 and 1.00) and this allows a simplification of the procedure by use of $D_2O/H_2O$ mixtures to vary the solvent density. If $M_P(1-\phi'\rho)$ is measured at $\rho = 1/\bar{v}_D$, or measured at

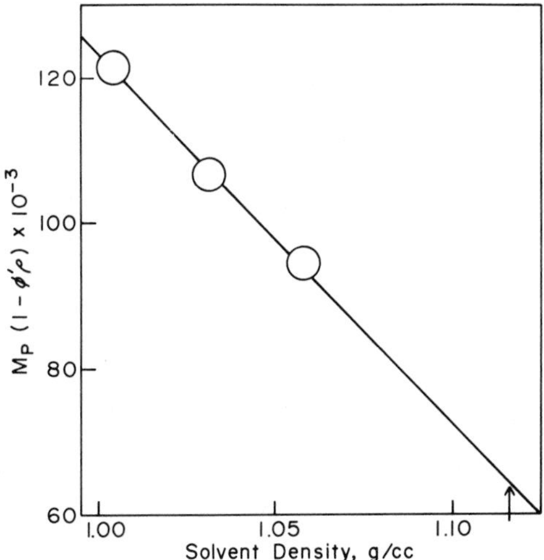


Figure 2. Determination of the molecular weight of the $Ca^{++}$-ATPase of sarcoplasmic reticulum in Tween 80. The protein contained 0.23 g bound phospholipid per g protein. The values of $M_P(1-\phi'\rho)$ are obtained directly from the experimental sedimenation equilibrium plots at different solvent densities. The partial specific volume of Tween 80 is 0.896 cc/g, and the arrow indicates the value of $\rho$ at which $\bar{v}_D = 1/\rho$. The protein molecular weight obtained from $M_P(1-\phi'\rho)$ at this point (with $\bar{v}_P = 0.738$ cc/g, $\bar{v}_L = 0.975$ cc/g) is 410,000. (The data are unpublished data of M. le Maire).

several densities and extrapolated to that point, the term $1-\bar{v}_D\rho$ of equation 1 vanishes, and $M_P$ can be determined <u>without knowledge of $\delta_D$</u>. Figure 2 provides an example. It shows data obtained for fully active $Ca^{++}$ ATPase from sarcoplasmic reticulum, solubilized in Tween 80. The experimental error is larger than appears at first glance because the uncertainty in the sedimentation equilibrium data is about 7%, and because of the long extrapolation required to reach $\rho = 1/\bar{v}_D$. The uncertainty in $M_P$ is probably close to 20%. The ATPase was prepared from purified vesicles (Meissner et al., 1973) that contained only a single kind of polypeptide chain, with molecular weight 120,000 (Rizzolo et al, 1976). The value of $M_P$ of 410,000, obtained by the extrapolation of Figure 2, thus suggests that the protein in the detergent is a trimer (true $M_P$ = 360,000) or possibly a tetramer (true $M_P$ = 480,000). Better accuracy would have been obtained with a detergent with a higher $\bar{v}_D$. Unfortunately the ATPase was slowly inactivated in several detergents with $\bar{v}_D$ in the range 0.93 to 0.97 cc/g, whereas activity was maintained for several days in Tween 80.

## EXPERIMENTAL RESULTS

Some of our early studies were done using deoxycholate as solubilizing detergent, because this substance is readily available in pure form and extensive studies of its self-association to form micelles have been made. There is no structural similarity between the hydrophobic moiety of deoxycholate and the hydrocarbon chains of the acyl groups of phospholipids, which creates a relatively high risk that the solubilized protein will not be in its authentic native state. Most of our more recent work has utilized nonionic detergents with long unbranched hydrocarbon chains, e.g., Lubrol WX (mixture of $C_{16}$ and $C_{18}$ saturated chains) and Tween 80 (oleyl chain).

We have found that microsomal cytochrome $b_5$ exists in the monomeric state in deoxycholate solution and in Triton X-100 (Robinson and Tanford, 1975; Visser et al., 1975). The major erythrocyte sialoglycoprotein is monomeric in deoxycholate (Grefrath, 1974) and in the nonionic detergent Lubrol WX (J.A. Reynolds, unpublished results). This protein does not have an assayable activity, so that one cannot be certain that the detergent-solubilized protein is representative of the native state, but there is other evidence to suggest that this glycoprotein has no tendency to form oligomeric structures with itself or with "band 3" polypeptides. Such interactions have been postulated on the basis of electron microscopic studies (Pinto da Silva and Nicolson, 1974).

The coat protein of the filamentous bacteriophages of the fd class, which is bound to the plasma membrane of the host bacterium in the course of phage biosynthesis, exists as a dimer in deoxycholate solution (Makino et al, 1975), and we have spectral evidence to indicate that the protein has the same conformation in this detergent as in phospholipid vesicles, which suggests that the observed state of association is representative of the *in vivo* membrane-bound protein (Nozaki et al., 1976).

As mentioned above, the $Ca^{++}$-stimulated ATPase from sarcoplasmic reticulum has a molecular weight in Tween 80 that corresponds to a trimer (or tetramer) of the ATPase polypeptide chain of molecular weight 120,000. Preparations of this protein in Tween 80 are actually not uniform in size and the results of Figure 2 apply to the major fraction obtained after chromatography on Sepharose 4 B (Le Maire et al., 1976). The enzyme in this fraction has a specific activity comparable to that of purified $Ca^{++}$ ATPase vesicles, whereas fractions containing protein at a lower molecular weight were found to have a lower specific activity, even though bound lipid is retained in all fractions. Experiments with the ATPase dissolved in deoxycholate lead to quite different results: monomer and dimer fractions are obtained, delipidation is complete, and all activity is lost (le Maire et al., 1976; Hardwicke and Green, 1974). The conclusion from these preliminary results and studies from other laboratories (Meissner et al., 1973; Warren et al., 1974) are that both the presence of phospholipid and a trimeric (or tetrameric) state are required for retention of ATPase activity.

The physiological function of the sarcoplasmic reticulum ATPase is to provide a mechanism for active trans-membrane transport of $Ca^{++}$ ions. We have not transferred the protein from the Tween 80 solution in which full ATPase activity was retained to a vesicular state in which $Ca^{++}$ transport can be studied, but this experiment has been done in other laboratories (Warren et al., 1974; Martonosi, 1968; Meissner and Fleischer, 1973), and reconstitution of the ability to transport $Ca^{++}$ has been demonstrated even after solubilization in detergents in which ATPase activity is retained for a much shorter length of time than in Tween 80.

(It should be noted parenthetically that the reconstitution experiments cited above, as well as our own work with the ATPase, were done, as previously indicated, with purified protein which by the usual criteria contained only a single kind of polypeptide chain. One recent study, however, has suggested that restoration of transport capability requires the presence of an additional polypeptide of low molecular weight (Racker and Eytan, 1975). This illustrates the tentative nature of almost all work with membrane proteins at the present early stage of development of this subject).

## DISCUSSION

It was stated at the beginning of this paper that there should be a close relation between the state of association of a protein and its biological function. This is borne out by the results given above. The monomeric state of cytochrome $b_5$, for example, is entirely consistent with its well-established function as an electron carrier in the microsomal electron transport chain (Strittmatter et al., 1972). There is no conceivable functional need for an oligomeric state: on the contrary, the rate of electron transfer from NADH-cytochrome $b_5$ reductase to cytochrome $b_5$ is determined by the rate of collision between these molecules in the membrane, which in turn is determined by the rate of lateral diffusion (Rogers and Strittmatter, 1972), and thus aided by small size.

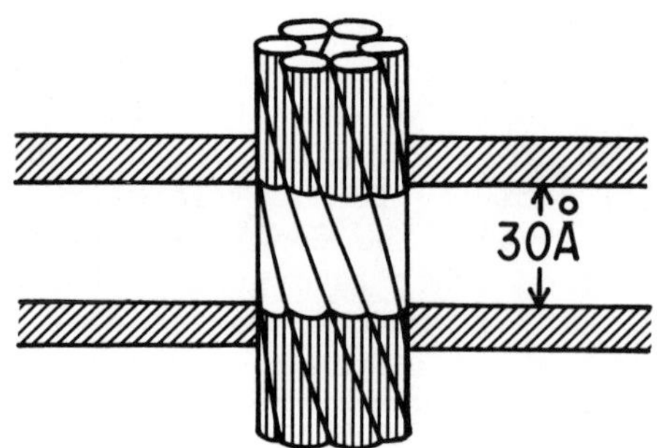


Figure 3. Hypothetical model for the formation of an aqueous channel across a phospholipid bilayer. The twisted cylinders are helical segments of 6 non-covalently associated polypeptide chains. Amino acids residues facing the internal channel are hydrophilic, whereas those in contact with lipid hydrocarbon chains (unshaded portion of outer surface) are hydrophobic. This model was proposed by Inouye (1974) for the major cell wall protein of _E. coli_. The complete amino acid sequence of this protein is known, and the arrangement of hydrophobic and hydrophilic amino acids is appropriate for the model structure. Whether this protein is in fact a trans-membrane protein has not been established.

The oligomeric state of the $Ca^{++}$ATPase of sarcoplasmic reticulum is especially interesting from a functional point of view because there is as yet no knowledge at all of the structural requirements for the creation of a channel, pore or other passage for the transport of ions across the hydrophobic core of a phospholipid bilayer, which by itself acts as a permeability barrier for ions and other strongly hydrophilic substances. Our results, indicating that the $Ca^{++}$ATPase is likely to be a trimer (or tetramer) of identical polypeptide chains, suggests that the structure traversing the membrane may be composed of symmetrically arranged elements, such as have been observed in bacteriorhodopsin (Henderson and Unwin, 1975) and may be related to the proton-pumping capability of that protein. An additional piece of information regarding this possibility is provided by studies of the products of mild proteolysis of the ATPase, which have shown that the ATPase polypeptide chain contains two halves of similar composition, both of which have hydrophobic regions that are presumably embedded in the membrane and are joined to each other by an external link (Rizzolo et al., 1976; Thorley-Lawson and Green, 1975). This suggests that the trans-membrane portion of the ATPase may actually have six (or eight) symmetrically-placed elements. Such an arrangement would lend itself well to the formation of an aqueous channel across the membrane, as is illustrated by Figure 3. There are no data at the present time that give any indication of how large a portion of each ATPase chain is actually embedded in the membrane, or whether membrane-embedded elements have an α-helical structure. Recent work (Shamoo et al., 1976) has shown, however, that a $Ca^{++}$-translocating channel can be formed from very small segments of the polypeptide chain, compatible with the individual helical elements of Figure 3.

ACKNOWLEDGMENTS

Some of the work described in this paper was done in the laboratory of my colleague, Dr. J.A. Reynolds, and all of it has benefitted from her generous advice. The experimental results have been obtained with the collaboration of S.P. Grefrath, M. le Maire, S. Makino, J.V. Møller, L.J. Rizzolo, N.C. Robinson and L. Visser. The work has been supported by research grants from the National Science Foundation and from the National Institutes of Health.

## REFERENCES


Casassa, E.F. and H. Eisenberg, Advan. Protein Chem. 19, 287, (1964).

Grefrath, S.P., Ph.D. thesis, Duke University, (1974).

Grefrath, S.P. and J.A. Reynolds, Proc. Natl. Acad. Sci. USA 71, 3913, (1974).

Hardwicke, P.M.D., and N.M. Green, Eur. J. Biochem. 42, 183, (1974).

Helenius, A., and K. Simons, Biochim. Biophys. Acta 415, 29, (1975).

Henderson, R. and P.N.T. Unwin, Nature 257, 28, (1975).

Huang, C., Biochemistry 8, 344 (1969).

Inouye, M., Proc. Natl. Acad. Sci. USA 71, 2396, (1974).

Klotz, I.M., D.W. Darnall and N.R. Langerman, The Proteins, edited by H. Neurath and R.L. Hill, vol. 1, ch.5, Academic Press, New York (1975).

le Maire, M., J.V. Møller and C. Tanford, Biochemistry 15, in press (1976).

Makino, S., J.A. Reynolds and C. Tanford, J. Biol. Chem. 248, 4926, (1973).

Makino, S., J.L. Woolford, C. Tanford, R.E. Webster, J. Biol. Chem. 250, 4327, (1975).

Martonosi, A., J. Biol. Chem. 243, 71, (1968).

Meissner, G. and S. Fleischer, Biochem. Biophys. Res. Commun. 52, 913, (1973).

Meissner, G., G.E. Conner, and S. Fleischer, Biochim. Biophys. Acta 298, 246, (1973).

Nozaki, Y., B.K. Chamberlain, R.E. Webster and C. Tanford, Nature 259, 335, (1976).

Pinto da Silva, P. and G.L. Nicolson, Biochim. Biophys. Acta 363, 311, (1974).

Racker, E. and E. Eytan, J. Biol. Chem. 250, 7533, (1975).

Rizzolo, L.J., M. le Maire, J.A. Reynolds and C. Tanford, Biochemistry 15, in press, (1976).

Robinson, N.C. and C. Tanford, Biochemistry 14, 369, (1975).

Rogers, M.J. and P. Strittmatter, J. Biol. Chem. 250, 5713, (1975).

Shamoo, A.E., T.E. Ryan, P.S. Stewart and D.H. MacLennan, Biophys. J. 16, 109a, (1976).

Strittmatter, P., M.J. Rogers and L. Spatz, J. Biol. Chem. 247 7188, (1972).

Tanford, C., J. Phys. Chem. 78, 2469, (1974).

Tanford, C., Y. Nozaki, J.A. Reynolds and S. Makino, Biochemistry 13, 2369, (1974).

Thorley-Lawson, D.A. and N.M. Green, Eur. J. Biochem. 59, 193, (1975).

Visser, L., N.C. Robinson and C. Tanford, Biochemistry 14, 1194, (1975).

Warren, G.B., P.A. Toon, N.J.M. Birdsall, A.G. Lee, and J.C. Metcalfe, Biochemistry 13, 5501, (1974).

Warren, G.B., P.A. Toon, N.J.M. Birdsall, A.G. Lee and J.C. Metcalfe, Proc. Natl. Acad. USA 71, 622, (1974).

MEMBRANE ELECTROSTATICS

Hermann Träuble

Max-Planck -Institut für biophysikalische Chemie

D-34 Göttingen-Nikolausberg, Germany

CONTENTS



## 1. INTRODUCTION

Suspensions of erythrocytes and of many other cells migrate toward the anode in an electrical field, indicating that the cell membranes posess a negative net surface charge. The origin of these charges are ionizable groups like carboxylic acid and phosphoric acid attached to membrane proteins, sugars or lipids. Among the lipids carrying a net negative charge between pH 3 and 9 are cardiolipin, phosphatidylserine, phosphatidylinositol, phosphatidylglycerol and phosphatidic acid (cf. Gurr and James,1971). The charge-carrying lipid in the human red cell membrane is phosphatidylserine, which comprises about 15-20 per cent of the total lipids. According to Verkleij et al. (1973) this lipid resides predominantly on the inner side of the membrane, demonstrating the possibility of asymmetric charge distribution on the two membrane sides. In the central nervous tissue the main acidic lipids are cerebroside sulphate and serine phosphoglycerides (cf. London and Vossenberg, 1973). A charge density as high as one elementary charge per 100 $Å^2$ was postulated by Hille et al. (1975) for nerve fibres in the vicinity of sodium channels.

The question arises: What is the significance of surface charge? Are there general principles applicable to any membrane, or is the rôle of surface charge different from one membrane to another?

Many biochemical reactions on the cell surface involve charged reactants. The surface charge will modify biochemical activity (Goldstein et al., 1964). Ionic species attempting to cross a charged membrane sense the surface electrostatic potential (McLaughlin et al., 1970). The conduction of nerve pulses is influenced by electrical charges at the axon membrane. The critical importance of a constant overall membrane charge has been demonstrated recently by Hubbard and Brody (1975) for Neurospora crassa. Using cultures of different mutant strains they succeeded in varying the relative proportions of the neutral and zwitterionic phospholipids within a wide range, whereas the proportion of the negatively charged cardiolipin remained always at a constant level.

In the following we shall try to work out some general principles relating the presence of negative surface charges to membrane structure and function on the basis of physico-chemical studies with charged lipid membranes. In broad terms the conclusion will be that surface charges provide a basis for the interaction of membranes with their electrolyte environment and vice versa. The subject can be discussed in terms of a few principles of thermodynamics and electrochemistry, and readers familiar with the polyelectrolyte field will recognize the close parallelism to this field.

Many of the experimental results presented in this article were obtained with bilayers of the synthetic phospholipids dimyristoyl phosphatidic acid (PA) and dimyristoyl methylphosphatidic acid (MPA), which carry two and one ionizable protons at their polar heads (cf. Fig. 1). MPA was synthesized with the aim of obtaining a lipid which has a small polar head, so that steric factors do not play a rôle, and which carries only one elementary charge per polar head, so that interferences between different ionizable groups can be excluded (Träuble et al., 1976).

Section Two introduces some of the results of the Gouy-Chapman theory of the diffuse electrical double layer. We require some understanding of the electrical surface potential $\Psi^o$ and its rôle in acid-base reactions at charged surfaces and, as a major point,

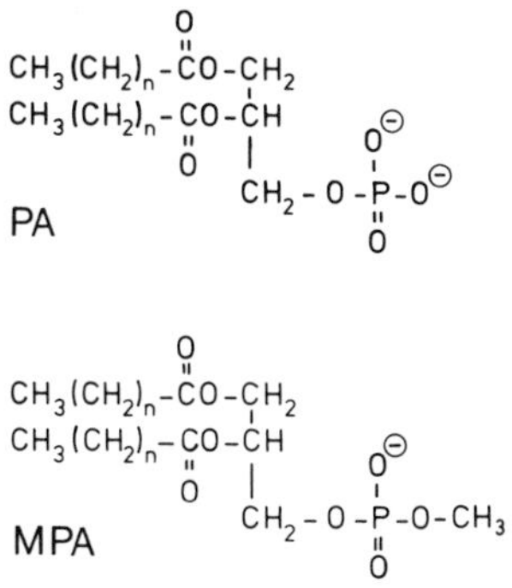

$CH_3(CH_2)_n$-CO-$CH_2$
$CH_3(CH_2)_n$-CO-CH
PA $CH_2$-O-P-$O^{\ominus}$

$CH_3(CH_2)_n$-CO-$CH_2$
$CH_3(CH_2)_n$-CO-CH
MPA $CH_2$-O-P-O-$CH_3$

Fig. 1 Chemical structure of the synthetic lipids Phosphatidic Acid and Methyl Phosphatidic Acid.

an expression for the free energy of the diffuse double layer. In Section Three we shall address ourselves to the question of how structural changes in membranes can be brought about by changes in the electrolyte environment; besides non-specific effects (changes in pH and ionic strength) we shall discuss the more specific interaction between divalent cations and negatively charged membranes. Section Four provides examples of the reverse effect, namely, the influence of membrane structural changes on the electrolyte environment; they comprise changes in pH and pulses of divalent cations. In Section Five we consider the rôle of surface charges for phase separation phenomena in heterogeneous membranes. Finally, Section Six is devoted to the electrostatic coupling between the two layers of a charged membrane. This leads us to propose a new mechanism for the transmission of signals through charged membranes.

We do not wish to provide a detailed theoretical background for all of these phenomena; this will be done in separate papers. Rather, it is our intention to discuss some phenomena which have a bearing on the general significance of negative surface charges for membrane structure and function.

## 2. SURFACE ELECTROSTATICS

We consider the surface charges as smeared out over the plane of the membrane and define a surface charge density $\sigma$ as a measure of the number of elementary charges per $cm^2$ of the surface.

As is known from the Gouy-Chapman theory (cf. Verwey and Overbeek, 1948; Davies and Rideal, 1961) a diffuse layer of counterions with cation build-up and anion depletion is established near a negatively charged surface. The "thickness" of this layer is given by $1/\kappa$, where the "Debye length" $\kappa$ is defined by

$$\kappa^2 = \frac{4\pi e^2}{\varepsilon kT}\sum_i n_i z_i^2$$

$$= \frac{8\pi}{\varepsilon}\frac{e^2}{kT} n \quad \text{for a 1:1 electrolyte.} \qquad (1)$$

Here $n_i$ is the bulk concentration of ions of kind $i$ in molecules per $cm^3$; $z_i$ is the valency. $\varepsilon$, $e$, $k$ and $T$ have their usual meaning. $1/\kappa$ has values of 400 Å, 96 Å, 31 Å and 10 Å for salt concentrations of $10^{-4}$, $10^{-3}$, $10^{-2}$ and 0.1 M; i.e. increasing salt concentration compresses the diffuse layer. For a 1:1 electrolyte the electrical potential in the plane of the fixed, negative charges is

$$\psi^o = 2\frac{kT}{e}\sinh^{-1}\frac{\sigma}{c} , \qquad (2)$$

where

$$c = \frac{\varepsilon}{2\pi}\frac{kT}{e}\kappa . \qquad (3)$$

The electrical potential decreases with increasing distance from the surface, and $1/\kappa$ defines the distance where $\Psi$ has decreased to about the $1/e$ fraction of its value at the surface. Fig. 2 illustrates the dependence of $\psi^o$ on salt concentration and charge density. The validity of eq.(2) for charged lipid membranes has been checked experimentally by McLaughlin et al. (1971) and Fromherz and Masters (1974) and excellent agreement with the theory was found. Therefore it appears that the Gouy-Chapman theory - despite all its known simplifications - gives reliable values of the surface potential of charged lipid layers.

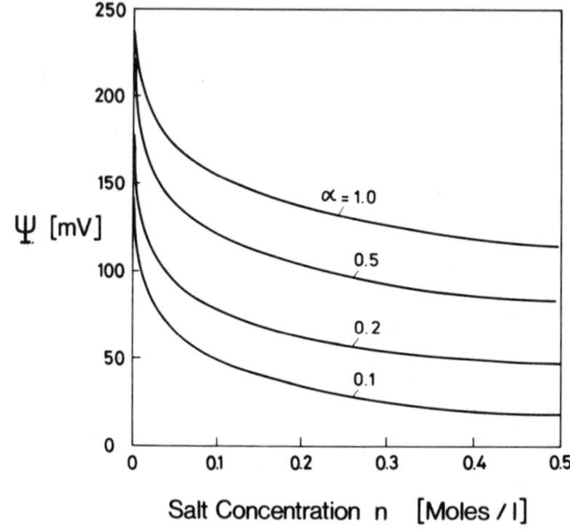


Fig. 2 Surface potential $\psi^o$ of a lipid membrane as a function of salt concentration n (1:1 electrolyte) and charge density at T = 300 K. $\alpha$ is the degree of dissociation of the lipid molecules with a molecular area of 50 $Å^2$.

One well-known consequence of the negative surface potential is the accumulation of protons at the membrane surface, and therefore the surface pH is lower than the bulk pH. The proton concentration at the surface $[H^+]_s$ is related to the bulk concentration $[H^+]_b$ by a Boltzmann factor:

$$[H^+]_s = [H^+]_b \exp(e\Psi^o/kT). \tag{4}$$

The surface potential can also markedly affect the ionization of surface-attached groups, resulting in a shift of the apparent pK. For example, we usually ascribe a pK of about 2 to the ionization of the first proton of a phosphate group in bulk solution. At a negatively charged surface the apparent pK can be around 6. The reason is that additional electrostatic work $(e\Psi^o)$ must be done to remove a proton from a negatively charged surface. The resulting change in Gibbs free energy in the dissociation process is therefore

$$\begin{aligned} \Delta G &= \Delta G^o + Le\Psi^o \\ &= - RT \ln K \end{aligned} \tag{5}$$

where $K$ is the apparent equilibrium dissociation constant. $\Delta G^o = - RT \ln K_o$ is the change in Gibbs free energy for dissociation in the bulk and $L$ equals Avogadro's number. The degree of dissociation, $\alpha$, is then given by

$$\alpha = \frac{K_o}{K_o + [H^+]_b \exp(e\Psi^o/kT)} \tag{6}$$

instead of the "bulk value"

$$\alpha_o = \frac{K_o}{K_o + [H^+]_b} .$$

For the application of eq.(6) to membranes one substitutes $\Psi^o$ according to eq.(2) and expresses $\sigma$ in terms of the number of charged groups per $cm^2$ of the surface. For a lipid membrane containing only ionizable lipids with a surface area $f$ per molecule

$$\sigma = \frac{e\alpha}{f} , \tag{7}$$

where $e\alpha$ $(0 \leq \alpha \leq 1)$ equals the charge per polar group.

As was shown elsewhere (Träuble et al. 1976) eq.(6) leads to the following expression for the apparent pK (i.e. the value of pH where $\alpha = 0.5$) of phosphate groups at a negatively charged lipid layer with one ionizable proton per lipid and a molecular area of 50 $Å^2$ immersed in a 1:1 electrolyte:

$$pK = pK_o + 0.86 - \log_{10} n. \tag{8}$$

Here $n$ is the molar salt concentration and $pK_o$ the "bulk pK". For $pK_o = 1.75$ and $n = 10^{-3}$M the apparent pK is about 6. Increasing salt concentration lowers the apparent pK. It should be noted that eq.(8) is valid in the so-called "high-potential region" where the surface potential can be approximated by $\Psi^o = 2\ (kT/e)\ \ln\ (2\sigma/c)$ (cf. Träuble et al., 1976).

The interaction between the surface charges and the counterions in the diffuse layer gives rise to a characteristic surface free energy which can be calculated from the charging integral (Träuble et al., 1976; Jähnig, 1976)

$$\phi = \int_0^\sigma \Psi^o(\sigma)d\sigma. \tag{9}$$

This is the work done in a hypothetical charging process whereby one $cm^2$ of the surface is charged from zero to the final value along a pathway of equilibrium states (cf. Harned and Owen, 1943). During this process the surface potential increases according to eq.(2). Substituting $\Psi^o$ according to eq.(2) we obtain after integration

$$\phi\ (\text{per } cm^2) = \sigma\Psi^o - \frac{\varepsilon}{\pi}\left(\frac{kT}{e}\right)^2 \kappa \left\{\cosh \frac{e\Psi^o}{2kT} - 1\right\}. \tag{10}$$

It can be shown that $\phi$ represents the additional work that must be done to ionize the polar groups of the lipid molecules in a bilayer compared with the ionization of isolated polar groups (Träuble et al. (1976).

Expression (10) differs from the one derived by Verwey and Overbeek (1948) by the additional term $\sigma\Psi^o$ and by its positive sign. The reasons for this difference are not trivial and are discussed in detail elsewhere (Träuble et al., 1976; Jähnig, 1976). May it suffice here to state that the use of Verwey and Overbeek's expression in the case of ionizable lipids leads to wrong predictions with regard both to the sign and to the order of magnitude of the effects to be discussed in Sections 3 and 5.

Fig. 3 shows the variation of $\phi$ with the charge density for different salt concentrations. The value of $\phi$ increases with increasing charge density and decreasing salt concentration (or increasing surface potential). Thus $\phi$ depends on two parameters which are important in biological systems.

For a lipid membrane with a molecular area of 50 $Å^2$ and one charge per molecule the calculated value of the double layer free energy amounts to about 7 kcal/mole (or 98 $erg/cm^2$) at $10^{-3}$M salt, and 3 kcal/mole (or 42 $erg/cm^2$) at 0.1 M salt. For comparison, the average energy of kinetic motion at 25 °C is 0.6 kcal/mole and the energy of the hydrogen bond ranges between 1 and 10 kcal/mole.

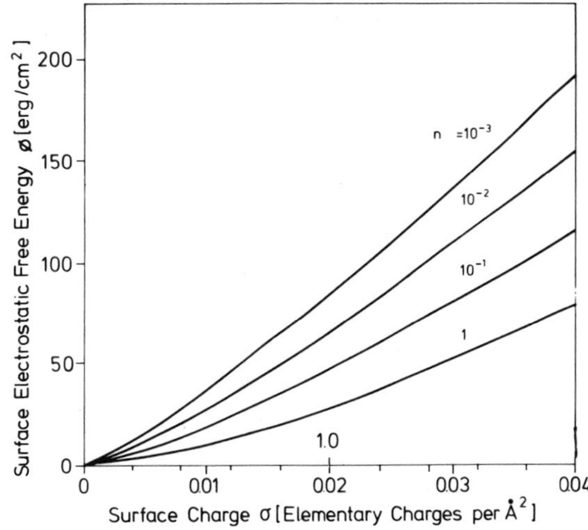


Fig. 3 Electrostatic free energy as a function of the surface charge density for different molar salt concentrations (1:1 electrolyte).

## 3. ELECTROSTATIC EFFECTS ON MEMBRANE STRUCTURE

Qualitatively one expects that surface charges tend to expand the membrane or "fluidize" the membrane structure, because of their lateral repulsion. Under appropriate conditions this tendency might suffice to trigger an ordered → fluid phase transition in the bilayer.

The most thorough way to quantify this influence would be to include the electrostatic interactions in the total free energy of the membrane and to calculate the new equilibrium state. Our aim here is rather to provide a general idea of the effects of surface charges on membrane structure and to develop a semiquantitative theory of their influence on the ordered ↔ fluid lipid phase transition.

We start by considering the change of the electrostatic free energy on expansion of a membrane containing L lipid molecules with an initial molecular area f. At the start the electrostatic free energy is

$$G^{el} = L\, f\, \phi. \tag{11}$$

In a linearized theory the change in $G^{el}$ resulting from an area change, $\Delta f$, is

$$\Delta G^{el} = L \left\{ \phi + f \frac{d\phi}{df} \right\} \Delta f\,. \tag{12}$$

Using eq.(7) we find from eq.(10)

$$\frac{d\phi}{df} = - \frac{\Psi^o \sigma}{f}\,. \tag{13}$$

The first term in eq.(12) describes an increase of $G^{el}$ due to the increase in membrane area, the second term describes a decrease in $G^{el}$ due to the decrease in charge density.

Substituting $\phi$ according to eq.(10) and $d\phi/df$ according to eq.(13) into eq.(12) yields

$$\Delta G^{el} = -\frac{\varepsilon}{\pi}\left(\frac{kT}{e}\right)^2 L\,\kappa\left\{\cosh\frac{e\Psi^o}{2kT} - 1\right\}\Delta f\,. \qquad (14)$$

This shows that the net effect of a membrane expansion is always a decrease in $G^{el}$, as one expects from the tendency of the electrostatic charge to spread itself out over the greatest possible area.

This tendency of the surface charges to expand a membrane can be demonstrated by pressure-area diagrams of monolayers of an ionizable lipid at different states of ionization, i.e. at different

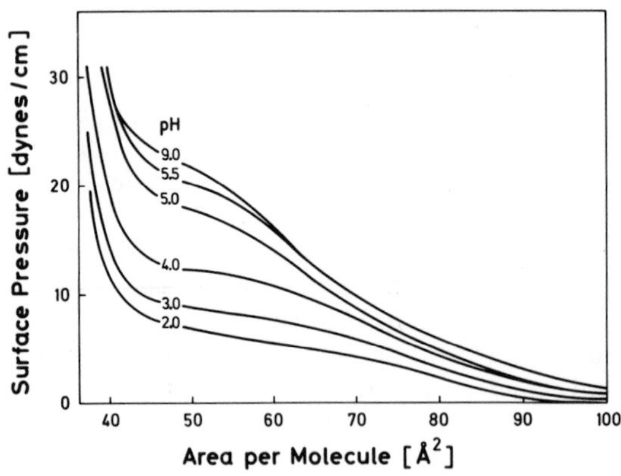


Fig. 4 Surface Pressure-Area diagrams of MPA-monolayers at different pH, 20 °C. Subphase $10^{-2}$M NaCl. These measurements were kindly supplied by Dipl. Math. M. Teubner.

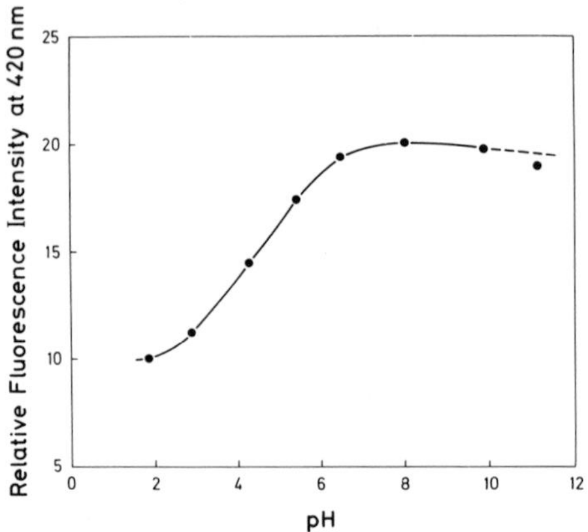


Fig. 5 Expansion of MPA bilayers during ionization of the phosphate groups leads to an increase in NPN fluorescence intensity. 50 °C, wavelength of excitation 340 nm. 2 x $10^{-4}$M MPA, 2 x $10^{-6}$M NPN , 5 x$10^{-3}$ M NaCl.

pH values (cf. Fig. 4). The corresponding effect with lipid bilayers is illustrated in Fig. 5. Here a pH titration was performed with a suspension of MPA at a temperature far above its transition in the presence of N-phenyl-naphthylamine (NPN). NPN is a fluorescence indicator which distributes itself between membrane and water according to the free volume present in the membrane, and which has a much higher fluorescence inside the membrane (Träuble and Overath, 1973). Increasing the pH leads to a twofold increase in the fluorescence intensity in the course of the ionization of the phosphate groups. This is interpreted as an additional incorporation of NPN into the electrostatically expanded membrane.

A nice example to illustrate further and to quantify the effect of electrostatic interactions on membrane structure is that of lipid phase transitions of the type first studied by D. Chapman. In this case it is possible to describe the interactions within the membrane by only two parameters, namely the transition temperature $T_t^*$ of the uncharged system and the entropy change $\Delta S^*$ at this transition.

The structural change at the ordered → fluid transition can be considered for our purpose simply as an expansion of the membrane, or an increase in the molecular area, $\Delta f$. According to the above considerations surface charges will support the action of increasing temperature in bringing about this membrane expansion. We expect a lower transition temperature for the ionized lipid.

A simple thermodynamic consideration (cf. Record, 1967) based on the equality of the Gibbs free energies of ordered and fluid membrane phases at $T_t$ allows us to express the expected shift in transition temperature as

$$\Delta T_t = \frac{\Delta G^{el}}{\Delta S^*} = \frac{G_{fl}^{el} - G_{ord}^{el}}{\Delta S^*} . \tag{15}$$

Here $\Delta G^{el}$ is the difference in the electrostatic free energy between the fluid (or expanded) and the ordered (or condensed) membrane. Thus $\Delta G^{el}$ can be directly substituted from eq.(14) to give an explicit expression for the effect of charge density and ionic strength on the lipid phase transition.

$\Delta S^*$ is the entropy difference between the fluid and the ordered states of the uncharged membrane. [In two recent studies on dipalmitoylphosphatidic acid (Jacobson and Papahadjopoulos, 1975) and dipalmitoylphosphatidyl serine (MacDonald et al., 1976) $\Delta S^*$ was erroneously interpreted as the entropy change for the specific state of ionization at which $T_t$ was measured.]

For a comparison with experiments on MPA suspensions we apply the high-potential approximation $\cosh x \approx 1/2 \exp x$ which is valid for most practical purposes when charged lipid membranes contain more than a few per cent ionized groups (cf.Träuble et al., 1976). This gives

$$\Delta T_t = -2\,\frac{kT}{e}\,\frac{L}{\Delta S^*}\,\sigma\,\Delta f + \frac{\varepsilon}{\pi}\left(\frac{kT}{e}\right)^2 \frac{L}{\Delta S^*}\,\kappa\,\Delta f\ . \tag{16}$$

The first term predicts a linear decrease in $T_t$ with increasing charge density; the second term contains the salt dependence, and predicts an increase in $T_t$ proportional to $\sqrt{n}$ . (There is, however, yet another, much stronger effect of the salt concentration on $T_t$ that operates via a change in the degree of dissociation at constant pH.) If values are inserted into eq.(16) one finds that the second term is small compared with the first one, at least for $n \lessapprox 0.2$ M and $\alpha \gtrapprox 0.2$.

The predictions made by eq.(16) were checked experimentally using suspensions of dimyristoyl-MPA, with one ionizable proton at its polar group (cf. Fig. 1). Transition temperatures were measured using NPN as a fluorescent indicator (Träuble and Overath, 1973). The strategy was (1) to measure $T_t$ for different values of $\sigma$ at constant ionic strength by changing the pH and (2) to alter the ionic strength at constant $\sigma$, which can be done at high pH where the lipid is fully ionized.

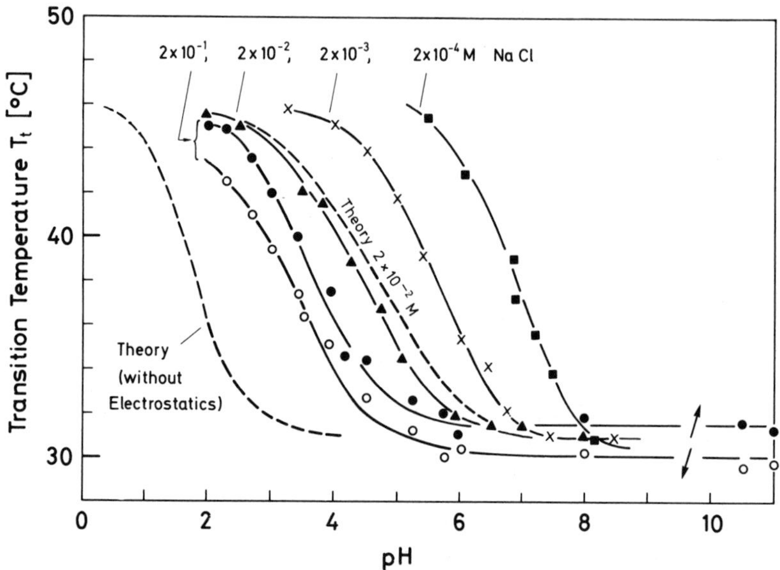


Fig. 6 Transition temperature of dimyristoyl-MPA as a function of pH at different salt concentrations. Ionization of the lipid lowers the value of $T_t$ by about 17 °C; the apparent pK decreases with increasing salt concentration. The $T_t$ values shown were measured at increasing temperature; only for 2 x $10^{-1}$M NaCl are shown also the values measured at decreasing temperature. The theoretical curve on the left was calculated without the influence of electrostatic forces on the dissociation using $pK_o$ = 1.75.

Fig. 6 shows the transition temperature as a function of pH at different salt concentrations. Ionization of the lipid lowers $T_t$ from about 47 $^oC$ to about 30 $^oC$ (fully charged state). The net decrease in $T_t$ is $\Delta T_t^{max} \approx 17$ $^oC$, and this value is calculated from eq.(16) if we use the measured value of $\Delta S^* = 17.5$ cal/mole deg (Blume, 1976) and take $\Delta f/f = 0.241$.

The $T_t$(pH)-curves in Fig. 6 resemble dissociation curves and, indeed, eq.(16) predicts that at constant ionic strength $\Delta T_t$ is proportional to $\sigma$ which in turn is proportional to the degree of dissociation $\alpha$. Thus the midpoints of the $T_t$(pH)-curves determine the apparent pK; the observed increase in pK with decreasing ionic strength is in full accord with eq.(8). Combining eqs.(16) and (6) and inserting $pK_o = 1.75$ and $T_t^* = 48$ $^oC$ leads to the theoretical curve in Fig. 6 calculated for $n = 2 \times 10^{-2}$M. This curve reproduces accurately the shape and the position of the measured curve. The agreement is equally good for the other salt concentrations, demonstrating the consistency between the theoretical approach and the experimental behavior.

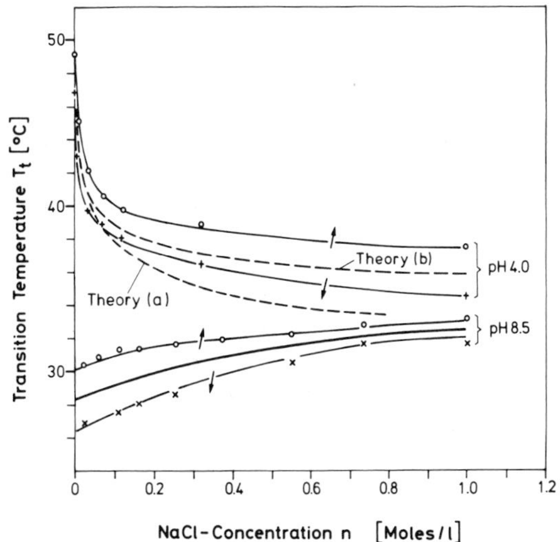


Fig. 7 Salt-dependence of the transition temperature of MPA at pH 8.5 (fully ionized state) and at pH 4 where the addition of salt increases the degree of dissociation. Symbols ↑ and ↓ denote measurements at increasing and decreasing temperature. Theoretical curves acc. to Eq.(16) with (b) and without (a) the second term.

The salt dependence of $T_t$ was studied at pH 8.5 and pH 4 and is illustrated in Fig. 7. The observed increase of $T_t$ with increasing salt at pH 8.5 is described by the last term in eq.(16). Surprisingly, at pH 4 $T_t$ shows a sharp initial decrease with increasing ionic strength. This effect can be explained quantitatively in terms of an increase in the degree of dissociation with increasing salt on the basis of eq.(6).

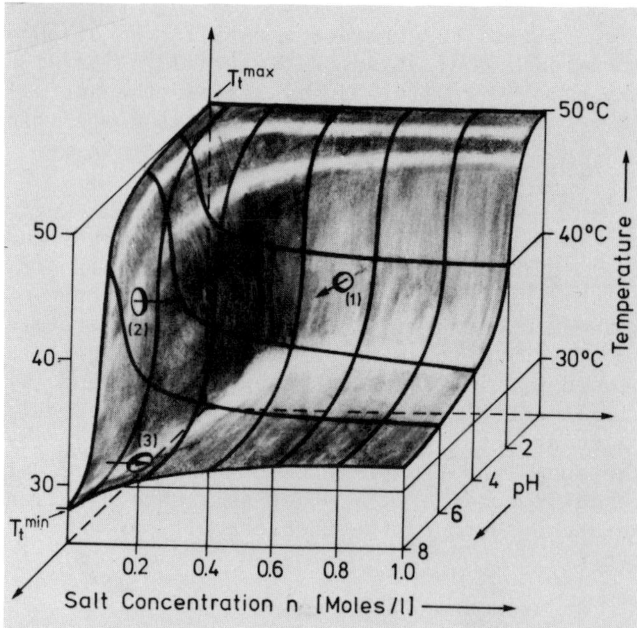


Fig. 8 Effects of pH and added salt (NaCl) on the transition temperature of MPA. Fluid state above, ordered state below the surface. The arrows show various ways to induce phase transitions at constant temperature.

A synopsis of the experimental results is given in Fig. 8 which shows $T_t$ as a function of pH and the molar salt concentration. The surface shown separates the regions of fluid and ordered membrane structure. This profile is in full accord with the predictions made by eq.(16). The arrows ((1) to (3)) indicate that in the temperature range between about 28 and 48 °C the lipid phase transition can be induced electrostatically at constant temperature. Fluidization of the membrane can be induced either by an increase in pH (arrow (1)) or by an increase in salt concentration (arrow (2)). Membrane condensation can be achieved by a decrease in pH or, when the lipid is fully ionized, by an increase in ionic strength (arrow (3)). All of these transitions have been verified experimentally (Träuble et al., 1976).

Without going into further detail we may draw the following conclusions: Surface charges tend to fluidize or expand lipid membranes. The relevant parameter is the electrostatic free energy, which makes a significant contribution to the free energy of the system and can be varied within a wide range by alteration either of the membrane's surface charge density or the salt concentration in the surrounding medium. Ordered ↔ fluid phase transitions can thus be induced at constant temperature by variations in pH and salt concentration. The possible biological significance is evident in view of the known dependence of several membrane enzymes and transport systems on the physical state of the surrounding lipid layer (cf. Overath et al., 1976).

Divalent Cations. The effects of pH and ionic strength on membrane structure are unspecific. More interesting in this respect are divalent cations, such as $Mg^{++}$ and $Ca^{++}$, which bind strongly and in a stoichiometric way to acidic phospholipids, whereby two elementary charges provide one binding site for a divalent cation. Of course, the apparent binding "constant" K depends on the surface potential in a similar way to that discussed for protons:

$$K = K_o e^{2e\Psi^o/kT} \tag{17}$$

where $K_o$ is the binding constant in the absence of surface charges.

The association of $Ca^{++}$ with phosphatidic acid and methylphosphatidic acid bilayers has been studied using murexide as an indicator for the free $Ca^{++}$ (Träuble and Eibl, 1975). When MPA is added to a $Ca^{++}$-containing solution (in the absence of additional electrolyte and at pH 8 so that the lipid is fully ionized) the amount of bound $Ca^{++}$ increases approximately linearly with the lipid content and reaches a saturation value when 1 $Ca^{++}$ is bound per two lipid molecules. As one would expect from eq.(17) the binding is stronger in the ordered state ($T < T_t$) where the surface potential is larger. The apparent binding constant decreases with increasing ionic strength, and therefore the addition of salt causes a gradual release of $Ca^{++}$ from the membrane surface.

The binding of $Ca^{++}$ reduces the effective surface charge density according to

$$\sigma = \sigma^o - 2 e N^* \tag{18}$$

where $\sigma^o$ denotes the initial surface charge density and $N^*$ is the surface density of bound $Ca^{++}$. The expected effect on membrane structure is (1) a reduction of the membrane area and (2) according to eq.(16) a linear increase in the transition temperature with increasing number of occupied binding sites.

A qualitative demonstration of the condensation effect is given in Fig. 9 which shows the decrease in NPN fluorescence intensity upon addition of $Ca^{++}$ to a suspension of MPA.

As was shown in a previous study (Träuble and Eibl, 1975), $Ca^{++}$ produces the expected linear increase in transition temperature in the case of dimyristoyl-phosphatidic acid. With MPA the effect is more complicated. Over a relatively wide range of $Ca^{++}$ concentration, when the molar ratio $m = [Ca^{++}]/[MPA]$ is $\lesssim 3$, the transition curve is biphasic as shown in Fig. 10. The first step of the transition occurs at a temperature characteristic of the $Ca^{++}$-free system ($\approx$30 °C), the second is characteristic of the $Ca^{++}$-saturated system ($\approx$50 °C). It therefore appears that, for low $Ca^{++}$ concentrations,

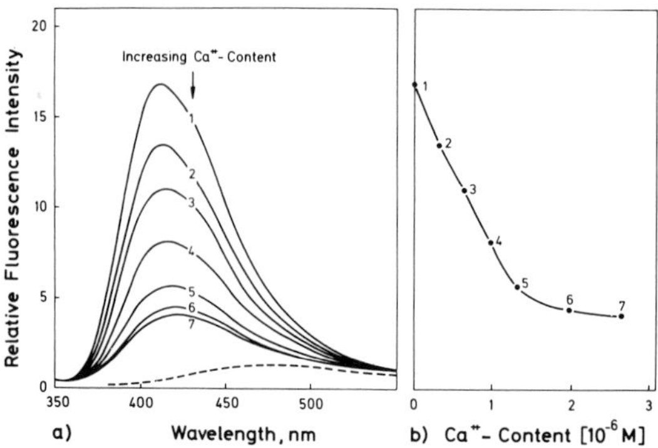


Fig. 9 Demonstration of $Ca^{++}$-binding to MPA by the decrease in NPN fluorescence intensity. a) Effect on spectra; Wavelength of excitation 340 nm. Dashed curve: NPN in $10^{-2}$M Tris, pH 8. 1: After addition of MPA ($3.5 \times 10^{-5}$M). b) Decrease in fluorescence intensity at the maximum with increasing $Ca^{++}$-content. 20 °C, $2 \times 10^{-6}$M NPN.

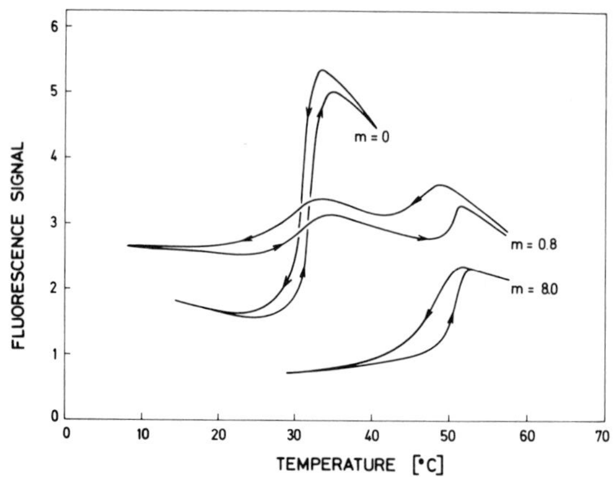


Fig. 10 Effect of $Ca^{++}$ on the transition of MPA ($2.5 \times 10^{-4}$M lipid, pH 7.8). m = molar ratio $[Ca^{++}]/[MPA]$. Continuous recording of NPN fluorescence intensity; $2 \times 10^{-6}$M NPN; $\lambda_{ex}$ = 340 nm, $\lambda_{em}$ = 420 nm.

a certain fraction of the lipids does not bind $Ca^{++}$, whereas another part binds the ion. $Ca^{++}$ appears to induce cluster formation and phase separation in these single-component model membranes. As one would expect the "amplitude" of the upper transition increases with increasing $Ca^{++}$-concentration, while that of the lower transition decreases and finally disappears.

In summary, the stoichiometric binding of divalent cations and the resulting membrane condensation (which allows one to trigger the fluid → ordered transition at constant temperature) make divalent cations potent regulatory factors of membrane structure.

## 4. ION PULSES INDUCED BY MEMBRANE STRUCTURAL CHANGES

A cell membrane bearing ionizable groups at its surface may be regarded as a reservoir (sink or source) for cations that can bind to the membrane surface. In the case of acidic lipids such cations are protons and divalent cations like $Ca^{++}$ and $Mg^{++}$. The "capacity" of the "membrane reservoir" can be estimated by considering a spherical cell with radius $r$ carrying $4\pi r^2/f$ ionizable groups on its inner surface where $f$ denotes the area per ionizable group. The value of $f$ is about 50 $Å^2$ in the case of ionizable lipids. If every group releases one cation this gives a cation concentration in the cell interior of $10^{-6}/r$ molar. For $r = 1\mu$ this corresponds to a $10^{-2}$molar solution. Pulses of protons or divalent cations even one or two orders of magnitude smaller could effectively alter the cell medium.

One way to activate this cation reservoir are membrane structural changes that alter the surface potential $\Psi^o$. Since the binding constants for protons or divalent cations to a charged surface depend sensitively on the value of $\Psi^o$ any change in $\Psi^o$ will cause a release or adsorption of ions. Relevant membrane structural changes are: (1) Alterations in the lateral distribution of neutral and charged lipids in the plane of the membrane (phase separation); (2) (Local) changes in membrane area as a result of conformational changes of membrane proteins or lipids.

To illustrate this mechanism we consider the effect of the area change at the thermal phase transition of MPA bilayers on the ionization of the phosphate groups. It is easy to predict qualitatively what will happen. An increase in membrane area reduces the charge density and consequently the surface potential $\Psi^o$. The expected effect is a release of protons from the membrane surface, or an increase in the degree of dissociation $\alpha$ at the ordered → fluid transition.

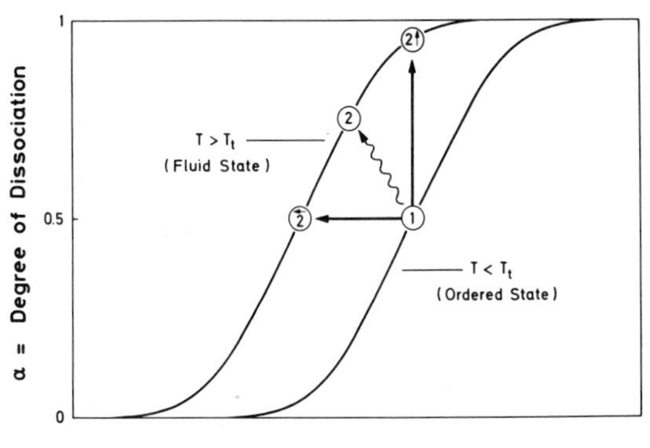


Fig. 11 Effect of a change in membrane area on the dissociation of the lipid phosphate groups; schematic.

Fig. 11 shows schematically the difference in the dissociation curve $\alpha$(pH) between the ordered or condensed state and the fluid or expanded state of the membrane. Starting from a value $\alpha_1$ in the ordered state there exist two extreme possibilities for the transition to a new $\alpha$-value on the dissociation curve of the expanded membrane. If the bulk pH is maintained constant there will be a large increase in $\alpha$ (transition 1 to 2↑); If $\alpha$ is maintained constant we obtain a relatively large decrease in pH (transition 1 to $\overleftarrow{2}$). The actual transition in an unbuffered or weakly buffered medium will be somewhere intermediate and the detailed theory will be developed in a forthcoming paper (Jähnig and Träuble, 1976). Here it suffices to describe the general line of thought.

Let us first estimate the shift in pK resulting from an increase in molecular area, $\Delta f$. Using the high-potential approximation $\psi^o = 2(kT/e)\ \ln\ (2\sigma/c)$ we obtain from eq.(6) the following representation for the dissociation curve $\alpha$(pH):

$$\alpha = \frac{K_o}{K_o + [H^+]_b\ (2\sigma/c)^2} \tag{19}$$

where $\sigma = e\alpha/f$. Setting $\alpha = 0.5$ we find for the shift in pK resulting from a change in molecular area, $\Delta f$:

$$\Delta pK = -\frac{2}{\ln 10}\ \frac{\Delta f}{f}\ . \tag{20}$$

For an area increase of $\Delta f/f = 0.3$ this yields $\Delta pK = -0.26$. The measured value of $\Delta pH$ will depend on the lipid concentration [L], on the initial value of $\alpha$ or on the bulk pH and on the ionic strength. At high pH and in an unbuffered medium small amounts of proton-accepting "impurities" will have a marked influence.

Starting point of the theory is the condition of electroneutrality for the whole system containing: ionizable lipid, electrolyte, $H^+$ and $OH^-$. This leads to the following relation between $\Delta\alpha$ and $\Delta[H^+]$ in an unbuffered medium

$$\Delta\alpha = \frac{1}{[L]}\left(1 + 10^{-14}/[H^+]_b^2\right)\Delta[H^+]_b \tag{21}$$

where [L] is the lipid concentration. The main step is now to calculate the change in $\alpha$ that results from a given change in membrane area. This is done on the basis of eq.(6). After a lengthy calculation to be presented elesewhere (Jähnig and Träuble, 1976), one finds

$$\Delta pH = \frac{F[L]}{\left([H^+] + [OH^-] + F[L]\right)}\ \frac{\alpha}{\sqrt{\ldots}}\ \frac{2}{\ln 10}\ \frac{\Delta N}{N} \tag{22}$$

where $$F = \frac{1}{\frac{1}{\alpha(1-\alpha)} + \frac{2N}{\sqrt{\cdots}}} \qquad (23)$$

and $$\sqrt{\cdots} = \sqrt{\alpha^2 N^2 + \left(\frac{\varepsilon}{2\pi}\frac{kT}{e^2}\kappa\right)^2} .$$

N is the number of ionizable groups per $cm^2$. Membrane expansion is equivalent to a decrease in N ($\Delta N$ negative) and should lead to a decrease in pH. For $\alpha = 0$ and $\alpha = 1$ F equals zero and thus $\Delta pH = 0$. Therefore the value of $/\Delta pH/$ is expected to go through a maximum as a function of bulk pH.

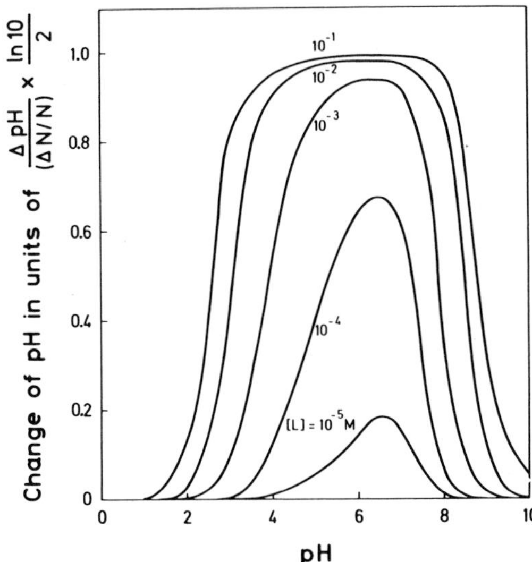


Fig. 12 Theoretical prediction for the change in pH at the ordered ↔ fluid transition of MPA bilayers for different pH. Bottom to top: increasing lipid concentration. Conditions: pK = 3.5, 0.2 M salt, T ≈ 300 K.

Fig. 12 shows the result of a numerical evaluation of equation (22) for pK = 3.5, n = 0.2 M and for different lipid concentrations. For $[L] \lesssim 10^{-3}$M $\Delta pH$ shows a maximum at around pH 6.5, the height of which increases with increasing lipid concentration. For larger values of [L] an extended plateau appears between pH 4 and pH 8.5 with a value of $\Delta pH \simeq -0.25$ for $\Delta N/N = -0.2$.

Fig. 13 shows a continuous recording of the pH in an unbuffered MPA-dispersion as a function of temperature. The ordered → fluid transition at about 35 °C produces the expected drop in the bulk pH with a value of $\Delta pH = -0.45$ under the conditions specified in the Caption. If such measurements are performed over the whole pH range one obtains the $\Delta pH$-pH dependence shown in Fig. 14 with a maximum around pH 4. The region below pH 3.5 - 4 represents the range of relatively large changes in proton concentration. This region is in satisfactory agreement with the predictions of the theory. The observed decrease in ΔpH for pH $\gtrsim$ 4.5 must have reasons not yet included in the theory. Systematic studies of this effect are in progress and will be published separately (Teubner and Träuble, in preparation).

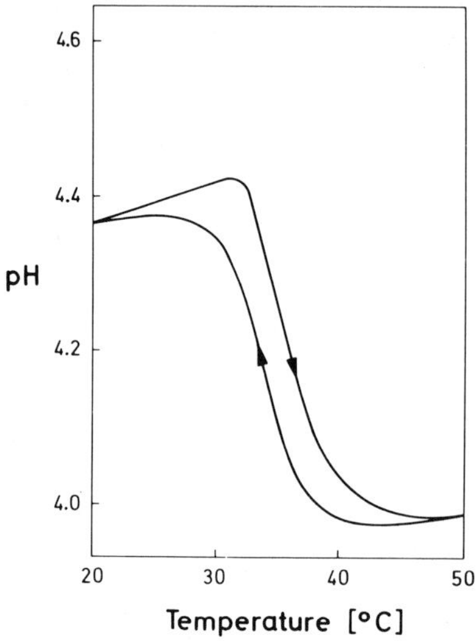


Fig. 13 Change in pH at the thermal transition of MPA bilayers. 0.2 M CsCl, 5 x $10^{-3}$M MPA. Continuous reading using a pH electrode. Protons are released when the membrane expands. This measurement was kindly provided by Dipl. Math. M. Teubner.

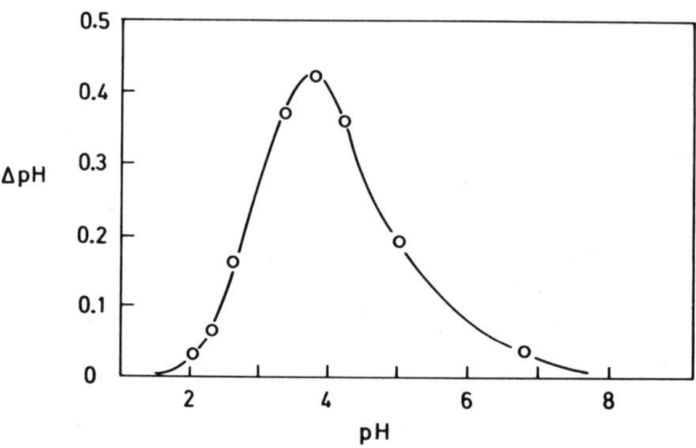


Fig. 14 Dependence of the pH-jump (ΔpH) at the ordered ↔ fluid transition of MPA on the bulk pH. Conditions: $10^{-2}$M MPA, 0.2 M NaCl. Apparent pK around 3.5.

A completely analogous treatment is possible for the case of $Ca^{++}$ binding or $Ca^{++}$ release resulting from membrane structural changes. In the theory one has to replace the proton dissociation curve, $\alpha$(pH), by the binding curve for $Ca^{++}$ to acidic lipids. The theoretical result can be written especially simply if we introduce a parameter $q$ denoting the ratio of free binding sites to total binding sites for $Ca^{++}$. The final result for the change in free $Ca^{++}$ content (symbol $Ca_o^{++}$) is

$$\frac{\Delta Ca_o^{++}}{Ca_o^{++}} = - \frac{4\,[L]}{Ca_o^{++}\,\frac{(5-4q)}{q(1-q)} + [L]}\;\frac{\Delta N}{N} \tag{24}$$

where [L] is the concentration of a lipid bearing one negative charge per polar head. The influence of the electrostatics on $Ca^{++}$ binding is implicit in the value of q. Again the theory predicts a release of calcium from the membrane as a result of an area increase (ordered → fluid transition). And according to eq.(24) this effect goes through a maximum when about 3/4 of the binding sites are occupied.

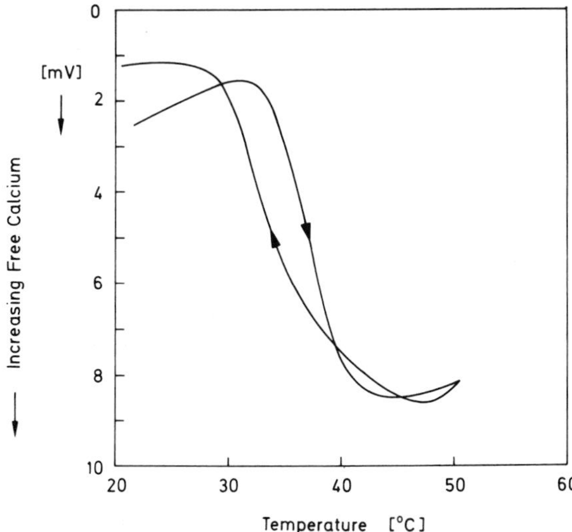


Fig. 15 Release of $Ca^{++}$ from the surface of MPA bilayers at the ordered→ fluid transition. Increasing readings in mV correspond to increasing free $Ca^{++}$ concentration. Conditions: Orion Ca-electrode, pH 9, 2 x $10^{-3}$M MPA, 0.2 M NaCl. The added calcium (2 x $10^{-4}$M) is quantitatively bound below the transition. Release: about 1 x $10^{-4}$M.

Fig. 15 shows the result of preliminary measurements of the free $Ca^{++}$ concentration in a MPA dispersion as a function of temperature using a $Ca^{++}$-sensitive electrode. The membrane expansion at around 38 °C causes the expected release of $Ca^{++}$ from the membrane surface.

The conclusion from these studies is that a membrane with ionizable groups can in fact function as a reservoir for protons and divalent cations. Changes in membrane structure lead to pulses of ions either from or to the membrane surface. This mechanism may be especially important in the case of divalent cations in view of the known regulatory function of these ions in many biological processes (cf.Cluthbert, 1970).

## 5. ELECTROSTATIC REGULATION OF MEMBRANE PHASE SEPARATIONS

Phase separation phenomena have been studied by metallurgists for more than hundred years. The theoretical foundation for much of our present understanding of phase stability was laid by J.W. Gibbs ninety years ago in his monumemtal work "On the Equilibrium of Heterogeneous Phases". Membrane phase separations are being studied since about three years and they are regarded as a new, exciting phenomenon (Shimshick and McConnell, 1973; Wu and McConnell, 1975; Ohnishi,1975; Lee, 1975; Galla and Sackmann, 1975).

If two components A and B coexist within a two-dimensional lattice their behavior depends on the difference between the interaction of AB pairs compared with AA and BB pairs. The energy gained by the creation of one AB nearest neighbour pair may be written as

$$w = U_{AB} - 1/2\,(U_{AA} + U_{BB}) \qquad (25)$$

where $U$ designates the energy per bond from one component to its $z$ neighbours. $w < 0$ means that unequal neighbours are preferred, i.e. the mixed state is favoured; $w > 0$ means that equal neighbours are preferred, i.e. separate phases are favoured. For disaturated phospholipids with identical polar heads a difference in chain length of more than 2 methylene groups seems to be necessary for the formation of separate phases (Phillips et al., 1972). Only if $w > 0$ and if the value of $w$ is large enough compared with the decrease in entropy that accompanies phase separation does the system form separate phases. Both the enthalpy and the entropy term depend on the proportions of the two components, which will be designated in the following by the fraction of A-molecules $\mu = N_A/N_T$.

As is shown in standard texts (Becker, 1966; Rao, 1970), enthalpy and entropy contributions may be written in the following way to give the molar change in Gibbs free energy on mixing:

$$G^m = RT\,\{\mu\ln\mu + (1-\mu)\ln(1-\mu)\} + Lzw\,\mu(1-\mu) \qquad (26)$$

where the first term accounts for the change in entropy and the second represents the mixing enthalpy (cf. eq.(25)); $z$ denotes the number of nearest neighbours.

Fig. 16a shows these two terms as a function of $\mu$ for $w > 0$ together with $G^m$-curves for different values of $r = Lzw/RT$. The combined curve exhibits two minima if $r > r_c$, or if $w$ is large enough, or if the temperature is below a critical value

$$T_c = \frac{Lz}{2R}\,w\,. \qquad (27)$$

For $T < T_c$ and for concentrations between those of the two minima the system separates spontaneously into two phases of different chemical position. The relation to the phase diagram in Fig. 16b is self-explanatory.

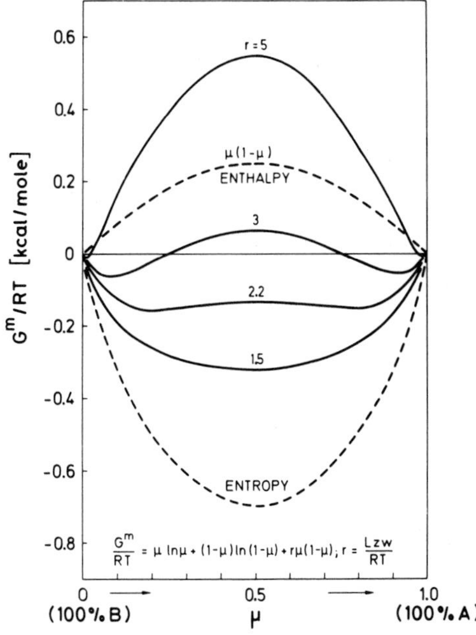


Fig. 16a Change in free energy on mixing two components A and B, μ = fraction of A. Solid lines: Combinations of enthalpy and entropy with increasing weight (r) of the enthalpy term.

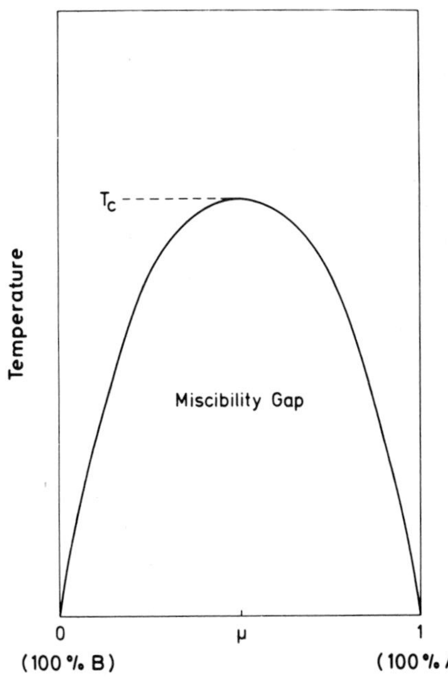


Fig. 16b Corresponding phase diagram with miscibility gap for $T < T_c$. The solid line is determined by the minima in Fig. 16a.

The occurence of phase separations in membranes is definitely an interesting phenomenon, because it implies physical and chemical inhomogeneities in the plane of the membrane, segregation of proteins, differences in surface pH etc. Even more interesting is the question of which parameters can be used to regulate membrane phase separations. In this context electrostatic interactions come into play. The question arises how the phase diagram of a two-component system is altered if one of the lipids is charged.

We know that surface charges give rise to an extra electrostatic free energy (eq.(10)), and from Section 3 it is clear that for a two-component system this energy has a minimum if the lipids are randomly distributed. Therefore the electrostatic free energy acts as a kind of "mixing affinity" that supports the entropy in producing a random distribution.

To quantify this statement we consider the effect of the electrostatic free energy on the phase diagram of a two-component membrane containing neutral and charged lipids. We calculate the electrostatic free energy (1) for the completely separated phases and (2) for the mixed membrane. We denote by $N_A$ the number of charged lipids per cm² and by $N_B$ the number of neutral lipids ($N_T = N_A+N_B$); thus $\mu = N_A/N_T$ is the fraction of charged lipids. If the two species have equal molecular areas $f$, then the average area over which the charges are smeared out is $F_s = f\,N_A$ in case 1 (s for separate phases), and $F_m = f\,N_T$ in case 2 (m for mixed). The corresponding charge densities are $\sigma_s = e\alpha/f$ and $\sigma_m = \mu e\alpha/f$. Denoting with $\phi_s$ and $\phi_m$ the corresponding electrostatic free energies per cm² we obtain for the difference in the molar electrostatic free energy between the two cases

$$\Delta G^{m,el} = G^{el}_s - G^{el}_m = \frac{L}{N_T}\,(\phi_s F_s - \phi_m F_m)\,, \qquad (28)$$

where $\phi$ has to be substituted according to eq.(10). This yields

$$\frac{\Delta G^{m,el}}{L} = 2\,kT\alpha\,\mu\left(\sinh^{-1}(\sigma_s/c) - \sinh^{-1}(\sigma_m/c)\right) - \frac{f\varepsilon}{\pi}(kT/e)^2\,\kappa(1-\mu) - \frac{f\varepsilon}{\pi}(kT/e)^2\,\kappa\left(\mu\cosh(\sinh^{-1}\sigma_s/c) - \cosh(\sinh^{-1}\sigma_m/c)\right) \qquad (29)$$

where $\kappa$ and $c$ are given by eqs.(1) and (3). Recalling that $c$ and $\kappa$ are proportional to $\sqrt{n}$ ($n$ = ionic strength), and $\sigma = e\alpha/f$ we see that $\Delta G^{m,el}$ depends on the ionic strength $n$, the charge per polar head $e\alpha$ (which in turn depends on the pH and the ionic strength) and the mixing ratio expressed by the fraction $\mu$ of A-molecules.

To obtain some feeling for these dependences we have evaluated eq.(29) numerically. Fig. 17 shows $\Delta G^{m,el}$ as a function of $\mu$ and $n$ for $\alpha = 0.5$ and $\alpha = 1$. This graph is drawn for $T = 300$ K, $\varepsilon = 80$ and $f = 50$ Å$^2$. $\Delta G^{m,el}$ is always positive which means that the charges favour the mixed state, or that $\Delta G^{m,el}$ supports the entropy in producing mixing. As one would expect $\Delta G^{m,el}$ is equal to zero for $\mu = 0$ and $\mu = 1$ and has a maximum for $\mu = 0.5$ when charged and neutral lipids have equal proportions. $\Delta G^{m,el}$ increases

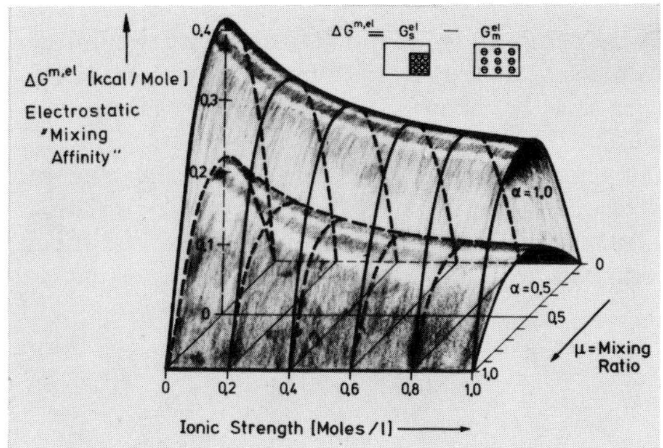


Fig. 17 Effect of the electrostatic free energy on the mixing behavior of neutral and charged lipids. The separate phases have always a higher electrostatic free energy ($\Delta G^{m,el} > 0$). $\Delta G^{m,el}$ is shown as a function of ionic strength and mixing ratio for one and 0.5 elementary charges on the ionizable lipids.

with decreasing ionic strength. For $\alpha = 1$, 0.1 M salt and a molar ratio 1:1 the electrostatic free energy of the mixed state is 0.35 kcal/mole lower than for separated phases.

Consider now the effect of this electrostatic term on the phase behavior of two lipids A and B which differ sufficiently in chain length, degree of unsaturation, or polar group structure so that they form separate phases in the absence of charges. If we "switch on" the charges on species A (for example by an increase in pH) the resulting electrostatic mixing affinity tends to produce mixing; a subsequent increase in ionic strength reduces the value of $\Delta G^{m,el}$, and this may reverse the process, leading again to separate phases. This shows that the electrostatic energy is an effective parameter to regulate the lateral organization of charged membranes.

Since the influence of $\Delta G^{m,el}$ is comparable with that of the mixing entropy its effect on the phase diagram will be to lower the critical temperature $T_c$ of the miscibility gap (cf. Fig. 16). To quantify this we add $\Delta G^{m,el}$ to eq.(26) and calculate the effective value of $T_c$ which may be written as

$$T_c^* = T_c - \Delta T_c$$

where $T_c$ is the critical temperature in the absence of electrostatic effects ($T_c = Lzw/2R$). To facilitate the calculation we replace the exact expression for $\Delta G^{m,el}$ (eq.(29)) by a simpler analytical form

$$\Delta G^{m,el} = 22.914 \; f \; L \; (1 - 0.396\sqrt{n}) \; \alpha^2 \mu \, (1-\mu) \tag{30}$$

which was deduced in a purely empirical way by curve-fitting and which yields a good description of the profile in Fig. 17 over a broad temperature range around $T = 300$ K. Eq.(30) gives the value of $\Delta G^{m,el}$ in cal if n is inserted as molar salt concentration and the molecular area f is taken in $cm^2$. L is Avogadro's number, $\alpha$ the degree of dissociation and $\mu$ the mixing ratio.

The new value of $T_c$ is found from the condition that the two minima and the point of inflection in the $G^m(\mu)$-curve coincide (cf. Fig. 16a). This is the standard procedure for the calculation of critical temperatures of mixed systems (cf. Rao, 1970). Thus we apply

$$\frac{\partial\ (G^m + \Delta G^{m,el})}{\partial \mu} = 0, \qquad \frac{\partial^2\ (G^m + \Delta G^{m,el})}{\partial \mu^2} = 0$$

with $G^m$ according to eq.(26) and $\Delta G^{m,el}$ according to eq.(30). We find

$$\Delta T_c = -\frac{Lf}{2R}\, 23\ (1 - 0.4\sqrt{n})\ \alpha^2\ . \tag{31}$$

For $\alpha = 1$ and $n \to 0$ we obtain as extremum $\Delta T_c = -\ 105\ ^oC$ which shows that the surface electrostatics may have a marked effect on the phase diagram. Since $\Delta G^{m,el}$ and consequently $\Delta T_c$ depend on pH and ionic strength, it is possible to regulate the lateral organization of the system at a given temperature by changing these parameters.

The expected effects of salt and pH may be illustrated by plotting $T_c$ as a function of pH for different salt concentrations. Here it is important to take into account that the degree of dissociation $\alpha$ depends not only on the pH but also on the salt concentration as discussed in Section 2. On the basis of eq.(6) we may approximate the dissociation curve of MPA by

$$\alpha = 0.5 \tanh 0.8\ (pH - pK) + 0.5,$$
and using eq.(7) for the apparent pK (32)
$$\alpha = 0.5 \tanh 0.8\ (pH - 2.61 - \log_{10} n) + 0.5\ .$$

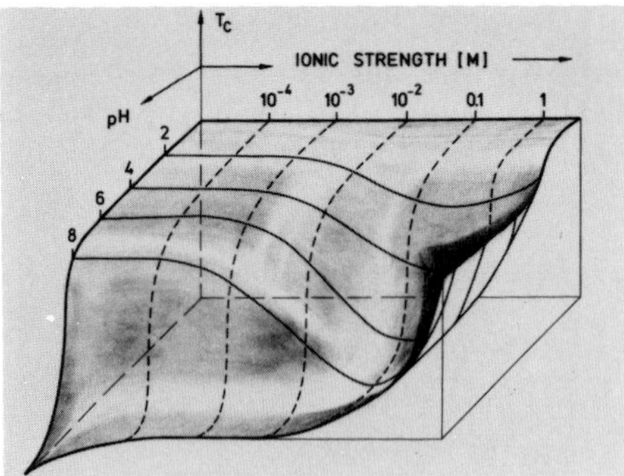


Fig. 18 Effect of surface electrostatics (pH and ionic strength) on the critical temperature $T_c$ of the miscibility gap of a two-component system (cf. Fig. 16b).

Using this relation we have evaluated the dependence of $T_c$ on pH and salt according to eq.(31) for MPA, and this is shown in Fig. 18. An increase in pH leads to a decrease of $T_c$ in the region of the pK which is around 8 for $n = 10^{-5}$M. With increasing ionic strength the pK decreases and the net decrease in $T_c$ becomes smaller because of the factor $(1 - 0.4\sqrt{n})$ in eq.(31). This behavior is to be expected as a result of "switching-on" of the charges. If we consider the effect of increasing salt, for example at pH 5, there is an initial decrease in $T_c$ followed by an increase at higher salt concentration. The decrease is a result of the salt-induced increase in the degree of dissociation; the subsequent increase in $T_c$ is related to the decrease of the electrostatic free energy with increasing ionic strength. Therefore if a system shows two phases at low salt concentration at the given temperature, increasing salt concentration may first induce mixing, and a further increase in salt concentration may cause phase separation again.

The essence of this chapter is that the electrostatic free energy represents a "negative" regulatory factor for the phase behavior of heterogeneous membranes. "Negative" has the sense that the lipids must be sufficiently different to form separate phases in their neutral form. Switching on of the charges on lipid A may then induce mixing of the two components. Conversely, if a membrane contains neutral and charged lipids randomly distributed it is possible to induce phase separation by reducing the value of $\Delta G^{m,el}$, for example by an increase in ionic strength or a reduction in pH.

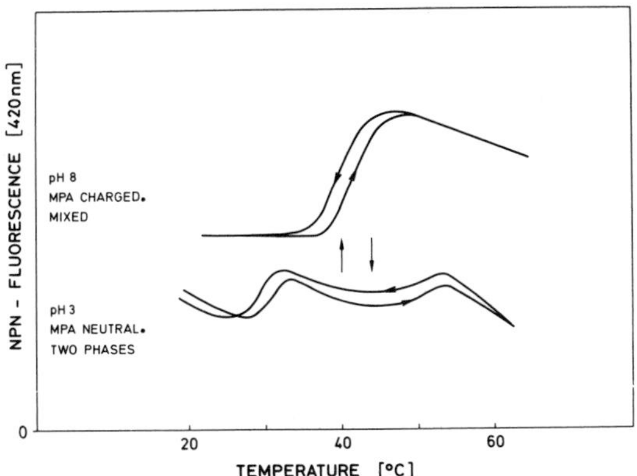


Fig. 19 Effect of pH on the phase behavior of an equimolar mixture of distearoyl-lecithin and dimyristoyl-MPA. Two phase transitions are seen in the NPN fluorescence intensity at low pH in the absence of charges. Switching-on of the charges on MPA (pH 8) induces mixing. $5 \times 10^{-4}$M lipid, $2 \times 10^{-3}$M NaCl, $5 \times 10^{-6}$M NPN.

The validity of this concept has been checked experimentally with mixtures of distearoyl-lecithin and dimyristoyl-methyl phosphatidic acid (MPA). Fig. 19 illustrates that at high pH, where MPA is negatively charged,the system shows only one thermal phase transition (ideal mixture of the two components) whereas at low pH, where MPA is electrically neutral, the system exhibits two distinct transitions indicating the existence of two separate phases.

## 6. ELECTROSTATIC COUPLING BETWEEN THE TWO LAYERS OF A MEMBRANE

The lipid bilayer as a structural unit of biological membranes has become a sanctuary which is worshipped by most membrane biologists. However, some authors treat the bilayer as a structural entity whereas others tacitly assume that it consists of two more or less independent monolayers. Are these two layers coupled? Do changes in the electrostatics on one side influence the surface electrostatics on the other side? Can the two layers independently undergo lipid phase transitions? Does phase separation in one layer alter the lateral distribution in the other layer? These are the questions to be discussed in the following.

As a starting point we consider the potential profile and the electrostatic free energy of a lipid membrane with charge densities $\sigma_1$ and $\sigma_2$ on the two sides (cf. Fig. 20). The membrane is assumed

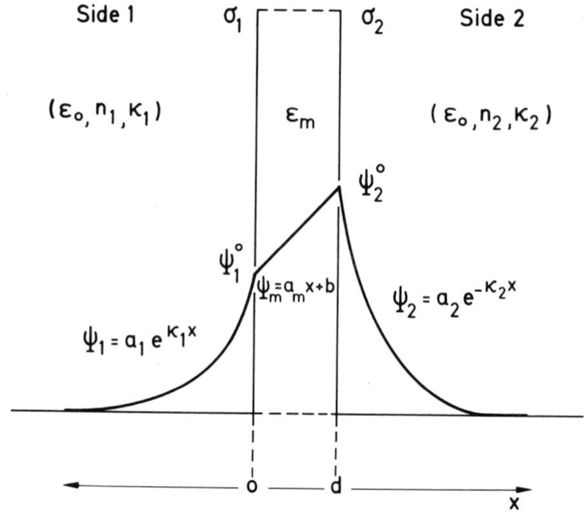


Fig. 20 Asymmetric lipid bilayer with charge densities $\sigma_1$ and $\sigma_2$ on the two surfaces. In the linearized case the electrical potential $\Psi$ follows an exponential law in the regions of the diffuse layers on the two membrane surfaces. A potential difference $\Delta\Psi^o = \Psi_2^o - \Psi_1^o$ and a corresponding electrical field $E = \Delta\Psi^o/d$ are present in the membrane.

to be impermeable to ions so that the potential profile is determined by the surface charges and the electrolytes on the two sides. The potential profile of such a membrane has been calculated by Ohki (Ohki, 1971, 1976) by solving Poisson's equation for the situation shown in Fig. 20.

For low potentials the Poisson equation can be linearized and the electrical potential in the region of the diffuse layers follows an exponential law. A linear potential profile exists within the membrane interior. Solution of Poisson's equation with the boundary conditions corresponding to the situation in Fig. 20 leads to the following expressions for the surface potentials $\Psi_1^o$ and $\Psi_2^o$:

$$\Psi_1^o = \frac{4\pi}{\varepsilon_o \kappa_1}\left\{\overbrace{\frac{(1+\kappa_2\, d\, \varepsilon_o/\varepsilon_m)}{(1+\kappa_2/\kappa_1+\kappa_2\, d\, \varepsilon_o/\varepsilon_m)}}^{a_1}\sigma_1 + \overbrace{\frac{1}{(1+\kappa_2/\kappa_1+\kappa_2\, d\, \varepsilon_o/\varepsilon_m)}}^{b_1}\sigma_2\right\} \tag{33a}$$

$$\Psi_2^o = \frac{4\pi}{\varepsilon_o \kappa_2}\left\{\underbrace{\frac{(1+\kappa_1\, d\, \varepsilon_o/\varepsilon_m)}{(1+\kappa_1/\kappa_2+\kappa_1\, d\, \varepsilon_o/\varepsilon_m)}}_{a_2}\sigma_2 + \underbrace{\frac{1}{(1+\kappa_1/\kappa_2+\kappa_1\, d\, \varepsilon_o/\varepsilon_m)}}_{b_2}\sigma_1\right\} \tag{33b}$$

where $\kappa_1$ and $\kappa_2$ are the Debye constants (cf. eq.(1)) for the two sides and $\varepsilon_o$ and $\varepsilon_m$ are the dielectric constants of the bulk solution and the membrane interior, respectively.

For identical electrolytes on the two sides ($n_1 = n_2$; $\kappa_1 = \kappa_2 = \kappa$) the surface potentials are

$$\Psi_1^o = \frac{4\pi}{\varepsilon_o \kappa}(a\sigma_1 + b\sigma_2), \tag{34a}$$

$$\Psi_2^o = \frac{4\pi}{\varepsilon_o \kappa}(a\sigma_2 + b\sigma_1), \tag{34b}$$

with

$$a = \frac{(1 + \kappa\, d\varepsilon_o/\varepsilon_m)}{(2 + \kappa\, d\varepsilon_o/\varepsilon_m)}, \tag{35}$$

$$b = \frac{1}{(2 + \kappa\, d\varepsilon_o/\varepsilon_m)}. \tag{36}$$

We note that for $\varepsilon_o = 80$, $\varepsilon_m \approx 10$, $d = 30$ Å and $1/\kappa = 30$ Å, $b$ is small compared to $a$ ($b = 0.1$, $a = 0.9$).

For comparison, the surface potential of an isolated surface is given by

$$\psi^o = \frac{4\pi}{\varepsilon_o \kappa} \sigma$$

in the linear case instead of by eq.(2).

The important result in this context is that the surface potential on one side depends also on the charge density on the other side. The low value of $b$ can be compensated by an asymmetry in the charge densities $\sigma_1$ and $\sigma_2$ which seems to exist for real membranes. In other words, even in the absence of charges on side two there exists a surface potential $\psi_2^o = (4\pi/\varepsilon_o\kappa)\, b\, \sigma_1$. An electrical field $E = \Delta\psi^o/d$ is present in the membrane interior, where the potential difference $\Delta\psi^o$ is given by

$$\Delta\psi^o = \psi_2^o - \psi_1^o = \frac{4\pi d}{\varepsilon_m(2 + \kappa\, d\, \varepsilon_o/\varepsilon_m)} (\sigma_2 - \sigma_1). \tag{37}$$

We note that in the general case, when $n_1 \neq n_2$ (or $\kappa_1 \neq \kappa_2$) a potential difference exists even when $\sigma_2 = \sigma_1$.

Because of the influence of the charge density $\sigma_1$ on the surface potential on the opposite side it is clear that a change in $\sigma_1$ , for example as a consequence of $Ca^{++}$-binding, influences the ionization reactions on the opposite side according to eq.(6). This means that in a cell the surface processes in the cell interior (and also field-dependent processes within the membrane) are directly affected by the surface electrostatics on the outside. Whether this effect is important in biological systems has still to be established.

Further insight into the electrostatic coupling between the two layers is obtained by calculating the electrostatic free energy $G^{el}$ of the membrane shown in Fig. 20. For this purpose we apply the charging process (cf. eq.(9)) to both membrane surfaces whereby we assume that the surface charges on the two sides are simultaneously increased from zero to their final values $\sigma_1$ and $\sigma_2$, respectively. A parameter $\lambda$ is introduced which increases from zero to unity in the course of the charging process.

This gives

$$G^{el} = \int_0^1 \Psi_1^o(\lambda\sigma_1, \lambda\sigma_2)\, d(\lambda\sigma_1) + \int_0^1 \Psi_2^o(\lambda\sigma_2, \lambda\sigma_1)\, d(\lambda\sigma_2), \tag{38}$$

and expressing $\Psi_1^o$ and $\Psi_2^o$ by eqs.(33a) and (33b) we obtain

$$G^{el} = \underbrace{\frac{2\pi}{\varepsilon_o\kappa_1} a_1\sigma_1^2}_{\phi_1} + \underbrace{\frac{2\pi}{\varepsilon_o\kappa_2} a_2\sigma_2^2}_{\phi_2} + \underbrace{\frac{2\pi}{\varepsilon_o}\left(\frac{b_1}{\kappa_1} + \frac{b_2}{\kappa_2}\right)\sigma_1\sigma_2}_{G^c} \tag{39}$$

$$= \phi_1 + \phi_2 + G^c.$$

For identical 1:1 electrolytes on the two sides ($\kappa_1 = \kappa_2 = \kappa$, $a_1 = a_2 = a$, $b_1 = b_2 = b$) this reads

$$G^{el} = \frac{2\pi a}{\varepsilon_o\kappa}(\sigma_1^2 + \sigma_2^2) + \frac{4\pi b}{\varepsilon_o\kappa}\sigma_1\sigma_2 . \tag{40}$$

The two terms $\phi_1$ and $\phi_2$ in eq.(39) represent the surface energies of the two sides; the term $G^c$ is denoted in the following as "coupling energy". This notation becomes clearer in the following paragraph where we investigate the influence of a phase separation in layer one on the lateral distribution in layer two. We shall see that the coupling energy $G^c$ has the effect that a charge separation on side one creates a similar charge pattern on side two of the membrane.

We consider again a planar lipid bilayer separating two compartments 1 and 2 with solutions of a 1:1 electrolyte of molar ion concentrations $n_1$ and $n_2$. The dielectric constants of the membrane interior and of the aqueous solution are $\varepsilon_m$ and $\varepsilon_o$, respectively. We assume that layer one contains two lipids A and B where A carries an ionizable proton at its polar moiety. Lipid A is electrically neutral at low pH (symbol A) and negatively charged at high pH (symbol $A^-$). The two lipids shall differ sufficiently in their chemical structure (chain length, degree of unsaturation, polar groups) so that they form separate phases in the plane of the membrane when A is uncharged. Layer two contains also charged and neutral lipid molecules A′ and B′ which may or may not have an "intrinsic" tendency to form separate phases. All of the lipid molecules can move in the plane of the membrane (lateral diffusion).

Suppose now that by one of the mechanisms discussed in Section 5 phase separation is induced in layer one so that after phase separation the membrane exhibits two areas U and L (upper and lower in Fig. 21) with different charge densities on side one. In Fig. 21b it is assumed that during phase separation on side one charged molecules have migrated from area U1 to area L1 in exchange for neutral molecules. Let $\sigma_1^o$ be the charge density on side one in the beginning (cf. Fig. 21a) then the charge density will decrease in area U1 and increase in area L1. Fig. 21b shows schematically the resulting charge density profile on side one.

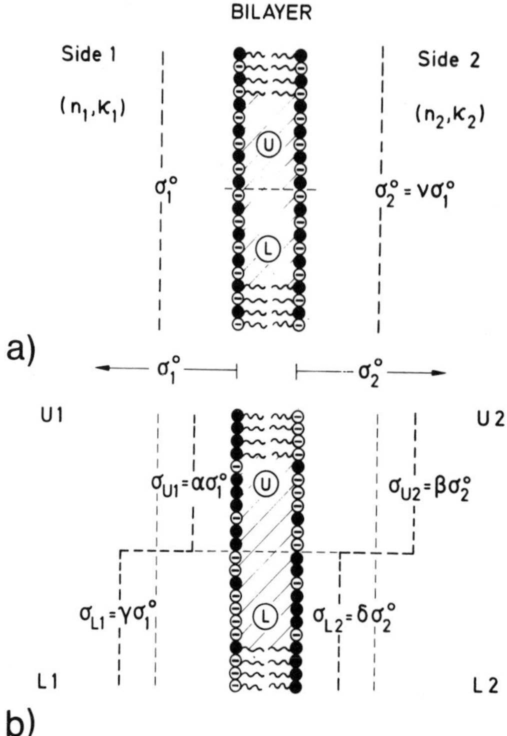


Fig. 21 Phase separation on side one of a lipid bilayer containing charged and neutral molecules (schematic). The vertical lines (--) represent the charge density in the plane of the membrane.

a) Homogeneous charge distribution on both sides before phase separation; charge densities $\sigma_1^o$ and $\sigma_2^o$.

b) Migration of charged lipids from area U1 to area L1 leads to a decrease of the charge density in area U1 and a corresponding increase in area L1. It is assumed that side two responds with a migration of charges from area L2 to area U2.

How does this process alter the electrostatic free energy of the rest of the membrane? The decrease of $\sigma$ in area U1 and the corresponding increase in area L1 lead to a decrease of the coupling energy $G^c$ in part U and a corresponding increase of $G^c$ in part L of the membrane. This is an equivalent of the decreased (or increased) repulsion between the charges on the opposite membrane sides in part U (or L) of the membrane. One now expects that a transfer of charges on side two from area L2 to area U2 can lower the total coupling energy. At the same time, however, the surface energy $\phi_2$ which tends to spread the charges homogeneously over the surface will increase.

The charge pattern that is eventually established in layer two can be calculated by minimization of the total free energy of the system. For this purpose we shall consider the charge distribution on side one as given (external constraint) and calculate the charge distribution on side two which corresponds to the lowest total free energy. For the sake of simplicity we assume that the two parts U and L of the membrane have equal areas of 1 cm² each. To derive an explicit expression for the free energy we introduce the following notations for the charge densities in the different parts (U1, L1, U2, L2) of the membrane (cf. Fig. 21 ). The charge densities on side 1 and 2 before phase separation (Fig. 21a) are denoted by

$$\sigma_1^o \text{ and } \sigma_2^o = \nu\sigma_1^o \tag{41}$$

where $\nu$ accounts for an asymmetry in the initial charge density. The charge densities in the areas U1 and L1 after phase separation (Fig. 21b) are

$$\begin{aligned} \sigma_{U1} &= \alpha\sigma_1^o, \\ \text{and } \sigma_{L1} &= \gamma\sigma_1^o. \end{aligned} \tag{42}$$

If during phase separation charged molecules ($A^-$) migrate from area U1 to L1 in exchange for neutral molecules (B) then the resulting charge density will be smaller in area U1 ($\alpha < 1$) and larger in area L1 ($\gamma > 1$). For a constant number of charged molecules

$$\gamma = 2 - \alpha. \tag{43}$$

Thus the value of $\alpha$ defines the charge pattern on side one which is considered as constant (external constraint).

Corresponding notations are used for the charge densities on side two:

$$\begin{aligned} \sigma_{U2} &= \beta\sigma_2^o, \\ \text{and } \sigma_{L2} &= \delta\sigma_2^o. \end{aligned} \tag{44}$$

For a constant number of charged molecules

$$\delta = 2 - \beta. \tag{45}$$

Obviously, the charge distribution on side two is described by the value of $\beta$. $\beta = 1$ is the case of homogeneous charge distribution and $\beta > 1$ means that charges have been transferred from L2 to U2. Our aim is now to calculate the "response function" $\beta(\alpha)$ by minimization of the total free energy of the membrane with respect to $\beta$ for a given value of $\alpha$.

Using the notations (41) through (45) and eq.(39) we may now express the electrostatic free energy of the membrane in the low-potential approximation as

$$G^{el} = G_U^{el} + G_L^{el}$$

$$= \underbrace{\phi_{U1} + \phi_{U2} + G_U^c}_{} + \underbrace{\phi_{L1} + \phi_{L2} + G_L^c}_{}, \quad \text{or}$$

$$G^{el} = \underbrace{\frac{2\pi}{\varepsilon_o\kappa_1} a_1(\alpha\sigma)^2}_{\phi_{U1}} + \underbrace{\frac{2\pi}{\varepsilon_o\kappa_2} a_2(\beta\nu\sigma)^2}_{\phi_{U2}} + \underbrace{\frac{2\pi}{\varepsilon_o}\left(\frac{b_1}{\kappa_1} + \frac{b_2}{\kappa_2}\right)\alpha\sigma\ \beta\nu\sigma}_{G_U^c} \tag{46}$$

$$+ \underbrace{\frac{2\pi}{\varepsilon_o\kappa_1} a_1(\gamma\sigma)^2}_{\phi_{L1}} + \underbrace{\frac{2\pi}{\varepsilon_o\kappa_2} a_2(\delta\nu\sigma)^2}_{\phi_{L2}} + \underbrace{\frac{2\pi}{\varepsilon_o}\left(\frac{b_1}{\kappa_1} + \frac{b_2}{\kappa_2}\right)\gamma\sigma\ \delta\nu\sigma}_{G_L^c}.$$

To this expression we have to add a term that accounts for the change in entropy during the redistribution of the molecules on areas U2 and L2. And perhaps we wish to include an intrinsic tendency of the molecules A' and B' towards phase separation. The proper expression that accounts for these influences is already given in Section 5, eq.(26). For a lipid layer of 1 cm² and a molecular area $f$ of the individual lipid molecules this expression reads

$$G^m = \frac{kT}{f}\left\{\mu\ln\mu + (1-\mu)\ln(1-\mu)\right\} + \frac{kT}{f} r\mu(1-\mu) \tag{47}$$

where $r = \frac{zw}{kT}$.

We recall that $G^m$ is the change in Gibbs free energy if two components A and B are mixed and $\mu$ denotes the fraction of A molecules.

Applied to a membrane with a total area of 2 cm² (one cm² for each area U2 and L2) we obtain

$$\Delta G^m = -2G^m(\mu_2^o) + G^m(\mu_{U2}) + G^m(\mu_{L2}) \tag{48}$$

as an expression for the change in $G^m$ that results from an unequal distribution of the charges on the two areas. Here $\mu_2^o$ denotes the concentration of the charged lipids on side two before phase separation, and $\mu_{U2}$ and $\mu_{L2}$ are the respective concentrations after phase separation. These quantities are related to the respective charge densities by

$$\mu_2^o = \frac{f}{e}\,\sigma_2^o \qquad \text{for the initial state,} \tag{49}$$

and

$$\mu_{U2} = \beta\mu_2^o \tag{50}$$

after phase separation.

$$\mu_{L2} = (2-\beta)\,\mu_2^o \tag{51}$$

Combining eqs.(46) through (51) we obtain the total free energy of the system as

$$G^T = G^{el} + \Delta G^m \, . \tag{52}$$

The response function $\beta(\alpha)$ is calculated from the condition

$$\partial G^T/\partial\beta = 0 \, . \tag{53}$$

Assuming identical electrolytes on the two sides, i.e. $\kappa_1 = \kappa_2 = \kappa$, $a_1 = a_2 = a$, $b_1 = b_2 = b$, we obtain

$$\alpha = 1 + \nu\left(\frac{a}{b}\right)(1-\beta) - r\,\frac{\varepsilon_o}{2\pi}\frac{kT}{e^2}\,f\,\frac{\kappa\nu}{b}\,(1-\beta) - \frac{\varepsilon_o}{8\pi}\frac{kT}{e}\frac{\kappa}{b\sigma}\,\ln\frac{\beta[1-(2-\beta)\mu]}{(2-\beta)\,[1-\beta\mu]} \tag{54}$$

where $\mu \equiv \mu_2^o$ (cf. eq.(49)) and $\sigma \equiv \sigma_2^o$. The values of $a$ and $b$ are defined in eqs.(35) and (36).

The last term in eq.(54) stems from the "mixing entropy" and the term with the prefactor $r$ from the intrinsic tendency towards phase separation in layer two. The main parameters are: the charge density $\sigma$; the value of $\nu$ describing the membrane asymmetry with respect to charge density; the Debye-Hückel parameter $\kappa$ describing the influence of the ionic strength $(\kappa \propto \sqrt{n})$, and the value of $r$ characterizing an intrinsic phase-separation-tendency in layer two.

To get a first impression of the meaning of eq.(54) we neglect the entropy term and set $r = 0$. This gives

$$\beta = 1 + \frac{1}{\nu}\left(\frac{b}{a}\right)(1-\alpha) \tag{55}$$

$$= 1 + \frac{1}{\nu\,(1+\kappa\,d\,\varepsilon_o/\varepsilon_m)}\,(1-\alpha)$$

which shows that in the simplest case the response of side two depends on two parameters: the salt concentration $(\kappa \propto \sqrt{n})$ and the asymmetry parameter $\nu$. A special solution of eq.(55) is $\beta = 1$, $\alpha = 1$ which is the case of homogeneous charge distribution on both sides.

Eq.(55) states that a decrease in $\alpha$ (describing a decrease in charge density in area U1) causes an increase in $\beta$ or an increase in the charge density on the opposite side of the membrane. This is in accord with the naive picture that charges of equal sign on opposite sides of the membrane repel each other.

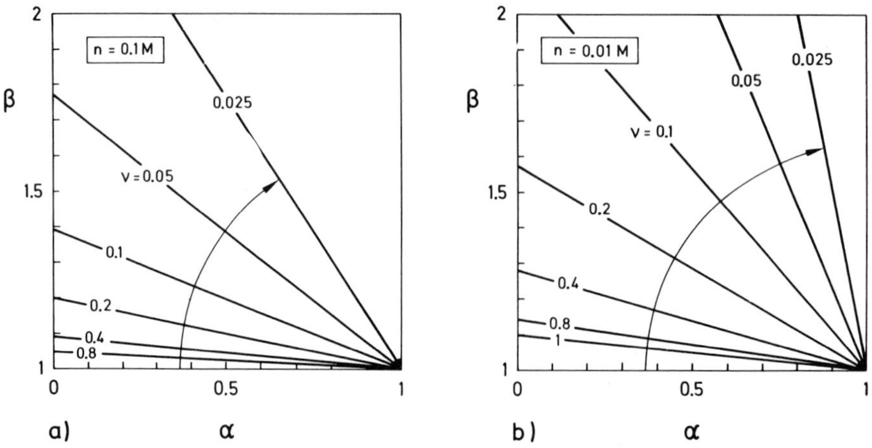


Fig. 22 Response of lipid layer two to a phase separation in layer one. Linearized theory without entropy. $\alpha = 1$, $\beta = 1$ is the case of homogeneous charge distribution on both membrane sides (cf. Fig. 21b). Charge depletion in area U1 on side one ($\alpha < 1$) causes charge concentration on the opposite side ($\beta > 1$) in area U2. Arrows indicate increasing membrane asymmetry (relative higher charge density on side one compared to side two).

Fig. 22 illustrates the response function $\beta(\alpha)$ according to eq.(55) for $n = 0.1$ M and 0.01 M and for different values of $\nu$. The sensitivity $|d\beta/d\alpha|$ increases with increasing membrane asymmetry (decreasing $\nu$) and decreasing salt concentration. For small values of $n$ and $\nu$ a small decrease in $\alpha$ from $\alpha = 1$ causes complete response ($\beta \to 2$) whereas for larger values of $n$ and $\nu$ side two shows only an incomplete response ($\beta < 2$ for $\alpha = 0$). Inspection of eq.(39) shows the reason for the increasing response with increasing membrane asymmetry or decreasing values of $\nu$. The surface energy $\phi_2$ shows a quadratic dependence on $\nu$ whereas the coupling energy $G^c$ depends linearly on $\nu$. Therefore decreasing $\nu$ leads to a relative increase of the coupling energy $G^c$.

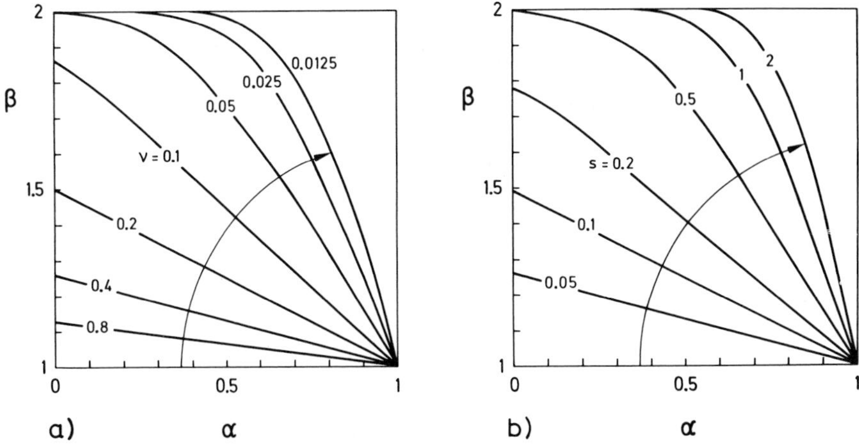


Fig. 23 Response function $\beta(\alpha)$ according to eq.(54); entropy term included. a) Effect of increasing membrane asymmetry or decreasing values of $\nu$; $n = 0.01$ M; $\sigma = 10^5$ esu (one elementary charge per lipid); $r = 0$. b) Effect of increasing charge density $\sigma = 10^5$ s; $\nu = 0.1$; $n = 0.01$ M; $r = 20$.

On the basis of the complete equation (54) the response becomes nonlinear and dependent on the values of $r$ and $\sigma$ (cf. Fig. 23 and 24). The characteristic influence of salt concentration and membrane asymmetry remains, however, the same. Fig. 23a shows the response function $\beta(\alpha)$ for increasing membrane asymmetry in the presence of the entropy influence. In contrast to Fig. 22 the response shows now the expected asymptotic behaviour for $\beta \to 2$. From Fig. 23b one sees that increasing charge density has a similar influence as increasing membrane asymmetry. The different curves in this Figure are drawn for different values of $s$, where $\sigma = 10^5 s$ esu; $s = 1$ corresponds to a charge density of one elementary charge per lipid molecule with a molecular area $f = 50$ Å$^2$.

The influence of an increasing tendency towards phase separation in layer 2, or increasing values of $r$ is illustrated in Fig. 24. Increasing values of $r$ produce a clockwise rotation of the $\beta(\alpha)$ curves around the central point $\alpha = 1$, $\beta = 1$ of the diagram. If $r$ exceeds a critical value $r_c$ layer two exhibits spontaneous phase separation i.e. $\beta \neq 1$ for $\alpha = 1$. For this case the complete $\beta(\alpha)$ curve (range $0 \leq \alpha \leq 2$, $0 \leq \beta \leq 2$) assumes an S-shaped form which is typical for systems showing phase transitions. The system shows instabilities, metastable states and hysteresis. The significance of these curves is as follows: For $\alpha < 1$ the value of $\beta$ is

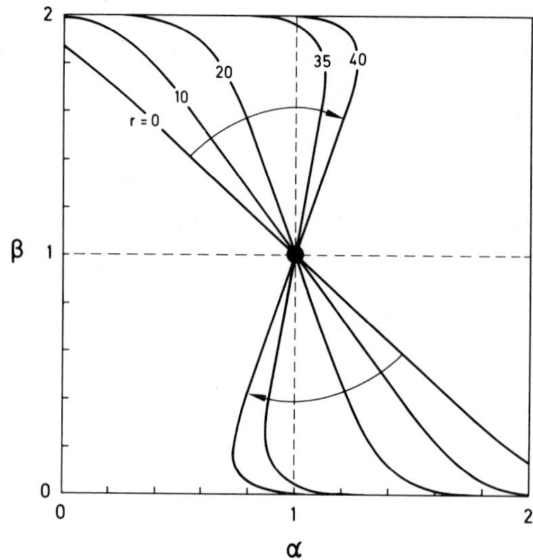


Fig. 24 Response function $\beta(\alpha)$ for different values of r, or different tendencies towards phase separation in layer two. For r 30 layer two exhibits spontaneous phase separation, i.e. $\beta = 1$ for $\alpha = 1$. When $\alpha$ reaches critical values layer two shows abrupt transitions in its lateral organization. $\sigma = 10^5$ esu; $\nu = 0.1$; $n = 0.01$; entropy included.

close to two, which means that side two carries charged patches in area U2 opposite to the area of low-charge density on side one (cf. Fig. 21b). With increasing $\alpha$ the charge density in area U1 increases whereas that in area L1 decreases. For a critical value of $\alpha$ layer two shows an abrupt translocation of charged patches from part U to part L. This transfer shows the characteristic features of a phase transition with metastable states and hysteresis.

To obtain a better understanding of the physical nature of the system we discuss now how the minima in the free energy arise for selected charge distributions on side two. For this purpose we consider the $\beta$-dependence of the individual energy terms in eq.(46) for a given value of $\alpha$, or for a given charge distribution on side one. Since the minima in the free energy arise from the electrostatic free energy, $G^{el}$, we ignore for this discussion the entropy and the r-term. As an example we take $\alpha = 0.5$ i.e. a situation with charge depletion in area U1 and charge concentration in area L1 on side one. Fig. 25 shows the dependence of the individual energy terms in eq.(46) on $\beta$. As one would expect from the tendency of the surface charge to spread itself over the largest possible area the surface energy $\phi_2$ has a minimum for $\beta = 1$, or for homo-

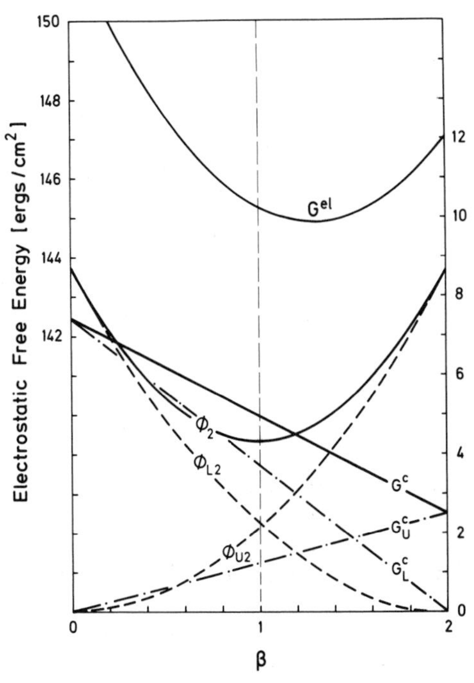


Fig. 25 Dependence of the different energy terms in eq. (46) on $\beta$, or on the charge distribution in layer two for a given state of phase separation in layer one ($\alpha = 0.5$). The surface energy of side two, $\phi_2 = \phi_{U2} + \phi_{L2}$, has a minimum for $\beta = 1$, or for homogeneous charge distribution. The coupling energy $G^c = G^c_U + G^c_L$ decreases linearly with increasing $\beta$. Summation of $G^c$ and $\phi_2$ results in a minimum of $G^{el}$ for $\beta \approx 1.3$ (cf. Fig. 22b). A large $\beta$-independent contribution to $G^{el}$ stems from the surface energy $\phi_1 = \phi_{U1} + \phi_{L1}$ on side one. The ordinate scale on the left hand side is valid for $G^{el}$, that on the right hand side for the other energy terms. Conditions: $\sigma = 5\text{x}10^4$esu; $\alpha = 0.5$; $n = 0.01$ M; $\nu = 0.2$; $\varepsilon_m = 10$; $\varepsilon_o = 80$; $d = 30$ Å; $T = 300$ K.

geneous charge distribution. In contrast, the coupling energy $G^c$ is proportional to $\sigma_1 \times \sigma_2$ (cf. eq.(39)) therefore $G^c_L$ equals zero for $\beta = 2$ where $\sigma_{L2} = 0$ and $G^c_U$ equals zero for $\beta = 0$ where $\sigma_{U2} = 0$. With increasing $\beta$ $G^c_U$ increases and $G^c_L$ decreases. However, because of the higher charge density in area L1 compared to U1 the value of $G^c_L$ decreases faster than $G^c_U$ increases. This gives a net linear decrease of $G^c$ with increasing $\beta$. As a result we obtain a minimum in $G^{el}$ for $\beta > 1$. In order that the coupling energy be comparable with the value of $\phi_2$ we need small values of $\nu$ or a relatively high charge density on side one compared to that on side two. This has the effect that the $\beta$-independent surface energy of side one is large. This is the reason why the value of $G^{el}$ is relatively large ( 145 erg/cm²) compared to the surface energy $\phi_2$ of side two and the coupling energies $G^c_L$ and $G^c_U$ which are in the range below 10 erg/cm².

Before discussing the possible biological implications of our model we consider briefly the approximations made and the questions that deserve further investigation. Firstly, our treatment is not self-consistent because we have neglected the feedback between the charge distribution on side two and the charge distribution on side one. The assumption $\alpha$ = const. implies that the forces inducing phase separation in layer one are strong compared with the processes on side two of the membrane. In order to treat the problem on a more sophisticated level one would have to combine the principles described in Section 5 concerning the influence of external parameters on the phase separation with the present model.

Secondly, we have ignored boundary effects at the boundary between regions of different surface charge density in the plane of the membrane. This approximation can be made only if the domains created by the phase separation are relatively large.

Thirdly, it was assumed that the number of charges in each layer of the membrane is constant. Eq.(54) is valid for $\gamma = 2 - \alpha$ and $\sigma = 2 - \beta$, i.e. for a constant number of charges on either side of the membrane. In view of Section 5 it seems probable that the process which induces phase separation in layer one (for example a decrease in pH or binding of $Ca^{++}$) causes a partial charge neutralization. This could be taken into account by setting $\gamma = 1$ which means that the total number of charges on side one decreases during phase separation and the remaining charged lipids concentrate in area L1 so that the charge density in this area remains on the initial value $\sigma_1^o$. It turns out that in the approximation leading to eq.(55) the two cases $\gamma = 1$ and $\gamma = 2 - \alpha$ are qualitatively not different and that only the sensitivity $|d\beta/d\alpha|$ is somewhat smaller for $\gamma = 1$. In a more refined version of the theory one would have to take into account that the degree of dissociation of the charge-carrying groups depends on the surface potential according to eq.(6) and that charge separation on side two involves a change in the charge per polar group.

Finally, the low-potential approximation (linearized Poisson equation) used in this Section is valid only for small charge density and high salt concentration. As is known from Section 2 and 3 the situation in biological membranes is better described by the so-called high potential approximation. Such a treatment will be presented in a forthcoming paper and it is shown that the major conclusions drawn here are qualitatively correct. An interesting feature that appears in this treatment is that the discussed mechanism depends on an externally applied potential difference. This means that the lateral organization of the membrane depends on an externally applied trans-membrane potential.

In this Section we have discussed two possibilities for an influence of the electrostatics on one side of a membrane on the surface electrostatics on the opposite side: (1) The first relies on the dependence of the surface potential $\psi^o$ (which regulates ionic processes at the menbrane surface) on the charge density on the opposite membrane side (cf. eqs.(33) and (34)). (2) The second involves an inductive effect of a phase separation on side one of the membrane on the lateral distribution on the opposite side. Both mechanisms may be regarded as possibilities for the transmission of signals across a membrane without the necessity of trans-membrane transport.

For the second case the question remains to be discussed how a change in surface charge density can be translated into biologically useful signals. As we have seen in the previous Sections a change in charge density creates local changes in surface pH; it affects dissociation or association reactions at the membrane surface according to eq.(6) which leads to pulses of protons or divalent cations. It can influence the electrostatic binding of enzymes or other ligands to the membrane surface. And as was shown in Section 3 a change in charge density affects the fluidity of the membrane with the possibility of triggering lipid phase transitions which in turn may influence the mobility and/or lateral distribution of proteins in the plane of the membrane. Finally, a change in charge density alters the transmembrane potential or the membrane interior electrical field and this may affect the conformation (and function) of intrinsic membrane proteins.

In summary, the sequence of events involved in the second mechanism is: First an "electrostatic signal" arrives on side one of the membrane.This can be a change in pH, ionic strength or a pulse of divalent cations etc. This signal alters the surface charge distribution on side one, for example by local charge neutralization or ionization of lipids leading to phase separation. The opposite layer of the membrane assumes a complementary charge pattern by way of electrostatic coupling across the membrane. Finally, these changes become biologically effective by one of the effects discussed. For example, a pulse of divalent cations on side one can lead to a similar pulse on the other side, or it can be translated into a change in surface pH or a change in some enzymatic activity that depends on surface pH or on the presence of $Ca^{++}$.

As we have seen a strong response requires a certain degree of membrane asymmetry. Therefore the mechanism is unidirectional in the sense that signal transmission is more effective from the side with high charge density to the side with low charge density than vice versa.

It is remarkable that the lipid composition of many biological membranes fulfills the criteria required for the functioning of the proposed mechanism. Biological membranes have in general a heterogenous lipid composition which allows phase separation. The fraction of charged lipids is generally about 10-30%. And as mentioned in the Introduction (see also the article by van Deenen et al., this volume) there is evidence that biological membranes are asymmetric with respect to their lipid composition and their charge densitiy.

## 7. SUMMARY

Surface charges provide a basis for the coupling between membrane structure and environmental electrolyte. The charges and the related electrostatic free energy tend to expand or fluidize the membrane structure. Under certain conditions this influence suffices to induce membrane phase separations and/or lipid phase transitions at constant temperature by changes in pH, ionic strength or divalent cations.

A membrane with ionizable groups can be regarded as a reservoir (sink or source) for cations like protons, $Ca^{++}$ or $Mg^{++}$. Membrane structural changes resulting from conformational changes of proteins or lipids lead to pulses of protons or divalent cations into the environment.

The presence of ionizable groups allows to regulate membrane phase separations by electrostatic means.

The two layers of a charged membrane are electrostatically coupled because of a dependence of the surface potential on the charge density on the opposite side. This leads to a "coupling term" in the free energy of a charged lipid bilayer with the consequence that phase separation on side one produces a complementary charge pattern on the other side. This is equivalent to the transmission of a signal across the membrane.

I am greatly indebted to Prof. M. Eigen for valuable criticism of this work. It is a pleasure to thank my colleagues, particularly Dipl. Math. M Teubner, Dr. P. Woolley and Dr. F. Jähnig for helpful discussions. This work has greatly benefitted from the collaboration with Dr. H. Eibl. The excellent technical assistance of Mrs. U. Sievers and the skill of Mrs. G. Daude in the preparation of the manuscript are gratefully acknowledged.

## REFERENCES

Becker, R., Theorie der Wärme, Springer Verlag, Berlin, Heidelberg, New York, 1966.

Blume, A., Doctoral Dissertation, University of Freiburg, 1976.

Cuthbert, A.W., Calcium and Cellular Function, London, 1970.

Davies, J.T. and E.K. Rideal, Interfacial phenomena, Academic Press, New York, 1961.

MacDonald, R.C., S.A. Simon and E. Baer, Biochemistry 15, 885, 1976.

Fromherz P. and B. Masters, Biochim. Biophys. Acta 356, 270, 1974.

Galla, H.J. and E. Sackmann, Biochim. Biophys. Acta 401, 509, 1975.

Goldstein, L., Y. Levine and E. Katchalski, Biochemistry 3, 1913, 1964.

Gurr, M.J. and A.T. James, Lipid Biochemistry: An introduction, Chapman and Hall, London, 1971.

Harned, H.S. and B.B. Owen, The physical chemistry of electrolytic solutions, Reinhold Publ. Corp., New York, 1943.

Hille, B., A.M. Woodhull and B.I. Shapiro, Phil. Trans. R. Soc. Lond. B. 270, 301, 1975.

Hubbard, S.C. and S. Brody, J. Biol. Chem. 250, 7173, 1975.

Jacobson, K. and D. Papahadjopoulos, Biochemistry 14, 152, 1975.

Jähnig, F., Biophys. Chem., in press.

Jähnig, F. and H. Träuble, in preparation.

McLaughlin, S.G.A., G. Szabo, G. Eisenmann and S.M. Ciani, Proc. Nat. Acad. Sci. (USA) 67, 1268, 1970.

McLaughlin, S.G.A., G. Szabo and G. Eisenmann, J. Gen. Physiol. 58, 667, 1971.

Lee, A.G., Biochim. Biophys. Acta 413, 11, 1975.

London, Y. and F.G.A. Vossenberg, Biochim. Biophys. Acta 178, 478, 1973.

Ohki, S., J. Coll. and Interface Sci. 37, 318, 1971.

Ohki, S., in: Progress in surface and membrane science, V. 10, edited by D.A. Cadenhead and J.F. Danielli, p. 117, Academic Press, New York, San Francisco, London, 1976.

Ohnishi, S., Adv. Biophys. 8, 35, 1975.

Overath, P., L. Thilo and H. Träuble, Trends in Biochem. Sci., in press

Phillips, M.C., H. Hauser and F. Paltauf, Chem. Phys. Lipids 8, 127, 1972.

Rao, Y.K., in: Phase diagrams, materials science and technology, V. I, edited by M. Alper, p. 1, Academic Press, New York and London, 1970.

Record, M.Th., Jr., Biopolymers 5, 975 and 993, 1967.

Shimshick, E.J. and H.M. McConnell, Biochemistry 12, 2351, 1973.

Teubner, M. and H. Träuble, in preparation.

Träuble, H. and P. Overath, Biochim. Biophys. Acta 307, 491, 1973.

Träuble, H. and H. Eibl, Proc. Nat. Acad. Sci. (USA) 71, 214, 1974.

Träuble, H. and H. Eibl, in: Funktional linkage in biomolecular systems, edited by F.O. Schmitt, D.M. Schneider and D.M. Crothers, p. 59, Raven Press, New York, 1975.

Träuble, H., M. Teubner, P. Woolley and H. Eibl, Biophys. Chem., in press.

Verkleij, A.J., R.F.A. Zwaal, B. Roelofsen, P. Comfurius, D. Kaselijn and L.L.M. van Deenen, Biochim. Biophys. Acta 323, 178, 1973.

Verwey, E.J.W. and J.Th.G. Overbeek, Theory of the stability of lyophobic colloids, Elsevier Publishing Co. Inc., New York, Amsterdam, London, Brussels, 1948.

Wu, S.H. and H.M. McConnell, Biochemistry 14, 847, 1975.

# DYNAMICS AND THERMODYNAMICS OF LIPID - PROTEIN INTERACTIONS IN MEMBRANES

Garret Vanderkooi and John T. Bendler*

Department of Chemistry

Northern Illinois University, DeKalb, IL 60115 USA

## ABSTRACT

A thermodynamic analysis of the process of protein aggregation in membranes is presented. Three factors have been identified as the principal determinants of the state of aggregation or dispersal of the intrinsic proteins in simple membranes: The entropy of mixing, the equilibrium between boundary lipid and bilayer lipid, and protein-protein interactions. The Hamaker method was used to estimate the strength of the nonbonded dispersion energy between proteins embedded in lipid. It was found that this energy gives a net attractive force between the proteins and for large proteins may be several times the thermal energy, even when the proteins are separated by one or more lipid molecules. A limiting law equation for the athermal mixing of proteins and lipids in a membrane was derived using a lattice solution theory model. It is shown that the entropy of mixing varies inversely with the size of the proteins. Increasing the effective size of the proteins, as through dimerization by crosslinking agents, may induce a further nonspecific aggregation of proteins to occur, on account of the combined effect of an increased nonbonded attraction and a decreased entropy of mixing.

* Present address: Midland Macromolecular Institute, 1910 West St. Andrews Drive, Midland, Michigan 48640 USA.

## INTRODUCTION

The purpose of this paper is to analyze the principal thermodynamic factors which determine the manner of distribution of proteins in simple membranes. Changes of the properties or composition of the lipid are known to affect the protein distribution. In several membranes, the proteins were found to be aggregated at temperatures below the thermal phase transition ($T_c$) of the lipids, but were dispersed when the lipids were in the liquid crystalline state (1-4). These experiments have been interpreted in terms of a "driving out" of the proteins from the orderly array of lipids in the gel state (1). Haest et al (2) showed, however, that the proteins in some bacterial membranes remained dispersed even when the lipids were in the gel state, if branched chain lipids were present. Thus more subtle factors may be involved than simply the state of the lipid.

Relatively little attention has been paid to the role of the proteins themselves in clustering phenomena. Hubbell et al. (3,4), using recombined rhodopsin membranes, showed that dark-adapted rhodopsin followed the general rule of being driven out of gel phase lipids, but was dispersed at temperatures above $T_c$. Bleaching of the rhodopsin caused it to be dispersed at all temperatures, however, regardless of the state of the lipid. If bleaching was carried out on the aggregated, dark-adapted rhodopsin at low temperature, the temperature had to be monentarily raised above $T_c$ to permit the redistribution of the proteins. This appears to be an example of a kinetic factor, i.e., low membrane fluidity in the gel state, preventing the attainment of the thermodynamically more stable dispersed state of the bleached rhodopsin. It is clear from these experiments that the condition of the protein itself in part dictates its mode of distribution in the membrane.

Erythrocyte glycophorin behaves in model membrane systems similarly to bleached rhodopsin, being dispersed both above and below the phase transition temperature (5). The sarcoplasmic reticulum ATPase, on the other hand, follows the general rule of being aggregated below $T_c$, and dispersed above $T_c$ in recombined membranes (6). From these examples on model membrane systems, we can see that it is a mistake to ignore the role of the proteins themselves, and the interactions between them, in protein distribution problems.

Protein aggregation in plasma membranes induced by lectins has been extensively studied. We do not propose to rationalize this type of aggregation in the present account. We are primarily concerned with thermodynamic effects of a general, nonspecific nature, whereas lectin-induced aggregation evidently involves some highly specific effects, including interaction between the intrinsic membrane proteins and the microtubule and microfilament systems (7,8). The potential physiological importance of the state of distribution of proteins in a membrane is emphasized by the lectin experiments, however.

In the following sections, we will first give a qualitative thermodynamic analysis of the factors which may be expected to govern the distribution of proteins in simple membranes. This represents a further development of our earlier work on the subject (9,10). Secondly, we will give an order of magnitude calculation of the nonbonded attractive energy between the proteins in a membrane. And finally, an equation will be given for calculating the entropy of mixing of lipids and proteins in a membrane, in the athermal limit.

## FREE ENERGY OF MIXING OF LIPIDS AND PROTEINS IN A MEMBRANE

We would like to identify the major enthalpic and entropic changes which accompany protein cluster formation in a simple membrane consisting only of intrinsic protein and lipid. Two processes will be considered: first, the transfer of a protein molecule from a constant external reference state into an expanse of lipid bilayer. (Since we are concerned with free energy differences, the nature of the reference state is unimportant; it could be a detergent micellar solution of the protein.) And second, the process of protein dimer formation and subsequent cluster formation in the plane of the membrane.

### Introduction of a Protein into a Lipid Bilayer

This process may be divided into two virtual steps: the creation of a hole in the lipid bilayer, and the filling of the hole with a protein. For this thought experiment, we assume that the perimeter of the hole is impenetrable to lipid, and also that the protein which will be inserted into the hole is an impenetrable mass. The motional freedom of the lipid aliphatic chains nearest to the perimeter of the hole will therefore be considerably restricted, since they can only move in directions parallel to or away from the hole, but not into the hole. Thus the entropy of these lipids will be decreased relative to that of the lipids in continuum bilayer, which can move in all directions. The enthalpy of the lipids at the perimeter of the hole will be increased, since they now have fewer nearest neighbors with which to interact. The ease with which a hole can be formed will depend upon the free energy change which accompanies the conversion of a continuum bilayer lipid molecule into a hole perimeter molecule.

The lipid molecules which form the perimeter of a hole in

the bilayer may be treated as a surface phase in equilibrium with continuum lipid, by analogy to the way in which the surface phase of a bulk liquid is defined. The thickness of this surface phase is unspecified; clearly, the physical properties of the first layer of lipid nearest the hole will be most affected, and the perturbing effect of the hole will decay as a function of distance away from the hole.

The bilayer analog of surface active agents should make hole formation easier, whereas other agents or conditions which increase the cohesive energy of the lipid bilayer will make hole formation more difficult. Changing from the liquid crystalline state to the gel state is in the latter category. Molecules which act in two dimensions as surface active agents will in general be those which are observed to increase membrane fluidity. An increase in fluidity parallels an increase in the specific volume of the lipids and a decrease in their cohesive energy density (10). Since general anesthetics are known to increase bilayer fluidity (11) and to decrease the density of membranes (12), they may be described in the present context as two dimensional surface active agents.

Once the hole has been formed in the lipid bilayer, introduction of a protein may take place. We will, for convenience, assume that no protein conformational changes accompany this process. The enthalpy increase which occurred upon hole formation will now be compensated for by the newly formed lipid-protein interactions that replace the broken lipid-lipid interactions. The net enthalpy change will therefore be the difference between these quantities. The lipid entropy decreased upon hole formation, and may decrease even more upon filling the hole with a protein, on account of the possibly attractive energy of interaction between the lipid and protein.

Fig. 1 illustrates the effect of introducing a protein into a lipid bilayer. For the purpose of illustration only, two layers of lipid around the monomeric protein are indicated as being perturbed by the presence of the protein. The term "boundary lipid" has been used to describe the partially immobilized lipid molecules surrounding the proteins in a membrane (13).

We can see from the above analysis that the free energy change which accompanies the introduction of a protein into a lipid bilayer from a constant reference state is wholly dependent upon the free energy difference between continuum bilayer lipid and boundary lipid. As will become evident in what follows, cluster formation of proteins results in a decrease in the amount of boundary lipid; for this reason, all lipid - dependent changes in the degree of dispersal of proteins in a membrane may be related

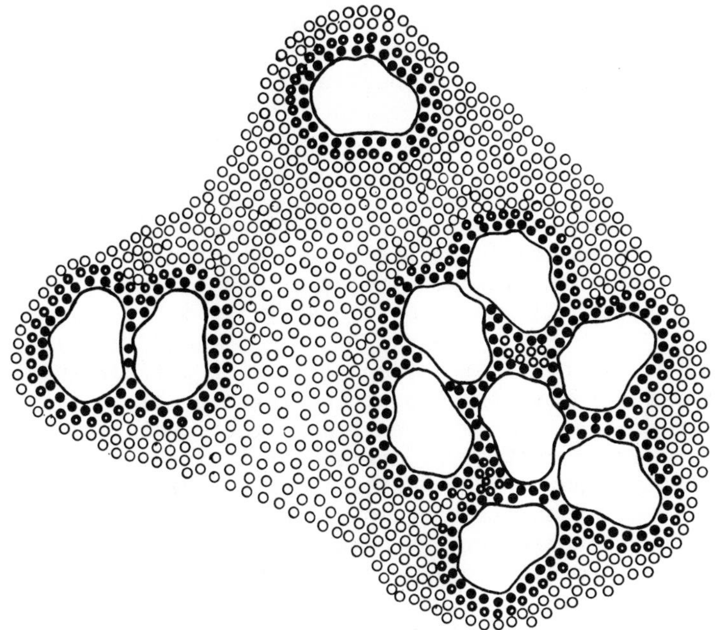

Fig. 1. Surface view of a membrane showing the relationship between boundary lipid and protein aggregation. Two layers of boundary lipid are shown; the first layer (filled circles) are the most strongly perturbed by the proteins, and the second layer (heavy circles) are less strongly perturbed. The open circles represent essentially unperturbed bilayer lipid. The monomeric protein (top) is completely surrounded by boundary lipid. In the protein dimer (left), much of the boundary lipid is excluded from between the proteins, and in the protein cluster (right), even more boundary lipid is removed from around the proteins. The number of boundary lipid layers shown here, and the amount removed upon protein cluster formation, are for illustrative purposes only and are strictly hypothetical.

directly to the energetics of the boundary lipid - bilayer lipid equilibrium.

## Protein Dimer and Cluster Formation

Having introduced the proteins into a lipid bilayer in a state of infinite dilution, the next step is to study the free energy changes which accompany the bringing together of a pair of protein molecules to form a dimer. If nonspecific dimer formation is energetically favored, the same forces may lead to larger cluster formation as well. We would ultimately like to know how the energy varies as a function of the distance of separation of proteins in a pair, and to derive from this information the radial distribution function.

Three types of contributions to the free energy of dimer or cluster formation may be identified: (a) changes in the entropy of mixing; (b) the protein-protein energy of interaction; and (c) change in the total amount of boundary lipid.

Entropy of mixing. The entropy of mixing will ordinarily decrease upon the formation of protein dimers or clusters. An equation is obtained in a later section for computing the athermal, positional or cratic, entropy of mixing of lipids and proteins in the plane of a membrane. The equation derived pertains to the idealized case of noninteracting molecules, which is obviously incorrect for membranes. A more complete treatment should include consideration of protein-protein interactions, and the difference in conformational entropy between boundary and bilayer lipid. The latter quantity may be either positive or negative, depending upon whether the bilayer is in the gel or fluid state.

Protein-protein interactions. The interactions between the proteins may be divided into two types: short and long range. We are concerned here with the long range, nonspecific contributions to the energy between proteins. These include chiefly the electrostatic (Coulombic and dipole) energy, and the nonbonded dispersion energy. Should the degree of exposure of the membrane components to water be dependent upon the mode of distribution of the proteins in the membrane, then there would also be a hydrophobic contribution. If short range specific interactions exist between the proteins (e.g., covalent bonds, hydrogen bonds, or salt linkages), their contribution to the free energy of dimer formation will ordinarily dominate the situation. We believe, however, that in the absence of such short range effects, the long range nonspecific interactions will not be negligible and may contribute significantly to the free energy of dimerization and cluster formation.

We will show in a subsequent section that the nonbonded energy between a pair of proteins embedded in a lipid medium is always negative, yielding an attractive force, and can be of an appreciable magnitude out to a distance of separation of many Angstroms. This nonbonded energy is strongly dependent on both the size and shape of the proteins involved, as well as on their relative orientations.

The long range electrostatic force between a pair of identical proteins may be either weakly attractive or strongly repulsive. Attraction may result from a favorable dipolar interaction if the net charge on the proteins is small. Of course, for nonidentical proteins bearing unlike net charges, there may be a strong Coulombic electrostatic attraction.

The electrostatic repulsion between a pair of charged proteins may be of such a magnitude as to completely overshadow all other long range attractive forces. The charge on the proteins may be increased by pH changes or by the removal of multivalent counterions which bind to and effectively neutralize the proteins. Decreasing the ionic strength of the medium will also increase the effective interprotein electrostatic repulsion through a reduction in ionic shielding. In complex membrane systems, the binding of foreign proteins to the membrane proteins may have the effect of either increasing or decreasing the net charge. The binding of antibodies or lectins to membranes may, for example, result in charge neutralization and thus eliminate the major repulsive force between the proteins.

Boundary lipid-bilayer lipid equilibrium. When two proteins approach each other so that their regions of influence overlap, the total amount of lipid perturbed by them will decrease, although the lipid which remains between the two proteins may experience an increased perturbation. Thus the amount of continuum bilayer lipid in the system will increase at the expense of the boundary lipid. This is illustrated in Fig. 1. Any influences which decrease the free energy of bilayer lipid relative to the boundary lipid will therefore favor dimerization and further aggregation of the proteins.

We would like to know the signs of the average enthalpy and entropy changes for the process of converting boundary lipid to bilayer lipid. The enthalpy and entropy values of the boundary lipid relative to the liquid crystalline or gel state bilayer may be deduced for the typical membranes in which the proteins are dispersed at temperatures above $T_c$, but are aggregated below that temperature. It is known that both the enthalpy and the entropy of bilayer lipids decrease upon going from the liquid crystalline state to the gel state (14). Since boundary lipid has decreased motional freedom relative to liquid crystalline bilayer lipid (13), we may safely say that the entropy of boundary lipid is lower than that of the bilayer. Since, however, in the typical membranes we are considering the proteins remain dispersed in the liquid crystalline bilayer, the enthalpy of the boundary lipid must also be lower than that of the bilayer, so that the net free energy difference ($\Delta G = \Delta H - T\Delta S$) favors the boundary lipid.

This situation must change when the temperature is brought

below $T_c$, causing the conversion of the lipids to gel form, and the concommitant aggregation of the proteins. It is probable that the entropy of boundary lipid is higher than that of gel bilayer lipid, since in the gel the aliphatic chains of the lipids are in a regular array, but in the boundary they are inevitably in a somewhat disordered state due to the irregular contour of the protein surface. For there to be a net negative free energy change for the boundary lipid to gel bilayer transition, we must assume that the enthalpy of the boundary lipid is also higher than that of the gel bilayer. Thus, we conclude that the average enthalpy and entropy of boundary lipid are both intermediate between those of liquid crystalline and gel state bilayer lipid. Since the enthalpy and entropy have opposing contributions to the free energy difference between boundary and bilayer lipid, a delicate thermodynamic control of the equilibrium is possible.

The presumed relative enthalpy and entropy values for boundary lipid are summarized in Fig. 2. This figure indicates not only that the average boundary lipid properties are intermediate between those of the liquid crystalline and gel states, but also that a progression of values may be expected as a function of the distance from the protein surface. As the distance from the surface increases, the values of the enthalpy and entropy must asymptotically approach those of the continuum bilayer. The figure shows the progression of values to be expected when the bilayer is in the liquid crystalline state. If the temperature is lowered so as to convert the bilayer lipid to the gel state, the outer layers of boundary lipid would also be converted to gel. How much of the boundary lipid is removed in this manner will depend upon the detailed free energy balance.

The amount of boundary lipid removed determines how tightly the proteins will be packed in a cluster. This characteristic may explain why the tightness of packing of the proteins in clusters is at times observed to be a function of temperature (4). As the temperature is lowered the free energy difference between that of more and more boundary lipid and bilayer lipid becomes negative, so that progressively more boundary lipid molecules are converted to bilayer, leaving less space between the proteins.

Anomalous protein aggregation effects may be caused by changes in the properties or composition of the lipid, if the magnitudes or signs of the boundary lipid-bilayer lipid entropy or enthalpy differences are affected. The nonaggregation of bacterial proteins in membranes enriched with branched chain lipids(2) may be explained in these terms. Since they cannot pack together as tightly as straight chain lipids in the gel state, the enthalpy difference between the boundary and gel states may be much smaller for the branched chains, so that $\Delta G$ for the boundary to bilayer conversion is no longer negative.

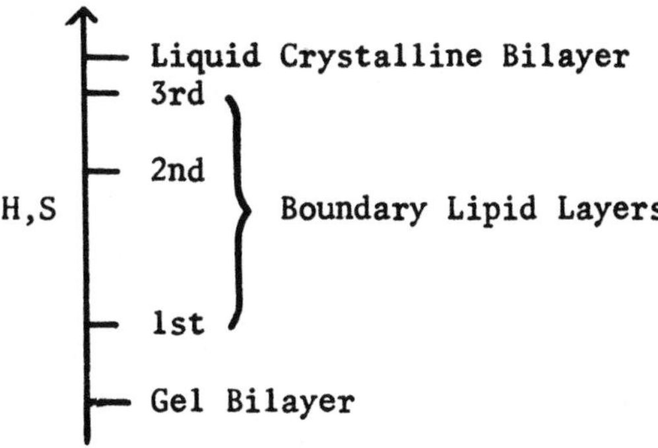


Fig. 2. Presumed relative enthalpy (H) and entropy (S) of boundary lipid layers as compared to the liquid crystalline and gel states of bilayer lipid. This diagram pertains to the case in which the bilayer is in the liquid crystalline state. The properties of the first boundary lipid layer differ the most from those of the bilayer, and the properties of the layers further away asymptotically approach those of unperturbed bilayer. No quantitative significance should be attached to the spacings shown on the graph.

Protein clusters. Is a protein cluster necessarily a rigid entity? If the major driving force for cluster formation is either a long range nonspecific attraction between the proteins, and/or a boundary lipid-bilayer lipid equilibrium which favors the bilayer, the proteins within the cluster may retain a considerable degree of rotational and lateral mobility, since some boundary lipid will likely remain between the proteins and act as a lubricant. If, on the other hand, short range specific linkages are formed directly between the proteins, then the cluster may become a solid, two dimensional crystal, in which translational and rotational mobility has been lost. The cytochrome oxidase membrane (15) and the purple membrane of *Halobacterium halobium* (16) are good examples of crystalline membranes.

## THE NONBONDED DISPERSION ENERGY BETWEEN MEMBRANE PROTEINS

In this section we will use the Hamaker integration method to provide an estimate of the nonbonded dispersion energy of interaction between proteins embedded in a lipoidal medium. The possibility that the nonbonded energy may play a role in determining the distribution of proteins in a membrane has been largely ignored. Because of its inverse sixth power dependence on the distance, it is usually considered to be strictly a short range energy. It was Hamaker (17) who originally showed, however, that the nonbonded energy between large particles may be appreciable and of long range. Hamaker obtained an explicit formula for computing the nonbonded energy between a pair of spheres as a function of the ratio of their separation distance and diameter, by integrating the energy function over the two volumes. The spheres were treated as having uniform atom density throughout. This gives an acceptable approximation for the pairwise summation of the nonbonded energy, provided that both the spheres and the distance between them are large by comparison to atomic dimensions.

We considered it worthwhile to use the Hamaker method to find out whether the nonbonded energy between the proteins in a membrane may be significant. Because of the several simplifying assumptions which had to be made, the numerical results obtained can only be considered to give an order of magnitude estimate. Our results showed that the magnitude may amount to several times RT for the nonbonded energy of interaction between a pair of proteins, and therefore it deserves further study and consideration.

Parsegian and Ninham (18) used the Lifshitz theory to compute the nonbonded dispersion energy between layers of decane separated by a water layer.They found that because of the large differences in the dielectric properties of the components in this system, the method of pairwise summation of the nonbonded energy considerably underestimated the more exact Lifshitz result. They showed however, that if the dielectric constant of all the components in a system are equal, the general Lifshitz equation actually reduces to the pairwise summation equation. In light of this, it did not appear to be necessary to use the more complicated Lifshitz theory to obtain an estimate of the nonbonded energy between proteins in a lipid bilayer, since protein and lipid have similar low dielectric constants. We consciously ignored the fact that a portion of most or all intrinsic membrane proteins extends beyond the confines of the lipid bilayer into water. The results of Parsegian and Ninham (18) show that the error introduced by this simplification is in the direction of underestimating the magnitude of the energy.

It was shown by Hamaker (17), and again by Flory (19) and others, that the proper way to compute the energy of interaction between a pair of solid particles immersed in a fluid medium involves the consideration of 3 types of interactions: (a) between the pair of solid particles; (b) between a solid particle and an equal volume of medium; and (c) between two volumes of medium having the same size, shape, and distance of separation as the pair of solid particles in question. The energy of bond formation, per particle, is given by:

$$\Delta w = \frac{1}{2}(w_{\ell\ell} + w_{pp}) - w_{\ell p} \qquad [1]$$

The subscripts $\ell$ and p denote lipid and protein, respectively, and the w's are the energies of interaction between the pair of real or virtual particles indicated by the subscripts. $\Delta w$ is the energy of formation of a pairwise contact, and will be negative for attractive interactions.

Hamaker (17) showed that for a pair of particles having the same composition but embedded in a fluid medium of different composition, the nonbonded dispersion energy will necessarily yield a negative value for $\Delta w$. The complicating effects of absorbed layers on this simple conclusion have been discussed at length by Vold (20) and Osmond et al (21), but their results do not appear to alter the original conclusion for the present problem.

## The Hamaker Equation

The Hamaker equation (17,21) may be written in the following form.

$$w_{12} = -\int_{V_1}\int_{V_2} dv_1\, dv_2 \sum_i \sum_j \frac{N_{i1} N_{j2} C_{ij}}{r^6} \qquad [2]$$

$V_1$, $V_2$, $dv_1$, and $dv_2$ are the volumes and differential volume elements of particles 1 and 2, respectively; r denotes the distance between the pair of volume elements; and $C_{ij}$ is the London-van der Waals constant for the pairwise interaction between atoms of type i and type j. The summations run over all atom types in particles 1 and 2; $N_{i1}$ and $N_{j2}$ are the atom densities of atom types i and j, respectively, in particles 1 and 2.

If the particles are assumed to be of homogeneous composition (but consisting of several kinds of atoms), Eq. 2 can readily be written as the product of a constant and a geometrical factor:

$$w_{12} = -A_{12}\, f(s) \qquad [3]$$

The quantity $A_{12}$ is known as the "Hamaker Constant", when expressed in the following form:

$$A_{12} = \pi^2 \sum_i \sum_j N_{i1}\, N_{j2}\, C_{ij} \qquad [4]$$

f(s) is the geometrical factor. Analytical expressions for f(s) are available for a limited number of geometries, which include equations for spheres of the same or different sizes (17), sphere and plane (17), and rectangular solids of various dimensions and orientations (22).

Values of $A_{ij}$ were computed for the three types of interaction: protein-protein, protein-lipid, and lipid- lipid. The atomic compositions of rhodopsin and of dipalmitoyl lecithin were used to obtain the values for $N_{i1}$ and $N_{i2}$ in Eq. 4. The density of the lipid was taken to be 1.019 gms/$cm^3$, on the basis of the partial specific volume determination by Huang and Charlton (23). The density of rhodopsin was computed from the residue molal volume tables of Conn and Edsall (24), using the amino acid composition given by Heller (25); a value of 1.308 gm/$cm^3$ was obtained.

The London-van der Waals constants, $C_{ij}$, were evaluated by means of the Slater-Kirkwood equation (26) using the values for the atomic polarizabilities and for $N_{eff}$ given by Momany et al (27).

The Hamaker constants obtained in this manner for lipids and proteins are given in the third column of Table I. These values are inversely dependent upon the densities. The fourth column in the table gives the Hamaker constants divided by the product of the densities, from which the Hamaker constants for a different set of assumed densities may readily be obtained. The product of $\Delta A$ (defined in Table I) and f(s) yields $-\Delta w$.

## Energy of Interaction As a Function of Protein Size and Shape

Fig. 3 gives f(s) and $-\Delta w$ as a function of protein volume for three geometries: pairs of equal spheres, cubes, and square plates. f(s) for spheres was computed with the equation given by Hamaker (17). For cubes and plates, Eq. 10 in the paper by Vold (22) was employed. The cubes were oriented with parallel opposing faces. The plates had dimensions of t x s x s, with t(thickness) being held constant at 50 Å. The interacting plates had t x s faces in parallel orientation. Curves are given in each case for closest approach distances (d) of 4.8 and 9.6 Å, which correspond respectively to the diameters of one and two lipid aliphatic chains. The volumes (V) are given in cubic Angstroms. These may readily be converted to molecular weight by the formula $MW = VN\rho \times 10^{-24}$, where N in Avagadro's number and $\rho$ is the density. For proteins, $MW \approx 0.8$ V.

The energy of interaction between the rectangular solids exceeds the thermal energy (RT) over a considerable portion of the range of volumes presented in Fig. 3. Taking RT as 0.6 kcal/mole, $-\Delta w$ is greater than RT at volumes larger than 30,000 to 40,000 $Å^3$

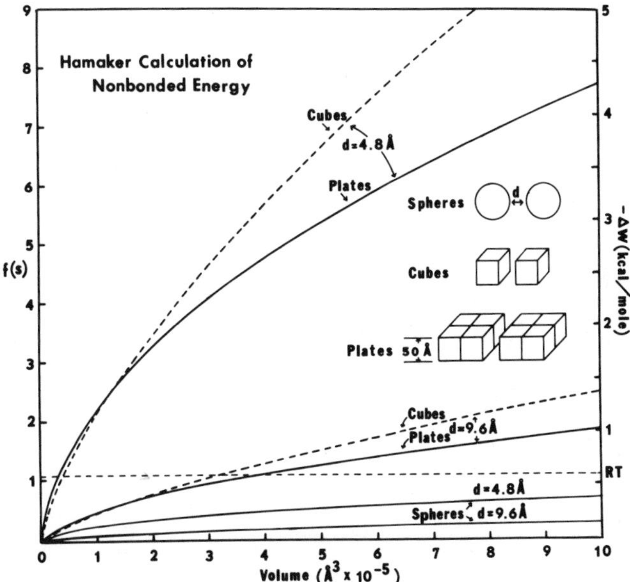


Fig. 3. Calculation of the nonbonded energy between proteins embedded in lipids. f(s) is the geometrical factor in the Hamaker equation, and Δw is the net energy of interaction per protein between a pair of proteins. The results for spheres, cubes and plates are given as a function of the protein volume. The plates had a constant thickness of 50 Å, and a square surface. In each case, d is the distance of closest approach between the proteins, and was held fixed at 4.8 Å or 9.6 Å. The thermal energy, RT, is indicated with a dashed line for comparison.

(24,000 to 32,000 Daltons) for plates and cubes at 4.8 Å separation; or at volumes larger than 300,000 to 400,000 Å$^3$ (240,000 to 320,000 Daltons) for plates and cubes at 9.6 Å separation.

Δw is strongly dependent upon shape; the energy of interaction between spheres is much weaker than between rectangular solids with juxtaposed faces. Therefore changes in protein conformation may have a marked effect on the nonbonded energy.

Fig. 4 gives f(s) and Δw for plates and cubes of constant volume, as a function of the distance of separation. The plates are of the size which one would obtain if 4 of the cubes (50 Å on a side) were joined to form a tetramer. For the plates, Δw is greater than RT out to a separation distance of over 10 Å.

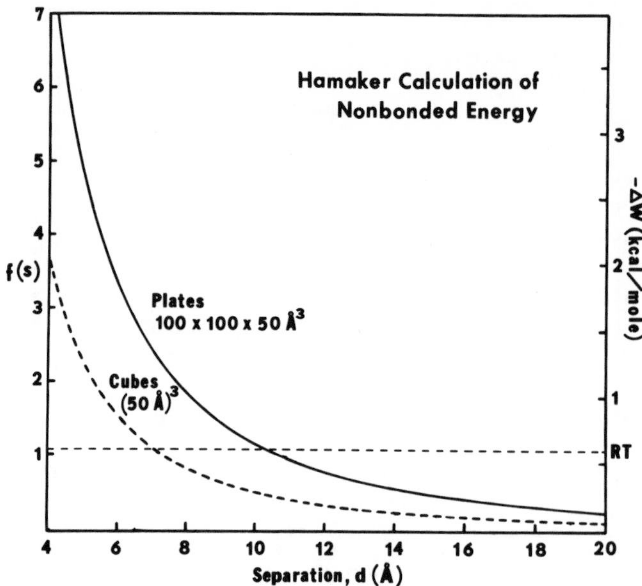


Fig. 4. The nonbonded energy between proteins embedded in lipid, expressed as a function of the distance of closest approach. Two cases are given: cubes of 50 $Å^3$, and plates having the size of a tetramer of these cubes. These volumes approximately correspond to protein molecular weights of 100,000 and 400,000 Daltons, respectively.

We conclude that for large proteins of favorable shape, the nonbonded energy of interaction is not negligible. Considering the dependence of the interaction energy on particle size, the nonbonded energy between pairs of dimers or multimers will be considerably greater than that between the free monomers. Thus if dimers are formed as a result of specific short range protein-protein interactions, or through being linked together by lectins or chemical crosslinking agents, the further aggregation of these dimers due to the nonbonded interactions will be favored. Dimerization simultaneously decreases the entropy of mixing and strengthens the energy of interaction between the remaining independent particles, and both of these factors favor nonspecific aggregation.

The plot of $-\Delta w$ given for plates in Fig. 3 may be used to estimate the energy of interaction between multimeric complexes as a function of the size of the complex. For example, if the monomer has a volume of 125,000$Å^3$, being a cube 50 Å sides, $\Delta w$ at 4.8 Å separation will be -1.38 kcal/mole, but if a tetramer is formed having dimensions of 50 x 100 x 100 $Å^3$, $\Delta w$ between the tetramers will be -2.97 kcal/mole, at the same separation distance. Even greater energies will be found between larger complexes, or between complexes of larger proteins.

## ATHERMAL ENTROPY OF MIXING

The entropy of mixing is one of the components of the free energy which favors the random dispersal of the proteins in a membrane. The other free energy components that we have identified which may favor dispersal of the proteins are electrostatic repulsion between similarly charged proteins, and an equilibrium between boundary and bilayer lipid which favors the boundary.

It would be a formidable undertaking to attempt to develop a complete expression for the entropy of mixing in membranes. Membranes are neither dilute solutions nor are their components noninteracting. Thus solution theories which make either of these assumptions are grossly inadequate. We nonetheless thought it of value to make a start in this direction, and to obtain the counterpart of an ideal entropy of mixing expression in two dimensions for molecules which differ greatly in size. We assume noninteraction between the molecules; this is the case of athermal mixing ($\Delta H = 0$), in the terminology of Guggenheim (28).

In the physical model we employ, the proteins and lipids are treated as rigid disks free to move about on a surface. Only two sizes of disks are considered: large disks representing the proteins and small ones for the lipids. It is assumed that the entire available surface is filled with disks. Since the disks are assumed to be noninteracting, all arrangements are equally probable. The entropy of mixing is then proportional to the logarithm of the number of posible arrangements. It is expressed as a function of the fractional surface area occupied by large disks, and the ratio of sizes of large to small disks.

We used two different approaches for deriving the expression for the entropy of mixing in the athermal limit. Both methods were based on a lattice model for two dimensional solutions. In the first, we employed a general equation given by Guggenheim (28) for computing the entropy of mixing of r-mers in a solution of monomers. By the second method, we obtained the same final expression as in the first case by writing out the combinatorial formula for the mixing of filled and empty sites on a lattice, from which the entropy of mixing was then derived.

## The Guggenheim method

Guggenheim (28) has reviewed the early work on the lattice model of computing the entropy of mixing for athermal mixtures of particles of different sizes. He gave explicit equations for computing the entropy of mixing of r-mers and monomers; an r-mer occupies r sites on a lattice, and a monomer occupies one site. r is therefore the ratio of areas between the large and small particles. The number of monomer molecules is denoted by $N_1$, and the number of r-mers by $N_r$. The total number of lattice sites, $N_s$, is just equal to $N_1 + r\,N_r$, so that all sites are filled. The fraction of sites occupied by r-mers is called $\phi$, and the fraction occupied by monomers is $1-\phi$. Clearly,

$$\phi = \frac{rN_r}{N_1 + rN_r} \qquad 1 - \phi = \frac{N_1}{N_1 + rN_r} \qquad [5]$$

Guggenheim (28) defines a quantity, called $\alpha$, as the ratio of the probability that a group of r sites, congruent with an r-mer, be wholly occupied by a single r-mer, to the probability that the group of sites by entirely occupied by monomers. He derives a general expression for the athermal entropy of mixing in terms of $\alpha$ and $\phi$:

$$\frac{\Delta S}{N_s k} = -\frac{1}{r}\int_0^{\phi} \ell n\ \alpha d\phi + \frac{\phi}{r}\int_0^{1} \ell n\ \alpha d\phi \qquad [6]$$

For the mixing of isotropic proteins with lipids, this equation takes on a particularly simple form. In that case, the probability that a group of r congruent sites (where r is now the ratio of the surface area of a protein to that of a lipid) be occupied by a protein is given by the area fraction, $\phi$, of protein; and the probability that those sites not be occupied by protein is $1-\phi$. Thus $\alpha = \frac{\phi}{1-\phi}$. Carrying out the integrations with this value of $\alpha$ yields:

$$\frac{\Delta S}{N_s k} = -\frac{1}{r}\left[\phi \ell n\phi + (1-\phi)\ \ell n(1-\phi)\right] \qquad [7]$$

In the case of r = 1 (particles of the same size) this equation reduces to the ideal entropy of mixing expression, since in that case the mole fraction, x, and the area or volume fraction $\phi$, become equivalent:

$$\frac{\Delta S}{N_s k} = -x\ \ell n\ x - (1-x)\ell n(1-x) \qquad [8]$$

$\Delta S$ is the athermal entropy of mixing of $N_1$ lipid molecules and $N_r$ protein molecules on $N_s$ lattice sites. $\Delta S/N_s$ gives the entropy of mixing per unit area, where the unit of area is defined as that of one lipid molecule. In practice, the ratio of protein to lipid surface area per molecule (r) can be a rather large number, causing $\Delta S/N_s$ to be very small compared to what it would be if all the molecules were the same size. For example, the surface area of one cytochrome oxidase protein complex is on the order of 3000 $Å^2$ (15), whereas the area of a phospholipid molecule is about 60 $Å^2$, giving r = 50. Should the effective value of r be increased, such as through the dimerization of the proteins by specific cross-linking agents, $\Delta S$ will correspondingly decrease.

## Lattice Statistics Method

The second derivation employs a lattice in which the mesh size is based upon the size of a protein molecule rather than a lipid molecule. Let the lattice contain $M_s$ sites, with the area of each site being equal to the surface area of one protein molecule. (In the first derivation, the lattice employed had $N_s$ sites, with one site having the area of one lipid molecule. Thus $r = N_s/M_s$). The number of ways in which $N_r$ proteins can be placed on $M_s$ sites ($N_r \leq M_s$) is given by the combinatorial formula:

$$Q = \frac{M_s!}{N_r!\ (M_s - N_r)!} \qquad [9]$$

The remaining sites are assumed to be filled with lipid molecules, but since there is just enough lipid to completely fill all these sites, the number of ways in which this filling can be done is unity. Thus Q is also the partition function for the athermal mixing of proteins and lipids on a lattice, and $\Delta S/k = \ell n Q$. The equivalence of this expression with Eq. 7 can be shown by replacing $M_s$ with $N_s/r$, and using the relationship $N_s = N_1 + rN_r$. Applying Stirling's approximation and rearranging terms gives:

$$\frac{\Delta S}{N_s k} = -\frac{1}{r}\left[\frac{N_1}{N_s}\,\ell n\,\frac{N_1}{N_s} + \frac{rN_r}{N_s}\,\ell n\,\frac{rN_r}{N_s}\right] \qquad [10]$$

Eqs. 10 and 7 are identical, since $N_1/N_s = (1-\phi)$ and $rN_r/N_s = \phi$.

## DISCUSSION

The finding that nonbonded, long range attractive forces exist between proteins in a membrane is important for understanding the dynamics of membranes. It means that if electrostatic repulsions are eliminated by one means or another, the residual force will not be zero, but attractive. This force may give rise to protein aggregation. The proteins in aggregates formed in this manner may be expected to retain rotational and translational mobility. The thermodynamic condition under which the proteins exist as loose clusters or aggregates corresponds to the state of partial miscibility of lipids and proteins described previously (9,10).

We can now understand why a limited degree of crosslinking of membrane proteins may result in a generalized aggregation. Crosslinking increases the effective size of the independent particles. This simultaneously decreases the entropy and increases the attractive force between the independent particles, both of which factors will favor protein aggregation. Charge neutralization and protein crosslinking may in some instances be effected simultaneously, as through the binding of extrinsic proteins.

The analysis of the boundary lipid-bilayer equilibrium presented here provides a possible explanation for the mode of action of certain classes of membrane-active drugs, especially the general anesthetics. If a drug is known to bind to a protein, or to change the permeability of a membrane, its effect can be rationalized (correctly or otherwise) in those terms. Neither of these explanations appear to apply in the case of the nonpolar, hydrocarbon-soluble general anesthetics. We propose that these molecules affect the equilibrium between boundary and bilayer lipid, and thus in turn between the aggregated and dispersed states of the proteins. For this mechanism to be effective, no massive shift in the state of protein aggregation is needed, but only a subtle shift in the monomer-multimer equilibrium of some critical membrane-bound allosteric enzyme system.

Table I

Hamaker constants for lipids and proteins

| Pairwise Interaction i | j | $A_{ij}$ (kcal/mole of interactions)[a] | $A_{ij}/\rho_i\rho_j$ |
|---|---|---|---|
| lipid | lipid | 15.1749 | 14.6143 |
| protein | protein | 21.8701 | 12.7831 |
| lipid | protein | 17.9666 | 13.4798 |
| $\Delta A$[b] | | 0.5559 | |

a. $\rho_{lipid} = 1.019$ gm/cm$^3$ and $\rho_{protein} = 1.308$ gm/cm$^3$

b. $\Delta A = \frac{1}{2}(A_{\ell\ell} + A_{pp}) - A_{\ell p}$

REFERENCES

1. Verkleij, A. J., Ververgaert, P. H. J., Van Deenen, L. L. M., and Elbers, P. F. (1972) Biochim. Biophys. Acta 288, 326-332.

2. Haest, C. W. M., Verkleij, A. J., De Gier, J., Scheek, R., Ververgaert, P. H. J., and Van Deenen, L. L. M. (1974) Biochim. Biophys. Acta 356, 17-26

3. Hong, K., and Hubbell, W. L. (1973) Biochemistry 12, 4517-4523.

4. Chen, Y. S., and Hubbell, W. L. (1973) Exp. Eye Res. 17, 517-532.

5. Kleemann, W., Grant, C. W. M., and McConnell, H. M. (1974) J. Supramol. Struct. 2, 609-616.

6. Kleemann,W., and McConnell, H. M. (1976) Biochim. Biophys. Acta 419, 206-222.

7. Wang, J. L., Gunther, G. R., and Edelman, G. M. (1975) J. Cell Biol. 66, 128-144.

8. DePetris, S. (1975) J. Cell Biol. 65, 123-146.

9. Vanderkooi, G. (1974) Biochim. Biophys. Acta 344, 307-345.

10. Vanderkooi, G. (1975) Int. J. Quantum Chem: Quantum Biology Symp. No. 2, 209-219.

11. Trudell, J. R., Hubbell, W. L., and Cohen, E. N. (1973) Biochim. Biophys. Acta 291, 321-327.

12. Seeman, P. (1974) Experientia 30, 759-760.

13. Jost, P. C., Griffith, O. H., Capaldi, R. A., and Vanderkooi, G. (1973) Proc. Nat. Acad. Sci. USA 70, 480-484.

14. Ladbrooke, B. D., and Chapman, D. (1969) Chem. Phys. Lipids 3, 304-367.

15. Vanderkooi, G., Senior, A. E., Capaldi, R. A., and Hayashi, H. (1972) Biochim. Biophys. Acta 274, 38-48.

16. Blaurock, A. E., and Stoeckenius, W. (1971) Nature New Biol. 233, 152-154.

17. Hamaker, H. C. (1937) Physica 4, 1058-1072.

18. Parsegian, V. A., and Ninham, B. W. (1971) J. Col. Interface Sci. 27, 332-341.

19. Flory, P. J. (1953) Principles of Polymer Chemistry, pp. 497-511, Cornell University Press, Ithaca.

20. Vold, M. J. (1961) J. Colloid Sci. 16, 1-12.

21. Osmond, D. W. J., Vincent, B., and Waite, F. A. (1973) J. Col. Interface Sci. 42, 262-269.

22. Vold, M. J. (1954) J. Colloid Sci. 9, 451-459.

23. Huang, C. H., and Charlton, J. P. (1971) J. Biol. Chem. 246, 2555-2560.

24. Cohn, E. J., and Edsall, J. T. (1943) Proteins, Amino Acids, and Peptides, p. 372, Hafner Publishing Co., New York.

25. Heller, J. (1968) Biochemistry, 7 2906-2913.

26. Scott, R. A., and Scheraga, H. A. (1966) J. Chem. Phys. 45, 2091-21

27. Momany, F. A., Carruthers, L. M., McGuire, R. F., and Scheraga, H. A. (1974) J. Phys. Chem. 78 1595-1630.

28. Guggenheim, E. A. (1952) Mixtures, pp. 185-214, Clarenden Press, Oxford.

# PARTICIPANTS

| | |
|---|---|
| Abrahamsson, S | Dept. of Structural Chemistry, University of Göteborg, Göteborg, Sweden. |
| Albertsson, P-A | Dept. of Biochemistry, University of Lund, Lund, Sweden. |
| Avron, M | Dept. of Biochemistry, Weizmann Institute of Science, Rehovot, Israel. |
| Baltscheffsky, H | Dept. of Biochemistry, University of Stockholm, Stockholm, Sweden. |
| Baltscheffsky, M | Dept. of Biochemistry, University of Stockholm, Stockholm, Sweden. |
| Cadenhead, D A | Dept. of Chemistry, State University of New York, Buffalo, N. Y., USA. |
| Chapman, D | Dept. of Chemistry, Chelsea College, University of London, London, UK. |
| Dahleń, B | Dept. of Structural Chemistry, University of Göteborg, Göteborg, Sweden. |
| Dallner, G | Dept. of Biochemistry, University of Stockholm, Stockholm, Sweden. |
| van Deenen, L | Laboratory of Biochemistry and Laboratory of Veterinary Biochemistry, University of Utrecht, Utrecht, The Netherlands. |
| DePierre, J W | Dept. of Biochemistry, University of Stockholm, Stockholm, Sweden. |
| Ehrenberg, A | Dept. of Biophysics, University of Stockholm, Stockholm, Sweden. |

| | |
|---|---|
| Ernster, L | Dept. of Biochemistry, University of Stockholm, Stockholm, Sweden. |
| Erkell, L | Dept. of Zoophysiology, University of Göteborg, Göteborg, Sweden. |
| Eylar, E H | Playfair Neuroscience Institute and Dept. of Biochemistry, University of Toronto, Toronto, Canada. |
| Forsén, S | Dept. of Physical Chemistry 2, University of Lund, Lund Institute of Technology, Lund, Sweden. |
| Fredga, A | Dept. of Organic Chemistry, University of Uppsala, Uppsala, Sweden. |
| de Haas, G H | Laboratory of Biochemistry, University of Utrecht, Utrecht, The Netherlands. |
| Hackenbrock, C R | Dept. of Cell Biology, University of Texas, Southwestern Medical School at Dallas, Texas, USA. |
| Hawthorne, J N | Dept. of Biochemistry, University Hospital and Medical School, Nottingham, UK. |
| Karlsson, K-A | Dept. of Medical Biochemistry, University of Göteborg, Göteborg, Sweden. |
| Larsson, K | Dept. of Food Science, Chemical Center, University of Lund, Lund, Sweden. |
| Liljenberg, C S | Dept. of Plant Physiology, University of Göteborg, Göteborg, Sweden. |
| Lindblom, G | Dept. of Physical Chemistry, University of Lund, Lund, Sweden. |
| Lucy, J A | Dept. of Biochemistry and Chemistry, Royal Free Hospital School of Medicine, University of London, London, UK. |
| Lundgren, G | Dept. of Inorganic Chemistry, University of Göteborg, Göteborg, Sweden. |

| | |
|---|---|
| Lundström, I | Research Laboratory of Electronics, Chalmers University of Technology, Göteborg, Sweden. |
| Luzzati, V | Centre de Génétique Moléculaire du C.N.R.S., Gif-sur-Yvette, France. |
| Löfgren, H | Dept. of Structural Chemistry, University of Göteborg, Göteborg, Sweden. |
| McConnell, H M | Stauffer Laboratory for Physical Chemistry, Stanford University, Stanford, Calif., USA. |
| Metcalfe, J C | Dept. of Biochemistry, University of Cambridge, Cambridge, UK. |
| Ödberg, L | Dept. of Physical Chemistry, Royal Institute of Technology, Stockholm, Sweden. |
| Olivecrona, T | Dept. of Chemistry, Section of Physiological Chemistry, University of Umeå, Umeå, Sweden. |
| Ovchinnikov, Y A | Shemyakin Institute of Bioorganic Chemistry, USSR Academy of Sciences, Moscow, USSR. |
| Pascher, I | Dept. of Structural Chemistry, University of Göteborg, Göteborg, Sweden. |
| Peterson, P A | Institute of Medical and Physiological Chemistry, University of Uppsala, Uppsala, Sweden. |
| Racker, E | Dept. of Biochemistry, Molecular and Cell Biology, Cornell University, Ithaca, N.Y., USA. |
| Radda, G K | Dept. of Biochemistry, University of Oxford, Oxford, UK. |
| Renkonen, O | Dept. of Biochemistry, University of Helsinki, Helsinki, Finland. |

| | |
|---|---|
| Samuelsson, B E | Dept. of Medical Biochemistry, University of Göteborg, Göteborg, Sweden. |
| Scanu, A M | Depts. of Medicine and Biochemistry, University of Chicago, Chicago, Ill., USA. |
| Singer, S J | Dept. of Biology, University of California at San Diego, La Jolla, Calif., USA. |
| Skou, J C | Institute of Physiology, University of Aarhus, Aarhus, Denmark. |
| Stoeckenius, W | University of California, School of Medicine, San Francisco, Calif., USA. |
| Sundell, S | Dept. of Structural Chemistry, University of Göteborg, Göteborg, Sweden. |
| Tanford, C | Dept. of Biochemistry, Duke University Medical Ctr., Durham, N.C., USA. |
| Träuble, H | Max-Planck-Institut für biophysikalische Chemie, Göttingen-Nikolausberg, Germany. |
| Vanderkooi, G | Dept. of Chemistry, Northern Illinois University, DeKalb, Ill., USA. |
| Virgin, H | Dept. of Plant Physiology, University of Göteborg, Göteborg, Sweden. |

# SUBJECT INDEX